Frontiers in Clinical Drug Research-Anti-Cancer Agents

(Volume 4)

Edited by

Atta-ur-Rahman, *FRS*
Honorary Life Fellow, Kings College, University of Cambridge
Cambridge, UK

General:

1. Any dispute or claim arising out of or in connection with this License Agreement or the Work (including non-contractual disputes or claims) will be governed by and construed in accordance with the laws of the U.A.E. as applied in the Emirate of Dubai. Each party agrees that the courts of the Emirate of Dubai shall have exclusive jurisdiction to settle any dispute or claim arising out of or in connection with this License Agreement or the Work (including non-contractual disputes or claims).
2. Your rights under this License Agreement will automatically terminate without notice and without the need for a court order if at any point you breach any terms of this License Agreement. In no event will any delay or failure by Bentham Science Publishers in enforcing your compliance with this License Agreement constitute a waiver of any of its rights.
3. You acknowledge that you have read this License Agreement, and agree to be bound by its terms and conditions. To the extent that any other terms and conditions presented on any website of Bentham Science Publishers conflict with, or are inconsistent with, the terms and conditions set out in this License Agreement, you acknowledge that the terms and conditions set out in this License Agreement shall prevail.

Bentham Science Publishers Ltd.
Executive Suite Y - 2
PO Box 7917, Saif Zone
Sharjah, U.A.E.
Email: subscriptions@benthamscience.org

BENTHAM
SCIENCE

CONTENTS

PREFACE

Frontiers in Clinical Drug Research - Anti-Cancer Agents presents on recent developments in various therapeutic approaches against different types of cancer. The book is a valuable resource for pharmaceutical scientists, postgraduate students and researchers seeking updated and critical information for devising research plans in anti-cancer research and developing clinical trials.

The chapters are written by authorities in the field. The contents of this volume represent exciting recent researches on advanced gastrointestinal cancers, breast cancer, cancer chemo-prevention and treatment, clinical anticancer nanodrug, adiponectin-based therapeutics, cancer vaccines, and chemotherapy medication. I hope that the readers will find these reviews valuable and thought provoking so that they may trigger further research in the quest for the new and novel therapies against cancers.

I am grateful for the timely efforts made by the editorial personnel, especially Mr. Mahmood Alam (Director Publications), and Mr. Shehzad Naqvi (Senior Manager Publications) at Bentham Science Publishers.

Atta-ur-Rahman, *FRS*
Honorary Life Fellow
Kings College
University of Cambridge
UK

List of Contributors

Abid H. Banday	Department of Chemistry and Biochemistry, University of Arizona, Tucson 85721, USA Department of Chemistry, Islamia College of science and Commerce, Hawal, Srinagar 190002, India
Angelos Evangelopoulos	Biologist, Evangelismos General Hospital, Athens, Greece
Andrea Lampis	Centre for Molecular Pathology, The Institute of Cancer Research London & Sutton, Sutton, United Kingdom
Andrea Nicolini	Department of Oncology, Transplantations and New Technologies in Medicine, University of Pisa, Italy
Christos Kazazis	Honorary Lecturer at Leicester University, Leicester, UK
Debra L. Zynger	Department of Pathology, Division of Genitourinary Pathology, The Ohio State University, Columbus, OH 43210, USA
Eleni Geladari	Internist, Evangelismos General Hospital, Athens, Greece
Handi Cao	State Key Laboratory of Pharmaceutical Biotechnology and Department of Pharmacology and Pharmacy, The University of Hong Kong, Hong Kong, China
Iskren Kotzev	University Hospital St. Marina, Varna, Bulgaria
Jens C. Hahne	Centre for Molecular Pathology, The Institute of Cancer Research London & Sutton, Sutton, United Kingdom
Jie Chen	State Key Laboratory of Pharmaceutical Biotechnology and Department of Pharmacology and Pharmacy, The University of Hong Kong, Hong Kong, China
Jing Liu	Beijing Key Lab of CryoBiomedical Engineering and Key Lab of Cryogenics, Technical Institute of Physics and Chemistry, Chinese Academy of Sciences, Beijing, 100190, China Department of Biomedical Engineering, Tsinghua University, Beijing, 100084, China
Mengjia Dou	Beijing Key Lab of CryoBiomedical Engineering and Key Lab of Cryogenics, Technical Institute of Physics and Chemistry, Chinese Academy of Sciences, Beijing, 100190, China Department of Thermal energy and Power engineering, University of Science & Technology, Beijing, 100083, China
Milko Mirchev	University Hospital St. Marina, Varna, Bulgaria
Ming Liu	State Key Laboratory of Pharmaceutical Biotechnology and Department of Pharmacology and Pharmacy, The University of Hong Kong, Hong Kong, China
Natalia G. Vallianou	Consultant in Internal Medicine, Evangelismos General Hospital, Athens, Greece

Nicola Valeri	Centre for Molecular Pathology, The Institute of Cancer Research London & Sutton, Sutton, United Kingdom Department of Gastrointestinal Oncology and Lymphoma, Royal Marsden Hospital London & Sutton, Sutton, United Kingdom
Paola Ferrari	Department of Oncology, Transplantations and New Technologies in Medicine, University of Pisa, Italy
Pier Mario Biava	Scientific Institute of Research and Care Multimedica, Milan, Italy
Syed Irtiqa Urfi	Department of Chemistry, Islamia College of science and Commerce, Hawal, Srinagar 190002, India
Victor J. Hruby	Department of Chemistry and Biochemistry, University of Arizona, Tucson 85721, USA
Wei Rao	Beijing Key Lab of CryoBiomedical Engineering and Key Lab of Cryogenics, Technical Institute of Physics and Chemistry, Chinese Academy of Sciences, Beijing, 100190, China
Yi Hou	Beijing Key Lab of CryoBiomedical Engineering and Key Lab of Cryogenics, Technical Institute of Physics and Chemistry, Chinese Academy of Sciences, Beijing, 100190, China Department of Thermal energy and Power engineering, University of Science & Technology, Beijing, 100083, China
Yu Wang	State Key Laboratory of Pharmaceutical Biotechnology and Department of Pharmacology and Pharmacy, The University of Hong Kong, Hong Kong, China

CHAPTER 1

Biomarkers for Monitoring Response to Therapies and Detection of Acquired Resistance in Advanced Gastrointestinal Cancers

Jens C. Hahne[1,*,°], **Milko Mirchev**[2,°], **Iskren Kotzev**[2], **Andrea Lampis**[1] and **Nicola Valeri**[1,3]

[1] *Centre for Molecular Pathology, The Institute of Cancer Research London & Sutton, Sutton, United Kingdom*

[2] *University Hospital St. Marina, Varna, Bulgaria*

[3] *Department of Gastrointestinal Oncology and Lymphoma, Royal Marsden Hospital London & Sutton, Sutton, United Kingdom*

Abstract: Gastrointestinal cancers are among the most common causes of cancer-related death in the world. Alone the subgroup of colorectal cancer is the third most common cancer in men (746,000 new diagnosed cases in 2012, 10.0% of all cancer cases) and the second in women (614,000 new diagnosed cases in 2012, 9.2% of all cancer cases) worldwide. Beside different treatment strategies including surgery, radiotherapy and/or chemotherapy have improved the survival of patients in the last years, there is still an urgent need to improve prognosis in advanced disease by means of targeted therapies as well as an early detection of development of acquired resistance. For this purpose, molecular biomarkers are a promising option especially in the light of increasing amount of personalized medicine.

In this chapter, we will review classic biomarkers for prognosis already in clinical use (*e.g.* carcinoembriogenic antigen, cancer antigen 19-9, serum pepsinogen I, α-fetoprotein, CD117) and new ones with their usefulness following combined therapeutic regimens. Some new technologies like the very sensitive digital-drople--PCR, next-generation sequencing and high-throughput screening methods for RNA expression patterns will be discussed shortly regarding their application for detecting and discovering potential new biomarkers. Finally, possible new potential mechanisms of resistance related to targeted pathways will be discussed as genetic and epigenetic alterations (hypermethylation of CpG-islands, copy-number variation, lncRNAs and single nucleotide polymorphisms). Furthermore, the potential of circulating tumour DNA and non-coding RNAs, with a special focus on microRNAs as new biomarkers will be addressed.

[°] Both authors contributed equally to this work

[*] **Corresponding author Jens C. Hahne:** Centre for Molecular Pathology, The Institute of Cancer Research, London & Sutton, Cotswold Road, Sutton, Surrey, SM2 5NG, United Kingdom; Tel: +44 0208 915 6633; E-mail: jens.hahne@icr.ac.uk

Atta-ur-Rahman (Ed.)

Keywords: α-fetoprotein, Anal cancer, Billary tract cancer, Biomarker, CA19-9, CD117, CEA, Cft-DNA, Cholangiocarcinoma, Colorectal cancer, Digital-droplet PCR, Gallbladder cancer, Gastric cancer, Gastrointestinal cancer, Liver cancer, Liquid biopsy, lncRNA, Microarrays, MicroRNA, Microsatellite instability, NanoString, Next-generation sequencing, Oesophagus cancer, Pancreatic cancer, SEPT9, Serum pepsinogen I, Small intestine cancer, TAG72, TPS.

INTRODUCTION

Gastrointestinal (GI) cancer summarizes a diverse group of cancers that affect the digestive tract system [1, 2]. This includes cancers of the oesophagus, gallbladder, liver and biliary tract, pancreas, stomach, small intestine, large bowel (colon and rectum) and anus (Table **1**).

GI cancer is the most common form of cancer responsible for nearly 25% of all new cancer diagnosis and responsible for most of cancer related death (around 30% of all cancer related death) worldwide [3, 4].

Clinical approaches for the treatment of GI cancers depend not only of cancer site but also on its clinicopathological stage or its development, and other patient-related factors. Treatment commonly includes surgery, chemotherapy and radiation therapy [5 - 13]. Even though these treatments over the last years have helped to prolong patients` survival, there is still an urgent need to provide a better prognosis particularly in advanced disease by means of targeted therapies as well as identify early biomarkers to overcome development of acquired resistance to treatment. Furthermore, the number of patients with metachronous metastatic disease, especially brain metastasis, is increasing caused by the therapeutical success and the longer survival of GI cancer patients [14 - 17]. This event represents another challenge that clinicians must deal for the treatment of GI cancer patients.

Molecular biomarkers might represent a promising and convenient option for patients` stratification that could help to provide a better treatment of GI cancer patients especially in the light of increasing amount of personalized medicine.

Table 1. Gastrointestinal cancer subtypes incidence.

GI cancer subtypes	Incidence
Colorectal cancer	Third most commonly diagnosed cancer; fourth leading cause of cancer death worldwide
Anal cancer	Makes up to 4% of anorectal malignancies and 1.5% of gastrointestinal malignancies; significant increased incidence rate particularly in HIV positive male
Small intestine cancer	Very uncommon cancer with increasing incidence over the last decades

(Table 1) contd.....

GI cancer subtypes	Incidence
Gastric cancer	Fourth most commonly diagnosed cancer; second leading cause of cancer death worldwide
Oesophagus cancer	Eighth most commonly diagnosed cancer with very high mortality
Pancreatic cancer	First leading cause of cancer death worldwide
Liver cancer	Sixth most commonly diagnosed cancer; third leading cause of cancer death worldwide
Cholangiocarcinoma	Very low incidence rate in western world and higher prevalence in Asia; account for 3% of all GI cancer
Gallbladder cancer	Sixth most frequent GI cancer with very high mortality

SOURCE AND TYPE OF BIOMARKERS

Tissue and Blood Derived Biomarkers

In the discovery phases for both diagnostic and prognostic setting, several studies have attempted to identify molecular biomarkers by using primary lesions as source of information of molecular aberrations (both genetic and genomic) and correlate this information with clinical assessment of the patients` [18, 19]. For instance in colorectal cancer (CRC) common genetic features as mutations in *Adenomatous Polyposis Coli* (*APC*), *Tumour Protein 53* (*TP53*), *Kirsten Rat Sarcoma* (*KRAS*), and *V-Raf Murine Sarcoma Viral Oncogene Homolog B* (*BRAF*) have been investigated for their potential usefulness as prognostic and predictive factors in early as well as advanced cancers [20, 21]. *KRAS* and *BRAF* are well known oncogenes mutated in a consistent proportion of CRC (40% and 8-10%, respectively) [22]. *APC* and *TP53* on the contrary are important tumour suppressors and they are also highly mutated in CRC [23]. Early studies focused on detection of genetic DNA-based biomarkers to potentially identify prognostic and predictive markers [24]. In addition, recently also RNA molecules (mRNAs, microRNAs, lncRNAs) have been investigated for the same purposes [25 - 27]. Availability of primary cancer tissues has also allowed to identify protein biomarkers mainly trough immunostaining of known deregulated proteins [28]. However, due to the heterogeneous mutational spectrum of cancers, multiple assays need to be tested and developed in order to improve sensitivity. Furthermore, it becomes evident that assessment of primary cancer does not necessary reflect the mutational status in metastatic disease [29]. Minimally invasive approaches as blood-based test to assess genomic, trascriptomic and proteomic markers on the contrary not only improve the patient's compliance over biopsy-based tissue recovery but also the genomic abnormalities characteristic of specific GI cancers might be better represented and thus allowing an improved cancer monitoring capability.

Protein Based *Versus* DNA and RNA Based Biomarkers

Proteomic technologies are generally used to discover and measure protein expression levels and have the potential to be used for developing new panels of biomarkers. In diagnostic settings several blood tests are presently used for monitoring cancer recurrence and prognosis. These tests detect the tumour markers carcinoembryonic antigen (CEA), carbohydrate antigen 19-9 (CA19-9), septin-9 (SEPT9), tissue polypeptide specific antigen (TPS) or tumour-associated glycoprotein-72 (TAG72) [30 - 33]. For instance CEA is knowingly elevated in blood of patients with colorectal cancer and others GI cancers [34]. However, its low sensitivity (5-10%) precludes its use for an early detection of cancer but blood over-expression has been used for monitor the disease and as prognostic factor [35, 36]. CA19-9 is a blood antigen that is found generally elevated in colorectal cancers but also other GI cancers [34, 37]. Nevertheless, it has been shown that CA19-9 is not very specific for tumour type and furthermore it has been found less sensitive than CEA [37]. Moreover, both biomarkers are significantly elevated in patients with metastatic disease but not in early cancer thus resulting not useful for diagnostic purposes [31]. Another possible biomarker test is the analyse of SEPT9 methylated DNA in serum [38] but once again this test has a moderate sensitivity (70%) and specificity (90%) [39]. TPS level in the blood indicates the tumour proliferation rate but with a diagnostic sensitivity around 58% it has only very limited importance as tumour marker [40]. The glycoprotein TAG72 can be measured in blood samples but it is not specific for GI cancer [41 - 43] and furthermore TAG72 protein level are found to be increased more frequently in poorly-differentiated tumours than in well-differentiated cancers [44].

New proteomic technologies like two-dimensional electrophoresis (2-DE) and two-dimensional differential in-gel electrophoresis (2D-DIGE) [45] but also the surface-enhanced laser desorption/ionization time of-flight mass spectrometry (SELDI-TOF-MS) or the matrix-assisted laser desorption/ionization time-of-flight mass spectrometry (MALDI-TOF-MS) [46] techniques are evolving as tools for new biomarker discovery. However, the main protein or metabolites derived biomarkers studied appear to be not enough sensitive or specific thus being unable to substitute already validated and used clinical tests is mainly for poor study design or experimental output validation [47]. The advantage of DNA and RNA, especially microRNA, based biomarkers is provided not only in higher stability in body fluids like blood, serum, plasma, urine but also in better specificity given the genomic make-up characteristic of many cancers [48, 49]. Evidence for nucleic acid based biomarkers is emerging during last decade through studies employing new technologies as next-generation sequencing (NGS) [50]. For instance, for colorectal cancer, most of the recent proposed prognostic tests are gene expression-based panels established on RNA from cancer tissue, for detection of potential recurrence like Oncotype DX Colon Cancer, GeneFx Colon, Coloprint, ColoGuideEx, OncoDefender-CRC, CologuidePro [51]. Nevertheless, another issue as preparation of the samples, storage and method of extraction for both proteomic and genomic-trancriptomic analyses are essential and might introduce technical bias that might affect a proper biomarker discovery and/or validation.

NEW POTENTIAL BIOMARKERS

Circulating Free Tumour DNA (cft-DNA)

In the last few years a lot of research has been focussed on the potential use of circulating DNA derived from tumour cells (cft-DNA). Fragmented DNA from cells is released into the blood stream following cell death [52 - 55]. Cft-DNA is relatively stable in blood and can be isolated as a derivation from serum or plasma both manually and with automated technologies [53]. Therefore, analysis of cft-DNA from blood -the so-called liquid biopsies- could represent a minimal-invasive method for monitoring the disease and implement the design of a therapeutic strategy as well as to assess the advent of a potential drug resistance [54]. Furthermore, it has the benefit to increase patients`compliance as well as reducing the tumour biopsies-related risks for the patient. Another important advantage is that cft-DNA isolated from the blood might represent better the intra-tumour heterogeneity than needle biopsy [56 - 58]. It has been proposed that a tumour mass can contain heterogeneous-cell populations with diverse molecular backgrounds and also between primary tumour and metastatic sites often genetic differences are observed [59, 60]. The use of cft-DNA isolated from blood overcomes the limitations of tumour biopsies which are also restricted to the sampling of a small part of a specific reachable tumour side.

For instance, by using ddPCR for analyzing cft-DNA resulted in a powerful method to monitor therapy response as well as disease progression of cancer patients [61, 62]. It has been shown that it was possible to detect gene specific mutations that are known to be associated with tumour progression several months before tumour progression was radiological assessed [63, 64].

Circulating Non-coding RNAs: MicroRNAs and lncRNAs

Another current focus of research is the use of non-coding RNA as new biomarkers [65 - 67]. Non-coding RNAs are a large family of RNA molecules that includes transfer RNAs (tRNAs), ribosomal RNAs (rRNAs), small nucleolar RNAs (snoRNAs), microRNA, small interfering RNA (siRNAs), small nuclear RNA (snRNAs), extracellular RNA (exRNAs), piwi-interacting RNA (piRNAs), small Cajal body RNA (scaRNAs) and long non-coding RNA (lncRNA), all of which are not coding for known proteins [66, 68 - 80]. Today lncRNAs are recognized and accepted as important regulators during development and pathological processes [81 - 85]. LncRNAs are involved in regulating gene expression by binding to chromatin regulatory proteins and they are able to alter chromatin modification as well as transcriptional or posttranscriptional gene regulation by interacting with other RNAs and proteins [70]. Recently, a crosstalk and strong linkage between lncRNA and microRNAs has been identified [86]. Sometimes lncRNA stability can be reduced by interaction with specific microRNAs but it is also possible that lncRNAs act as microRNA decoys resulting in sequestering microRNAs and leading by this to expression of the microRNA target genes [86]. Furthermore, lncRNAs can favour gene expression by competing with microRNAs for specific binding sites in the non-coding regions of mRNAs and prevent the transcriptional repression caused by microRNA binding [86]. Surprisingly some lncRNAs have been identified that can be processed into microRNAs [86].

Of special interest in the light of biomarkers are microRNAs since it has been well established that microRNAs can regulate gene expression acting on target proteins at translational level [80, 87]. Each microRNA can have several target mRNAs; thus interaction of microRNA with the target mRNAs results in direct deregulation of different target proteins acting simultaneously in regulation of diverse cellular pathways [88, 89]. Therefore, variation in microRNA expression can result in reduced mRNA level of the targets and by this direct leading to change in protein levels within the cell [89, 90]. MicroRNAs expression patterns are tissue-specific [91] and often define the physiological nature of the cell [92]. Several publications show that altered microRNA expression in the context of several diseases (*e.g.* cancer, viral diseases, neurodegenerative disease, immune-related diseases) and some pathological conditions are caused by an aberrant expression level of microRNAs [93 - 101]. Therefore, altered microRNA expression patterns are most probably well suited as biomarkers not only for supporting cancer diagnosis but also as predictive tools to anti-cancer therapy as well as for monitoring drug resistance. Another feature is that microRNAs seem to be more robust biomarkers, having demonstrated a higher sensitivity and specificity in several studies aiming to identify cancer specific microRNAs [102].

MicroRNAs are short non-coding RNAs (18-22 nucleotides) and this characteristic offers the advantage of an increased stability in body fluids (over a gene-based mRNA detection approach). Furthermore, microRNAs seem to be associated with RNA-binding multiprotein complexes in blood [88, 103, 104] and it has been shown that their stability is not affected by sample storage conditions [105], an important issue for exploratory studies aiming to identify novel biomarkers. Therefore, microRNAs are more suitable then other RNA based markers.

Furthermore, microRNAs could overcome the low sensitivity of traditional biomarkers -especially for early stage cancers- that is known for most of the blood-based biomarker currently in use, *e.g.* CA19-9 and CEA [36]. Furthermore, advances in the field of biosensor technology allowing today the linear detection of microRNA directly over a wide range of concentrations down to attomolar concentration in human serum background [106 - 110]. The proof-of-principle was made for detection of the cancer biomarker miR-155 in breast cancer patients [106, 111]. So it will soon be possible to develop biosensor specific for any microRNA and enable the fast and accurate measurement of the microRNA in question even if it is present in a low concentration in the blood.

THE ADVENT OF NEW PLATFORMS AND TECHNOLOGIES FOR PERSONALIZED MEDICINE

New technologies and platforms are evolving for the discovery and detection of potential new prognostic and predictive biomarkers. In this paragraph an overview will be given about some of the main technologies that have been tested over last years and their relative applications in different GI cancers for monitoring disease progression [64, 112 - 114], like digital-droplet PCR, next-generation sequencing, transcriptome-based platforms.

Digital-Droplet PCR (ddPCR)

The third generation of PCR, the so called ddPCR, is a PCR method for absolute quantification by counting nucleic acid molecules encapsulated in discrete, volumetrically defined, water-in-oil droplet partitions. The nucleic acid template sample is randomly distributed into these partitions, such that some droplets contain no nucleic acid template and others contain one or more copies of the template defined by a Poisson distribution [115 - 117]. As an end-point measurement, following PCR amplification in single droplets, ddPCR does neither require parallel amplification of any housekeeping gene for normalization for RNA based detections nor the need of standard curves generation for detection and quantification of a target DNA [115, 118 - 122]. Based on its high sensitivity ddPCR is often used for detecting gene copy number variations or minority targets detection such as allelic imbalance (like single-nucleotide polymorphism (SNP) and loss of heterozygosity (LOH)) and identification of rare or low level mutations in cancer [61, 62, 115, 121 - 124] and as independent method for verification of next-generation sequencing [61, 62, 125].

Next-Generation Sequencing (NGS)

Nowadays for NGS several different sequencing platforms have been rapidly evolving (*e.g.* Ion Torrent's PGM, Pacific Biosciences' RS and Illumina MiSeq) and the preparation of sequence libraries is specific to each platform [126, 127]. In the aforementioned case DNA pre-processing is mandatory and can be subdivided into several steps. First, extracted genomic DNA is randomly fragmented into a library of small DNA sequences (fragment size is platform specific and varies between 100 bp and 20 kbp) [128]. Then blunt-ended DNA fragments are created by an enzymatic end polishing step. Finally, specific adapters (still platform specific) are ligated to the fragments at the 3′ and 5′ends. Depending on the sequencing platform the libraries must be pre-amplified prior to sequencing or can be used directly without any amplification step [126, 129 - 132]. Sequencing on the generated libraries is performed in parallel by a stepwise repetition of four reactions: addition of nucleotides; washing step to remove non-incorporated nucleotides; detection of the identity of the incorporated nucleotides on each library fragment and finally a washing step that include removal of fluorescent labels or blocking reactive groups [126, 127]. Individual sequences are assembled to the reference genome and the whole-genome sequence is derived from the consensus of several aligned reads [126, 127]. The applications for NGS can span from gene expression analysis by full RNA sequencing [133 - 135], characterization of RNA structures [136, 137], miRNA expression [138, 139], to analysing chromatin immune precipitation (ChIP)-enriched sequences for identification of transcription factor or miRNA binding sites [140 - 142], DNA methylation studies [142 - 144] and identification of disease related genes and identification of tumour mutations for cancer personalized medicine [126, 145 - 152]. Of special interest is genome-wide sequencing of RNA and DNA in plasma samples for predicting responsiveness or drug resistance to anti-cancer therapy [153].

Other High-Throughput Platforms

Several commercial available MicroArrays for detection of expression of RNAs, microRNAs and long non-coding RNAs (lncRNAs) have been often used for genome-wide analysis of expression differences between normal and disease samples or to study RNA expression changes caused by substances like cytokines, growth-factors or drugs in a cell [154 - 158].

However, a limitation of all microarrays is that an amplification step is always necessary before the chip based measurement takes place.

Another high-throughput method to study RNA, microRNA and lncRNA expression patterns is based on the NanoString nCounter technology. This multiplexed technology detects directly the gene target expression levels without any enzymatic amplification reaction [159 - 161]. NanoString nCounter allows counting molecules in a given sample directly by using barcoded target 5′-end sequence-specific probes for capturing and purification as well as a barcoded 3′-end target sequence-specific fluorescent-labelled probe for detection [159 - 161]. Furthermore, NanoString nCounter technology can also be used for quantification of DNA, CNV and proteins [162, 163]. The advantage of this technology is that RNA and DNA, beside isolated from fresh-frozen material, can also be derived from crude tissue lysates, plasma or serum samples and formalin fixed paraffin embedded tissues [159, 160, 164].

As already mentioned GI cancer includes a heterogenous group of cancers that affect the digestive system [1, 2]. In the following part the different subgroups of GI cancer will be addressed.

COLORECTAL CANCER

Incidence, Diagnosis and Risk Factors

Colorectal cancer is one of the leading causes of death from cancer in Western countries [165 - 168]. It is the third most commonly diagnosed cancer in men and women and the fourth leading cause of cancer death worldwide [167, 169 - 171]. With over 63% of all cancer cases colorectal cancer is mainly a disease of developed countries with a Western culture [171, 172]. Colorectal cancer survival is highly dependent upon stage of disease at diagnosis, and the 5-year survival rate ranges from 90% for cancers detected at the localized stage to 10% for patients with metastatic disease at time of diagnosis [169, 173]. Common risk factors for colorectal cancer development are represented by: age, hereditary factors (familial adenomatous polyposis (FAP) and hereditary nonpolyposis colorectal cancer (HNPCC)), inflammatory bowel disease; low physical activity and excessive body weight, smoking, heavy alcohol consumption and poor nutritional practices [168 - 170, 172, 174 - 182].

Chromosomal and microsatellite instability are also involved in developing colorectal cancer especially in the case of FAP and HNPCC, respectively [182 - 187]. Both are hereditary genetic disorders – in the case of FAP mutations in the *APC* gene are present and in the case of HNPCC mutations in mismatch-repair genes (*e.g. hMSH2, hMSH3, hMSH6, hMLH1, hMLH3, hPMS1, hPMS2*) result in microsatellite instability [182, 188]. Microsatellite instability can be caused by a high-level CpG-island methylation phenotype and *hMLH1* promoter methylation. Given the fact that it is very often associated with *BRAF V600E* mutation the simultaneous analysis of microsatellite instability and *BRAF* mutation confers an inferior prognosis [189 - 192]. These genetic alterations result in an increased risk to develop colorectal cancer [182, 186, 187]. Even in the case of HNPCC microRNA deregulation seems to be the key; as discussed later in this chapter over-expression of miR-155 was identified as reason for microsatellite instability by significantly down-regulating core mismatch-repair proteins [193]. This finding underlines once again the importance of microRNAs for physiological and pathological conditions as well as their potential to function as biomarkers.

Despite being an age-related disease commonly arising after 50 years, over the last years colorectal cancer diagnosed in younger people is dramatically increasing [169, 194 - 196].

Current Biomarkers in Use

For colorectal cancer a panel of tumour markers (CEA, CA19-9, p53, RAS, thymidine synthase, dihydropyrimidine dehydrogenase, thymidine phosphorylase, deleted in colon cancer (DCC) protein and microsatellite instability) has been studied and its role in pre-operative staging and post-operative follow-up has been evaluated [36].

In the majority of colorectal cancer cases the glycoprotein CEA is over-expressed and secreted into the blood [34, 197]. Therefore the plasma level of CEA can be measured quantitatively but unfortunately due to its lack of sensitivity in early disease stages it is not recommended as a screening test for colorectal cancer [34, 36, 197]. Nevertheless, the pre-operative measurement of CEA level in patients is supportive for staging and surgical treatment planning [36]. Especially in patients who undergo resection for metastases CEA levels have prognostic value [198] and the lack of return of CEA to normal levels after surgical treatment is a marker for inadequate resection [197, 199]. The post-operative monitoring of CEA level is a useful diagnostic marker because CEA is the most frequent indicator of recurrence in asymptomatic patients and is sensitive for detecting metastases [200, 201]. Therefore, the ASCO guidelines suggest that CEA testing should be performed every 3 months in patients with stage II or III disease for at least 3 years after diagnosis given that the patient is a candidate for surgery or systemic therapy [36]. Also a systemic therapy of metastatic colorectal cancer should be monitored by measuring the level of CEA at the beginning and every 1 to 3 months during active treatment [36]. Persistently rising values above baseline suggest disease progression and require restaging [36]. However, early rise of CEA (4 to 6 weeks after treatment initiation) should not always be interpreted as an alarming sign since it can occur as treatment induced changes in liver function, especially during oxaliplatin treatment, or can be related to other non-cancer related conditions such as gastritis, peptic ulcer disease, liver diseases and chronic inflammation [202, 203].

The role of CA19-9 as a prognostic marker in stage IV colorectal cancer patients who have undergone curative resection is supported by several studies, mainly from Asia [204 - 207]. Post-operative CA19-9 level seems to be an independent predictor for overall survival and the combination of elevated CEA and CA19-9 levels were positively associated with disease recurrence and lymphatic invasion [208]. But in some other studies the prognostic and predictive value of CA19-9 level was not always confirmed [205, 209, 210] and it is obviously that CA19-9 is a less sensitive biomarker than CEA [37].

Current Treatment Modalities

It is well established that aberrant epidermal growth factor receptor (EGFR) signalling is involved in the progression of colorectal cancer [211 - 213]. EGFR is an upstream receptor of the RAS-RAF-MEK- as well as the AKT-PI3K-pathway; both pathways resulting in cell proliferation and increased cell survival after activation [214, 215]. Therefore, blocking of EGFR is an attractive and efficient therapeutic option [214]. Cetuximab and panitumumab, are two monoclonal antibodies that prevent the activation of signal transduction pathways *via* EGFR binding and are approved for colorectal cancer treatment [216 - 219]. Unfortunately this targeted therapy against EGFR has only limited success in a specific subset of patients with metastatic colon cancer [214, 220 - 223].

A standard chemotherapeutic regime in the presence of distant metastases (disease stage IV) is a combined chemotherapy such as FOLFOX (folonioc acid, 5`-fluorouracil, oxaliplatin) or FOLFIRI (folonic acid, 5`-fluorouracil, irinotecan) associated with antibodies directed against EGFR (cetuximab, panitumumab) or anti-angiogenic drugs that target VEGF pathway (bevacizumab, aflibercept) [214, 224 - 227]. However, patients` selection for cetuximab and panitumumab treatment is based on activating mutations in the KRAS and NRAS genes (that result in permanent activation of the RAS-MEK-ERK signalling pathway) [214, 228 - 232]. Therefore, *KRAS* and *NRAS* are being used as biomarkers for patients` selection for the use of anti-EGFR monoclonal antibodies treatment [233 - 235]. Thus, only patients who are reported wild-type for mutations in *RAS* genes are suited for this targeted therapy [219, 236, 237].

Several clinical studies (*e.g.* CRYSTAL, PRIME and OPUS) have shown that patients with metastatic colorectal cancer, whose tumours contain activating mutations of *RAS* genes, do not derive any benefit from EGFR antibody therapy and this therapy may in fact have a detrimental effect in such patients especially if it is combined with oxaliplatin [238]. Retrospective analyses of tumours from these studies with newer methods such as next generation sequencing or BEAMing technology resulted in identification of additional mutations in 15-31%

of tumours that were initially considered as RAS-wild-type and those patients had worse outcomes [238 - 240].

Beside mutations in the RAS genes also mutation in *BRAF* and *PIK3CA* might represent a reason for failure of the antibody based anti-EGFR therapy [228, 241, 242]. Furthermore, amplification of *HER-2*, *MET* and *KRAS* as well as silencing of *PTEN* have been recently identified as additional mechanisms that lead to failure of cetuximab and panitumumab based treatments [243 - 248].

Nevertheless, also patients who initially respond to the anti-EGFR antibody therapy are most likely to develop resistance within 3-12 months against the treatment [214, 220, 223, 249]. However, by using liquid biopsies it has been shown that it is possible to detect *KRAS* mutations 10 months before disease progression is radiologically detected [64, 214, 250 - 252]. This further demonstrates the power of new detection systems. Therefore, therapy can be altered earlier to prevent the progression and increase patient's survival. Some reports have shown that it is possible to overcome the acquired resistance to anti-EGFR antibodies with a combined therapy [244, 253]. A multidrug approach that involves the use of cetuximab with the MEK inhibitor pimasertib or the MET inhibitors crizotinib resulted in restoring sensitivity to resistant tumours in some patients [244, 253]. Combination therapies however might bear the risk of higher toxicity and they could not *per se* improve the treatment in every patient [254, 255] and therefore combination therapies should be single patient tailored. Liquid biopsies are one possibility to monitor such treatments. First studies are underway to use cft-DNA present in the blood of colorectal cancer patients as marker for treatment response and for detection of resistance development [256, 257]. Changes in cft-DNA are in good correlation to later tumour response and therefore liquid biopsies seem to be a good prognostic tool [256].

Potential Biomarkers Currently Under Further Investigation

The first microRNAs found to be reduced in colorectal cancer patients are miR-143 and miR-145 [258]. In agreement with the outstanding role of *KRAS* for colorectal cancer development, *KRAS* has been identified as one of the miR-143 targets [259]. In the meantime, large-scale microRNA expression analyses have been performed in human colorectal cells and colon cancer patients resulting in a microRNA expression pattern for colon cancer [260, 261]. As a logical consequence, the potential of microRNAs as diagnostic markers has been suggested in the last years [154, 262 - 264]. Given the amount of deregulated microRNA in colon cancer it is not surprisingly that several microRNAs have been investigated as biomarkers in the context of this disease. A closer look into recent discoveries shows that microRNAs probably could allow a very distinct discrimination amongst colon cancer stages [154, 265] thus the importance of different expressed microRNAs as diagnostic markers increases further. MiR-92 has been found highly expressed in plasma of colorectal cancer patients and seems to be a promising biomarker for colorectal cancer screening tests [262 - 264]. For instance, in plasma of advanced colorectal cancer patients, miR-92a and miR-29a might be of diagnostic value [262, 263]. The proof-of-concept that microRNAs are also useful for monitoring the therapeutic response was provided in a cohort of colon cancer patients treated with 5′-fluorouracil and the next-generation oral 5′-fluorouracil compound S-1 [262, 266]. In this study expression of miR-181b and let-7g was significantly correlated with successful therapy [262, 266]. In HNPCC regulation of mismatch-repair proteins by miR-155 was proofed [193]. High expression of miR-155 resulted in increased microsatellite instability by reducing the expression level of MLH1, MSH2 and MSH6 [193]. A high-throughput microRNA screening revealed that miR-145 is down-regulated in tumour samples with different microsatellite status compared to healthy colorectal tissue [267]. Based on this data it was possible to associate microRNA expression levels to specific microsatellite status as well as recurrence of colorectal cancer [267]. In other studies the expression of miR-320 and miR-498 has been shown to correlate with disease-free survival [267], whereas high expression of miR-200c and miR-21 appear to be associated with poor prognosis for colorectal cancer patients [262, 268 - 270]. Like in many types of cancers miR-21 over-expression has been associated with low sensitivity and poor response to chemotherapy in colorectal cancer [269 - 271].

Recently it has been shown that a single nucleotide polymorphism in microRNA or microRNA target can be used as biomarker for survival and response to treatment for colorectal cancer patients [272, 273]. For prognosis of early-stage colorectal cancer, the single-nucleotide polymorphism rs61764370 in the let-7 complementary site 6 of KRAS mRNA was identified [273]. Furthermore, it seems possible that this single-nucleotide polymorphism could be a biomarker for the benefit from anti-EGFR antibodies based therapies in metastatic colorectal cancer [273]. In high-risk locally advanced colorectal cancer, the single nucleotide polymorphism rs4919510 in miR-608 was identified as potential diagnostic marker for disease prognosis [272].

Also lncRNAs could be useful biomarkers for colorectal cancer [274]. Two lncRNAs (HOTAIR and CRNDE) were found to be up-regulated in tumour tissue and blood of colorectal cancer patients [275, 276]. The expression level of HOTAIR correlates with the tumour stage and overall survival [275]. Therefore, detection of HOTAIR level in the blood of colorectal cancer patients could serve as additional prognostic marker [275] but it must be kept in mind that altered expression of HOTAIR is also observed in the context of other diseases *e.g.* gastric and pancreatic cancer [277, 278].

The most important points of the above discussion about colorectal cancer are summarized in Table **2**.

Table 2. Colorectal cancer.

Risk factors	- Age - Low physical activity - Excessive body weight - Smoking - Heavy alcohol consumption - Poor nutritional practices - Inflammatory bowel disease - Hereditary factors especially chromosomal and microsatellite instability in the case of familial adenomatous polyposis (FAP) and hereditary nonpolyposis colorectal cancer (HNPCC)
Current treatment modalities	- Surgery - FOLFOX (folonioc acid, 5`-fluorouracil, oxaliplatin) - FOLFIRI (folonic acid, 5`-fluorouracil, irinotecan) - Antibodies directed against EGFR (cetuximab and panitumumab) - Anti-angiogenic drugs that target VEGF pathway (bevacizumab, aflibercept)
Current biomarkers	- CEA - CA19-9 - p53 - RAS mutations - Thymidine synthase - Dihydropyrimidine dehydrogenase - Thymidine phosphorylase - Deleted in colon cancer (DCC) protein - Microsatellite instability
Potential biomarkers currently under investigation	MicroRNAs: -MiR-143 - MiR-145 - MiR-92 -MiR-92a - MiR-29a - MiR-181b - Let-7g - Mir-155 - MiR-320 - MiR-498 - MiR-200c

(Table 2) contd.....

Potential biomarkers currently under investigation	- MiR-21 - MiR-608 Long non-coding RNAs: - HOTAIR - CRNDE Single-nucleotide polymorphism: -*KRAS* (rs61764370) - miR-608 (rs4919510)

ANAL CANCER

Incidence, Diagnosis and Risk Factors

Cancer of anal canal makes up to 4% of all anorectal malignancies and 1.5% of gastrointestinal malignancies [165, 279 - 281]. But anal cancer has a significant increased incidence rate (*e.g.* 130% since 1970 in UK according to Cancer Research UK) particularly in human immunodeficiency virus (HIV) positive male [281]. As high risk factors causing anal cancer human papillomavirus (HPV) infection, anoreceptive intercourse, cigarette smoking, and immunosuppression, have been identified [282, 283].

Current Biomarkers in Use

Different biomarkers have been studied in squamous cell anal cancer which elucidated to some extend the cancer pathogenesis but none of them are sufficient to determine the prognosis or guide the treatment [284].

Current Treatment Modalities

In different clinical trials it was proved that anal cancer responded well to standard treatment with concurrent radio- and chemo-therapy resulting in a 60% cure rate [285 - 289]. Anal cancer metastasizes preferential through the lymphatic system into liver, lung and lymph nodes [290, 291].

Neither EGFR mutation nor amplification is observed in anal cancer even if sometimes EGFR is overexpressed [292, 293]. In the context of this cancer EGFR over-expression is not dictating the treatment regime [294, 295]. Surprisingly also *KRAS* mutations are an exception in anal cancer patients [294, 296].

Given the fact that anal cancer is often caused by HPV-infection [283, 297] it is not surprisingly that therapies used for the treatment of other HPV-related cancers (*e.g.* head and neck or cervical cancer [298, 299]) are also effective in anal cancer patients. Furthermore, typical consequences of HPV-infection like genomic instability, gene alterations and increased DNA damage are observed in anal cancer [300 - 302].

Phosphorylation and subsequent activation of AKT resulting in an increased activation of the central PI3K/AKT-signal transduction pathway also play an important role in anal cancer like in most of HPV-caused tumours [297, 303, 304].

Potential Biomarkers Currently Under Further Investigation

Anal cancer is poorly examined on molecular and biological level most probably due to the fact that it is a rare tumour with low tissue availability. No data exist about aberrant microRNA expression in anal cancer patients.

Since anal cancer is in most cases HPV-related, the cell cycle regulator p16 may serve as a biomarker for detecting HPV-DNA. Tumours with moderate to high p16 expression may respond better to chemoradiotherapy and have a lower relapse rate compared to those with low expression [290]. In a systematic review which analyzed 29 different biomarkers, only p53 and p21 were prognostic in more than one study and only p53 mutations were associated with poorer survival [305].

Up to now only one study address epigenetic differences in the context of anal cancer [306]. It was possible to identify 16 CpG sites with differentially methylation status in high-risk anal cancer patients compared to low-risk patients and by this to discriminate locally advanced from early anal cancer [306]. Therefore, it seems possible to use epigenetic alterations as potential biomarkers for stratification of anal cancer patients.

The most important points of the above discussion about anal cancer are summarized in Table **3**.

Table 3. Anal cancer.

Risk factors	- Human papillomavirus (HPV) infection - Anoreceptive intercourse - Cigarette smoking - Immunosuppression
Current treatment modalities	- Surgery - Chemoradiation therapy based on mitomycin C and 5′-fluorouracil 5′ fluorouracil and cisplatin - Radiotherapy
Current biomarkers	- Not available

(Table 3) contd.....

Potential biomarkers currently under investigation	Proteins: - Cell cycle regulator p16 - p53 Epigenetic alterations: - Methylation status of 16 CpG sites

SMALL INTESTINE CANCER

Incidence, Diagnosis and Risk Factors

Cancer of the small intestine is very uncommon but the incidence of small intestine cancer has increased over the last decades [165, 307 - 310]. Patients with Crohn's disease, Celiac disease, Familial Adenomatous Polyposis (FAP) and Peutz-Jeghers syndrome have increased risk for developing small intestine cancer [307, 309 - 313]. Furthermore, some risk factors like high consumption of red or processed meat, saturated fat, obesity and smoking have been connected to this type of GI cancer [307, 311]. The prognosis for carcinomas of the small intestine is poor (5 years relative survival $< 30\%$) [165, 307, 308].

Current Biomarkers in Use

Up to now no specific biomarkers are available for predicting prognosis and treatment decision for small intestinal cancer.

Current Treatment Modalities

Because of the low prevalence the efficacy of chemotherapy for treating small intestine cancer is only addressed in limited studies with small patient numbers [314 - 320]. Treatment with 5-fluorouracil of advanced small intestine cancer patients results in median survival of 9-13 months [314 - 316]. Recently it was demonstrated that curative resection and definitive chemotherapy for unresectable cases of small intestine cancer can be an effective treatment approach [317, 321].

Potential Biomarkers Currently Under Further Investigation

Furthermore, only few studies have been performed to analyze genetic differences and aberrant microRNA expression in small intestine cancer.

One study based on 46 patients showed that small intestine cancer can be subdivided into microsatellite instable and microsatellite stable tumours [322]. Patients with chromosomal stable small intestine cancer were found to have low frequencies of altered CpG island methylation, rarely *BRAF* mutations but often mutations in *KRAS* gen [322].

The most important points of the above discussion about small intestine cancer are summarized in Table **4**.

Table 4. Small intestine cancer.

Risk factors	- Crohn's disease - Celiac disease - Familial Adenomatous Polyposis (FAP) - Peutz-Jeghers syndrome - High consumption of red or processed meat - High consumption of saturated fat - Obesity - Smoking
Current treatment modalities	- Surgery - 5′-fluorouracil
Current biomarkers	- Not available
Potential biomarkers currently under investigation	Epigenetic alterations: - Microsatellite instability

GASTRIC CANCER

Incidence, Diagnosis and Risk Factors

Gastric cancer also known as stomach cancer was the most common cancer less than a century ago [323, 324]. Since the widespread use of refrigerators the rate of gastric cancer decreased [325]. Beside increased consumption of fresh vegetables and fruits the intake of salt, which had been used as a food preservative, decreased. Furthermore also food contaminations caused by carcinogenic compounds arising from unrefrigerated meat products have been strongly decreased [325]. Although it is no longer the most common cancer worldwide, every year approximately 990,000 people are still diagnosed with gastric cancer worldwide, of whom about 738,000 die from cancer [323, 324, 326]. Currently, as incidence, gastric cancer is the fourth most common cancer and the second leading cause of cancer death worldwide [165, 323, 327]. The high mortality is caused by the fact that gastric cancer often forms metastasis [328]. The highest incidence rates are observed in East Asia, East Europe, and South America, whereas the lowest rates are observed in North America and most parts of Africa [323, 329].

Gastric cancer can be subdivided into two subgroups: an intestinal and a diffuse type [330 - 332]. Intestinal tumours progress in sequential steps starting with atrophic gastritis followed by intestinal metaplasia then dysplasia and finally cancer [331, 332]. Diffuse gastric cancer is only related on chronic gastritis associated with *Helicobacter pylori* (*H. pylori*) infection [331, 332].

In a small subgroup of gastric cancer patients (up to 3%) hereditary diffuse gastric cancer (HDGC) is diagnosed [333]. Mutations in the *E-cadherin* gene are known to cause HDGC [334].

Current Treatment Modalities

Surprisingly incidence rates of gastric cancer are found to be 2- to 3-folds higher in men than in women [323, 326, 335]. The prognosis for carcinomas of the stomach is poor (5 years relative survival < 30%) [165]. Nevertheless, in patients' receiving perioperative chemotherapy before operation 5-year survival was found to be 36% [336] and for advanced or metastatic gastric cancer the 5-year survival is around 5-20% with median overall survival being less than 12 months [337 - 339].

Best survival results are achieved with a combined chemotherapy composed of 5`-fluorouracil and cis-platin in combination with an anthracycline [340, 341].

Current Biomarkers in Use

Based on several studies showing over-expression or amplification of human epidermal growth factor receptor-2 (HER-2 or ERBB2) in up to 30% of gastric tumours it becomes more and more reasonable to recognize HER-2 as an important biomarker and key driver in gastric cancer [340, 342 - 349]. The HER-2 positivity rate in gastric tumours is very similar to HER-2 positive breast cancer cases [350, 351]. Based on the good tolerability profile of the HER-2 antibody trastuzumab in patients with breast cancer [352, 353] trastuzumab was also introduced into the treatment of gastric cancer patients [340, 345]. In one large international multicenter study it was shown that addition of trastuzumab to the standard combination chemotherapy significantly improved overall survival compared with chemotherapy alone in advanced gastric cancer [345]. But up to now it is not finally sorted out if the observed positive trastuzumab effect is merely based on the fact that HER-2 expression is more common found in patients with intestinal-type tumours that are known to have a better clinical outcome compared to gastric cancer patients with diffuse-type tumours [344, 354 - 357]. Nevertheless, in several studies were demonstrated that HER-2-positive gastric tumours are a more aggressive cancer subgroup and resulting in poor outcome like it is the case in HER-2-positive breast cancer [342, 344, 358, 359]. Therefore, further studies are an urgent need to clarify if HER-2 expression has a good or poor prognosis in gastric cancer.

Potential Biomarkers Currently Under Further Investigation

Also in gastric cancer the first attempts have been done to use cft-DNA for monitoring the disease status [360, 361]. In a small gastric cancer patient cohort the concentration of cft-DNA was found to increase in parallel with established serological biomarkers but cft-DNA has not had advantage over serological biomarkers already in use [360].

Gastric cancer is characterized by numerous genetic and epigenetic alterations. A variety of different genetic and epigenetic changes are involved in and favour development of gastric cancer [362, 363]. Loss of the *APC* tumour suppressor gene, *ß-catenin* and *K-ras* mutations as well as inactivation of *TP53* tumour suppressor gene are often found in gastric cancer patients [362, 364 - 366]. Also *c-met* and *HER-2* (already discussed above) gene amplifications are often observed [362, 367 - 371]. Furthermore CpG-hypermethylation of *hMLH1* mismatch repair gene resulting in microsatellite instability is a common event in gastric cancer [330, 362, 372 - 377]. Beside hypermethylation, also increased rate of hypomethylation can be a reason for cancer as it was shown for the oncogenes *ELK1*, *FRAT2*, *r-RAS*, *RHOB*, and *RHO6* that are found to be activated by demethylation of specific CpG sites in more than half of gastric cancers [378, 379].

Nevertheless, overexpression of cyclin E, reduced expression of p27KIP1 by deranged degradation in proteasome and reactivation of telomerase are often observed in gastric cancer patients [380, 381]. A prerequisite for progression of poorly differentiated or scirrhous gastric carcinomas is the loss of E-cadherin (especially in the context of *H. pylori* infection) and catenins by mutation or CpG hypermethylation [382 - 385]. Additional some reports describe loss of heterozygosity (LOH) and mutation of tumour suppressor gene *PTEN* in gastric cancer [386, 387].

The role of non-coding RNA (*e.g.* microRNA and lncRNA) in gastric cancer has been extensively analyzed during the last years [328, 388 - 392]. Up to now the role of lncRNAs in context of gastric cancer is not fully understood and a lot of work is still necessary. Nevertheless, it is obviously that the altered expression of some lncRNAs (*e.g.* H19, HOTAIR, ANRIL) are connected to gastric cancer and formation of metastasis but not limited to gastric cancer and therefore these lncRNAs have only limited use as specific biomarkers [277, 278, 393 - 396]. Recently it was demonstrated that the c-Myc induced lncRNA CCAT-1 (colon cancer-associated transcript-1) has a high expression in early gastric cancer and is involved in cancer proliferation, migration and metastasis [362, 397]. But once again the lncRNA CCAT-1 is not specific for gastric cancer and is found to be over-expressed also in colon cancer, hepatocellular carcinoma and breast cancer [398 - 401]. In contrast microRNAs seem to be much better biomarkers. It was possible to discriminate gastric cancer caused by *H. pylori* from *H. pylori* negative cancer based on the microRNA expression pattern [402]. In general, despite miR-21 that is very often connected to different cancer types [271] miR-106, miR-17, and miR-221 are discussed as potential biomarkers for gastric cancer in several studies [403 - 408]. Furthermore, increased expression of miR-10 and miR-21 as well as low level of miR-125a and miR-146a were identified as indication for high metastasis risk in gastric cancer patients [49, 409, 410]. Using liquid biopsies from gastric cancer patients a signature based on five microRNAs (miR-1, miR-20a, miR-27a, miR-34, miR-423) was identified as diagnostic fingerprint for gastric cancer [411]. Additional the microRNA expression level correlated well with the tumour stage [411] and therefore this microRNA fingerprint will be a useful diagnostic tool in the future. This is an excellent example for the use of microRNA isolated from blood as diagnostic parameter.

The most important points of the above discussion about gastric cancer are summarized in Table **5**.

Table 5. Gastric cancer.

Risk factors	-*Helicobacter pylori* infection - High intake of salt - Consumption of enormous amounts of smoked foods, salted fish and meat, and pickled vegetables - Low intake of fresh vegetables and fruits - Intake of food contaminations caused by unrefrigerated meat products - Gender
Current treatment modalities	- Perioperative chemotherapy - 5`-fluorouracil and cis-platin in combination with an anthracycline and trastuzumab
Current biomarkers	- CEA - CA19-9 - HER-2
Potential biomarkers currently under investigation	Epigenetic alterations: - Microsatellite instability Proteins: - Cyclin E - p27KIP1 - Telomerase - E-cadherin Long non-coding RNAs: - HOTAIR - H19

(Table 5) contd.....

Potential biomarkers currently under investigation	- ANRIL - CCAT-1 MicroRNAs: -Fingerprint based on miR-1, miR-20a, miR-27a, miR-34 and miR-423 - MiR-21 - MiR-106 - MiR-17 - MiR-221 - MiR-10 - MiR-125a - MiR-146a

OESOPHAGEAL CANCER

Incidence, Diagnosis and Risk Factors

The incidence of oesophageal cancer is rapidly increasing during the last years [412] and it is currently the eighth most common incident cancer in the world with a very high mortality [413 - 416]. Risk factors are represented by smoking, hot tea drinking, red meat consumption, poor oral health, low intake of fresh fruits and vegetables, obesity, nass use (a chewing tobacco product), opium consumption and low socioeconomic status [413, 414, 417 - 425].

Oesophageal cancer is subdivided into two groups according to histological appearance: oesophageal squamous cell carcinoma (predominant in western countries) and oesophageal adenocarcinoma (most common form in Asia) [426, 427].

Current Biomarkers in Use

No molecular markers for prediction of prognosis or treatment effect have been discovered so far for squamous cell carcinoma. For oesophagus adenocarcinoma the HER-2 status should be determined by immunohistochemistry for protein expression or *in-situ* hybridization for HER2 gene amplification [428].

Current Treatment Modalities

HER-2 positive oesophageal cancer patients can be treated with trastuzumab-containig regimens [345, 429].

Nevertheless, oesophageal cancer is mostly treated by radiation with concurrent chemotherapy or surgery but the 5-year overall survival rates is with 5% to 10% very poor [416, 430, 431]. The most commonly used drugs for oesophagus cancer treatment are 5′-fluorouracil and cisplatin [431].

One reason for the high mortality of oesophageal cancer is the late state of diagnosis [413] another reason is the role of tumour microenvirement [432]. Especially in the case of oesophageal cancer the surrounding cells seem to have an important influence on the tumour cells [432]. Beside cancer-associated fibroblasts that are able to support tumour growth and metastasis by altering the extracellular matrix by secreting growth factors and cytokines [432] also different immune cells (*e.g.* myeloid-derived suppressor cells, tumour-associated macrophages and regulatory T-cells) are involved in the development of oesophageal cancer [432]. Most probably therapies targeting the tumour microenvirement might be able to increase the survival of oesophageal cancer patients and currently different clinical studies are underway that target the cells in the tumour environment [432].

Potential Biomarkers Currently Under Further Investigation

Recently, a study using cft-DNA sequencing for analysing tumour mutations in oesophageal cancer patients was published [433]. It becomes obviously by comparing pre-surgery and post-surgery cft-DNA sequencing results that mutations specific for oesophageal cancer [434] had reduced frequency or disappeared after surgery [433]. Therefore cft-DNA could be used for detection and monitor therapy of oesophageal cancer [433].

As one frequent reason for oesophageal cancer the inactivation of *ECRG 4* by hypermethylation was identified [435]. Other hypermethylation hotspots in oesophageal cancer are *p16*, *RASSF1A* and *retinoic acid receptor-beta2* gene [436 - 439].

MicroRNAs are found to be deregulated in oesophageal cancer but most of these miRNAs are also dysregulated in other cancers [440]. MiR-21 together with miR-375 could have prognostic relevance but up to now a detailed analysis of circulating miRNAs as parameter for diagnosis and treatment response for oesophageal cancer is missing [440, 441]. In contrast, lncRNAs have already been evaluated as important players in oesophageal cancer progression as well as biomarkers [442 - 444]. The expression pattern of three lncRNAs (ENST00000435885.1, XLOC_013014 and ENST00000547963.1) was identified as independent prognostic factor for oesophageal tumour patients with different disease stages [444].

Furthermore, increased copy numbers of several *ABC transporter* genes resulting in higher protein level of these transporters were found in oesophageal cancer patients [445]. Especially high expression of ABCC4 was identified as poor prognostic factor for oesophageal cancer [445].

Up to now studies focussing on the identification of single-nucleotide polymorphism in the context of oesophageal cancer are only performed in Asian populations [446 - 449]. Single-nucleotide polymorphism in alcohol dehydrogenase 1B (*ADH1B*) (rs1229984), aldehyde dehydrogenase 2 family (*ALDH2*) (rs671), phospholipase C, epsilon 1 (*PLCE1*) (rs2274223), in chromosome 20 open reading frame (*C20orf54*) (rs13042395), phosphodiesterase 4D (*PDE4D*) (rs10052657), runt-related transcription factor 1 (*RUNX1*) (rs2014300), near to unc-5 homolog C (C. elegans)-like (*UNC5CL*) (rs10484761) and in vascular endothelial growth factor (*VEGF*) gen has been linked with oesophageal cancer [446 - 450].

The most important points of the above discussion about oesophagus cancer are summarized in Table **6**.

Table 6. Oesophagus cancer.

Risk factors	- Smoking - Hot tea drinking - Red meat consumption - Poor oral health - Obesity - Nass use - Opium consumption - Low socioeconomic status - Low intake of fresh vegetables and fruits
Current treatment modalities	- Radiation with concurrent chemotherapy or surgery - 5′-fluorouracil and cisplatin - Trastuzumab for HER-2 positive oesophagus adenocarcinoma
Current biomarkers	- Not available for squamous cell carcinoma - HER-2 for oesophagus adenocarcinoma
Potential biomarkers currently under investigation	Epigenetic alterations: - Hypermethylation of genes coding for *ECRG 4*, *p16*, *RASSF1A* and *retinoic acid receptor-beta2* Single-nucleotide polymorphism: -*ADH1B* (rs1229984) -*ALDH2* (rs671) -*PLCE1* (rs2274223) -*C20orf54* (rs13042395) -*PDE4D* (rs10052657) -*RUNX1* (rs2014300) -*UNC5CL* (rs10484761) Proteins: - ABCC4 MicroRNAs: - Fingerprint of miR-21 and miR-375

(Table 6) contd.....

Potential biomarkers currently under investigation	Long non-coding RNAs: - ENST00000435885.1 - XLOC_013014 - ENST00000547963.1

PANCREATIC CANCER

Incidence, Diagnosis and Risk Factors

Pancreatic cancer has an extremely high mortality and is the leading cause of death from cancer [165, 451] partially due to the fact that it is often diagnosed at an advanced stage. The burden of pancreatic disorders is expected to increase over time [451, 452]. Pancreatic cancer affects people with Africans origin more than any other race [451, 452]. Heavy drinking, smoking and non-O blood groups have been evaluated as significant risk factors for pancreatic cancer [451, 453 - 455]. More than 80% of pancreatic cancer cases are due to sporadically occurring mutations, approximately 10% are related to inherited conditions and 10-15% have a positive familial history for pancreatic cancer [456].

Current Biomarkers in Use

The tumor-associated CA19-9 antigen is the only classical biomarker that has been extensively studied in pancreatic cancer [457, 458]. Due to its low specificity and sensitivity it is not useful as screening marker for pancreatic cancer [458]. Furthermore, CA19-9 level is elevated in many other conditions such as tumours of the upper gastrointestinal tract, ovarian cancer, hepatocellular carcinoma, colorectal cancer, inflammation of the hepatobiliary system and benign diseases such as thyroid disease or biliary obstruction [458]. Nevertheless, the pre-operative CA19-9 level has been shown to predict patient`s outcome; a preoperative level of CA19-9 $\geq$500 IU/ml indicates a worse prognosis after surgery [457]. In the post-operative setting the level of CA19-9 may predict disease recurrence, which is helpful in patient's follow-up and for the decision who is treated with surgery alone or with adjuvant chemo- and/or radiotherapy [457]. However, a combination of imaging studies, biopsy and CA19-9 levels is a better option for disease follow-up rather than either methodology alone [457].

Current Treatment Modalities

For pancreatic cancer traditional chemotherapy is ineffective because pancreatic carcinoma has one of the densest stroma of all epithelial tumours and this dense stroma protects cancer cells against the drugs [459, 460]. Therefore surgery remains the only curative treatment for pancreatic cancer [461]. But even after curative surgical treatment the 5-year survival rate is only 20% [462].

Chemotherapy for pancreatic cancer is mostly restricted to 5′-fluorouracil or gemcitabine monotherapy [463, 464] or a combination therapy containing one of these drugs [461, 465 - 468]. The combination therapy FOLFIRINOX (5′-fluorouracil, leucovorin, irinotecan and oxaliplatin) enables patients to have a progression-free survival of 6.4 months and an overall survival rate of 11.1 months [459, 469].

Based on the finding that EGFR is over-expressed in 90% of pancreatic cancer patients [470, 471] a combination therapy with gemcitabine and the EGFR targeting drug erlotinib was established [467].

Potential Biomarkers Currently Under Further Investigation

In pancreatic cancer the most commonly mutated genes are *KRAS, TP53, CDKN2A* and *SMAD4* [457, 472] leading to an increased cell proliferation, survival, motility and remodelling. One subgroup of pancreatic cancer (medullary pancreatic carcinoma) that has a better prognosis compared to other pancreatic cancer subgroups is characterized by epigenetic silenced *MLH1* gene *via* promoter methylation and the missing *MLH1* gene expression results in microsatellite instability [457, 473 - 475]. In general hypermethylation of promoters are a common epigenetic deregulation found in pancreatic cancer [476]. In the course of malignant progression *FOXA1/2* gene expression is lost due to increasing promoter hypermetylation [477] and as a result *E-cadherin* expression is reduced and epithelial-mesenchymal transformation is increased. More often genomic instability in pancreatic cancer is caused by hypomethylation resulting in over-expression of genes and proteins [476]. One example is *MUC4* gene expression that is increased from normal tissue over pancreatic lesion to pancreatic cancer caused by increasing promoter hypomethylation [478].

The use of cft-DNA in the context of pancreatic cancer for clinical purpose is limited due to the low sensitivity and high variability observed in different studies up to now [479 - 481].

In contrast microRNA expression has been found to be deregulated in pancreatic cancer patients [482 - 486]. A panel of four microRNAs (miR-145, miR-150, miR-223 and miR-636) was identified as diagnostic marker for pancreatic cancer which together with the classical biomarker CA19-9 enhanced significantly the specificity and sensitivity for prediction [487]. According to another study the expression level of miR-1290 in serum and tumour tissue of pancreatic cancer patients have a higher diagnostic accuracy than CA19-9 [482, 488, 489]. It is not surprisingly that even in pancreatic cancer miR-21 has prognostic potential and a high miR-21 expression level correlates with poor clinical outcome [490]. But also a down-regulation of miR-34 was associated with a decrease in overall survival of pancreatic cancer patients [491]. Furthermore, over-expression of the lncRNAs HOTAIR and MALAT-1 were also identified as marker for poor clinical outcome in the context of pancreatic cancer [492, 493]. Therefore, non-coding RNAs are most probably best suited to increase detection and therapy of pancreatic cancer patients in the near future.

The most important points of the above discussion about pancreatic cancer are summarized in Table **7**.

Table 7. Pancreatic cancer.

Risk factors	- Heavy drinking - Smoking - Non-O blood groups - African origin
Current treatment modalities	- Surgery - Monotherapy with 5′-fluorouracil or gemcitabine - FOLFIRINOX (5′-fluorouracil, leucovorin, irinotecan and oxaliplatin) - Combination therapy with gemcitabine and erlotinib
Current biomarkers	- CA19-9
Potential biomarkers currently under investigation	Epigenetic alterations: - Microsatellite instability MicroRNAs: - Fingerprint of miR-145, miR-150, miR-223 and miR-636 - MiR-1290 - MiR-21 - MiR-34 Long non-coding RNAs: - HOTAIR - MALAT-1

LIVER AND BILARY TRACT CANCER

Incidence, diagnosis and risk factors

Liver cancer is the sixth most commonly diagnosed cancer worldwide [165] and primary liver cancer is the third largest cause of cancer mortality in the world [326, 494]. This GI cancer type is primarily associated with chronic hepatitis B virus (HBV) and hepatitis C virus (HCV) infections, but is also linked to other risk factors such as dietary aflatoxin exposure, alcohol-related cirrhosis, fatty liver disease, obesity, smoking, diabetes, and iron overload [495 - 499].

Cholangiocarcinoma also known as bile duct cancer has an extremely low incidence rate in western world (1-2 case per 100,000 people) and a higher prevalence in Asia caused by endemic chronic parasitic infestation [500]. But over the past decades incidence have rising worldwide and account for 3% of all GI cancer [501, 502]. This epithelial cell based tumour is in nearly all cases an incurable and rapidly lethal cancer because up to now no curative treatment exists except surgery [503].

Current Biomarkers in Use

Serum alpha-fetoprotein (AFP) levels together with abdominal ultrasound are the most commonly used methods for screening and diagnosis of hepatocellular carcinoma [504]. High levels of AFP are associated with larger tumour size, bilobar involvement and portal vein thrombosis [505]. Even survival has been shown to be markedly reduced in patients with AFP ≥400 IU/ml [505]. Also, very good correlations between tumour differentiation as well as tumour stage according to the BCLC classification and the AFP level have been discovered [504]. Despite these data, the diagnostic value of AFP is considered controversial since its sensitivity and specificity are unstable and current guidelines recommend its use for diagnosis of liver carcinoma but only in combination with imaging modalities [506].

Current Treatment Modalities

Given the fact that most cholangiocarcinoma patients have advanced stage disease and are inoperable at the time of diagnosis, they generally receive palliative care treatments such as radiation therapy and chemotherapy [503]. 5′-fluorouracil with leucovorin, gemcitabine as a single agent or gemcitabine plus cisplatin, irinotecan, or capecitabine are used in clinical practice [507 - 511]. These chemotherapeutic agents are also used as additional therapies after surgery in cases where resection has apparently been successful [512 - 516].

For liver cancer the only curative treatment is liver resection or liver transplantation. Certain signaling pathways and molecular alterations have been identified as driver of liver cancer development by promoting cell growth and survival [517, 518]. This knowledge offers the possibility to develop targeted therapies [518]. Especially the multi tyrosin-kinase inhibitor sorafenib has been shown to be effective against liver cancer [519] and recently sorafenib was approved as first-line treatment for this disease [519, 520]. Combined therapy is much more efficient and a potent treatment for advanced liver cancer patients is based on a combination of sorafenib with transarterial chemoembolization [521]. Furthermore, one clinical study demonstrated a synergistic effect for a combination therapy based on sorafenib and rapamycin, an mTOR inhibitor, in liver cancer patients [522]. Another multi-kinase inhibitor, regorafenib, has been successfully tested in clinical studies as a second-line drug in liver cancer patients who progress after prior sorafenib treatment [523] and a phase 3 trial was successful completed (presented at the 18th World Congress on Gastrointestinal Cancer (Barcelona, Spain, 2016), Abstract LBA-03). Therefore, regorafenib is currently under approval for second-line therapy for liver cancer by FDA.

Another new approach for treating liver cancer is immunotherapy [524 - 526] and the first results from clinical trials are promising. Some studies (ClinicalTrials.gov Identifier: NCT02576509, NCT01658878, NCT02423343, NCT01853618, NCT02519348, NCT01968109) target molecules that are involved in regulation of the immune response. By this the anti-cancer immune response is activated by either blocking inhibitory molecules or activating stimulatory molecules [527]. Another possibility is to use monoclonal antibodies directed against specific tumour antigens for initiating an immune response (ClinicalTrials.gov Identifier: NCT02435433, NCT01631552, NCT01306058, NCT02560779) [528].

In other studies, genetic modified cells or virus are used. In the so called "adoptive T-cell therapy" T-cells from the patients are genetically modified to enhance their activity and then re-introduced into the patient [529]. By this, the anti-cancer response of the immune system is improved (ClinicalTrials.gov Identifier: NCT01174121). Some modified virus strains are used to specific infect tumour cells (ClinicalTrials.gov Identifier: NCT02562755, NCT01071941). The oncolytic viruses cause self-destruction of tumour cells and result in an increased immune response due to the higher amount of released tumour specific antigens [530 - 532].

Potential Biomarkers Currently Under Further Investigation

Deregulated microRNA expression pattern that modulate signalling pathways has been observed in liver cancer. MiR-122 is often reduced expressed in liver cancer [533, 534]. This microRNA is known to target cyclin G1 as well as the cell-cycle inhibitors CDKN1B (p27kip1) and CDKN1C (p57kip2) resulting in abnormal cell-cycle in tumour cells [535, 536]. Furthermore, as in most human cancers miR-21 is found to be over-expressed in liver cancer resulting in an increased activity of PI3K/AKT pathway leading to an increased invasion and proliferation [534, 537]. Beside the fact that these microRNAs could most probably be used as therapeutic target for new treatments of liver cancer the deregulated microRNAs are helpful for diagnosis, therapy and subgrouping of liver cancers [534]. For example miR-26 down-regulation results in activation of NF-κB and IL-6 signaling pathways and is an indication for inflammation-related liver cancer [538, 539]. Patients with low miR-26 expression are responding better to interferon therapy than do other patients [538, 540]. Even for prediction of resistance to gemcitabine treatment in cholangiocarcinoma patients microRNA level of miR-21 and miR-200b in the blood can be used [541].

MiR-126 down-regulation is an indication for alcohol-related liver carcinoma and mir-96 is found to be over-expressed specifically in HBV-related cancers [534]. Theras the up-regulation of miR-21, miR-122 and miR-223 in serum is detected in liver cancer patients with HBV- or HCV-infection [542, 543]. This set of up-regulated microRNAs was proofed to be a better biomarker for liver cancer compared to α-fetoprotein [542, 543]. MiR-15b and miR-130b showed good potential as useful biomarkers for early-stage liver cancer patients because these microRNAs are strongly up-regulated in an early phase of the disease when α-fetoprotein level is still extremely low [544].

Also some lncRNAs are highly up-regulated in liver cancer and detection of some lncRNAs can be used as independent diagnostic marker for this disease like MALAT-1 for liver cancer recurrence after liver transplantation [545].

Another biomarker for detection and treatment success as well as prognosis of liver carcinoma is cft-DNA [546]. The increased DNA methylation present in liver cancer patients can be detected in cft-DNA [547 - 550] and this might be used as a blood based test for early detection of liver cancer in the near future.

The most important points of the above discussion about liver and biliary tract cancer are summarized in Table **8**.

Table 8. Liver and biliary tract cancer.

Risk factors	- Chronic hepatitis B virus infection - Chronic hepatitis C virus infection - Dietary aflatoxin exposure - Alcohol-related cirrhosis - Fatty liver disease - Obesity - Smoking - Diabetes - Iron overload - Chronic parasitic infestation
Current treatment modalities	- Surgery - Liver transplantation - Radiation - Chemotherapy (5′-fluorouracil with leucovorin, gemcitabine as a single agent or gemcitabine plus cisplatin, irinotecan, or capecitabine) - Sorafenib alone or in combination with transarterial chemoembolization - Regorafenib - Immunotherapy

(Table 8) contd.....

Current biomarkers	- AFP
Potential biomarkers currently under investigation	MicroRNAs: - Fingerprint of miR-21, miR-122 and miR-223 - MiR-26 - MiR-200b - MiR-126 - MiR-96 - MiR-130b - MiR-15b Long non-coding RNA: - MALAT-1

GALLBLADDER CANCER

Incidence, Diagnosis and Risk Factors

Gallbladder cancer is the sixth most frequent GI cancer and the most common malignancy of the biliary tract, representing 80%–95% of biliary tract cancers worldwide [551, 552]. Overall mean survival is only 6 months, while 5-year survival rate is only 5% [551, 553]. Risk factors are chronic inflammation in the gallbladder, gallstones, obesity and chronic salmonella infection. Surprisingly incidence rates of gallbladder cancer are found to be 2- to 6-folds higher in women than in men [554].

Current Biomarkers in Use

CA19-9 and CEA are most commonly used in the diagnosis of gallbladder cancer [555, 556].

Current Treatment Modalities

Surgical resection of gallbladder cancer is the only chance for complete cure but only very few patients are diagnosed in an early stage so that surgery is still possible [557, 558]. Nevertheless the recurrence rate is high among surgical treated patients most probably due to the high metastasis rate of gallbladder cancer [557]. Advanced stage disease can be treated only palliative with gemcitabine, platin-based or fluoropyrimidine-based chemotherapy often in combination with radiotherapy [559 - 563]. In general chemotherapy has little effect on this kind of cancer and up to now a standard chemotherapy for gallbladder cancer has not been established [511]. Unfortunately a targeted therapy has only a very limited role in the management of gallbladder cancer mostly caused by limited understanding of the molecular carcinogenesis of adenocarcinomas of the gallbladder [558, 559, 564 - 566].

Potential Biomarkers Currently Under Further Investigation

At least first steps have been made to identify aberrant expressed microRNAs in gallbladder cancer [567 - 571]. An increased expression of miR-20a, miR-21, miR-155 and miR-182 was found to be associated with gallbladder cancer [569, 572 - 575]. Especially for miR-20a a central role in gallbladder cancer development was proofed and high expression level of this microRNA is correlated to local tumour invasion, increased epithelial-to-mesenchymal transformation as well as distant metastatic sides and therefore indication of poor patient survival [575]. MiR-155 over-expression in gallbladder cancer patients seems to be a marker for lymph node metastasis and vascular invasion [569]. In agreement with these observations gallbladder patients with high level of miR-155 have a lower survival rate compared to patients with low level of miR-155 [569]. Further studies are needed but it seems probably that miR-155 could become a prognostic marker for gallbladder cancer.

Nevertheless, most deregulated microRNAs are down-regulated in gallbladder tumours [567] as it is the case for miR-1, miR-26a, miR-34a, miR-130a, miR-135a-5p, miR-145, miR-146b-5p, miR-218-5p and miR-335 [570, 576 - 582]. By multivariate analysis the expression level of miR-335 was identified as independent prognostic factor for the overall survival of gallbladder cancer patients [570]. MiR-26 expression level was found to be in good agreement with pathological TMN stage and also expression level of miR-135a-5p correlated with the histological tumour grade [576, 577]. Interestingly, some of these microRNAs are down-regulated by lncRNAs, *e.g.* miR-218-5p by CCAT-1; miR-130a by HOTAIR [579, 581].

Recently it has been discovered that also a panel of circulating microRNAs (let-7a, miR-21, miR-143, miR-187, miR-202 and miR-335) isolated from blood can be used to discriminate gallbladder cancer patients from healthy donors [583]. Furthermore, the level of microRNAs corresponded with tumour pathological features and it seems to become a promising method for early detection of gallbladder cancer [583].

The most important points of the above discussion about gallbladder cancer are summarized in Table **9**.

Conclusions and Perspectives

Even though in recent years, the survival of patients affected by the diverse group of cancers of the digestive tract and commonly summarized as GI tumours, has improved [14 - 17], there is still an urgent need of discovery of biomolecular tools in order to implement and support traditional detection methods for an earlier and

Table 9. Gallbladder cancer.

Risk factors	- Chronic inflammation in the gallbladder - Gallstones - Obesity - Chronic salmonella infection - Gender
Current treatment modalities	- Surgery - Chemotherapy with gemcitabine, platin-compounds or fluoropyrimidine-compounds - Radiotherapy
Current biomarkers	- CEA - CA19-9
Potential biomarkers currently under investigation	MicroRNAs: - Fingerprint of let-7a, miR-21, miR-143, miR-187, miR-202 and miR-335 - MiR-20a - MiR-21 - MiR-155 - MiR-182 - MiR-1 - MiR-26a - MiR-34a - MiR-130a - MiR-135a-5p - MiR-145 - MiR-146b-5p - MiR-218-5p - MiR-335

more specific diagnosis. Moreover, biomarkers for cancer detection could be used for monitoring the disease in case of chemotherapeutic treatment either with standard or targeted therapies but also as follow-up post surgery or in event of recurrence. However, this objective could be reached only by the discovery and validation of biomarkers with demonstrated high sensitivity and specificity in regard to the cancer-type. Blood based biomarkers are emerging not only as more suitable for detecting the more complex cancer heterogeneity than in primary cancer tissue source [584, 585], but also the so called "liquid biopsy", is overcoming the lower compliance on patients on traditional biopsy methods and reducing the procedure associated risks [113, 586, 587]. Among new biomarkers for cancer the monitoring of cft-DNA and non-coding RNA (*e.g.* microRNA and lncRNA) have been deeply studied in last decade for their high stability in body fluids [588, 589] thus representing new highly promising biomarkers. Furthermore, these biomarkers (or a combination of biomarkers) could be of pivotal importance in the light of increasing amount of personalized medicine. Therefore, it might be of extreme importance to pursue studies able to explore and validate the potentiality of high sensitive methods (ddPCR, NGS platforms, nCounter) in order to implement traditional biomarkers for diagnostic and prognostic settings but also importantly for prediction of response to therapeutic drugs in cancer treatment. Thus, the identification of new molecular biomarkers could represent a promising tool for guiding a better treatment of cancer patients not only in the field of GI cancer.

CONFLICT OF INEREST

The authors (editor) declares no conflict of interest, financial or otherwise.

ACKNOWLEDGEMENT

Study concept and design (JCH, MM, NV); drafting the manuscript (JCH, MM, AL); critical revision of the manuscript (AL, IK, NV).

REFERENCES

[1] Pourhoseingholi MA, Vahedi M, Baghestani AR. Burden of gastrointestinal cancer in Asia; an overview. Gastroenterol Hepatol Bed Bench 2015; 8(1): 19-27.
[PMID: 25584172]

[2] Ajani A, Lynch PM, Janjan NA, Curley SA, Eds. Gastrointestinal Cancer. 2005.

[3] Siegel RL, Miller KD, Jemal A. Cancer statistics, 2015. CA Cancer J Clin 2015; 65(1): 5-29.
[http://dx.doi.org/10.3322/caac.21254] [PMID: 25559415]

[4] Torre LA, Bray F, Siegel RL, Ferlay J, Lortet-Tieulent J, Jemal A. Global cancer statistics, 2012. CA Cancer J Clin 2015; 65(2): 87-108.
[http://dx.doi.org/10.3322/caac.21262] [PMID: 25651787]

[5] Olcina MM, Giaccia AJ. Reducing radiation-induced gastrointestinal toxicity - the role of the PHD/HIF axis. J Clin Invest 2016; 126(10): 3708-15.
[http://dx.doi.org/10.1172/JCI84432] [PMID: 27548524]

[6] Rautio T, Kairaluoma M, Sand J. [Novel techniques in the treatment of rectal cancer]. Duodecim 2016; 132(12): 1160-4. [Novel techniques in the treatment of rectal cancer].
[PMID: 27483632]

[7] Ristamäki R, Ålgars A. [Principles of oncologic drug therapy following surgery for bowel cancer]. Duodecim 2016; 132(12): 1155-9. [Principles of oncologic drug therapy following surgery for bowel cancer].
[PMID: 27483631]

[8] Murphy MB. Adjunctive Therapy of Gastric Cancer: Moving the Field Forward. Oncology (Williston Park) 2016; 30(7): 646-7.
[PMID: 27422111]

[9] Chan BA, Jang RW, Wong RK, Swallow CJ, Darling GE, Elimova E. Improving Outcomes in Resectable Gastric Cancer: A Review of Current and Future Strategies. Oncology (Williston Park) 2016; 30(7): 635-45.
[PMID: 27422110]

[10] Ismael HN, Denbo J, Cox S, *et al.* Biologic mesh spacer placement facilitates safe delivery of dose-intense radiation therapy: A novel treatment option for unresectable liver tumors. Eur J Surg Oncol 2016; 42(10): 1591-6.

[http://dx.doi.org/10.1016/j.ejso.2016.05.021] [PMID: 27296729]

[11] Jakhetiya A, Garg PK, Prakash G, Sharma J, Pandey R, Pandey D. Targeted therapy of gastrointestinal stromal tumours. World J Gastrointest Surg 2016; 8(5): 345-52. [http://dx.doi.org/10.4240/wjgs.v8.i5.345] [PMID: 27231512]

[12] Rutkowski P, Hompes D. Combined Therapy of Gastrointestinal Stromal Tumors. Surg Oncol Clin N Am 2016; 25(4): 735-59. [http://dx.doi.org/10.1016/j.soc.2016.05.006] [PMID: 27591496]

[13] Hung AY, Canning CA, Patel KM, Holland JM, Kachnic LA. Radiation therapy for gastrointestinal cancer. Hematol Oncol Clin North Am 2006; 20(2): 287-320. [http://dx.doi.org/10.1016/j.hoc.2006.01.016] [PMID: 16730296]

[14] Go PH, Klaassen Z, Meadows MC, Chamberlain RS. Gastrointestinal cancer and brain metastasis: a rare and ominous sign. Cancer 2011; 117(16): 3630-40. [http://dx.doi.org/10.1002/cncr.25940] [PMID: 21319152]

[15] Esmaeilzadeh M, Majlesara A, Faridar A, *et al.* Brain metastasis from gastrointestinal cancers: a systematic review. Int J Clin Pract 2014; 68(7): 890-9. [http://dx.doi.org/10.1111/ijcp.12395] [PMID: 24666726]

[16] Da Silva AN, Nagayama K, Schlesinger DJ, Sheehan JP. Gamma Knife surgery for brain metastases from gastrointestinal cancer. J Neurosurg 2009; 111(3): 423-30. [http://dx.doi.org/10.3171/2008.9.JNS08281] [PMID: 19722810]

[17] Minn YK, Lee DH, Hyung WJ, *et al.* MicroRNA-200 family members and ZEB2 are associated with brain metastasis in gastric adenocarcinoma. Int J Oncol 2014; 45(6): 2403-10. [PMID: 25270520]

[18] Roth AD, Tejpar S, Delorenzi M, *et al.* Prognostic role of KRAS and BRAF in stage II and III resected colon cancer: results of the translational study on the PETACC-3, EORTC 40993, SAKK 60-00 trial. J Clin Oncol 2010; 28(3): 466-74. [http://dx.doi.org/10.1200/JCO.2009.23.3452] [PMID: 20008640]

[19] Hutchins G, Southward K, Handley K, *et al.* Value of mismatch repair, KRAS, and BRAF mutations in predicting recurrence and benefits from chemotherapy in colorectal cancer. J Clin Oncol 2011; 29(10): 1261-70. [http://dx.doi.org/10.1200/JCO.2010.30.1366] [PMID: 21383284]

[20] Jass JR. Classification of colorectal cancer based on correlation of clinical, morphological and molecular features. Histopathology 2007; 50(1): 113-30. [http://dx.doi.org/10.1111/j.1365-2559.2006.02549.x] [PMID: 17204026]

[21] Sinicrope FA, Shi Q, Smyrk TC, *et al.* Molecular markers identify subtypes of stage III colon cancer associated with patient outcomes. Gastroenterology 2015; 148(1): 88-99. [http://dx.doi.org/10.1053/j.gastro.2014.09.041] [PMID: 25305506]

[22] Fariña-Sarasqueta A, van Lijnschoten G, Moerland E, *et al.* The BRAF V600E mutation is an independent prognostic factor for survival in stage II and stage III colon cancer patients. Ann Oncol 2010; 21(12): 2396-402. [http://dx.doi.org/10.1093/annonc/mdq258] [PMID: 20501503]

[23] Fearon ER, Vogelstein B. A genetic model for colorectal tumorigenesis. Cell 1990; 61(5): 759-67. [http://dx.doi.org/10.1016/0092-8674(90)90186-I] [PMID: 2188735]

[24] De Rosa M, Rega D, Costabile V, *et al.* The biological complexity of colorectal cancer: insights into biomarkers for early detection and personalized care. Therap Adv Gastroenterol 2016; 9(6): 861-86. [http://dx.doi.org/10.1177/1756283X16659790] [PMID: 27803741]

[25] Hollis M, Nair K, Vyas A, Chaturvedi LS, Gambhir S, Vyas D. MicroRNAs potential utility in colon cancer: Early detection, prognosis, and chemosensitivity. World J Gastroenterol 2015; 21(27): 8284-92.

[http://dx.doi.org/10.3748/wjg.v21.i27.8284] [PMID: 26217080]

[26] Weng M, Wu D, Yang C, *et al.* Noncoding RNAs in the development, diagnosis, and prognosis of colorectal cancer. Transl Res 2016.
[PMID: 27810413]

[27] Zhang JX, Song W, Chen ZH, *et al.* Prognostic and predictive value of a microRNA signature in stage II colon cancer: a microRNA expression analysis. Lancet Oncol 2013; 14(13): 1295-306.
[http://dx.doi.org/10.1016/S1470-2045(13)70491-1] [PMID: 24239208]

[28] O'Dwyer D, Ralton LD, O'Shea A, Murray GI. The proteomics of colorectal cancer: identification of a protein signature associated with prognosis. PLoS One 2011; 6(11): e27718.
[http://dx.doi.org/10.1371/journal.pone.0027718] [PMID: 22125622]

[29] Klein CA. Parallel progression of primary tumours and metastases. Nat Rev Cancer 2009; 9(4): 302-12.
[http://dx.doi.org/10.1038/nrc2627] [PMID: 19308069]

[30] deVos T, Tetzner R, Model F, *et al.* Circulating methylated SEPT9 DNA in plasma is a biomarker for colorectal cancer. Clin Chem 2009; 55(7): 1337-46.
[http://dx.doi.org/10.1373/clinchem.2008.115808] [PMID: 19406918]

[31] Vukobrat-Bijedic Z, Husic-Selimovic A, Sofic A, *et al.* Cancer Antigens (CEA and CA 19-9) as Markers of Advanced Stage of Colorectal Carcinoma. Med Arh 2013; 67(6): 397-401.
[http://dx.doi.org/10.5455/medarh.2013.67.397-401] [PMID: 25568506]

[32] Nicholson BD, Shinkins B, Pathiraja I, *et al.* Blood CEA levels for detecting recurrent colorectal cancer. Cochrane Database Syst Rev 2015; (12): CD011134.
[PMID: 26661580]

[33] Svobodova S, Topolcan O, Holubec L Jr, Levy M, Pecen L, Svacina S. Parameters of biological activity in colorectal cancer. Anticancer Res 2011; 31(1): 373-8.
[PMID: 21273626]

[34] Polat E, Duman U, Duman M, *et al.* Diagnostic value of preoperative serum carcinoembryonic antigen and carbohydrate antigen 19-9 in colorectal cancer. Curr Oncol 2014; 21(1): e1-7.
[http://dx.doi.org/10.3747/co.21.1711] [PMID: 24523606]

[35] Duffy MJ. Carcinoembryonic antigen as a marker for colorectal cancer: is it clinically useful? Clin Chem 2001; 47(4): 624-30.
[PMID: 11274010]

[36] Locker GY, Hamilton S, Harris J, *et al.* ASCO. ASCO 2006 update of recommendations for the use of tumor markers in gastrointestinal cancer. J Clin Oncol 2006; 24(33): 5313-27.
[http://dx.doi.org/10.1200/JCO.2006.08.2644] [PMID: 17060676]

[37] Bagaria B, Sood S, Sharma R, Lalwani S. Comparative study of CEA and CA19-9 in esophageal, gastric and colon cancers individually and in combination (ROC curve analysis). Cancer Biol Med 2013; 10(3): 148-57.
[PMID: 24379990]

[38] Warren JD, Xiong W, Bunker AM, *et al.* Septin 9 methylated DNA is a sensitive and specific blood test for colorectal cancer. BMC Med 2011; 9: 133.
[http://dx.doi.org/10.1186/1741-7015-9-133] [PMID: 22168215]

[39] Molnár B, Tóth K, Barták BK, Tulassay Z. Plasma methylated septin 9: a colorectal cancer screening marker. Expert Rev Mol Diagn 2015; 15(2): 171-84.
[http://dx.doi.org/10.1586/14737159.2015.975212] [PMID: 25429690]

[40] Sun SS, Hsieh JF, Tsai SC, Ho YJ, Kao CH. Tissue polypeptide specific antigen (TPS) as a tumor marker in nasopharyngeal carcinoma. Anticancer Res 2000; 20(6C): 4661-3.
[PMID: 11205197]

[41] Galietta A, Pizzi C, Pettinato G, *et al.* Differential TAG-72 epitope expression in breast cancer and lymph node metastases: a marker of a more aggressive phenotype. Oncol Rep 2002; 9(1): 135-40. [PMID: 11748471]

[42] Anastasi E, Manganaro L, Granato T, Benedetti Panici P, Frati L, Porpora MG. Is CA72-4 a useful biomarker in differential diagnosis between ovarian endometrioma and epithelial ovarian cancer? Dis Markers 2013; 35(5): 331-5. [http://dx.doi.org/10.1155/2013/984641] [PMID: 24191126]

[43] Hareyama H, Sakuragi N, Makinoda S, Fujimoto S. Serum and tissue measurements of CA72-4 in patients with endometrial carcinoma. J Clin Pathol 1996; 49(12): 967-70. [http://dx.doi.org/10.1136/jcp.49.12.967] [PMID: 9038731]

[44] Pasquali C, Sperti C, D'Andrea AA, Costantino V, Filipponi C, Pedrazzoli S. Clinical value of serum TAG-72 as a tumor marker for pancreatic carcinoma. Comparison with CA 19-9. Int J Pancreatol 1994; 15(3): 171-7. [PMID: 7930777]

[45] Derijks-Engwegen JY, Cats A, Smits ME, Schellens JH, Beijnen JH. Improving colorectal cancer management: the potential of proteomics. Biomarkers Med 2008; 2(3): 253-89. [http://dx.doi.org/10.2217/17520363.2.3.253] [PMID: 20477414]

[46] Palmblad M, Tiss A, Cramer R. Mass spectrometry in clinical proteomics - from the present to the future. Proteomics Clin Appl 2009; 3(1): 6-17. [http://dx.doi.org/10.1002/prca.200800090] [PMID: 21136932]

[47] Zolg W. The proteomic search for diagnostic biomarkers: lost in translation? Mol Cell Proteomics 2006; 5(10): 1720-6. [http://dx.doi.org/10.1074/mcp.R600001-MCP200] [PMID: 16546995]

[48] Ma R, Jiang T, Kang X. Circulating microRNAs in cancer: origin, function and application. J Exp Clin Cancer Res 2012; 31: 38. [http://dx.doi.org/10.1186/1756-9966-31-38] [PMID: 22546315]

[49] Wu HH, Lin WC, Tsai KW. Advances in molecular biomarkers for gastric cancer: miRNAs as emerging novel cancer markers. Expert Rev Mol Med 2014; 16: e1. [http://dx.doi.org/10.1017/erm.2013.16] [PMID: 24456939]

[50] Khodakov D, Wang C, Zhang DY. Diagnostics based on nucleic acid sequence variant profiling: PCR, hybridization, and NGS approaches. Adv Drug Deliv Rev 2016; 105(Pt A): 3-19. [http://dx.doi.org/10.1016/j.addr.2016.04.005] [PMID: 27089811]

[51] Lopez NE, Weiss AC, Robles J, Fanta P, Ramamoorthy SL. A systematic review of clinically available gene expression profiling assays for stage II colorectal cancer: initial steps toward genetic staging. Am J Surg 2016; 212(4): 700-14. [http://dx.doi.org/10.1016/j.amjsurg.2016.06.019] [PMID: 27637150]

[52] Elshimali YI, Khaddour H, Sarkissyan M, Wu Y, Vadgama JV. The clinical utilization of circulating cell free DNA (CCFDNA) in blood of cancer patients. Int J Mol Sci 2013; 14(9): 18925-58. [http://dx.doi.org/10.3390/ijms140918925] [PMID: 24065096]

[53] Haber DA, Velculescu VE. Blood-based analyses of cancer: circulating tumor cells and circulating tumor DNA. Cancer Discov 2014; 4(6): 650-61. [http://dx.doi.org/10.1158/2159-8290.CD-13-1014] [PMID: 24801577]

[54] Newman AM, Bratman SV, To J, *et al.* An ultrasensitive method for quantitating circulating tumor DNA with broad patient coverage. Nat Med 2014; 20(5): 548-54. [http://dx.doi.org/10.1038/nm.3519] [PMID: 24705333]

[55] Qin Z, Ljubimov VA, Zhou C, Tong Y, Liang J. Cell-free circulating tumor DNA in cancer. Chin J Cancer 2016; 35: 36. [http://dx.doi.org/10.1186/s40880-016-0092-4] [PMID: 27056366]

[56] Fedyanin M, Popova A, Polyanskaya E, Tjulandin S. Role of stem cells in colorectal cancer progression and prognostic and predictive characteristics of stem cells markers in colorectal cancer. Curr Stem Cell Res Ther 2016.
[http://dx.doi.org/10.2174/1574888X11666160905092938] [PMID: 27593447]

[57] Rothwell DG, Smith N, Morris D, *et al.* Genetic profiling of tumours using both circulating free DNA and circulating tumour cells isolated from the same preserved whole blood sample. Mol Oncol 2016; 10(4): 566-74.
[http://dx.doi.org/10.1016/j.molonc.2015.11.006] [PMID: 26639657]

[58] Chabon JJ, Simmons AD, Lovejoy AF, *et al.* Circulating tumour DNA profiling reveals heterogeneity of EGFR inhibitor resistance mechanisms in lung cancer patients. Nat Commun 2016; 7: 11815.
[http://dx.doi.org/10.1038/ncomms11815] [PMID: 27283993]

[59] Popławski AB, Jankowski M, Erickson SW, *et al.* Frequent genetic differences between matched primary and metastatic breast cancer provide an approach to identification of biomarkers for disease progression. Eur J Hum Genet 2010; 18(5): 560-8.
[http://dx.doi.org/10.1038/ejhg.2009.230] [PMID: 20051991]

[60] Sylvester BE, Vakiani E. Tumor evolution and intratumor heterogeneity in colorectal carcinoma: insights from comparative genomic profiling of primary tumors and matched metastases. J Gastrointest Oncol 2015; 6(6): 668-75.
[PMID: 26697200]

[61] Bardelli A. Liquid Biopsies: Genotyping Circulating Tumour DNA in Colorectal Cancer. Ann Oncol 2014; 25.

[62] Diaz LA Jr, Bardelli A. Liquid biopsies: genotyping circulating tumor DNA. J Clin Oncol 2014; 32(6): 579-86.
[http://dx.doi.org/10.1200/JCO.2012.45.2011] [PMID: 24449238]

[63] Gao F, Pfeifer E, Farah H, *et al.* Microdroplet digital PCR: detection and quantitation of biomarkers in archived tissue and serial plasma samples in patients with lung cancer. J Thorac Oncol 2015; 10(1): 212-7.
[http://dx.doi.org/10.1097/JTO.0000000000000349] [PMID: 25654729]

[64] Crowley E, Di Nicolantonio F, Loupakis F, Bardelli A. Liquid biopsy: monitoring cancer-genetics in the blood. Nat Rev Clin Oncol 2013; 10(8): 472-84.
[http://dx.doi.org/10.1038/nrclinonc.2013.110] [PMID: 23836314]

[65] Jiang C, Li X, Zhao H, Liu H. Long non-coding RNAs: potential new biomarkers for predicting tumor invasion and metastasis. Mol Cancer 2016; 15(1): 62.
[http://dx.doi.org/10.1186/s12943-016-0545-z] [PMID: 27686732]

[66] Qi P, Zhou XY, Du X. Circulating long non-coding RNAs in cancer: current status and future perspectives. Mol Cancer 2016; 15(1): 39.
[http://dx.doi.org/10.1186/s12943-016-0524-4] [PMID: 27189224]

[67] Schmitt AM, Chang HY. Long Noncoding RNAs in Cancer Pathways. Cancer Cell 2016; 29(4): 452-63.
[http://dx.doi.org/10.1016/j.ccell.2016.03.010] [PMID: 27070700]

[68] Eddy SR. Non-coding RNA genes and the modern RNA world. Nat Rev Genet 2001; 2(12): 919-29.
[http://dx.doi.org/10.1038/35103511] [PMID: 11733745]

[69] Choudhuri S. Small noncoding RNAs: biogenesis, function, and emerging significance in toxicology. J Biochem Mol Toxicol 2010; 24(3): 195-210.
[http://dx.doi.org/10.1002/jbt.20325] [PMID: 20143452]

[70] Wilusz JE, Sunwoo H, Spector DL. Long noncoding RNAs: functional surprises from the RNA world. Genes Dev 2009; 23(13): 1494-504.
[http://dx.doi.org/10.1101/gad.1800909] [PMID: 19571179]

[71] Azlan A, Dzaki N, Azzam G. Argonaute: The executor of small RNA function. J Genet Genomics 2016; 43(8): 481-94.
[http://dx.doi.org/10.1016/j.jgg.2016.06.002] [PMID: 27569398]

[72] Geiger J, Dalgaard LT. Interplay of mitochondrial metabolism and microRNAs. Cell Mol Life Sci 2016.
[PMID: 27563705]

[73] An Y, Furber KL, Ji S. Pseudogenes regulate parental gene expression *via* ceRNA network. J Cell Mol Med 2016.
[PMID: 27561207]

[74] Beermann J, Piccoli MT, Viereck J, Thum T. Non-coding RNAs in Development and Disease: Background, Mechanisms, and Therapeutic Approaches. Physiol Rev 2016; 96(4): 1297-325.
[http://dx.doi.org/10.1152/physrev.00041.2015] [PMID: 27535639]

[75] de Almeida RA, Fraczek MG, Parker S, Delneri D, O'Keefe RT. Non-coding RNAs and disease: the classical ncRNAs make a comeback. Biochem Soc Trans 2016; 44(4): 1073-8.
[http://dx.doi.org/10.1042/BST20160089] [PMID: 27528754]

[76] Evans JR, Feng FY, Chinnaiyan AM. The bright side of dark matter: lncRNAs in cancer. J Clin Invest 2016; 126(8): 2775-82.
[http://dx.doi.org/10.1172/JCI84421] [PMID: 27479746]

[77] Granados-Riveron JT, Aquino-Jarquin G. The complexity of the translation ability of circRNAs. Biochim Biophys Acta 2016; 1859: 1245-51.

[78] Khurana E, Fu Y, Chakravarty D, Demichelis F, Rubin MA, Gerstein M. Role of non-coding sequence variants in cancer. Nat Rev Genet 2016; 17(2): 93-108.
[http://dx.doi.org/10.1038/nrg.2015.17] [PMID: 26781813]

[79] Quinn JJ, Chang HY. Unique features of long non-coding RNA biogenesis and function. Nat Rev Genet 2016; 17(1): 47-62.
[http://dx.doi.org/10.1038/nrg.2015.10] [PMID: 26666209]

[80] He L, Hannon GJ. MicroRNAs: small RNAs with a big role in gene regulation. Nat Rev Genet 2004; 5(7): 522-31.
[http://dx.doi.org/10.1038/nrg1379] [PMID: 15211354]

[81] Guttman M, Donaghey J, Carey BW, *et al.* lincRNAs act in the circuitry controlling pluripotency and differentiation. Nature 2011; 477(7364): 295-300.
[http://dx.doi.org/10.1038/nature10398] [PMID: 21874018]

[82] Sauvageau M, Goff LA, Lodato S, *et al.* Multiple knockout mouse models reveal lincRNAs are required for life and brain development. eLife 2013; 2: e01749.
[http://dx.doi.org/10.7554/eLife.01749] [PMID: 24381249]

[83] Ounzain S, Pezzuto I, Micheletti R, *et al.* Functional importance of cardiac enhancer-associated noncoding RNAs in heart development and disease. J Mol Cell Cardiol 2014; 76: 55-70.
[http://dx.doi.org/10.1016/j.yjmcc.2014.08.009] [PMID: 25149110]

[84] Herriges MJ, Swarr DT, Morley MP, *et al.* Long noncoding RNAs are spatially correlated with transcription factors and regulate lung development. Genes Dev 2014; 28(12): 1363-79.
[http://dx.doi.org/10.1101/gad.238782.114] [PMID: 24939938]

[85] Li G, Zhang H, Wan X, *et al.* Long noncoding RNA plays a key role in metastasis and prognosis of hepatocellular carcinoma. Biomed Res Int 2014; 2014: 780521.
[http://dx.doi.org/10.1155/2014/780521]

[86] Yoon JH, Abdelmohsen K, Gorospe M. Functional interactions among microRNAs and long noncoding RNAs. Semin Cell Dev Biol 2014; 34: 9-14.
[http://dx.doi.org/10.1016/j.semcdb.2014.05.015] [PMID: 24965208]

[87] Cannell IG, Kong YW, Bushell M. How do microRNAs regulate gene expression? Biochem Soc Trans 2008; 36(Pt 6): 1224-31.
[http://dx.doi.org/10.1042/BST0361224] [PMID: 19021530]

[88] Macfarlane LA, Murphy PR. MicroRNA: Biogenesis, Function and Role in Cancer. Curr Genomics 2010; 11(7): 537-61.
[http://dx.doi.org/10.2174/138920210793175895] [PMID: 21532838]

[89] Pasquinelli AE. MicroRNAs and their targets: recognition, regulation and an emerging reciprocal relationship. Nat Rev Genet 2012; 13(4): 271-82.
[PMID: 22411466]

[90] von Schack D, Agostino MJ, Murray BS, *et al.* Dynamic changes in the microRNA expression profile reveal multiple regulatory mechanisms in the spinal nerve ligation model of neuropathic pain. PLoS One 2011; 6(3): e17670.
[http://dx.doi.org/10.1371/journal.pone.0017670] [PMID: 21423802]

[91] Lagos-Quintana M, Rauhut R, Yalcin A, Meyer J, Lendeckel W, Tuschl T. Identification of tissue-specific microRNAs from mouse. Curr Biol 2002; 12(9): 735-9.
[http://dx.doi.org/10.1016/S0960-9822(02)00809-6] [PMID: 12007417]

[92] Lim LP, Lau NC, Garrett-Engele P, *et al.* Microarray analysis shows that some microRNAs downregulate large numbers of target mRNAs. Nature 2005; 433(7027): 769-73.
[http://dx.doi.org/10.1038/nature03315] [PMID: 15685193]

[93] Murakami Y, Yasuda T, Saigo K, *et al.* Comprehensive analysis of microRNA expression patterns in hepatocellular carcinoma and non-tumorous tissues. Oncogene 2006; 25(17): 2537-45.
[http://dx.doi.org/10.1038/sj.onc.1209283] [PMID: 16331254]

[94] Ha TY. MicroRNAs in Human Diseases: From Lung, Liver and Kidney Diseases to Infectious Disease, Sickle Cell Disease and Endometrium Disease. Immune Netw 2011; 11(6): 309-23.
[http://dx.doi.org/10.4110/in.2011.11.6.309] [PMID: 22346770]

[95] Ha TY. MicroRNAs in Human Diseases: From Cancer to Cardiovascular Disease. Immune Netw 2011; 11(3): 135-54.
[http://dx.doi.org/10.4110/in.2011.11.3.135] [PMID: 21860607]

[96] Ha TY. The Role of MicroRNAs in Regulatory T Cells and in the Immune Response. Immune Netw 2011; 11(1): 11-41.
[http://dx.doi.org/10.4110/in.2011.11.1.11] [PMID: 21494372]

[97] O'Connell RM, Rao DS, Chaudhuri AA, Baltimore D. Physiological and pathological roles for microRNAs in the immune system. Nat Rev Immunol 2010; 10(2): 111-22.
[http://dx.doi.org/10.1038/nri2708] [PMID: 20098459]

[98] Esteller M. Non-coding RNAs in human disease. Nat Rev Genet 2011; 12(12): 861-74.
[http://dx.doi.org/10.1038/nrg3074] [PMID: 22094949]

[99] Iorio MV, Croce CM. microRNA involvement in human cancer. Carcinogenesis 2012; 33(6): 1126-33.
[http://dx.doi.org/10.1093/carcin/bgs140] [PMID: 22491715]

[100] Acunzo M, Romano G, Wernicke D, Croce CM. MicroRNA and cancer--a brief overview. Adv Biol Regul 2015; 57: 1-9.
[http://dx.doi.org/10.1016/j.jbior.2014.09.013] [PMID: 25294678]

[101] Balatti V, Pekarky Y, Croce CM. Role of microRNA in chronic lymphocytic leukemia onset and progression. J Hematol Oncol 2015; 8: 12.
[http://dx.doi.org/10.1186/s13045-015-0112-x] [PMID: 25886051]

[102] Ren A, Dong Y, Tsoi H, Yu J. Detection of miRNA as non-invasive biomarkers of colorectal cancer. Int J Mol Sci 2015; 16(2): 2810-23.
[http://dx.doi.org/10.3390/ijms16022810] [PMID: 25633103]

[103] Mitchell PS, Parkin RK, Kroh EM, *et al.* Circulating microRNAs as stable blood-based markers for cancer detection. Proc Natl Acad Sci USA 2008; 105(30): 10513-8. [http://dx.doi.org/10.1073/pnas.0804549105] [PMID: 18663219]

[104] Gardiner AS, Twiss JL, Perrone-Bizzozero NI. Competing Interactions of RNA-Binding Proteins, MicroRNAs, and Their Targets Control Neuronal Development and Function. Biomolecules 2015; 5(4): 2903-18. [http://dx.doi.org/10.3390/biom5042903] [PMID: 26512708]

[105] Grasedieck S, Schöler N, Bommer M, *et al.* Impact of serum storage conditions on microRNA stability. Leukemia 2012; 26(11): 2414-6. [http://dx.doi.org/10.1038/leu.2012.106] [PMID: 22504138]

[106] Cardoso AR, Moreira FT, Fernandes R, Sales MG. Novel and simple electrochemical biosensor monitoring attomolar levels of miRNA-155 in breast cancer. Biosens Bioelectron 2016; 80: 621-30. [http://dx.doi.org/10.1016/j.bios.2016.02.035] [PMID: 26901459]

[107] Bohunicky B, Mousa SA. Biosensors: the new wave in cancer diagnosis. Nanotechnol Sci Appl 2010; 4: 1-10. [PMID: 24198482]

[108] Campuzano S, Pedrero M, Pingarrón JM. Electrochemical genosensors for the detection of cancer-related miRNAs. Anal Bioanal Chem 2014; 406(1): 27-33. [http://dx.doi.org/10.1007/s00216-013-7459-z] [PMID: 24247551]

[109] Labib M, Berezovski MV. Electrochemical sensing of microRNAs: avenues and paradigms. Biosens Bioelectron 2015; 68: 83-94. [http://dx.doi.org/10.1016/j.bios.2014.12.026] [PMID: 25562735]

[110] Xia N, Zhang L. Nanomaterials-Based Sensing Strategies for Electrochemical Detection of MicroRNAs. Materials (Basel) 2014; 7(7): 5366-84. [http://dx.doi.org/10.3390/ma7075366] [PMID: 28788133]

[111] Liu J, Huang WH, Yang HX, Luo Y. Expression and function of miR-155 in breast cancer. Biotechnol Biotec Eq 2015; 29: 840-3. [http://dx.doi.org/10.1080/13102818.2015.1043946]

[112] Siravegna G, Mussolin B, Buscarino M, *et al.* Clonal evolution and resistance to EGFR blockade in the blood of colorectal cancer patients. Nat Med 2015; 21(7): 827. [http://dx.doi.org/10.1038/nm0715-827b] [PMID: 26151329]

[113] Siravegna G, Bardelli A. Blood circulating tumor DNA for non-invasive genotyping of colon cancer patients. Mol Oncol 2016; 10(3): 475-80. [http://dx.doi.org/10.1016/j.molonc.2015.12.005] [PMID: 26774880]

[114] Bardelli A, Siena S. Molecular mechanisms of resistance to cetuximab and panitumumab in colorectal cancer. J Clin Oncol 2010; 28(7): 1254-61. [http://dx.doi.org/10.1200/JCO.2009.24.6116] [PMID: 20100961]

[115] Pinheiro LB, Coleman VA, Hindson CM, *et al.* Evaluation of a droplet digital polymerase chain reaction format for DNA copy number quantification. Anal Chem 2012; 84(2): 1003-11. [http://dx.doi.org/10.1021/ac202578x] [PMID: 22122760]

[116] Sykes PJ, Neoh SH, Brisco MJ, Hughes E, Condon J, Morley AA. Quantitation of targets for PCR by use of limiting dilution. Biotechniques 1992; 13(3): 444-9. [PMID: 1389177]

[117] Vogelstein B, Kinzler KW. Digital PCR. Proc Natl Acad Sci USA 1999; 96(16): 9236-41. [http://dx.doi.org/10.1073/pnas.96.16.9236] [PMID: 10430926]

[118] Bhat S, Herrmann J, Armishaw P, Corbisier P, Emslie KR. Single molecule detection in nanofluidic digital array enables accurate measurement of DNA copy number. Anal Bioanal Chem 2009; 394(2):

457-67.
[http://dx.doi.org/10.1007/s00216-009-2729-5] [PMID: 19288230]

[119] Bhat S, Curach N, Mostyn T, Bains GS, Griffiths KR, Emslie KR. Comparison of methods for accurate quantification of DNA mass concentration with traceability to the international system of units. Anal Chem 2010; 82(17): 7185-92.
[http://dx.doi.org/10.1021/ac100845m] [PMID: 20690645]

[120] Dong LH, Meng Y, Sui ZW, Wang J, Wu LQ, Fu BQ. Comparison of four digital PCR platforms for accurate quantification of DNA copy number of a certified plasmid DNA reference material. Sci Rep-Uk 2015; p. 5.

[121] Sanders R, Huggett JF, Bushell CA, Cowen S, Scott DJ, Foy CA. Evaluation of digital PCR for absolute DNA quantification. Anal Chem 2011; 83(17): 6474-84.
[http://dx.doi.org/10.1021/ac103230c] [PMID: 21446772]

[122] Lo YM, Lun FM, Chan KC, *et al.* Digital PCR for the molecular detection of fetal chromosomal aneuploidy. Proc Natl Acad Sci USA 2007; 104(32): 13116-21.
[http://dx.doi.org/10.1073/pnas.0705765104] [PMID: 17664418]

[123] Pohl G, Shih IeM. Principle and applications of digital PCR. Expert Rev Mol Diagn 2004; 4(1): 41-7.
[http://dx.doi.org/10.1586/14737159.4.1.41] [PMID: 14711348]

[124] Wang J, Ramakrishnan R, Tang Z, *et al.* Quantifying EGFR alterations in the lung cancer genome with nanofluidic digital PCR arrays. Clin Chem 2010; 56(4): 623-32.
[http://dx.doi.org/10.1373/clinchem.2009.134973] [PMID: 20207772]

[125] White RA III, Blainey PC, Fan HC, Quake SR. Digital PCR provides sensitive and absolute calibration for high throughput sequencing. BMC Genomics 2009; 10: 116.
[http://dx.doi.org/10.1186/1471-2164-10-116] [PMID: 19298667]

[126] Buermans HP, den Dunnen JT. Next generation sequencing technology: Advances and applications. Biochim Biophys Acta 2014; 1842: 1932-41.

[127] Mardis ER. Next-generation sequencing platforms. Annu Rev Anal Chem (Palo Alto, Calif) 2013; 6: 287-303.
[http://dx.doi.org/10.1146/annurev-anchem-062012-092628] [PMID: 23560931]

[128] Head SR, Komori HK, LaMere SA, *et al.* Library construction for next-generation sequencing: overviews and challenges. Biotechniques 2014; 6: 61-64, 6, 8.
[http://dx.doi.org/10.2144/000114133]

[129] Dabney J, Meyer M. Length and GC-biases during sequencing library amplification: a comparison of various polymerase-buffer systems with ancient and modern DNA sequencing libraries. Biotechniques 2012; 52(2): 87-94.
[http://dx.doi.org/10.2144/000113809] [PMID: 22313406]

[130] Hafner M, Renwick N, Brown M, *et al.* RNA-ligase-dependent biases in miRNA representation in deep-sequenced small RNA cDNA libraries. RNA 2011; 17(9): 1697-712.
[http://dx.doi.org/10.1261/rna.2799511] [PMID: 21775473]

[131] Bentley DR, Balasubramanian S, Swerdlow HP, *et al.* Accurate whole human genome sequencing using reversible terminator chemistry. Nature 2008; 456(7218): 53-9.
[http://dx.doi.org/10.1038/nature07517] [PMID: 18987734]

[132] Dressman D, Yan H, Traverso G, Kinzler KW, Vogelstein B. Transforming single DNA molecules into fluorescent magnetic particles for detection and enumeration of genetic variations. Proc Natl Acad Sci USA 2003; 100(15): 8817-22.
[http://dx.doi.org/10.1073/pnas.1133470100] [PMID: 12857956]

[133] Lappalainen T, Sammeth M, Friedländer MR, *et al.* Geuvadis Consortium. Transcriptome and genome sequencing uncovers functional variation in humans. Nature 2013; 501(7468): 506-11.
[http://dx.doi.org/10.1038/nature12531] [PMID: 24037378]

[134] Westra HJ, Peters MJ, Esko T, *et al.* Systematic identification of trans eQTLs as putative drivers of known disease associations. Nat Genet 2013; 45(10): 1238-43.
[http://dx.doi.org/10.1038/ng.2756] [PMID: 24013639]

[135] Ingolia NT, Ghaemmaghami S, Newman JR, Weissman JS. Genome-wide analysis *in vivo* of translation with nucleotide resolution using ribosome profiling. Science 2009; 324(5924): 218-23.
[http://dx.doi.org/10.1126/science.1168978] [PMID: 19213877]

[136] Valen E, Pascarella G, Chalk A, *et al.* Genome-wide detection and analysis of hippocampus core promoters using DeepCAGE. Genome Res 2009; 19(2): 255-65.
[http://dx.doi.org/10.1101/gr.084541.108] [PMID: 19074369]

[137] de Klerk E, Venema A, Anvar SY, *et al.* Poly(A) binding protein nuclear 1 levels affect alternative polyadenylation. Nucleic Acids Res 2012; 40(18): 9089-101.
[http://dx.doi.org/10.1093/nar/gks655] [PMID: 22772983]

[138] Nolte-'t Hoen EN, Buermans HP, Waasdorp M, Stoorvogel W, Wauben MH, 't Hoen PA. Deep sequencing of RNA from immune cell-derived vesicles uncovers the selective incorporation of small non-coding RNA biotypes with potential regulatory functions. Nucleic Acids Res 2012; 40(18): 9272-85.
[http://dx.doi.org/10.1093/nar/gks658] [PMID: 22821563]

[139] Berezikov E, Thuemmler F, van Laake LW, *et al.* Diversity of microRNAs in human and chimpanzee brain. Nat Genet 2006; 38(12): 1375-7.
[http://dx.doi.org/10.1038/ng1914] [PMID: 17072315]

[140] Andersson R, Gebhard C, Miguel-Escalada I, *et al.* An atlas of active enhancers across human cell types and tissues. Nature 2014; 507(7493): 455-61.
[http://dx.doi.org/10.1038/nature12787] [PMID: 24670763]

[141] Blat Y, Kleckner N. Cohesins bind to preferential sites along yeast chromosome III, with differential regulation along arms *versus* the centric region. Cell 1999; 98(2): 249-59.
[http://dx.doi.org/10.1016/S0092-8674(00)81019-3] [PMID: 10428036]

[142] Consortium EP. ENCODE Project Consortium. The ENCODE (ENCyclopedia Of DNA Elements) Project. Science 2004; 306(5696): 636-40.
[http://dx.doi.org/10.1126/science.1105136] [PMID: 15499007]

[143] Herb BR, Wolschin F, Hansen KD, *et al.* Reversible switching between epigenetic states in honeybee behavioral subcastes. Nat Neurosci 2012; 15(10): 1371-3.
[http://dx.doi.org/10.1038/nn.3218] [PMID: 22983211]

[144] Laird PW. Principles and challenges of genomewide DNA methylation analysis. Nat Rev Genet 2010; 11(3): 191-203.
[http://dx.doi.org/10.1038/nrg2732] [PMID: 20125086]

[145] Norton ME, Brar H, Weiss J, *et al.* Non-Invasive Chromosomal Evaluation (NICE) Study: results of a multicenter prospective cohort study for detection of fetal trisomy 21 and trisomy 18. Am J Obstet Gynecol 2012; 207: 137 e1-8.

[146] Kitzman JO, Snyder MW, Ventura M, *et al.* Noninvasive whole-genome sequencing of a human fetus. Sci Transl Med 2012; 4(137): 137ra76.
[http://dx.doi.org/10.1126/scitranslmed.3004323] [PMID: 22674554]

[147] Hoischen A, van Bon BW, Gilissen C, *et al.* De novo mutations of SETBP1 cause Schinzel-Giedion syndrome. Nat Genet 2010; 42(6): 483-5.
[http://dx.doi.org/10.1038/ng.581] [PMID: 20436468]

[148] Vissers LE, de Ligt J, Gilissen C, *et al.* A de novo paradigm for mental retardation. Nat Genet 2010; 42(12): 1109-12.
[http://dx.doi.org/10.1038/ng.712] [PMID: 21076407]

[149] Ng SB, Buckingham KJ, Lee C, *et al.* Exome sequencing identifies the cause of a mendelian disorder. Nat Genet 2010; 42(1): 30-5.
[http://dx.doi.org/10.1038/ng.499] [PMID: 19915526]

[150] Bernards R. Finding effective cancer therapies through loss of function genetic screens. Curr Opin Genet Dev 2014; 24: 23-9.
[http://dx.doi.org/10.1016/j.gde.2013.11.007] [PMID: 24657533]

[151] Macaulay IC, Voet T. Single cell genomics: advances and future perspectives. PLoS Genet 2014; 10(1): e1004126.
[http://dx.doi.org/10.1371/journal.pgen.1004126] [PMID: 24497842]

[152] Chen R, Mias GI, Li-Pook-Than J, *et al.* Personal omics profiling reveals dynamic molecular and medical phenotypes. Cell 2012; 148(6): 1293-307.
[http://dx.doi.org/10.1016/j.cell.2012.02.009] [PMID: 22424236]

[153] Murtaza M, Dawson SJ, Tsui DW, *et al.* Non-invasive analysis of acquired resistance to cancer therapy by sequencing of plasma DNA. Nature 2013; 497(7447): 108-12.
[http://dx.doi.org/10.1038/nature12065] [PMID: 23563269]

[154] Slattery ML, Herrick JS, Mullany LE, *et al.* An evaluation and replication of miRNAs with disease stage and colorectal cancer-specific mortality. Int J Cancer 2015; 137(2): 428-38.
[http://dx.doi.org/10.1002/ijc.29384] [PMID: 25484364]

[155] Braconi C, Valeri N, Kogure T, *et al.* Expression and functional role of a transcribed noncoding RNA with an ultraconserved element in hepatocellular carcinoma. Proc Natl Acad Sci USA 2011; 108(2): 786-91.
[http://dx.doi.org/10.1073/pnas.1011098108] [PMID: 21187392]

[156] Wernert N, Kaminski A, Haddouti M, Hahne JC. Tumor-stroma interactions of metastatic prostate cancer cell lines: analyses using microarrays. Methods Mol Biol 2007; 382: 223-37.
[http://dx.doi.org/10.1007/978-1-59745-304-2_14] [PMID: 18220234]

[157] Liu CG, Calin GA, Volinia S, Croce CM. MicroRNA expression profiling using microarrays. Nat Protoc 2008; 3(4): 563-78.
[http://dx.doi.org/10.1038/nprot.2008.14] [PMID: 18388938]

[158] Li MH, Fu SB, Xiao HS. Genome-wide analysis of microRNA and mRNA expression signatures in cancer. Acta Pharmacol Sin 2015; 36(10): 1200-11.
[http://dx.doi.org/10.1038/aps.2015.67] [PMID: 26299954]

[159] Geiss GK, Bumgarner RE, Birditt B, *et al.* Direct multiplexed measurement of gene expression with color-coded probe pairs. Nat Biotechnol 2008; 26(3): 317-25.
[http://dx.doi.org/10.1038/nbt1385] [PMID: 18278033]

[160] Malkov VA, Serikawa KA, Balantac N, *et al.* Multiplexed measurements of gene signatures in different analytes using the Nanostring nCounter Assay System. BMC Res Notes 2009; 2: 80.
[http://dx.doi.org/10.1186/1756-0500-2-80] [PMID: 19426535]

[161] Valeri N, Braconi C, Gasparini P, *et al.* MicroRNA-135b promotes cancer progression by acting as a downstream effector of oncogenic pathways in colon cancer. Cancer Cell 2014; 25(4): 469-83.
[http://dx.doi.org/10.1016/j.ccr.2014.03.006] [PMID: 24735923]

[162] Ullal AV, Peterson V, Agasti SS, *et al.* Cancer cell profiling by barcoding allows multiplexed protein analysis in fine-needle aspirates. Sci Transl Med 2014; 6(219): 219ra9.
[http://dx.doi.org/10.1126/scitranslmed.3007361] [PMID: 24431113]

[163] Apicella M, Migliore C, Capeloa T, *et al.* Dual MET/EGFR therapy leads to complete response and resistance prevention in a MET-amplified gastroesophageal xenopatient cohort. Oncogene 2016.
[PMID: 27524418]

[164] Trevisani F, Ghidini M, Larcher A, *et al.* MicroRNA 193b-3p as a predictive biomarker of chronic

kidney disease in patients undergoing radical nephrectomy for renal cell carcinoma. Br J Cancer 2016; 115(11): 1343-50.
[http://dx.doi.org/10.1038/bjc.2016.329] [PMID: 27802451]

[165] Siegel RL, Miller KD, Jemal A. Cancer statistics, 2016. CA Cancer J Clin 2016; 66(1): 7-30.
[http://dx.doi.org/10.3322/caac.21332] [PMID: 26742998]

[166] Boyle P, Langman JS. ABC of colorectal cancer: Epidemiology. BMJ 2000; 321(7264): 805-8.
[http://dx.doi.org/10.1136/bmj.321.7264.805] [PMID: 11009523]

[167] Brenner H, Kloor M, Pox CP. Colorectal cancer. Lancet 2014; 383(9927): 1490-502.
[http://dx.doi.org/10.1016/S0140-6736(13)61649-9] [PMID: 24225001]

[168] Kotzev AI, Kamenova MA. Highlights in the prevention of sporadic colorectal cancer. memo -. Mag Eur Med Oncol 2014; 7: 111-4.
[http://dx.doi.org/10.1007/s12254-014-0140-1]

[169] Haggar FA, Boushey RP. Colorectal cancer epidemiology: incidence, mortality, survival, and risk factors. Clin Colon Rectal Surg 2009; 22(4): 191-7.
[http://dx.doi.org/10.1055/s-0029-1242458] [PMID: 21037809]

[170] Kotzev I, Mirchev M, Manevska B, Ivanova I, Kaneva M. Risk and protective factors for development of colorectal polyps and cancer (Bulgarian experience). Hepatogastroenterology 2008; 55(82-83): 381-7.
[PMID: 18613371]

[171] Boyle P, Ferlay J. Mortality and survival in breast and colorectal cancer. Nat Clin Pract Oncol 2005; 2(9): 424-5.
[http://dx.doi.org/10.1038/ncponc0288] [PMID: 16264992]

[172] Janout V, Kollárová H. Epidemiology of colorectal cancer. Biomed Pap Med Fac Univ Palacky Olomouc Czech Repub 2001; 145(1): 5-10.
[http://dx.doi.org/10.5507/bp.2001.001] [PMID: 12415635]

[173] Jemal A, Clegg LX, Ward E, *et al.* Annual report to the nation on the status of cancer, 1975-2001, with a special feature regarding survival. Cancer 2004; 101(1): 3-27.
[http://dx.doi.org/10.1002/cncr.20288] [PMID: 15221985]

[174] Labianca R, Beretta GD, Mosconi S, Milesi L, Pessi MA. Giog. Colorectal cancer: screening. Ann Oncol 2005; 16: 127-32.
[http://dx.doi.org/10.1093/annonc/mdi730]

[175] Boardman LA, Morlan BW, Rabe KG, *et al.* Colorectal cancer risks in relatives of young-onset cases: is risk the same across all first-degree relatives? Clin Gastroenterol Hepatol 2007; 5(10): 1195-8.
[http://dx.doi.org/10.1016/j.cgh.2007.06.001] [PMID: 17702662]

[176] Jackson-Thompson J, Ahmed F, German RR, Lai SM, Friedman C. Descriptive epidemiology of colorectal cancer in the United States, 1998-2001. Cancer 2006; 107(5) (Suppl.): 1103-11.
[http://dx.doi.org/10.1002/cncr.22007] [PMID: 16835911]

[177] Tsong WH, Koh WP, Yuan JM, Wang R, Sun CL, Yu MC. Cigarettes and alcohol in relation to colorectal cancer: the Singapore Chinese Health Study. Br J Cancer 2007; 96(5): 821-7.
[http://dx.doi.org/10.1038/sj.bjc.6603623] [PMID: 17311023]

[178] Johnson IT, Lund EK. Review article: nutrition, obesity and colorectal cancer. Aliment Pharmacol Ther 2007; 26(2): 161-81.
[http://dx.doi.org/10.1111/j.1365-2036.2007.03371.x] [PMID: 17593063]

[179] Papadopoulos N, Nicolaides NC, Wei YF, *et al.* Mutation of a mutL homolog in hereditary colon cancer. Science 1994; 263(5153): 1625-9.
[http://dx.doi.org/10.1126/science.8128251] [PMID: 8128251]

[180] Willett WC. Diet and cancer: an evolving picture. JAMA 2005; 293(2): 233-4.

[http://dx.doi.org/10.1001/jama.293.2.233] [PMID: 15644551]

[181] Zisman AL, Nickolov A, Brand RE, Gorchow A, Roy HK. Associations between the age at diagnosis and location of colorectal cancer and the use of alcohol and tobacco: implications for screening. Arch Intern Med 2006; 166(6): 629-34.
[http://dx.doi.org/10.1001/archinte.166.6.629] [PMID: 16567601]

[182] Kotzev AI, Mirchev M. Genetics of colorectal cancer. Scripta Scientifica Medica 2006; 38: 57-63.

[183] Ahlquist DA. Aggressive polyps in hereditary nonpolyposis colorectal cancer: targets for screening. Gastroenterology 1995; 108(5): 1590-2.
[http://dx.doi.org/10.1016/0016-5085(95)90711-4] [PMID: 7729653]

[184] Rustgi AK. Hereditary gastrointestinal polyposis and nonpolyposis syndromes. N Engl J Med 1994; 331(25): 1694-702.
[http://dx.doi.org/10.1056/NEJM199412223312507] [PMID: 7969362]

[185] Jasperson KW, Samowitz WS, Burt RW. Constitutional mismatch repair-deficiency syndrome presenting as colonic adenomatous polyposis: clues from the skin. Clin Genet 2011; 80(4): 394-7.
[http://dx.doi.org/10.1111/j.1399-0004.2010.01543.x] [PMID: 21039432]

[186] Carethers JM, Stoffel EM. Lynch syndrome and Lynch syndrome mimics: The growing complex landscape of hereditary colon cancer. World J Gastroenterol 2015; 21(31): 9253-61.
[http://dx.doi.org/10.3748/wjg.v21.i31.9253] [PMID: 26309352]

[187] Giglia MD, Chu DI. Familial Colorectal Cancer: Understanding the Alphabet Soup. Clin Colon Rectal Surg 2016; 29(3): 185-95.
[http://dx.doi.org/10.1055/s-0036-1584290] [PMID: 27582643]

[188] Quiroga D, Lyerly HK, Morse MA. Deficient Mismatch Repair and the Role of Immunotherapy in Metastatic Colorectal Cancer. Curr Treat Options Oncol 2016; 17(8): 41.
[http://dx.doi.org/10.1007/s11864-016-0414-4] [PMID: 27315067]

[189] Bettstetter M, Dechant S, Ruemmele P, *et al.* Distinction of hereditary nonpolyposis colorectal cancer and sporadic microsatellite-unstable colorectal cancer through quantification of MLH1 methylation by real-time PCR. Clin Cancer Res 2007; 13(11): 3221-8.
[http://dx.doi.org/10.1158/1078-0432.CCR-06-3064] [PMID: 17545526]

[190] Lochhead P, Kuchiba A, Imamura Y, *et al.* Microsatellite instability and BRAF mutation testing in colorectal cancer prognostication. J Natl Cancer Inst 2013; 105(15): 1151-6.
[http://dx.doi.org/10.1093/jnci/djt173] [PMID: 23878352]

[191] Goldstein J, Tran B, Ensor J, *et al.* Multicenter retrospective analysis of metastatic colorectal cancer (CRC) with high-level microsatellite instability (MSI-H). Ann Oncol 2014; 25(5): 1032-8.
[http://dx.doi.org/10.1093/annonc/mdu100] [PMID: 24585723]

[192] Venderbosch S, Nagtegaal ID, Maughan TS, *et al.* Mismatch repair status and BRAF mutation status in metastatic colorectal cancer patients: a pooled analysis of the CAIRO, CAIRO2, COIN, and FOCUS studies. Clin Cancer Res 2014; 20(20): 5322-30.
[http://dx.doi.org/10.1158/1078-0432.CCR-14-0332] [PMID: 25139339]

[193] Valeri N, Gasparini P, Fabbri M, *et al.* Modulation of mismatch repair and genomic stability by miR-155. Proc Natl Acad Sci USA 2010; 107(15): 6982-7.
[http://dx.doi.org/10.1073/pnas.1002472107] [PMID: 20351277]

[194] O'Connell JB, Maggard MA, Liu JH, Etzioni DA, Livingston EH, Ko CY. Rates of colon and rectal cancers are increasing in young adults. Am Surg 2003; 69(10): 866-72.
[PMID: 14570365]

[195] O'Connell JB, Maggard MA, Livingston EH, Yo CK. Colorectal cancer in the young. Am J Surg 2004; 187(3): 343-8.
[http://dx.doi.org/10.1016/j.amjsurg.2003.12.020] [PMID: 15006562]

[196] Fairley TL, Cardinez CJ, Martin J, *et al.* Colorectal cancer in U.S. adults younger than 50 years of age, 1998-2001. Cancer 2006; 107(5) (Suppl.): 1153-61.
[http://dx.doi.org/10.1002/cncr.22012] [PMID: 16862554]

[197] Goldstein MJ, Mitchell EP. Carcinoembryonic antigen in the staging and follow-up of patients with colorectal cancer. Cancer Invest 2005; 23(4): 338-51.
[http://dx.doi.org/10.1081/CNV-58878] [PMID: 16100946]

[198] Ueno H, Mochizuki H, Hatsuse K, Hase K, Yamamoto T. Indicators for treatment strategies of colorectal liver metastases. Ann Surg 2000; 231(1): 59-66.
[http://dx.doi.org/10.1097/00000658-200001000-00009] [PMID: 10636103]

[199] Hohenberger P, Schlag PM, Gerneth T, Herfarth C. Pre- and postoperative carcinoembryonic antigen determinations in hepatic resection for colorectal metastases. Predictive value and implications for adjuvant treatment based on multivariate analysis. Ann Surg 1994; 219(2): 135-43.
[http://dx.doi.org/10.1097/00000658-199402000-00005] [PMID: 8129484]

[200] Pietra N, Sarli L, Costi R, Ouchemi C, Grattarola M, Peracchia A. Role of follow-up in management of local recurrences of colorectal cancer: a prospective, randomized study. Dis Colon Rectum 1998; 41(9): 1127-33.
[http://dx.doi.org/10.1007/BF02239434] [PMID: 9749496]

[201] Arnaud JP, Koehl C, Adloff M. Carcinoembryonic antigen (CEA) in diagnosis and prognosis of colorectal carcinoma. Dis Colon Rectum 1980; 23(3): 141-4.
[http://dx.doi.org/10.1007/BF02587615] [PMID: 7379666]

[202] Sørbye H, Dahl O. Transient CEA increase at start of oxaliplatin combination therapy for metastatic colorectal cancer. Acta Oncol 2004; 43(5): 495-8.
[http://dx.doi.org/10.1080/02841860410032380] [PMID: 15360055]

[203] Moertel CG, Fleming TR, Macdonald JS, Haller DG, Laurie JA, Tangen C. An evaluation of the carcinoembryonic antigen (CEA) test for monitoring patients with resected colon cancer. JAMA 1993; 270(8): 943-7.
[http://dx.doi.org/10.1001/jama.1993.03510080047030] [PMID: 8141873]

[204] Choi YJ, Chang WJ, Shin SW, Park KH, Kim ST, Kim YH. The prognostic role of serum C-X-C chemokine receptor type 4 in patients with metastatic or recurrent colorectal cancer. Onco Targets Ther 2016; 9: 3307-12.
[http://dx.doi.org/10.2147/OTT.S104511] [PMID: 27330310]

[205] Ozawa T, Ishihara S, Kawai K, *et al.* Prognostic Significance of Preoperative Serum Carbohydrate Antigen 19-9 in Patients With Stage IV Colorectal Cancer. Clin Colorectal Cancer 2016; 15(4): e157-63.
[http://dx.doi.org/10.1016/j.clcc.2016.04.012] [PMID: 27245559]

[206] Huo YR, Huang Y, Liauw W, Zhao J, Morris DL. Prognostic Value of Carcinoembryonic Antigen (CEA), AFP, CA19-9 and CA125 for Patients with Colorectal Cancer with Peritoneal Carcinomatosis Treated by Cytoreductive Surgery and Intraperitoneal Chemotherapy. Anticancer Res 2016; 36(3): 1041-9.
[PMID: 26976996]

[207] Kawahara D, Fujita F, Hayashi H, Matsuyama T, Eguchi S. The Significance of Combined Measurement of p53 Antibody and other Tumor Markers for Colorectal Cancer after Curative Resection. Hepatogastroenterology 2015; 62(139): 624-8.
[PMID: 26897942]

[208] Abe S, Kawai K, Ishihara S, *et al.* Prognostic impact of carcinoembryonic antigen and carbohydrate antigen 19-9 in stage IV colorectal cancer patients after R0 resection. J Surg Res 2016; 205(2): 384-92.
[http://dx.doi.org/10.1016/j.jss.2016.06.078] [PMID: 27664887]

[209] Stiksma J, Grootendorst DC, van der Linden PW. CA 19-9 as a marker in addition to CEA to monitor

colorectal cancer. Clin Colorectal Cancer 2014; 13(4): 239-44.
[http://dx.doi.org/10.1016/j.clcc.2014.09.004] [PMID: 25442815]

[210] Burz C, Aziz BY, Bălăcescu L, *et al.* Tumor markers used in monitoring the tumor recurrence in patients with colorectal cancer. Clujul Med 2016; 89(3): 378-83.
[http://dx.doi.org/10.15386/cjmed-635] [PMID: 27547057]

[211] Nicholson RI, Gee JM, Harper ME. EGFR and cancer prognosis. Eur J Cancer 2001; 37 (Suppl. 4): S9-S15.
[http://dx.doi.org/10.1016/S0959-8049(01)00231-3] [PMID: 11597399]

[212] Seshacharyulu P, Ponnusamy MP, Haridas D, Jain M, Ganti AK, Batra SK. Targeting the EGFR signaling pathway in cancer therapy. Expert Opin Ther Targets 2012; 16(1): 15-31.
[http://dx.doi.org/10.1517/14728222.2011.648617] [PMID: 22239438]

[213] Regad T. Targeting RTK Signaling Pathways in Cancer. Cancers (Basel) 2015; 7(3): 1758-84.
[http://dx.doi.org/10.3390/cancers7030860] [PMID: 26404379]

[214] Van Emburgh BO, Sartore-Bianchi A, Di Nicolantonio F, Siena S, Bardelli A. Acquired resistance to EGFR-targeted therapies in colorectal cancer. Mol Oncol 2014; 8(6): 1084-94.
[http://dx.doi.org/10.1016/j.molonc.2014.05.003] [PMID: 24913799]

[215] Scaltriti M, Baselga J. The epidermal growth factor receptor pathway: a model for targeted therapy. Clin Cancer Res 2006; 12(18): 5268-72.
[http://dx.doi.org/10.1158/1078-0432.CCR-05-1554] [PMID: 17000658]

[216] Li S, Schmitz KR, Jeffrey PD, Wiltzius JJ, Kussie P, Ferguson KM. Structural basis for inhibition of the epidermal growth factor receptor by cetuximab. Cancer Cell 2005; 7(4): 301-11.
[http://dx.doi.org/10.1016/j.ccr.2005.03.003] [PMID: 15837620]

[217] Mendelsohn J, Baselga J. The EGF receptor family as targets for cancer therapy. Oncogene 2000; 19(56): 6550-65.
[http://dx.doi.org/10.1038/sj.onc.1204082] [PMID: 11426640]

[218] Weber J, McCormack PL. Panitumumab: in metastatic colorectal cancer with wild-type KRAS. BioDrugs 2008; 22(6): 403-11.
[http://dx.doi.org/10.2165/0063030-200822060-00006] [PMID: 18998757]

[219] Lo Nigro C, Ricci V, Vivenza D, *et al.* Evaluation of antibody-dependent cell-mediated cytotoxicity activity and cetuximab response in KRAS wild-type metastatic colorectal cancer patients. World J Gastrointest Oncol 2016; 8(2): 222-30.
[http://dx.doi.org/10.4251/wjgo.v8.i2.222] [PMID: 26909137]

[220] Cunningham D, Humblet Y, Siena S, *et al.* Cetuximab monotherapy and cetuximab plus irinotecan in irinotecan-refractory metastatic colorectal cancer. N Engl J Med 2004; 351(4): 337-45.
[http://dx.doi.org/10.1056/NEJMoa033025] [PMID: 15269313]

[221] Jonker DJ, O'Callaghan CJ, Karapetis CS, *et al.* Cetuximab for the treatment of colorectal cancer. N Engl J Med 2007; 357(20): 2040-8.
[http://dx.doi.org/10.1056/NEJMoa071834] [PMID: 18003960]

[222] Van Cutsem E, Köhne CH, Hitre E, *et al.* Cetuximab and chemotherapy as initial treatment for metastatic colorectal cancer. N Engl J Med 2009; 360(14): 1408-17.
[http://dx.doi.org/10.1056/NEJMoa0805019] [PMID: 19339720]

[223] Van Cutsem E, Peeters M, Siena S, *et al.* Open-label phase III trial of panitumumab plus best supportive care compared with best supportive care alone in patients with chemotherapy-refractory metastatic colorectal cancer. J Clin Oncol 2007; 25(13): 1658-64.
[http://dx.doi.org/10.1200/JCO.2006.08.1620] [PMID: 17470858]

[224] de Gramont A, Figer A, Seymour M, *et al.* Leucovorin and fluorouracil with or without oxaliplatin as first-line treatment in advanced colorectal cancer. J Clin Oncol 2000; 18(16): 2938-47.
[http://dx.doi.org/10.1200/JCO.2000.18.16.2938] [PMID: 10944126]

[225] Cremolini C, Loupakis F, Antoniotti C, *et al.* FOLFOXIRI plus bevacizumab *versus* FOLFIRI plus bevacizumab as first-line treatment of patients with metastatic colorectal cancer: updated overall survival and molecular subgroup analyses of the open-label, phase 3 TRIBE study. Lancet Oncol 2015; 16(13): 1306-15.
[http://dx.doi.org/10.1016/S1470-2045(15)00122-9] [PMID: 26338525]

[226] Adam R, Wicherts DA, de Haas RJ, *et al.* Patients with initially unresectable colorectal liver metastases: is there a possibility of cure? J Clin Oncol 2009; 27(11): 1829-35.
[http://dx.doi.org/10.1200/JCO.2008.19.9273] [PMID: 19273699]

[227] Peeters M, Price TJ, Cervantes A, *et al.* Final results from a randomized phase 3 study of FOLFIRI +/- panitumumab for second-line treatment of metastatic colorectal cancer. Ann Oncol 2014; 25(1): 107-16.
[http://dx.doi.org/10.1093/annonc/mdt523] [PMID: 24356622]

[228] De Roock W, Claes B, Bernasconi D, *et al.* Effects of KRAS, BRAF, NRAS, and PIK3CA mutations on the efficacy of cetuximab plus chemotherapy in chemotherapy-refractory metastatic colorectal cancer: a retrospective consortium analysis. Lancet Oncol 2010; 11(8): 753-62.
[http://dx.doi.org/10.1016/S1470-2045(10)70130-3] [PMID: 20619739]

[229] Amado RG, Wolf M, Peeters M, *et al.* Wild-type KRAS is required for panitumumab efficacy in patients with metastatic colorectal cancer. J Clin Oncol 2008; 26(10): 1626-34.
[http://dx.doi.org/10.1200/JCO.2007.14.7116] [PMID: 18316791]

[230] Benvenuti S, Sartore-Bianchi A, Di Nicolantonio F, *et al.* Oncogenic activation of the RAS/RAF signaling pathway impairs the response of metastatic colorectal cancers to anti-epidermal growth factor receptor antibody therapies. Cancer Res 2007; 67(6): 2643-8.
[http://dx.doi.org/10.1158/0008-5472.CAN-06-4158] [PMID: 17363584]

[231] Karapetis CS, Khambata-Ford S, Jonker DJ, *et al.* K-ras mutations and benefit from cetuximab in advanced colorectal cancer. N Engl J Med 2008; 359(17): 1757-65.
[http://dx.doi.org/10.1056/NEJMoa0804385] [PMID: 18946061]

[232] Lièvre A, Bachet JB, Le Corre D, *et al.* KRAS mutation status is predictive of response to cetuximab therapy in colorectal cancer. Cancer Res 2006; 66(8): 3992-5.
[http://dx.doi.org/10.1158/0008-5472.CAN-06-0191] [PMID: 16618717]

[233] Yokota T. Are KRAS/BRAF mutations potent prognostic and/or predictive biomarkers in colorectal cancers? Anticancer Agents Med Chem 2012; 12(2): 163-71.
[http://dx.doi.org/10.2174/187152012799014968] [PMID: 22043994]

[234] Douillard JY, Oliner KS, Siena S, *et al.* Panitumumab-FOLFOX4 treatment and RAS mutations in colorectal cancer. N Engl J Med 2013; 369(11): 1023-34.
[http://dx.doi.org/10.1056/NEJMoa1305275] [PMID: 24024839]

[235] Stintzing S, Jung A, Rossius L, *et al.* Analysis of KRAS/NRAS and BRAF mutations in FIRE-3: A randomized phase III study of FOLFIRI plus cetuximab or bevacizumab as first-line treatment for wild-type (WT) KRAS (exon 2) metastatic colorectal cancer (mCRC) patients. Eur J Cancer 2013; 49: S8-9.

[236] Schmoll HJ, Van Cutsem E, Stein A, *et al.* ESMO Consensus Guidelines for management of patients with colon and rectal cancer. a personalized approach to clinical decision making. Ann Oncol 2012; 23(10): 2479-516.
[http://dx.doi.org/10.1093/annonc/mds236] [PMID: 23012255]

[237] Wong R, Cunningham D. Using predictive biomarkers to select patients with advanced colorectal cancer for treatment with epidermal growth factor receptor antibodies. J Clin Oncol 2008; 26(35): 5668-70.
[http://dx.doi.org/10.1200/JCO.2008.19.5024] [PMID: 19001346]

[238] Van Cutsem E, Cervantes A, Adam R, *et al.* ESMO consensus guidelines for the management of

patients with metastatic colorectal cancer. Ann Oncol 2016; 27(8): 1386-422. [http://dx.doi.org/10.1093/annonc/mdw235] [PMID: 27380959]

[239] Van Cutsem E, Lenz HJ, Köhne CH, *et al.* Fluorouracil, leucovorin, and irinotecan plus cetuximab treatment and RAS mutations in colorectal cancer. J Clin Oncol 2015; 33(7): 692-700. [http://dx.doi.org/10.1200/JCO.2014.59.4812] [PMID: 25605843]

[240] Bokemeyer C, Köhne CH, Ciardiello F, *et al.* FOLFOX4 plus cetuximab treatment and RAS mutations in colorectal cancer. Eur J Cancer 2015; 51(10): 1243-52. [http://dx.doi.org/10.1016/j.ejca.2015.04.007] [PMID: 25937522]

[241] Sartore-Bianchi A, Martini M, Molinari F, *et al.* PIK3CA mutations in colorectal cancer are associated with clinical resistance to EGFR-targeted monoclonal antibodies. Cancer Res 2009; 69(5): 1851-7. [http://dx.doi.org/10.1158/0008-5472.CAN-08-2466] [PMID: 19223544]

[242] Di Nicolantonio F, Martini M, Molinari F, *et al.* Wild-type BRAF is required for response to panitumumab or cetuximab in metastatic colorectal cancer. J Clin Oncol 2008; 26(35): 5705-12. [http://dx.doi.org/10.1200/JCO.2008.18.0786] [PMID: 19001320]

[243] Bertotti A, Migliardi G, Galimi F, *et al.* A molecularly annotated platform of patient-derived xenografts ("xenopatients") identifies HER2 as an effective therapeutic target in cetuximab-resistant colorectal cancer. Cancer Discov 2011; 1(6): 508-23. [http://dx.doi.org/10.1158/2159-8290.CD-11-0109] [PMID: 22586653]

[244] Bardelli A, Corso S, Bertotti A, *et al.* Amplification of the MET receptor drives resistance to anti-EGFR therapies in colorectal cancer. Cancer Discov 2013; 3(6): 658-73. [http://dx.doi.org/10.1158/2159-8290.CD-12-0558] [PMID: 23729478]

[245] Smith G, Bounds R, Wolf H, Steele RJ, Carey FA, Wolf CR. Activating K-Ras mutations outwith 'hotspot' codons in sporadic colorectal tumours - implications for personalised cancer medicine. Br J Cancer 2010; 102(4): 693-703. [http://dx.doi.org/10.1038/sj.bjc.6605534] [PMID: 20147967]

[246] Valtorta E, Misale S, Sartore-Bianchi A, *et al.* KRAS gene amplification in colorectal cancer and impact on response to EGFR-targeted therapy. Int J Cancer 2013; 133(5): 1259-65. [http://dx.doi.org/10.1002/ijc.28106] [PMID: 23404247]

[247] Frattini M, Saletti P, Romagnani E, *et al.* PTEN loss of expression predicts cetuximab efficacy in metastatic colorectal cancer patients. Br J Cancer 2007; 97(8): 1139-45. [http://dx.doi.org/10.1038/sj.bjc.6604009] [PMID: 17940504]

[248] Yonesaka K, Zejnullahu K, Okamoto I, *et al.* Activation of ERBB2 signaling causes resistance to the EGFR-directed therapeutic antibody cetuximab. Sci Transl Med 2011; 3(99): 99ra86. [http://dx.doi.org/10.1126/scitranslmed.3002442] [PMID: 21900593]

[249] Misale S, Yaeger R, Hobor S, *et al.* Emergence of KRAS mutations and acquired resistance to anti-EGFR therapy in colorectal cancer. Nature 2012; 486(7404): 532-6. [PMID: 22722830]

[250] Kalikaki A, Politaki H, Souglakos J, *et al.* KRAS genotypic changes of circulating tumor cells during treatment of patients with metastatic colorectal cancer. PLoS One 2014; 9(8): e104902. [http://dx.doi.org/10.1371/journal.pone.0104902] [PMID: 25137394]

[251] Diaz LA Jr, Williams RT, Wu J, *et al.* The molecular evolution of acquired resistance to targeted EGFR blockade in colorectal cancers. Nature 2012; 486(7404): 537-40. [PMID: 22722843]

[252] Bettegowda C, Sausen M, Leary RJ, *et al.* Detection of circulating tumor DNA in early- and late-stage human malignancies. Sci Transl Med 2014; 6(224): 224ra24. [http://dx.doi.org/10.1126/scitranslmed.3007094] [PMID: 24553385]

[253] Misale S, Arena S, Lamba S, *et al.* Blockade of EGFR and MEK intercepts heterogeneous mechanisms of acquired resistance to anti-EGFR therapies in colorectal cancer. Sci Transl Med 2014;

6(224): 224ra26.
[http://dx.doi.org/10.1126/scitranslmed.3007947] [PMID: 24553387]

[254] Hecht JR, Mitchell E, Chidiac T, *et al.* A randomized phase IIIB trial of chemotherapy, bevacizumab, and panitumumab compared with chemotherapy and bevacizumab alone for metastatic colorectal cancer. J Clin Oncol 2009; 27(5): 672-80.
[http://dx.doi.org/10.1200/JCO.2008.19.8135] [PMID: 19114685]

[255] Tol J, Koopman M, Cats A, *et al.* Chemotherapy, bevacizumab, and cetuximab in metastatic colorectal cancer. N Engl J Med 2009; 360(6): 563-72.
[http://dx.doi.org/10.1056/NEJMoa0808268] [PMID: 19196673]

[256] Tie J, Kinde I, Wang Y, *et al.* Circulating tumor DNA as an early marker of therapeutic response in patients with metastatic colorectal cancer. Ann Oncol 2015; 26(8): 1715-22.
[http://dx.doi.org/10.1093/annonc/mdv177] [PMID: 25851626]

[257] Tie J, Wang Y, Tomasetti C, *et al.* Circulating tumor DNA analysis detects minimal residual disease and predicts recurrence in patients with stage II colon cancer. Sci Transl Med 2016; 8(346): 346ra92.
[http://dx.doi.org/10.1126/scitranslmed.aaf6219] [PMID: 27384348]

[258] Michael MZ, O' Connor SM, van Holst Pellekaan NG, Young GP, James RJ. Reduced accumulation of specific microRNAs in colorectal neoplasia. Mol Cancer Res 2003; 1(12): 882-91.
[PMID: 14573789]

[259] Chen X, Guo X, Zhang H, *et al.* Role of miR-143 targeting KRAS in colorectal tumorigenesis. Oncogene 2009; 28(10): 1385-92.
[http://dx.doi.org/10.1038/onc.2008.474] [PMID: 19137007]

[260] Cummins JM, He Y, Leary RJ, *et al.* The colorectal microRNAome. Proc Natl Acad Sci USA 2006; 103(10): 3687-92.
[http://dx.doi.org/10.1073/pnas.0511155103] [PMID: 16505370]

[261] Zhou JJ, Zheng S, Sun LF, Zheng L. MicroRNA regulation network in colorectal cancer metastasis. World J Biol Chem 2014; 5(3): 301-7.
[http://dx.doi.org/10.4331/wjbc.v5.i3.301] [PMID: 25225598]

[262] Ju J. miRNAs as biomarkers in colorectal cancer diagnosis and prognosis. Bioanalysis 2010; 2(5): 901-6.
[http://dx.doi.org/10.4155/bio.10.45] [PMID: 20543966]

[263] Huang Z, Huang D, Ni S, Peng Z, Sheng W, Du X. Plasma microRNAs are promising novel biomarkers for early detection of colorectal cancer. Int J Cancer 2010; 127(1): 118-26.
[http://dx.doi.org/10.1002/ijc.25007] [PMID: 19876917]

[264] Ng EK, Chong WW, Jin H, *et al.* Differential expression of microRNAs in plasma of patients with colorectal cancer: a potential marker for colorectal cancer screening. Gut 2009; 58(10): 1375-81.
[http://dx.doi.org/10.1136/gut.2008.167817] [PMID: 19201770]

[265] Schetter AJ, Okayama H, Harris CC. The role of microRNAs in colorectal cancer. Cancer J 2012; 18(3): 244-52.
[http://dx.doi.org/10.1097/PPO.0b013e318258b78f] [PMID: 22647361]

[266] Nakajima G, Hayashi K, Xi Y, *et al.* Non-coding MicroRNAs hsa-let-7g and hsa-miR-181b are Associated with Chemoresponse to S-1 in Colon Cancer. Cancer Genomics Proteomics 2006; 3(5): 317-24.
[PMID: 18172508]

[267] Schepeler T, Reinert JT, Ostenfeld MS, *et al.* Diagnostic and prognostic microRNAs in stage II colon cancer. Cancer Res 2008; 68(15): 6416-24.
[http://dx.doi.org/10.1158/0008-5472.CAN-07-6110] [PMID: 18676867]

[268] Xi Y, Formentini A, Chien M, *et al.* Prognostic Values of microRNAs in Colorectal Cancer. Biomark Insights 2006; 2: 113-21.

[PMID: 18079988]

[269] Schetter AJ, Leung SY, Sohn JJ, *et al.* MicroRNA expression profiles associated with prognosis and therapeutic outcome in colon adenocarcinoma. JAMA 2008; 299(4): 425-36.
[http://dx.doi.org/10.1001/jama.299.4.425] [PMID: 18230780]

[270] Slaby O, Svoboda M, Fabian P, *et al.* Altered expression of miR-21, miR-31, miR-143 and miR-145 is related to clinicopathologic features of colorectal cancer. Oncology 2007; 72(5-6): 397-402.
[http://dx.doi.org/10.1159/000113489] [PMID: 18196926]

[271] Valeri N, Gasparini P, Braconi C, *et al.* MicroRNA-21 induces resistance to 5-fluorouracil by down-regulating human DNA MutS homolog 2 (hMSH2). Proc Natl Acad Sci USA 2010; 107(49): 21098-103.
[http://dx.doi.org/10.1073/pnas.1015541107] [PMID: 21078976]

[272] Sclafani F, Chau I, Cunningham D, *et al.* Sequence variation in mature microRNA-608 and benefit from neo-adjuvant treatment in locally advanced rectal cancer patients. Carcinogenesis 2016; 37(9): 852-7.
[http://dx.doi.org/10.1093/carcin/bgw073] [PMID: 27381831]

[273] Sclafani F, Chau I, Cunningham D, *et al.* Prognostic role of the LCS6 KRAS variant in locally advanced rectal cancer: results of the EXPERT-C trial. Ann Oncol 2015; 26(9): 1936-41.
[http://dx.doi.org/10.1093/annonc/mdv285] [PMID: 26162609]

[274] Xie X, Tang B, Xiao YF, *et al.* Long non-coding RNAs in colorectal cancer. Oncotarget 2016; 7(5): 5226-39.
[http://dx.doi.org/10.18632/oncotarget.6446] [PMID: 26637808]

[275] Svoboda M, Slyskova J, Schneiderova M, *et al.* HOTAIR long non-coding RNA is a negative prognostic factor not only in primary tumors, but also in the blood of colorectal cancer patients. Carcinogenesis 2014; 35(7): 1510-5.
[http://dx.doi.org/10.1093/carcin/bgu055] [PMID: 24583926]

[276] Graham LD, Pedersen SK, Brown GS, *et al.* Colorectal Neoplasia Differentially Expressed (CRNDE), a Novel Gene with Elevated Expression in Colorectal Adenomas and Adenocarcinomas. Genes Cancer 2011; 2(8): 829-40.
[http://dx.doi.org/10.1177/1947601911431081] [PMID: 22393467]

[277] Geng YJ, Xie SL, Li Q, Ma J, Wang GY. Large intervening non-coding RNA HOTAIR is associated with hepatocellular carcinoma progression. J Int Med Res 2011; 39(6): 2119-28.
[http://dx.doi.org/10.1177/147323001103900608] [PMID: 22289527]

[278] Wu ZH, Wang XL, Tang HM, *et al.* Long non-coding RNA HOTAIR is a powerful predictor of metastasis and poor prognosis and is associated with epithelial-mesenchymal transition in colon cancer. Oncol Rep 2014; 32(1): 395-402.
[http://dx.doi.org/10.3892/or.2014.3186] [PMID: 24840737]

[279] Martin FT, Kavanagh D, Waldron R. Squamous cell carcinoma of the anal canal. Surgeon 2009; 7(4): 232-7.
[http://dx.doi.org/10.1016/S1479-666X(09)80091-7] [PMID: 19736891]

[280] Bendell JC, Ryan DP. Current perspectives on anal cancer. Oncology (Williston Park) 2003; 17: 492-7. 502-3; discussion 3, 7-9.

[281] Salati SA, Al Kadi A. Anal cancer - a review. Int J Health Sci (Qassim) 2012; 6(2): 206-30.
[http://dx.doi.org/10.12816/0006000] [PMID: 23580899]

[282] Uronis HE, Bendell JC. Anal cancer: an overview. Oncologist 2007; 12(5): 524-34.
[http://dx.doi.org/10.1634/theoncologist.12-5-524] [PMID: 17522240]

[283] Johnson LG, Madeleine MM, Newcomer LM, Schwartz SM, Daling JR. Anal cancer incidence and survival: the surveillance, epidemiology, and end results experience, 1973-2000. Cancer 2004; 101(2): 281-8.

[http://dx.doi.org/10.1002/cncr.20364] [PMID: 15241824]

[284] Das P, Crane CH, Eng C, Ajani JA. Prognostic factors for squamous cell cancer of the anal canal. Gastrointest Cancer Res 2008; 2(1): 10-4.
[PMID: 19259318]

[285] Ajani JA, Winter KA, Gunderson LL, *et al.* Fluorouracil, mitomycin, and radiotherapy *vs* fluorouracil, cisplatin, and radiotherapy for carcinoma of the anal canal: a randomized controlled trial. JAMA 2008; 299(16): 1914-21.
[http://dx.doi.org/10.1001/jama.299.16.1914] [PMID: 18430910]

[286] UKCCCR Anal Cancer Trial Working Party. UK Co-ordinating Committee on Cancer Research. Epidermoid anal cancer: results from the UKCCCR randomised trial of radiotherapy alone *versus* radiotherapy, 5-fluorouracil, and mitomycin. Lancet 1996; 348(9034): 1049-54.
[http://dx.doi.org/10.1016/S0140-6736(96)03409-5] [PMID: 8874455]

[287] Bartelink H, Roelofsen F, Eschwege F, *et al.* Concomitant radiotherapy and chemotherapy is superior to radiotherapy alone in the treatment of locally advanced anal cancer: results of a phase III randomized trial of the European Organization for Research and Treatment of Cancer Radiotherapy and Gastrointestinal Cooperative Groups. J Clin Oncol 1997; 15(5): 2040-9.
[http://dx.doi.org/10.1200/JCO.1997.15.5.2040] [PMID: 9164216]

[288] Flam M, John M, Pajak TF, *et al.* Role of mitomycin in combination with fluorouracil and radiotherapy, and of salvage chemoradiation in the definitive nonsurgical treatment of epidermoid carcinoma of the anal canal: results of a phase III randomized intergroup study. J Clin Oncol 1996; 14(9): 2527-39.
[http://dx.doi.org/10.1200/JCO.1996.14.9.2527] [PMID: 8823332]

[289] Bentzen AG, Guren MG, Wanderås EH, *et al.* Chemoradiotherapy of anal carcinoma: survival and recurrence in an unselected national cohort. Int J Radiat Oncol Biol Phys 2012; 83(2): e173-80.
[http://dx.doi.org/10.1016/j.ijrobp.2011.12.062] [PMID: 22436791]

[290] Glynne-Jones R, Nilsson PJ, Aschele C, *et al.* European Society for Medical Oncology (ESMO); European Society of Surgical Oncology (ESSO); European Society of Radiotherapy and Oncology (ESTRO). Anal cancer: ESMO-ESSO-ESTRO clinical practice guidelines for diagnosis, treatment and follow-up. Eur J Surg Oncol 2014; 40(10): 1165-76.
[http://dx.doi.org/10.1016/j.ejso.2014.07.030] [PMID: 25239441]

[291] Ghosn M, Kourie HR, Abdayem P, Antoun J, Nasr D. Anal cancer treatment: current status and future perspectives. World J Gastroenterol 2015; 21(8): 2294-302.
[http://dx.doi.org/10.3748/wjg.v21.i8.2294] [PMID: 25741135]

[292] Lukan N, Ströbel P, Willer A, *et al.* Cetuximab-based treatment of metastatic anal cancer: correlation of response with KRAS mutational status. Oncology 2009; 77(5): 293-9.
[http://dx.doi.org/10.1159/000259615] [PMID: 19923868]

[293] Van Damme N, Deron P, Van Roy N, *et al.* Epidermal growth factor receptor and K-RAS status in two cohorts of squamous cell carcinomas. BMC Cancer 2010; 10: 189.
[http://dx.doi.org/10.1186/1471-2407-10-189] [PMID: 20459770]

[294] Gnanajothy R, Warren GW, Okun S, Peterson LL. A combined modality therapeutic approach to metastatic anal squamous cell carcinoma with systemic chemotherapy and local therapy to sites of disease: case report and review of literature. J Gastrointest Oncol 2016; 7(3): E58-63.
[http://dx.doi.org/10.21037/jgo.2015.11.02] [PMID: 27284490]

[295] Paliga A, Onerheim R, Gologan A, *et al.* EGFR and K-ras gene mutation status in squamous cell anal carcinoma: a role for concurrent radiation and EGFR inhibitors? Br J Cancer 2012; 107(11): 1864-8.
[http://dx.doi.org/10.1038/bjc.2012.479] [PMID: 23093229]

[296] Serup-Hansen E, Linnemann D, Høgdall E, Geertsen PF, Havsteen H. KRAS and BRAF mutations in anal carcinoma. APMIS 2015; 123(1): 53-9.
[http://dx.doi.org/10.1111/apm.12306] [PMID: 25244542]

[297] Zhang L, Wu J, Ling MT, Zhao L, Zhao KN. The role of the PI3K/Akt/mTOR signalling pathway in human cancers induced by infection with human papillomaviruses. Mol Cancer 2015; 14: 87. [http://dx.doi.org/10.1186/s12943-015-0361-x] [PMID: 26022660]

[298] Marur S, D'Souza G, Westra WH, Forastiere AA. HPV-associated head and neck cancer: a virus-related cancer epidemic. Lancet Oncol 2010; 11(8): 781-9. [http://dx.doi.org/10.1016/S1470-2045(10)70017-6] [PMID: 20451455]

[299] Crosbie EJ, Einstein MH, Franceschi S, Kitchener HC. Human papillomavirus and cervical cancer. Lancet 2013; 382(9895): 889-99. [http://dx.doi.org/10.1016/S0140-6736(13)60022-7] [PMID: 23618600]

[300] Thomas MK, Pitot HC, Liem A, Lambert PF. Dominant role of HPV16 E7 in anal carcinogenesis. Virology 2011; 421(2): 114-8. [http://dx.doi.org/10.1016/j.virol.2011.09.018] [PMID: 21999991]

[301] Thomas LK, Bermejo JL, Vinokurova S, *et al.* Chromosomal gains and losses in human papillomavirus-associated neoplasia of the lower genital tract - a systematic review and meta-analysis. Eur J Cancer 2014; 50(1): 85-98. [http://dx.doi.org/10.1016/j.ejca.2013.08.022] [PMID: 24054023]

[302] Ojesina AI, Lichtenstein L, Freeman SS, *et al.* Landscape of genomic alterations in cervical carcinomas. Nature 2014; 506(7488): 371-5. [http://dx.doi.org/10.1038/nature12881] [PMID: 24390348]

[303] Patel H, Polanco-Echeverry G, Segditsas S, *et al.* Activation of AKT and nuclear accumulation of wild type TP53 and MDM2 in anal squamous cell carcinoma. Int J Cancer 2007; 121(12): 2668-73. [http://dx.doi.org/10.1002/ijc.23028] [PMID: 17721920]

[304] Cacheux W, Rouleau E, Briaux A, *et al.* Mutational analysis of anal cancers demonstrates frequent PIK3CA mutations associated with poor outcome after salvage abdominoperineal resection. Br J Cancer 2016; 114(12): 1387-94. [http://dx.doi.org/10.1038/bjc.2016.144] [PMID: 27219019]

[305] Lampejo T, Kavanagh D, Clark J, *et al.* Prognostic biomarkers in squamous cell carcinoma of the anus: a systematic review. Br J Cancer 2010; 103(12): 1858-69. [http://dx.doi.org/10.1038/sj.bjc.6605984] [PMID: 21063399]

[306] Siegel EM, Eschrich S, Winter K, *et al.* Epigenomic characterization of locally advanced anal cancer: a radiation therapy oncology group 98-11 specimen study. Dis Colon Rectum 2014; 57(8): 941-57. [http://dx.doi.org/10.1097/DCR.0000000000000160] [PMID: 25003289]

[307] Pan SY, Morrison H. Epidemiology of cancer of the small intestine. World J Gastrointest Oncol 2011; 3(3): 33-42. [http://dx.doi.org/10.4251/wjgo.v3.i3.33] [PMID: 21461167]

[308] Bilimoria KY, Bentrem DJ, Wayne JD, Ko CY, Bennett CL, Talamonti MS. Small bowel cancer in the United States: changes in epidemiology, treatment, and survival over the last 20 years. Ann Surg 2009; 249(1): 63-71. [http://dx.doi.org/10.1097/SLA.0b013e31818e4641] [PMID: 19106677]

[309] Wright NH, Pennazio M, Howe JR, *et al.* Pathology and Genetics of Tumours of the Digestive System.World Health Organization Classification of Tumours. Lyon: IARC Press 2000; pp. 69-92.

[310] Curado MP, Edwards B, Shin HR, *et al.* In: Lyon, Ed: Cancer Incidence in Five Continents, Vol. IX, IARC Scientific Publications, Lyon France.

[311] Chow WH, Linet MS, McLaughlin JK, Hsing AW, Chien HT, Blot WJ. Risk factors for small intestine cancer. Cancer Causes Control 1993; 4(2): 163-9. [http://dx.doi.org/10.1007/BF00053158] [PMID: 8481495]

[312] Neugut AI, Jacobson JS, Suh S, Mukherjee R, Arber N. The epidemiology of cancer of the small

bowel. Cancer Epidemiol Biomarkers Prev 1998; 7(3): 243-51.
[PMID: 9521441]

[313] Canavan C, Abrams KR, Mayberry J. Meta-analysis: colorectal and small bowel cancer risk in patients with Crohn's disease. Aliment Pharmacol Ther 2006; 23(8): 1097-104.
[http://dx.doi.org/10.1111/j.1365-2036.2006.02854.x] [PMID: 16611269]

[314] Jigyasu D, Bedikian AY, Stroehlein JR. Chemotherapy for primary adenocarcinoma of the small bowel. Cancer 1984; 53(1): 23-5.
[http://dx.doi.org/10.1002/1097-0142(19840101)53:1<23::AID-CNCR2820530106>3.0.CO;2-U]
[PMID: 6690001]

[315] Ouriel K, Adams JT. Adenocarcinoma of the small intestine. Am J Surg 1984; 147(1): 66-71.
[http://dx.doi.org/10.1016/0002-9610(84)90036-9] [PMID: 6691554]

[316] Crawley C, Ross P, Norman A, Hill A, Cunningham D. The Royal Marsden experience of a small bowel adenocarcinoma treated with protracted venous infusion 5-fluorouracil. Br J Cancer 1998; 78(4): 508-10.
[http://dx.doi.org/10.1038/bjc.1998.523] [PMID: 9716035]

[317] Nakanoko T, Koga T, Taketani K, *et al.* Characteristics and Treatment Strategies for Small Bowel Adenocarcinoma in Advanced-stage Cases. Anticancer Res 2015; 35(7): 4135-8.
[PMID: 26124367]

[318] Polyzos A, Kouraklis G, Giannopoulos A, Bramis J, Delladetsima JK, Sfikakis PP. Irinotecan as salvage chemotherapy for advanced small bowel adenocarcinoma: a series of three patients. J Chemother 2003; 15(5): 503-6.
[PMID: 14598944]

[319] Bettini AC, Beretta GD, Sironi P, Mosconi S, Labianca R. Chemotherapy in small bowel adenocarcinoma associated with celiac disease: a report of three cases. Tumori 2003; 89(2): 193-5.
[PMID: 12841670]

[320] Koo DH, Yun SC, Hong YS, *et al.* Adjuvant chemotherapy for small bowel adenocarcinoma after curative surgery. Oncology 2011; 80(3-4): 208-13.
[http://dx.doi.org/10.1159/000328506] [PMID: 21720183]

[321] Higashi D, Ishibashi Y, Tamura T, *et al.* Clinical features of and chemotherapy for cancer of the small intestine. Anticancer Res 2010; 30(8): 3193-7.
[PMID: 20871040]

[322] Warth A, Kloor M, Schirmacher P, Bläker H. Genetics and epigenetics of small bowel adenocarcinoma: the interactions of CIN, MSI, and CIMP. Mod Pathol 2011; 24(4): 564-70.
[http://dx.doi.org/10.1038/modpathol.2010.223] [PMID: 21297586]

[323] Karimi P, Islami F, Anandasabapathy S, Freedman ND, Kamangar F. Gastric cancer: descriptive epidemiology, risk factors, screening, and prevention. Cancer Epidemiol Biomarkers Prev 2014; 23(5): 700-13.
[http://dx.doi.org/10.1158/1055-9965.EPI-13-1057] [PMID: 24618998]

[324] Bosetti C, Bertuccio P, Malvezzi M, *et al.* Cancer mortality in Europe, 2005-2009, and an overview of trends since 1980. Ann Oncol 2013; 24(10): 2657-71.
[http://dx.doi.org/10.1093/annonc/mdt301] [PMID: 23921790]

[325] La Vecchia C, Negri E, D'Avanzo B, Franceschi S. Electric refrigerator use and gastric cancer risk. Br J Cancer 1990; 62(1): 136-7.
[http://dx.doi.org/10.1038/bjc.1990.245] [PMID: 2390474]

[326] Ferlay J, Shin HR, Bray F, Forman D, Mathers C, Parkin DM. Estimates of worldwide burden of cancer in 2008: GLOBOCAN 2008. Int J Cancer 2010; 127(12): 2893-917.
[http://dx.doi.org/10.1002/ijc.25516] [PMID: 21351269]

[327] Soerjomataram I, Lortet-Tieulent J, Parkin DM, *et al.* Global burden of cancer in 2008: a systematic

analysis of disability-adjusted life-years in 12 world regions. Lancet 2012; 380(9856): 1840-50. [http://dx.doi.org/10.1016/S0140-6736(12)60919-2] [PMID: 23079588]

[328] Yang ZG, Gao L, Guo XB, Shi YL. Roles of long non-coding RNAs in gastric cancer metastasis. World J Gastroenterol 2015; 21(17): 5220-30. [http://dx.doi.org/10.3748/wjg.v21.i17.5220] [PMID: 25954095]

[329] Forman D, Burley VJ. Gastric cancer: global pattern of the disease and an overview of environmental risk factors. Best Pract Res Clin Gastroenterol 2006; 20(4): 633-49. [http://dx.doi.org/10.1016/j.bpg.2006.04.008] [PMID: 16997150]

[330] Ushijima T, Sasako M. Focus on gastric cancer. Cancer Cell 2004; 5(2): 121-5. [http://dx.doi.org/10.1016/S1535-6108(04)00033-9] [PMID: 14998488]

[331] Smith MG, Hold GL, Tahara E, El-Omar EM. Cellular and molecular aspects of gastric cancer. World J Gastroenterol 2006; 12(19): 2979-90. [http://dx.doi.org/10.3748/wjg.v12.i19.2979] [PMID: 16718776]

[332] Tahara E. Genetic pathways of two types of gastric cancer. IARC Sci Publ 2004; (157): 327-49. [PMID: 15055305]

[333] Mirchev M, Kotzev A, Kotzev I. Hereditary gastric cancer. Medical Review 2016; 52: 5-10. [Medicinski pregled].

[334] Fitzgerald RC, Caldas C. Clinical implications of E-cadherin associated hereditary diffuse gastric cancer. Gut 2004; 53(6): 775-8. [http://dx.doi.org/10.1136/gut.2003.022061] [PMID: 15138199]

[335] Brown LM, Devesa SS. Epidemiologic trends in esophageal and gastric cancer in the United States. Surg Oncol Clin N Am 2002; 11(2): 235-56. [http://dx.doi.org/10.1016/S1055-3207(02)00002-9] [PMID: 12424848]

[336] Cunningham D, Allum WH, Stenning SP, *et al.* MAGIC Trial Participants. Perioperative chemotherapy *versus* surgery alone for resectable gastroesophageal cancer. N Engl J Med 2006; 355(1): 11-20. [http://dx.doi.org/10.1056/NEJMoa055531] [PMID: 16822992]

[337] Cunningham SC, Kamangar F, Kim MP, *et al.* Survival after gastric adenocarcinoma resection: eighteen-year experience at a single institution. J Gastrointest Surg 2005; 9(5): 718-25. [http://dx.doi.org/10.1016/j.gassur.2004.12.002] [PMID: 15862270]

[338] Khan FA, Shukla AN. Pathogenesis and treatment of gastric carcinoma: "an up-date with brief review". J Cancer Res Ther 2006; 2(4): 196-9. [http://dx.doi.org/10.4103/0973-1482.29830] [PMID: 17998703]

[339] Kamangar F, Dores GM, Anderson WF. Patterns of cancer incidence, mortality, and prevalence across five continents: defining priorities to reduce cancer disparities in different geographic regions of the world. J Clin Oncol 2006; 24(14): 2137-50. [http://dx.doi.org/10.1200/JCO.2005.05.2308] [PMID: 16682732]

[340] Park SC, Chun HJ. Chemotherapy for advanced gastric cancer: review and update of current practices. Gut Liver 2013; 7(4): 385-93. [http://dx.doi.org/10.5009/gnl.2013.7.4.385] [PMID: 23898376]

[341] Wagner AD, Grothe W, Haerting J, Kleber G, Grothey A, Fleig WE. Chemotherapy in advanced gastric cancer: a systematic review and meta-analysis based on aggregate data. J Clin Oncol 2006; 24(18): 2903-9. [http://dx.doi.org/10.1200/JCO.2005.05.0245] [PMID: 16782930]

[342] Gravalos C, Jimeno A. HER2 in gastric cancer: a new prognostic factor and a novel therapeutic target. Ann Oncol 2008; 19(9): 1523-9. [http://dx.doi.org/10.1093/annonc/mdn169] [PMID: 18441328]

[343] Hofmann M, Stoss O, Shi D, *et al.* Assessment of a HER2 scoring system for gastric cancer: results from a validation study. Histopathology 2008; 52(7): 797-805. [http://dx.doi.org/10.1111/j.1365-2559.2008.03028.x] [PMID: 18422971]

[344] Tanner M, Hollmén M, Junttila TT, *et al.* Amplification of HER-2 in gastric carcinoma: association with Topoisomerase IIalpha gene amplification, intestinal type, poor prognosis and sensitivity to trastuzumab. Ann Oncol 2005; 16(2): 273-8. [http://dx.doi.org/10.1093/annonc/mdi064] [PMID: 15668283]

[345] Bang YJ, Van Cutsem E, Feyereislova A, *et al.* ToGA Trial Investigators. Trastuzumab in combination with chemotherapy *versus* chemotherapy alone for treatment of HER2-positive advanced gastric or gastro-oesophageal junction cancer (ToGA): a phase 3, open-label, randomised controlled trial. Lancet 2010; 376(9742): 687-97. [http://dx.doi.org/10.1016/S0140-6736(10)61121-X] [PMID: 20728210]

[346] Mammano E, Belluco C, Sciro M, *et al.* Epidermal growth factor receptor (EGFR): mutational and protein expression analysis in gastric cancer. Anticancer Res 2006; 26(5A): 3547-50. [PMID: 17094480]

[347] Yakes FM, Chinratanalab W, Ritter CA, King W, Seelig S, Arteaga CL. Herceptin-induced inhibition of phosphatidylinositol-3 kinase and Akt Is required for antibody-mediated effects on p27, cyclin D1, and antitumor action. Cancer Res 2002; 62(14): 4132-41. [PMID: 12124352]

[348] Yano T, Doi T, Ohtsu A, *et al.* Comparison of HER2 gene amplification assessed by fluorescence *in situ* hybridization and HER2 protein expression assessed by immunohistochemistry in gastric cancer. Oncol Rep 2006; 15(1): 65-71. [PMID: 16328035]

[349] Gunturu KS, Woo Y, Beaubier N, Remotti HE, Saif MW. Gastric cancer and trastuzumab: first biologic therapy in gastric cancer. Ther Adv Med Oncol 2013; 5(2): 143-51. [http://dx.doi.org/10.1177/1758834012469429] [PMID: 23450234]

[350] Dybdal N, Leiberman G, Anderson S, *et al.* Determination of HER2 gene amplification by fluorescence *in situ* hybridization and concordance with the clinical trials immunohistochemical assay in women with metastatic breast cancer evaluated for treatment with trastuzumab. Breast Cancer Res Treat 2005; 93(1): 3-11. [http://dx.doi.org/10.1007/s10549-004-6275-8] [PMID: 16184453]

[351] Slamon DJ, Godolphin W, Jones LA, *et al.* Studies of the HER-2/neu proto-oncogene in human breast and ovarian cancer. Science 1989; 244(4905): 707-12. [http://dx.doi.org/10.1126/science.2470152] [PMID: 2470152]

[352] Piccart-Gebhart MJ, Procter M, Leyland-Jones B, *et al.* Herceptin Adjuvant (HERA) Trial Study Team. Trastuzumab after adjuvant chemotherapy in HER2-positive breast cancer. N Engl J Med 2005; 353(16): 1659-72. [http://dx.doi.org/10.1056/NEJMoa052306] [PMID: 16236737]

[353] Smith I, Procter M, Gelber RD, *et al.* HERA study team. 2-year follow-up of trastuzumab after adjuvant chemotherapy in HER2-positive breast cancer: a randomised controlled trial. Lancet 2007; 369(9555): 29-36. [http://dx.doi.org/10.1016/S0140-6736(07)60028-2] [PMID: 17208639]

[354] Yamashita K, Sakuramoto S, Katada N, *et al.* Diffuse type advanced gastric cancer showing dismal prognosis is characterized by deeper invasion and emerging peritoneal cancer cell: the latest comparative study to intestinal advanced gastric cancer. Hepatogastroenterology 2009; 56(89): 276-81. [PMID: 19453074]

[355] Miyahara R, Niwa Y, Matsuura T, *et al.* Prevalence and prognosis of gastric cancer detected by screening in a large Japanese population: data from a single institute over 30 years. J Gastroenterol Hepatol 2007; 22(9): 1435-42.

[http://dx.doi.org/10.1111/j.1440-1746.2007.04991.x] [PMID: 17573829]

[356] Bang Y, Chung H, Xu J, *et al.* Pathological features of advanced gastric cancer (GC): Relationship to human epidermal growth factor receptor 2 (HER2) positivity in the global screening programme of the ToGA trial. J Clin Oncol 2009; •••: 27.

[357] Zheng H, Takahashi H, Murai Y, *et al.* Pathobiological characteristics of intestinal and diffuse-type gastric carcinoma in Japan: an immunostaining study on the tissue microarray. J Clin Pathol 2007; 60(3): 273-7.
[http://dx.doi.org/10.1136/jcp.2006.038778] [PMID: 16714395]

[358] Browne BC, O'Brien N, Duffy MJ, Crown J, O'Donovan N. HER-2 signaling and inhibition in breast cancer. Curr Cancer Drug Targets 2009; 9(3): 419-38.
[http://dx.doi.org/10.2174/156800909788166484] [PMID: 19442060]

[359] Song Y, Huang J, Wang JW. [Relationship between HER2/neu gene amplification and protein expression and prognosis in patients with advanced gastric carcinoma]. Chin J Cancer 2010; 29(1): 76-81. [Relationship between HER2/neu gene amplification and protein expression and prognosis in patients with advanced gastric carcinoma].
[http://dx.doi.org/10.5732/cjc.009.10377] [PMID: 20038314]

[360] Hamakawa T, Kukita Y, Kurokawa Y, *et al.* Monitoring gastric cancer progression with circulating tumour DNA. Br J Cancer 2015; 112(2): 352-6.
[http://dx.doi.org/10.1038/bjc.2014.609] [PMID: 25490524]

[361] Kim K, Shin DG, Park MK, *et al.* Circulating cell-free DNA as a promising biomarker in patients with gastric cancer: diagnostic validity and significant reduction of cfDNA after surgical resection. Ann Surg Treat Res 2014; 86(3): 136-42.
[http://dx.doi.org/10.4174/astr.2014.86.3.136] [PMID: 24761422]

[362] Yokozaki H, Yasui W, Tahara E. Genetic and epigenetic changes in stomach cancer. Int Rev Cytol 2001; 204: 49-95.
[http://dx.doi.org/10.1016/S0074-7696(01)04003-7] [PMID: 11243597]

[363] Fu DG. Epigenetic alterations in gastric cancer (Review). Mol Med Rep 2015; 12(3): 3223-30. [Review].
[http://dx.doi.org/10.3892/mmr.2015.3816] [PMID: 25997695]

[364] Lee JH, Abraham SC, Kim HS, *et al.* Inverse relationship between APC gene mutation in gastric adenomas and development of adenocarcinoma. Am J Pathol 2002; 161(2): 611-8.
[http://dx.doi.org/10.1016/S0002-9440(10)64216-2] [PMID: 12163385]

[365] Park WS, Oh RR, Park JY, *et al.* Frequent somatic mutations of the beta-catenin gene in intestinal-type gastric cancer. Cancer Res 1999; 59(17): 4257-60.
[PMID: 10485468]

[366] Maesawa C, Tamura G, Suzuki Y, *et al.* The sequential accumulation of genetic alterations characteristic of the colorectal adenoma-carcinoma sequence does not occur between gastric adenoma and adenocarcinoma. J Pathol 1995; 176(3): 249-58.
[http://dx.doi.org/10.1002/path.1711760307] [PMID: 7674088]

[367] Rüschoff J, Hanna W, Bilous M, *et al.* HER2 testing in gastric cancer: a practical approach. Mod Pathol 2012; 25(5): 637-50.
[http://dx.doi.org/10.1038/modpathol.2011.198] [PMID: 22222640]

[368] Abrahao-Machado LF, Scapulatempo-Neto C. HER2 testing in gastric cancer: An update. World J Gastroenterol 2016; 22(19): 4619-25.
[http://dx.doi.org/10.3748/wjg.v22.i19.4619] [PMID: 27217694]

[369] Boku N. HER2-positive gastric cancer. Gastric Cancer 2014; 17(1): 1-12.
[http://dx.doi.org/10.1007/s10120-013-0252-z] [PMID: 23563986]

[370] Kawakami H, Okamoto I, Arao T, *et al.* MET amplification as a potential therapeutic target in gastric

cancer. Oncotarget 2013; 4(1): 9-17.
[http://dx.doi.org/10.18632/oncotarget.718] [PMID: 23327903]

[371] Lee HE, Kim MA, Lee HS, *et al.* MET in gastric carcinomas: comparison between protein expression and gene copy number and impact on clinical outcome. Br J Cancer 2012; 107(2): 325-33.
[http://dx.doi.org/10.1038/bjc.2012.237] [PMID: 22644302]

[372] Hirst M, Marra MA. Epigenetics and human disease. Int J Biochem Cell Biol 2009; 41(1): 136-46.
[http://dx.doi.org/10.1016/j.biocel.2008.09.011] [PMID: 18852064]

[373] Panani AD. Cytogenetic and molecular aspects of gastric cancer: clinical implications. Cancer Lett 2008; 266(2): 99-115.
[http://dx.doi.org/10.1016/j.canlet.2008.02.053] [PMID: 18381231]

[374] An C, Choi IS, Yao JC, *et al.* Prognostic significance of CpG island methylator phenotype and microsatellite instability in gastric carcinoma. Clin Cancer Res 2005; 11(2 Pt 1): 656-63.
[PMID: 15701853]

[375] Fleisher AS, Esteller M, Wang S, *et al.* Hypermethylation of the hMLH1 gene promoter in human gastric cancers with microsatellite instability. Cancer Res 1999; 59(5): 1090-5.
[PMID: 10070967]

[376] Leung SY, Yuen ST, Chung LP, Chu KM, Chan AS, Ho JC. hMLH1 promoter methylation and lack of hMLH1 expression in sporadic gastric carcinomas with high-frequency microsatellite instability. Cancer Res 1999; 59(1): 159-64.
[PMID: 9892201]

[377] Jung HY, Jung KC, Shim YH, Ro JY, Kang GH. Methylation of the hMLH1 promoter in multiple gastric carcinomas with microsatellite instability. Pathol Int 2001; 51(6): 445-51.
[http://dx.doi.org/10.1046/j.1440-1827.2001.01222.x] [PMID: 11422806]

[378] Nishigaki M, Aoyagi K, Danjoh I, *et al.* Discovery of aberrant expression of R-RAS by cancer-linked DNA hypomethylation in gastric cancer using microarrays. Cancer Res 2005; 65(6): 2115-24.
[http://dx.doi.org/10.1158/0008-5472.CAN-04-3340] [PMID: 15781621]

[379] Sadikovic B, Al-Romaih K, Squire JA, Zielenska M. Cause and consequences of genetic and epigenetic alterations in human cancer. Curr Genomics 2008; 9(6): 394-408.
[http://dx.doi.org/10.2174/138920208785699580] [PMID: 19506729]

[380] Alsina M, Landolfi S, Aura C, *et al.* Cyclin E amplification/overexpression is associated with poor prognosis in gastric cancer. Ann Oncol 2015; 26(2): 438-9.
[http://dx.doi.org/10.1093/annonc/mdu535] [PMID: 25403579]

[381] Niu YL, Li YJ, Wang JB, *et al.* CacyBP/SIP nuclear translocation regulates p27Kip1 stability in gastric cancer cells. World J Gastroenterol 2016; 22(15): 3992-4001.
[http://dx.doi.org/10.3748/wjg.v22.i15.3992] [PMID: 27099442]

[382] Becker KF, Atkinson MJ, Reich U, *et al.* E-cadherin gene mutations provide clues to diffuse type gastric carcinomas. Cancer Res 1994; 54(14): 3845-52.
[PMID: 8033105]

[383] Chan AO. E-cadherin in gastric cancer. World J Gastroenterol 2006; 12(2): 199-203.
[http://dx.doi.org/10.3748/wjg.v12.i2.199] [PMID: 16482618]

[384] Chan AO, Lam SK, Wong BC, *et al.* Promoter methylation of E-cadherin gene in gastric mucosa associated with Helicobacter pylori infection and in gastric cancer. Gut 2003; 52(4): 502-6.
[http://dx.doi.org/10.1136/gut.52.4.502] [PMID: 12631658]

[385] Terrés AM, Pajares JM, O'Toole D, Ahern S, Kelleher D. H pylori infection is associated with downregulation of E-cadherin, a molecule involved in epithelial cell adhesion and proliferation control. J Clin Pathol 1998; 51(5): 410-2.
[http://dx.doi.org/10.1136/jcp.51.5.410] [PMID: 9708215]

[386] Li YL, Tian Z, Wu DY, Fu BY, Xin Y. Loss of heterozygosity on 10q23.3 and mutation of tumor suppressor gene PTEN in gastric cancer and precancerous lesions. World J Gastroenterol 2005; 11(2): 285-8.
[http://dx.doi.org/10.3748/wjg.v11.i2.285] [PMID: 15633233]

[387] Sato K, Tamura G, Tsuchiya T, *et al.* Analysis of genetic and epigenetic alterations of the PTEN gene in gastric cancer. Virchows Arch 2002; 440(2): 160-5.
[http://dx.doi.org/10.1007/s004280100499] [PMID: 11964046]

[388] Song H, Sun W, Ye G, *et al.* Long non-coding RNA expression profile in human gastric cancer and its clinical significances. J Transl Med 2013; 11: 225.
[http://dx.doi.org/10.1186/1479-5876-11-225] [PMID: 24063685]

[389] Wang F, Sun GP, Zou YF, Hao JQ, Zhong F, Ren WJ. MicroRNAs as promising biomarkers for gastric cancer. Cancer Biomark 2012; 11(6): 259-67.
[http://dx.doi.org/10.3233/CBM-2012-00284] [PMID: 23248184]

[390] Liu HS, Xiao HS. MicroRNAs as potential biomarkers for gastric cancer. World J Gastroenterol 2014; 20(34): 12007-17.
[http://dx.doi.org/10.3748/wjg.v20.i34.12007] [PMID: 25232237]

[391] Pan HW, Li SC, Tsai KW. MicroRNA dysregulation in gastric cancer. Curr Pharm Des 2013; 19(7): 1273-84.
[PMID: 23092346]

[392] Ishiguro H, Kimura M, Takeyama H. Role of microRNAs in gastric cancer. World J Gastroenterol 2014; 20(19): 5694-9.
[http://dx.doi.org/10.3748/wjg.v20.i19.5694] [PMID: 24914330]

[393] Fellig Y, Ariel I, Ohana P, *et al.* H19 expression in hepatic metastases from a range of human carcinomas. J Clin Pathol 2005; 58(10): 1064-8.
[http://dx.doi.org/10.1136/jcp.2004.023648] [PMID: 16189152]

[394] Zhang L, Zhou XF, Pan GF, Zhao JP. Enhanced expression of long non-coding RNA ZXF1 promoted the invasion and metastasis in lung adenocarcinoma. Biomed Pharmacother 2014; 68(4): 401-7.
[http://dx.doi.org/10.1016/j.biopha.2014.03.001] [PMID: 24721325]

[395] Gutschner T, Hämmerle M, Eissmann M, *et al.* The noncoding RNA MALAT1 is a critical regulator of the metastasis phenotype of lung cancer cells. Cancer Res 2013; 73(3): 1180-9.
[http://dx.doi.org/10.1158/0008-5472.CAN-12-2850] [PMID: 23243023]

[396] Yap KL, Li S, Muñoz-Cabello AM, *et al.* Molecular interplay of the noncoding RNA ANRIL and methylated histone H3 lysine 27 by polycomb CBX7 in transcriptional silencing of INK4a. Mol Cell 2010; 38(5): 662-74.
[http://dx.doi.org/10.1016/j.molcel.2010.03.021] [PMID: 20541999]

[397] Mizrahi I, Mazeh H, Grinbaum R, *et al.* Colon Cancer Associated Transcript-1 (CCAT1) Expression in Adenocarcinoma of the Stomach. J Cancer 2015; 6(2): 105-10.
[http://dx.doi.org/10.7150/jca.10568] [PMID: 25561974]

[398] Alaiyan B, Ilyayev N, Stojadinovic A, *et al.* Differential expression of colon cancer associated transcript1 (CCAT1) along the colonic adenoma-carcinoma sequence. BMC Cancer 2013; 13: 196.
[http://dx.doi.org/10.1186/1471-2407-13-196] [PMID: 23594791]

[399] He X, Tan X, Wang X, *et al.* C-Myc-activated long noncoding RNA CCAT1 promotes colon cancer cell proliferation and invasion. Tumour Biol 2014; 35(12): 12181-8.
[http://dx.doi.org/10.1007/s13277-014-2526-4] [PMID: 25185650]

[400] Zhang XF, Liu T, Li Y, Li S. Overexpression of long non-coding RNA CCAT1 is a novel biomarker of poor prognosis in patients with breast cancer. Int J Clin Exp Pathol 2015; 8(8): 9440-5.
[PMID: 26464701]

[401] Deng L, Yang SB, Xu FF, Zhang JH. Long noncoding RNA CCAT1 promotes hepatocellular carcinoma progression by functioning as let-7 sponge. J Exp Clin Canc Res 2015; p. 34.

[402] Chang H, Kim N, Park JH, *et al.* Different microRNA expression levels in gastric cancer depending on Helicobacter pylori infection. Gut Liver 2015; 9(2): 188-96.
[http://dx.doi.org/10.5009/gnl13371] [PMID: 25167801]

[403] Komatsu S, Ichikawa D, Tsujiura M, *et al.* Prognostic impact of circulating miR-21 in the plasma of patients with gastric carcinoma. Anticancer Res 2013; 33(1): 271-6.
[PMID: 23267156]

[404] Zhou H, Guo JM, Lou YR, *et al.* Detection of circulating tumor cells in peripheral blood from patients with gastric cancer using microRNA as a marker. J Mol Med (Berl) 2010; 88(7): 709-17.
[http://dx.doi.org/10.1007/s00109-010-0617-2] [PMID: 20349219]

[405] Tsujiura M, Ichikawa D, Komatsu S, *et al.* Circulating microRNAs in plasma of patients with gastric cancers. Br J Cancer 2010; 102(7): 1174-9.
[http://dx.doi.org/10.1038/sj.bjc.6605608] [PMID: 20234369]

[406] Wang M, Gu H, Wang S, *et al.* Circulating miR-17-5p and miR-20a: molecular markers for gastric cancer. Mol Med Rep 2012; 5(6): 1514-20.
[PMID: 22406928]

[407] Zheng Y, Cui L, Sun W, *et al.* MicroRNA-21 is a new marker of circulating tumor cells in gastric cancer patients. Cancer Biomark 2011-2012; 10(2): 71-7.
[http://dx.doi.org/10.3233/CBM-2011-0231] [PMID: 22430134]

[408] Cai H, Yuan Y, Hao YF, Guo TK, Wei X, Zhang YM. Plasma microRNAs serve as novel potential biomarkers for early detection of gastric cancer. Med Oncol 2013; 30(1): 452.
[http://dx.doi.org/10.1007/s12032-012-0452-0] [PMID: 23307259]

[409] Wang YY, Ye ZY, Zhao ZS, *et al.* Clinicopathologic significance of miR-10b expression in gastric carcinoma. Hum Pathol 2013; 44(7): 1278-85.
[http://dx.doi.org/10.1016/j.humpath.2012.10.014] [PMID: 23351547]

[410] Kogo R, Mimori K, Tanaka F, Komune S, Mori M. Clinical significance of miR-146a in gastric cancer cases. Clin Cancer Res 2011; 17(13): 4277-84.
[http://dx.doi.org/10.1158/1078-0432.CCR-10-2866] [PMID: 21632853]

[411] Liu R, Zhang C, Hu Z, *et al.* A five-microRNA signature identified from genome-wide serum microRNA expression profiling serves as a fingerprint for gastric cancer diagnosis. Eur J Cancer 2011; 47(5): 784-91.
[http://dx.doi.org/10.1016/j.ejca.2010.10.025] [PMID: 21112772]

[412] Pennathur A, Gibson MK, Jobe BA, Luketich JD. Oesophageal carcinoma. Lancet 2013; 381(9864): 400-12.
[http://dx.doi.org/10.1016/S0140-6736(12)60643-6] [PMID: 23374478]

[413] Zhang Y. Epidemiology of esophageal cancer. World J Gastroenterol 2013; 19(34): 5598-606.
[http://dx.doi.org/10.3748/wjg.v19.i34.5598] [PMID: 24039351]

[414] Mao WM, Zheng WH, Ling ZQ. Epidemiologic risk factors for esophageal cancer development. Asian Pac J Cancer Prev 2011; 12(10): 2461-6.
[PMID: 22320939]

[415] Enzinger PC, Mayer RJ. Esophageal cancer. N Engl J Med 2003; 349(23): 2241-52.
[http://dx.doi.org/10.1056/NEJMra035010] [PMID: 14657432]

[416] Le Bras GF, Farooq MH, Falk GW, Andl CD. Esophageal cancer: The latest on chemoprevention and state of the art therapies. Pharmacol Res 2016; 113(Pt A): 236-44.
[http://dx.doi.org/10.1016/j.phrs.2016.08.021] [PMID: 27565381]

[417] Blot WJ. Invited commentary: more evidence of increased risks of cancer among alcohol drinkers. Am

J Epidemiol 1999; 150(11): 1138-40.
[http://dx.doi.org/10.1093/oxfordjournals.aje.a009939] [PMID: 10588074]

[418] D'Onofrio V, Bovero E, Iaquinto G. Characterization of acid and alkaline reflux in patients with Barrett's esophagus. G.O.S.P.E. Operative Group for the study of Esophageal Precancer. Dis Esophagus 1997; 10: 16-22. discussion -3

[419] Guanrei Y, Songliang Q. Endoscopic surveys in high-risk and low-risk populations for esophageal cancer in China with special reference to precursors of esophageal cancer. Endoscopy 1987; 19(3): 91-5.
[http://dx.doi.org/10.1055/s-2007-1018250] [PMID: 3608927]

[420] Lin J, Zeng R, Cao W, Luo R, Chen J, Lin Y. Hot beverage and food intake and esophageal cancer in southern China. Asian Pac J Cancer Prev 2011; 12(9): 2189-92.
[PMID: 22296354]

[421] Lagergren J, Lagergren P. Recent developments in esophageal adenocarcinoma. CA Cancer J Clin 2013; 63(4): 232-48.
[http://dx.doi.org/10.3322/caac.21185] [PMID: 23818335]

[422] Nieman KM, Romero IL, Van Houten B, Lengyel E. Adipose tissue and adipocytes support tumorigenesis and metastasis. Bba-Mol Cell Biol L 2013; 1831: 1533-41.

[423] Duggan C, Onstad L, Hardikar S, Blount PL, Reid BJ, Vaughan TL. Association between markers of obesity and progression from Barrett's esophagus to esophageal adenocarcinoma. Clin Gastroenterol Hepatol 2013; 11(8): 934-43.
[http://dx.doi.org/10.1016/j.cgh.2013.02.017] [PMID: 23466711]

[424] Carr JS, Zafar SF, Saba N, Khuri FR, El-Rayes BF. Risk factors for rising incidence of esophageal and gastric cardia adenocarcinoma. J Gastrointest Cancer 2013; 44(2): 143-51.
[http://dx.doi.org/10.1007/s12029-013-9480-z] [PMID: 23435833]

[425] Löfdahl HE, Lu Y, Lagergren P, Lagergren J. Risk factors for esophageal adenocarcinoma after antireflux surgery. Ann Surg 2013; 257(4): 579-82.
[http://dx.doi.org/10.1097/SLA.0b013e3182888384] [PMID: 23426349]

[426] Lin Y, Totsuka Y, He Y, *et al.* Epidemiology of esophageal cancer in Japan and China. J Epidemiol 2013; 23(4): 233-42.
[http://dx.doi.org/10.2188/jea.JE20120162] [PMID: 23629646]

[427] Napier KJ, Scheerer M, Misra S. Esophageal cancer: A Review of epidemiology, pathogenesis, staging workup and treatment modalities. World J Gastrointest Oncol 2014; 6(5): 112-20.
[http://dx.doi.org/10.4251/wjgo.v6.i5.112] [PMID: 24834141]

[428] Bartley AN, Washington MK, Ventura CB, *et al.* HER2 Testing and Clinical Decision Making in Gastroesophageal Adenocarcinoma: Guideline From the College of American Pathologists, American Society for Clinical Pathology, and American Society of Clinical Oncology. Arch Pathol Lab Med 2016; 140(12): 1345-63.
[http://dx.doi.org/10.5858/arpa.2016-0331-CP] [PMID: 27841667]

[429] Lordick F, Mariette C, Haustermans K, Obermannová R, Arnold D, Committee EG. ESMO Guidelines Committee. Oesophageal cancer: ESMO Clinical Practice Guidelines for diagnosis, treatment and follow-up. Ann Oncol 2016; 27 (Suppl. 5): v50-7.
[http://dx.doi.org/10.1093/annonc/mdw329] [PMID: 27664261]

[430] Cooper JS, Guo MD, Herskovic A, *et al.* Radiation Therapy Oncology Group. Chemoradiotherapy of locally advanced esophageal cancer: long-term follow-up of a prospective randomized trial (RTOG 85-01). JAMA 1999; 281(17): 1623-7.
[http://dx.doi.org/10.1001/jama.281.17.1623] [PMID: 10235156]

[431] Tepper J, Krasna MJ, Niedzwiecki D, *et al.* Phase III trial of trimodality therapy with cisplatin, fluorouracil, radiotherapy, and surgery compared with surgery alone for esophageal cancer: CALGB

9781. J Clin Oncol 2008; 26(7): 1086-92.
[http://dx.doi.org/10.1200/JCO.2007.12.9593] [PMID: 18309943]

[432] Lin EW, Karakasheva TA, Hicks PD, Bass AJ, Rustgi AK. The tumor microenvironment in esophageal cancer. Oncogene 2016; 35(41): 5337-49.
[http://dx.doi.org/10.1038/onc.2016.34] [PMID: 26923327]

[433] Luo H, Li H, Hu Z, *et al.* Noninvasive diagnosis and monitoring of mutations by deep sequencing of circulating tumor DNA in esophageal squamous cell carcinoma. Biochem Biophys Res Commun 2016; 471(4): 596-602.
[http://dx.doi.org/10.1016/j.bbrc.2016.02.011] [PMID: 26876573]

[434] Weaver JM, Ross-Innes CS, Shannon N, *et al.* OCCAMS consortium. Ordering of mutations in preinvasive disease stages of esophageal carcinogenesis. Nat Genet 2014; 46(8): 837-43.
[http://dx.doi.org/10.1038/ng.3013] [PMID: 24952744]

[435] Yue CM, Deng DJ, Bi MX, Guo LP, Lu SH. Expression of ECRG4, a novel esophageal cancer-related gene, downregulated by CpG island hypermethylation in human esophageal squamous cell carcinoma. World J Gastroenterol 2003; 9(6): 1174-8.
[http://dx.doi.org/10.3748/wjg.v9.i6.1174] [PMID: 12800218]

[436] Wu DL, Sui FY, Jiang XM, Jiang XH. Methylation in esophageal carcinogenesis. World J Gastroenterol 2006; 12(43): 6933-40.
[http://dx.doi.org/10.3748/wjg.v12.i43.6933] [PMID: 17109513]

[437] Qiu H, Zhang W, El-Naggar AK, *et al.* Loss of retinoic acid receptor-beta expression is an early event during esophageal carcinogenesis. Am J Pathol 1999; 155(5): 1519-23.
[http://dx.doi.org/10.1016/S0002-9440(10)65467-3] [PMID: 10550308]

[438] Kuroki T, Trapasso F, Yendamuri S, *et al.* Promoter hypermethylation of RASSF1A in esophageal squamous cell carcinoma. Clin Cancer Res 2003; 9(4): 1441-5.
[PMID: 12684417]

[439] Wong ML, Tao Q, Fu L, *et al.* Aberrant promoter hypermethylation and silencing of the critical 3p21 tumour suppressor gene, RASSF1A, in Chinese oesophageal squamous cell carcinoma. Int J Oncol 2006; 28(3): 767-73.
[PMID: 16465383]

[440] Gu J, Wang Y, Wu X. MicroRNA in the pathogenesis and prognosis of esophageal cancer. Curr Pharm Des 2013; 19(7): 1292-300.
[PMID: 23092349]

[441] Mathé EA, Nguyen GH, Bowman ED, *et al.* MicroRNA expression in squamous cell carcinoma and adenocarcinoma of the esophagus: associations with survival. Clin Cancer Res 2009; 15(19): 6192-200.
[http://dx.doi.org/10.1158/1078-0432.CCR-09-1467] [PMID: 19789312]

[442] Tang WW, Wu Q, Li SQ, *et al.* Implication of lncRNAs in pathogenesis of esophageal cancer. Onco Targets Ther 2015; 8: 3219-26.
[http://dx.doi.org/10.2147/OTT.S87856] [PMID: 26609239]

[443] Shen WJ, Zhang F, Zhao X, Xu J. LncRNAs and Esophageal Squamous Cell Carcinoma - Implications for Pathogenesis and Drug Development. J Cancer 2016; 7(10): 1258-64.
[http://dx.doi.org/10.7150/jca.14869] [PMID: 27390601]

[444] Li J, Chen Z, Tian L, *et al.* LncRNA profile study reveals a three-lncRNA signature associated with the survival of patients with oesophageal squamous cell carcinoma. Gut 2014; 63(11): 1700-10.
[http://dx.doi.org/10.1136/gutjnl-2013-305806] [PMID: 24522499]

[445] Sun Y, Shi N, Lu H, *et al.* ABCC4 copy number variation is associated with susceptibility to esophageal squamous cell carcinoma. Carcinogenesis 2014; 35(9): 1941-50.
[http://dx.doi.org/10.1093/carcin/bgu043] [PMID: 24510239]

[446] Cui R, Kamatani Y, Takahashi A, *et al.* Functional variants in ADH1B and ALDH2 coupled with alcohol and smoking synergistically enhance esophageal cancer risk. Gastroenterology 2009; 137(5): 1768-75.
[http://dx.doi.org/10.1053/j.gastro.2009.07.070] [PMID: 19698717]

[447] Abnet CC, Freedman ND, Hu N, *et al.* A shared susceptibility locus in PLCE1 at 10q23 for gastric adenocarcinoma and esophageal squamous cell carcinoma. Nat Genet 2010; 42(9): 764-7.
[http://dx.doi.org/10.1038/ng.649] [PMID: 20729852]

[448] Wang LD, Zhou FY, Li XM, *et al.* Genome-wide association study of esophageal squamous cell carcinoma in Chinese subjects identifies susceptibility loci at PLCE1 and C20orf54. Nat Genet 2010; 42(9): 759-63.
[http://dx.doi.org/10.1038/ng.648] [PMID: 20729853]

[449] Wu C, Hu Z, He Z, *et al.* Genome-wide association study identifies three new susceptibility loci for esophageal squamous-cell carcinoma in Chinese populations. Nat Genet 2011; 43(7): 679-84.
[http://dx.doi.org/10.1038/ng.849] [PMID: 21642993]

[450] Liu CY, Wu MC, Chen F, *et al.* A Large-scale genetic association study of esophageal adenocarcinoma risk. Carcinogenesis 2010; 31(7): 1259-63.
[http://dx.doi.org/10.1093/carcin/bgq092] [PMID: 20453000]

[451] Yadav D, Lowenfels AB. The epidemiology of pancreatitis and pancreatic cancer. Gastroenterology 2013; 144(6): 1252-61.
[http://dx.doi.org/10.1053/j.gastro.2013.01.068] [PMID: 23622135]

[452] Raimondi S, Maisonneuve P, Lowenfels AB. Epidemiology of pancreatic cancer: an overview. Nat Rev Gastroenterol Hepatol 2009; 6(12): 699-708.
[http://dx.doi.org/10.1038/nrgastro.2009.177] [PMID: 19806144]

[453] Amundadottir L, Kraft P, Stolzenberg-Solomon RZ, *et al.* Genome-wide association study identifies variants in the ABO locus associated with susceptibility to pancreatic cancer. Nat Genet 2009; 41(9): 986-90.
[http://dx.doi.org/10.1038/ng.429] [PMID: 19648918]

[454] Wolpin BM, Chan AT, Hartge P, *et al.* ABO blood group and the risk of pancreatic cancer. J Natl Cancer Inst 2009; 101(6): 424-31.
[http://dx.doi.org/10.1093/jnci/djp020] [PMID: 19276450]

[455] Wolpin BM, Kraft P, Gross M, *et al.* Pancreatic cancer risk and ABO blood group alleles: results from the pancreatic cancer cohort consortium. Cancer Res 2010; 70(3): 1015-23.
[http://dx.doi.org/10.1158/0008-5472.CAN-09-2993] [PMID: 20103627]

[456] Yeo TP. Demographics, epidemiology, and inheritance of pancreatic ductal adenocarcinoma. Semin Oncol 2015; 42(1): 8-18.
[http://dx.doi.org/10.1053/j.seminoncol.2014.12.002] [PMID: 25726048]

[457] Ducreux M, Cuhna AS, Caramella C, *et al.* ESMO Guidelines Committee. Cancer of the pancreas: ESMO Clinical Practice Guidelines for diagnosis, treatment and follow-up. Ann Oncol 2015; 26 (Suppl. 5): v56-68.
[http://dx.doi.org/10.1093/annonc/mdv295] [PMID: 26314780]

[458] Frebourg T, Bercoff E, Manchon N, *et al.* The evaluation of CA 19-9 antigen level in the early detection of pancreatic cancer. A prospective study of 866 patients. Cancer 1988; 62(11): 2287-90.
[http://dx.doi.org/10.1002/1097-0142(19881201)62:11<2287::AID-CNCR2820621103>3.0.CO;2-H] [PMID: 3179943]

[459] Teague A, Lim KH, Wang-Gillam A. Advanced pancreatic adenocarcinoma: a review of current treatment strategies and developing therapies. Ther Adv Med Oncol 2015; 7(2): 68-84.
[http://dx.doi.org/10.1177/1758834014564775] [PMID: 25755680]

[460] Li X, Ma Q, Xu Q, Duan W, Lei J, Wu E. Targeting the cancer-stroma interaction: a potential

approach for pancreatic cancer treatment. Curr Pharm Des 2012; 18(17): 2404-15. [http://dx.doi.org/10.2174/138161281120924041] [PMID: 22372501]

[461] Garrido-Laguna I, Hidalgo M. Pancreatic cancer: from state-of-the-art treatments to promising novel therapies. Nat Rev Clin Oncol 2015; 12(6): 319-34. [http://dx.doi.org/10.1038/nrclinonc.2015.53] [PMID: 25824606]

[462] Malvezzi M, Bertuccio P, Levi F, La Vecchia C, Negri E. European cancer mortality predictions for the year 2014. Ann Oncol 2014; 25(8): 1650-6. [http://dx.doi.org/10.1093/annonc/mdu138] [PMID: 24759568]

[463] Moertel CG. Chemotherapy of gastrointestinal cancer. N Engl J Med 1978; 299(19): 1049-52. [http://dx.doi.org/10.1056/NEJM197811092991906] [PMID: 360064]

[464] Burris HA III, Moore MJ, Andersen J, *et al.* Improvements in survival and clinical benefit with gemcitabine as first-line therapy for patients with advanced pancreas cancer: a randomized trial. J Clin Oncol 1997; 15(6): 2403-13. [http://dx.doi.org/10.1200/JCO.1997.15.6.2403] [PMID: 9196156]

[465] Conroy T, Gourgou-Bourgade S, Ducreux M. FOLFIRINOX *versus* Gemcitabine for Metastatic Pancreatic Cancer REPLY. N Engl J Med 2011; 365: 769. [PMID: 21864187]

[466] Von Hoff DD, Ramanathan RK, Borad MJ, *et al.* Gemcitabine plus nab-paclitaxel is an active regimen in patients with advanced pancreatic cancer: a phase I/II trial. J Clin Oncol 2011; 29(34): 4548-54. [http://dx.doi.org/10.1200/JCO.2011.36.5742] [PMID: 21969517]

[467] Moore MJ, Goldstein D, Hamm J, *et al.* National Cancer Institute of Canada Clinical Trials Group. Erlotinib plus gemcitabine compared with gemcitabine alone in patients with advanced pancreatic cancer: a phase III trial of the National Cancer Institute of Canada Clinical Trials Group. J Clin Oncol 2007; 25(15): 1960-6. [http://dx.doi.org/10.1200/JCO.2006.07.9525] [PMID: 17452677]

[468] Oettle H, Riess H, Stieler JM, *et al.* Second-line oxaliplatin, folinic acid, and fluorouracil *versus* folinic acid and fluorouracil alone for gemcitabine-refractory pancreatic cancer: outcomes from the CONKO-003 trial. J Clin Oncol 2014; 32(23): 2423-9. [http://dx.doi.org/10.1200/JCO.2013.53.6995] [PMID: 24982456]

[469] Gourgou-Bourgade S, Bascoul-Mollevi C, Desseigne F, *et al.* Impact of FOLFIRINOX compared with gemcitabine on quality of life in patients with metastatic pancreatic cancer: results from the PRODIGE 4/ACCORD 11 randomized trial. J Clin Oncol 2013; 31(1): 23-9. [http://dx.doi.org/10.1200/JCO.2012.44.4869] [PMID: 23213101]

[470] Fjällskog ML, Lejonklou MH, Oberg KE, Eriksson BK, Janson ET. Expression of molecular targets for tyrosine kinase receptor antagonists in malignant endocrine pancreatic tumors. Clin Cancer Res 2003; 9(4): 1469-73. [PMID: 12684421]

[471] Tobita K, Kijima H, Dowaki S, *et al.* Epidermal growth factor receptor expression in human pancreatic cancer: Significance for liver metastasis. Int J Mol Med 2003; 11(3): 305-9. [PMID: 12579331]

[472] Biankin AV, Waddell N, Kassahn KS, *et al.* Australian Pancreatic Cancer Genome Initiative. Pancreatic cancer genomes reveal aberrations in axon guidance pathway genes. Nature 2012; 491(7424): 399-405. [http://dx.doi.org/10.1038/nature11547] [PMID: 23103869]

[473] Yamamoto H, Itoh F, Nakamura H, *et al.* Genetic and clinical features of human pancreatic ductal adenocarcinomas with widespread microsatellite instability. Cancer Res 2001; 61(7): 3139-44. [PMID: 11306499]

[474] Nakata B, Wang YQ, Yashiro M, *et al.* Prognostic value of microsatellite instability in resectable

pancreatic cancer. Clin Cancer Res 2002; 8(8): 2536-40.
[PMID: 12171881]

[475] Ueki T, Toyota M, Sohn T, *et al.* Hypermethylation of multiple genes in pancreatic adenocarcinoma. Cancer Res 2000; 60(7): 1835-9.
[PMID: 10766168]

[476] Neureiter D, Jäger T, Ocker M, Kiesslich T. Epigenetics and pancreatic cancer: pathophysiology and novel treatment aspects. World J Gastroenterol 2014; 20(24): 7830-48.
[http://dx.doi.org/10.3748/wjg.v20.i24.7830] [PMID: 24976721]

[477] Song Y, Washington MK, Crawford HC. Loss of FOXA1/2 is essential for the epithelial-t--mesenchymal transition in pancreatic cancer. Cancer Res 2010; 70(5): 2115-25.
[http://dx.doi.org/10.1158/0008-5472.CAN-09-2979] [PMID: 20160041]

[478] Zhu Y, Zhang JJ, Zhu R, *et al.* The increase in the expression and hypomethylation of MUC4 gene with the progression of pancreatic ductal adenocarcinoma. Med Oncol 2011; 28 (Suppl. 1): S175-84.
[http://dx.doi.org/10.1007/s12032-010-9683-0] [PMID: 20922503]

[479] Lewis AR, Valle JW, McNamara MG. Pancreatic cancer: Are "liquid biopsies" ready for prime-time? World J Gastroenterol 2016; 22(32): 7175-85.
[http://dx.doi.org/10.3748/wjg.v22.i32.7175] [PMID: 27621566]

[480] Earl J, Garcia-Nieto S, Martinez-Avila JC, *et al.* Circulating tumor cells (Ctc) and kras mutant circulating free Dna (cfdna) detection in peripheral blood as biomarkers in patients diagnosed with exocrine pancreatic cancer. BMC Cancer 2015; 15: 797.
[http://dx.doi.org/10.1186/s12885-015-1779-7] [PMID: 26498594]

[481] Sausen M, Phallen J, Adleff V, *et al.* Clinical implications of genomic alterations in the tumour and circulation of pancreatic cancer patients. Nat Commun 2015; 6: 7686.
[http://dx.doi.org/10.1038/ncomms8686] [PMID: 26154128]

[482] Previdi MC, Carotenuto P, Zito D, Pandolfo R, Braconi C. Noncoding RNAs as novel biomarkers in pancreatic cancer: what do we know? Future Oncol 2016.
[PMID: 27841659]

[483] Cote GA, Gore AJ, McElyea SD, *et al.* A pilot study to develop a diagnostic test for pancreatic ductal adenocarcinoma based on differential expression of select miRNA in plasma and bile. Am J Gastroenterol 2014; 109(12): 1942-52.
[http://dx.doi.org/10.1038/ajg.2014.331] [PMID: 25350767]

[484] Roldo C, Missiaglia E, Hagan JP, *et al.* MicroRNA expression abnormalities in pancreatic endocrine and acinar tumors are associated with distinctive pathologic features and clinical behavior. J Clin Oncol 2006; 24(29): 4677-84.
[http://dx.doi.org/10.1200/JCO.2005.05.5194] [PMID: 16966691]

[485] Bloomston M, Frankel WL, Petrocca F, *et al.* MicroRNA expression patterns to differentiate pancreatic adenocarcinoma from normal pancreas and chronic pancreatitis. JAMA 2007; 297(17): 1901-8.
[http://dx.doi.org/10.1001/jama.297.17.1901] [PMID: 17473300]

[486] Lee EJ, Gusev Y, Jiang J, *et al.* Expression profiling identifies microRNA signature in pancreatic cancer. Int J Cancer 2007; 120(5): 1046-54.
[http://dx.doi.org/10.1002/ijc.22394] [PMID: 17149698]

[487] Schultz NA, Dehlendorff C, Jensen BV, *et al.* MicroRNA biomarkers in whole blood for detection of pancreatic cancer. JAMA 2014; 311(4): 392-404.
[http://dx.doi.org/10.1001/jama.2013.284664] [PMID: 24449318]

[488] Li A, Yu J, Kim H, *et al.* Serum miR-1290 as a marker of pancreatic cancer--response. Clin Cancer Res 2013; 19(18): 5252-3.
[http://dx.doi.org/10.1158/1078-0432.CCR-13-1899] [PMID: 23881921]

[489] Li A, Yu J, Kim H, *et al.* MicroRNA array analysis finds elevated serum miR-1290 accurately distinguishes patients with low-stage pancreatic cancer from healthy and disease controls. Clin Cancer Res 2013; 19(13): 3600-10.
[http://dx.doi.org/10.1158/1078-0432.CCR-12-3092] [PMID: 23697990]

[490] Frampton AE, Krell J, Jamieson NB, *et al.* microRNAs with prognostic significance in pancreatic ductal adenocarcinoma: A meta-analysis. Eur J Cancer 2015; 51(11): 1389-404.
[http://dx.doi.org/10.1016/j.ejca.2015.04.006] [PMID: 26002251]

[491] Jamieson NB, Morran DC, Morton JP, *et al.* MicroRNA molecular profiles associated with diagnosis, clinicopathologic criteria, and overall survival in patients with resectable pancreatic ductal adenocarcinoma. Clin Cancer Res 2012; 18(2): 534-45.
[http://dx.doi.org/10.1158/1078-0432.CCR-11-0679] [PMID: 22114136]

[492] Pang EJ, Yang R, Fu XB, Liu YF. Overexpression of long non-coding RNA MALAT1 is correlated with clinical progression and unfavorable prognosis in pancreatic cancer. Tumour Biol 2015; 36(4): 2403-7.
[http://dx.doi.org/10.1007/s13277-014-2850-8] [PMID: 25481511]

[493] Kim K, Jutooru I, Chadalapaka G, *et al.* HOTAIR is a negative prognostic factor and exhibits pro-oncogenic activity in pancreatic cancer. Oncogene 2013; 32(13): 1616-25.
[http://dx.doi.org/10.1038/onc.2012.193] [PMID: 22614017]

[494] Petrick JL, Braunlin M, Laversanne M, Valery PC, Bray F, McGlynn KA. International trends in liver cancer incidence, overall and by histologic subtype, 1978-2007. Int J Cancer 2016; 139(7): 1534-45.
[http://dx.doi.org/10.1002/ijc.30211] [PMID: 27244487]

[495] Perz JF, Armstrong GL, Farrington LA, Hutin YJ, Bell BP. The contributions of hepatitis B virus and hepatitis C virus infections to cirrhosis and primary liver cancer worldwide. J Hepatol 2006; 45(4): 529-38.
[http://dx.doi.org/10.1016/j.jhep.2006.05.013] [PMID: 16879891]

[496] Blonski W, Kotlyar DS, Forde KA. Non-viral causes of hepatocellular carcinoma. World J Gastroenterol 2010; 16(29): 3603-15.
[http://dx.doi.org/10.3748/wjg.v16.i29.3603] [PMID: 20677332]

[497] Bosch FX, Ribes J, Díaz M, Clèries R. Primary liver cancer: worldwide incidence and trends. Gastroenterology 2004; 127(5) (Suppl. 1): S5-S16.
[http://dx.doi.org/10.1053/j.gastro.2004.09.011] [PMID: 15508102]

[498] Chuang SC, La Vecchia C, Boffetta P. Liver cancer: descriptive epidemiology and risk factors other than HBV and HCV infection. Cancer Lett 2009; 286(1): 9-14.
[http://dx.doi.org/10.1016/j.canlet.2008.10.040] [PMID: 19091458]

[499] Parkin DM, Bray FI. International patterns of cancer incidence and mortality.Cancer epidemiology and prevention. 3rd ed. Oxford University Press 2006; pp. 101-38.
[http://dx.doi.org/10.1093/acprof:oso/9780195149616.003.0008]

[500] Tyson GL, El-Serag HB. Risk factors for cholangiocarcinoma. Hepatology 2011; 54(1): 173-84.
[http://dx.doi.org/10.1002/hep.24351] [PMID: 21488076]

[501] Patel T. Worldwide trends in mortality from biliary tract malignancies. BMC Cancer 2002; 2: 10.
[http://dx.doi.org/10.1186/1471-2407-2-10] [PMID: 11991810]

[502] Gatto M, Bragazzi MC, Semeraro R, *et al.* Cholangiocarcinoma: update and future perspectives. Dig Liver Dis 2010; 42(4): 253-60.
[http://dx.doi.org/10.1016/j.dld.2009.12.008] [PMID: 20097142]

[503] Vauthey JN, Blumgart LH. Recent advances in the management of cholangiocarcinomas. Semin Liver Dis 1994; 14(2): 109-14.
[http://dx.doi.org/10.1055/s-2007-1007302] [PMID: 8047893]

[504] Liu C, Xiao GQ, Yan LN, *et al.* Value of α-fetoprotein in association with clinicopathological features of hepatocellular carcinoma. World J Gastroenterol 2013; 19(11): 1811-9. [http://dx.doi.org/10.3748/wjg.v19.i11.1811] [PMID: 23555170]

[505] Tangkijvanich P, Anukulkarnkusol N, Suwangool P, *et al.* Clinical characteristics and prognosis of hepatocellular carcinoma: analysis based on serum alpha-fetoprotein levels. J Clin Gastroenterol 2000; 31(4): 302-8. [http://dx.doi.org/10.1097/00004836-200012000-00007] [PMID: 11129271]

[506] Verslype C, Rosmorduc O, Rougier P, Group EG. ESMO Guidelines Working Group. Hepatocellular carcinoma: ESMO-ESDO Clinical Practice Guidelines for diagnosis, treatment and follow-up. Ann Oncol 2012; 23 (Suppl. 7): vii41-8. [http://dx.doi.org/10.1093/annonc/mds225] [PMID: 22997453]

[507] Glimelius B, Hoffman K, Sjödén PO, *et al.* Chemotherapy improves survival and quality of life in advanced pancreatic and biliary cancer. Ann Oncol 1996; 7(6): 593-600. [http://dx.doi.org/10.1093/oxfordjournals.annonc.a010676] [PMID: 8879373]

[508] Choi CW, Choi IK, Seo JH, *et al.* Effects of 5-fluorouracil and leucovorin in the treatment of pancreatic-biliary tract adenocarcinomas. Am J Clin Oncol 2000; 23(4): 425-8. [http://dx.doi.org/10.1097/00000421-200008000-00023] [PMID: 10955877]

[509] Park JS, Oh SY, Kim SH, *et al.* Single-agent gemcitabine in the treatment of advanced biliary tract cancers: a phase II study. Jpn J Clin Oncol 2005; 35(2): 68-73. [http://dx.doi.org/10.1093/jjco/hyi021] [PMID: 15709089]

[510] Bhargava P, Jani CR, Savarese DM, O'Donnell JL, Stuart KE, Rocha Lima CM. Gemcitabine and irinotecan in locally advanced or metastatic biliary cancer: preliminary report. Oncology (Williston Park) 2003; 17(9) (Suppl. 8): 23-6. [PMID: 14569844]

[511] Knox JJ, Hedley D, Oza A, *et al.* Combining gemcitabine and capecitabine in patients with advanced biliary cancer: a phase II trial. J Clin Oncol 2005; 23(10): 2332-8. [http://dx.doi.org/10.1200/JCO.2005.51.008] [PMID: 15800324]

[512] Su CH, Tsay SH, Wu CC, *et al.* Factors influencing postoperative morbidity, mortality, and survival after resection for hilar cholangiocarcinoma. Ann Surg 1996; 223(4): 384-94. [http://dx.doi.org/10.1097/00000658-199604000-00007] [PMID: 8633917]

[513] Todoroki T, Ohara K, Kawamoto T, *et al.* Benefits of adjuvant radiotherapy after radical resection of locally advanced main hepatic duct carcinoma. Int J Radiat Oncol Biol Phys 2000; 46(3): 581-7. [http://dx.doi.org/10.1016/S0360-3016(99)00472-1] [PMID: 10701737]

[514] Alden ME, Mohiuddin M. The impact of radiation dose in combined external beam and intraluminal Ir-192 brachytherapy for bile duct cancer. Int J Radiat Oncol Biol Phys 1994; 28(4): 945-51. [http://dx.doi.org/10.1016/0360-3016(94)90115-5] [PMID: 8138448]

[515] González González D, Gouma DJ, Rauws EA, van Gulik TM, Bosma A, Koedooder C. Role of radiotherapy, in particular intraluminal brachytherapy, in the treatment of proximal bile duct carcinoma. Ann Oncol 1999; 10 (Suppl. 4): 215-20. [http://dx.doi.org/10.1093/annonc/10.suppl_4.S215] [PMID: 10436826]

[516] Takada T, Amano H, Yasuda H, *et al.* Study Group of Surgical Adjuvant Therapy for Carcinomas of the Pancreas and Biliary Tract. Is postoperative adjuvant chemotherapy useful for gallbladder carcinoma? A phase III multicenter prospective randomized controlled trial in patients with resected pancreaticobiliary carcinoma. Cancer 2002; 95(8): 1685-95. [http://dx.doi.org/10.1002/cncr.10831] [PMID: 12365016]

[517] Shibata T, Aburatani H. Exploration of liver cancer genomes. Nat Rev Gastro Hepat 2014; p. 11.

[518] Avila MA, Berasain C, Sangro B, Prieto J. New therapies for hepatocellular carcinoma. Oncogene 2006; 25(27): 3866-84.

[http://dx.doi.org/10.1038/sj.onc.1209550] [PMID: 16799628]

[519] Bruix J, Llovet JM. Major achievements in hepatocellular carcinoma. Lancet 2009; 373(9664): 614-6. [http://dx.doi.org/10.1016/S0140-6736(09)60381-0] [PMID: 19231618]

[520] Llovet JM, Ricci S, Mazzaferro V, *et al.* SHARP Investigators Study Group. Sorafenib in advanced hepatocellular carcinoma. N Engl J Med 2008; 359(4): 378-90. [http://dx.doi.org/10.1056/NEJMoa0708857] [PMID: 18650514]

[521] Zhao Y, Wang WJ, Guan S, *et al.* Sorafenib combined with transarterial chemoembolization for the treatment of advanced hepatocellular carcinoma: a large-scale multicenter study of 222 patients. Ann Oncol 2013; 24(7): 1786-92. [http://dx.doi.org/10.1093/annonc/mdt072] [PMID: 23508822]

[522] Gedaly R, Angulo P, Chen C, *et al.* The role of PI3K/mTOR inhibition in combination with sorafenib in hepatocellular carcinoma treatment. Anticancer Res 2012; 32(7): 2531-6. [PMID: 22753710]

[523] Bruix J, Tak WY, Gasbarrini A, *et al.* Regorafenib as second-line therapy for intermediate or advanced hepatocellular carcinoma: multicentre, open-label, phase II safety study. Eur J Cancer 2013; 49(16): 3412-9. [http://dx.doi.org/10.1016/j.ejca.2013.05.028] [PMID: 23809766]

[524] Pardee AD, Butterfield LH. Immunotherapy of hepatocellular carcinoma: Unique challenges and clinical opportunities. OncoImmunology 2012; 1(1): 48-55. [http://dx.doi.org/10.4161/onci.1.1.18344] [PMID: 22720211]

[525] Tsuchiya N, Sawada Y, Endo I, Uemura Y, Nakatsura T. Potentiality of immunotherapy against hepatocellular carcinoma. World J Gastroenterol 2015; 21(36): 10314-26. [http://dx.doi.org/10.3748/wjg.v21.i36.10314] [PMID: 26420958]

[526] Liu D, Staveley-O'Carroll KF, Li G. Immune-based Therapy Clinical Trials in Hepatocellular Carcinoma. J Clin Cell Immunol 2015; 6(6): 6. [http://dx.doi.org/10.4172/2155-9899.1000376] [PMID: 26877890]

[527] Kudo M. Immune Checkpoint Blockade in Hepatocellular Carcinoma. Liver Cancer 2015; 4(4): 201-7. [http://dx.doi.org/10.1159/000367758] [PMID: 26732472]

[528] Scott AM, Allison JP, Wolchok JD. Monoclonal antibodies in cancer therapy. Cancer Immun 2012; 12: 14. [PMID: 22896759]

[529] Willimsky G, Protzer U, Knolle P, Heikenwalder M. Adoptive T-cell therapy to treat liver cancer: is the liver microenvironment key? Oncotarget 2013; 4(8): 1117-8. [http://dx.doi.org/10.18632/oncotarget.1179] [PMID: 23907631]

[530] Chiocca EA, Rabkin SD. Oncolytic viruses and their application to cancer immunotherapy. Cancer Immunol Res 2014; 2(4): 295-300. [http://dx.doi.org/10.1158/2326-6066.CIR-14-0015] [PMID: 24764576]

[531] Vilarinho S, Taddei TH. New frontier in liver cancer treatment: oncolytic viral therapy. Hepatology 2014; 59(1): 343-6. [http://dx.doi.org/10.1002/hep.26605] [PMID: 23836532]

[532] Heo J, Reid T, Ruo L, *et al.* Randomized dose-finding clinical trial of oncolytic immunotherapeutic vaccinia JX-594 in liver cancer. Nat Med 2013; 19(3): 329-36. [http://dx.doi.org/10.1038/nm.3089] [PMID: 23396206]

[533] Kutay H, Bai S, Datta J, *et al.* Downregulation of miR-122 in the rodent and human hepatocellular carcinomas. J Cell Biochem 2006; 99(3): 671-8. [http://dx.doi.org/10.1002/jcb.20982] [PMID: 16924677]

[534] Nishida N, Goel A. Genetic and epigenetic signatures in human hepatocellular carcinoma: a systematic

review. Curr Genomics 2011; 12(2): 130-7.
[http://dx.doi.org/10.2174/138920211795564359] [PMID: 21966251]

[535] Gramantieri L, Ferracin M, Fornari F, *et al.* Cyclin G1 is a target of miR-122a, a microRNA frequently down-regulated in human hepatocellular carcinoma. Cancer Res 2007; 67(13): 6092-9.
[http://dx.doi.org/10.1158/0008-5472.CAN-06-4607] [PMID: 17616664]

[536] Fornari F, Gramantieri L, Ferracin M, *et al.* MiR-221 controls CDKN1C/p57 and CDKN1B/p27 expression in human hepatocellular carcinoma. Oncogene 2008; 27(43): 5651-61.
[http://dx.doi.org/10.1038/onc.2008.178] [PMID: 18521080]

[537] Meng F, Henson R, Wehbe-Janek H, Ghoshal K, Jacob ST, Patel T. MicroRNA-21 regulates expression of the PTEN tumor suppressor gene in human hepatocellular cancer. Gastroenterology 2007; 133(2): 647-58.
[http://dx.doi.org/10.1053/j.gastro.2007.05.022] [PMID: 17681183]

[538] Ji J, Shi J, Budhu A, *et al.* MicroRNA expression, survival, and response to interferon in liver cancer. N Engl J Med 2009; 361(15): 1437-47.
[http://dx.doi.org/10.1056/NEJMoa0901282] [PMID: 19812400]

[539] Naugler WE, Sakurai T, Kim S, *et al.* Gender disparity in liver cancer due to sex differences in MyD88-dependent IL-6 production. Science 2007; 317(5834): 121-4.
[http://dx.doi.org/10.1126/science.1140485] [PMID: 17615358]

[540] Kota J, Chivukula RR, O'Donnell KA, *et al.* Therapeutic microRNA delivery suppresses tumorigenesis in a murine liver cancer model. Cell 2009; 137(6): 1005-17.
[http://dx.doi.org/10.1016/j.cell.2009.04.021] [PMID: 19524505]

[541] Meng F, Henson R, Lang M, *et al.* Involvement of human micro-RNA in growth and response to chemotherapy in human cholangiocarcinoma cell lines. Gastroenterology 2006; 130(7): 2113-29.
[http://dx.doi.org/10.1053/j.gastro.2006.02.057] [PMID: 16762633]

[542] Xu J, Wu C, Che X, *et al.* Circulating microRNAs, miR-21, miR-122, and miR-223, in patients with hepatocellular carcinoma or chronic hepatitis. Mol Carcinog 2011; 50(2): 136-42.
[http://dx.doi.org/10.1002/mc.20712] [PMID: 21229610]

[543] Tomimaru Y, Eguchi H, Nagano H, *et al.* Circulating microRNA-21 as a novel biomarker for hepatocellular carcinoma. J Hepatol 2012; 56(1): 167-75.
[http://dx.doi.org/10.1016/j.jhep.2011.04.026] [PMID: 21749846]

[544] Liu AM, Yao TJ, Wang W, *et al.* Circulating miR-15b and miR-130b in serum as potential markers for detecting hepatocellular carcinoma: a retrospective cohort study. BMJ Open 2012; 2(2): e000825.
[http://dx.doi.org/10.1136/bmjopen-2012-000825] [PMID: 22403344]

[545] Lai MC, Yang Z, Zhou L, *et al.* Long non-coding RNA MALAT-1 overexpression predicts tumor recurrence of hepatocellular carcinoma after liver transplantation. Med Oncol 2012; 29(3): 1810-6.
[http://dx.doi.org/10.1007/s12032-011-0004-z] [PMID: 21678027]

[546] Tang JC, Feng YL, Guo T, Xie AY, Cai XJ. Circulating tumor DNA in hepatocellular carcinoma: trends and challenges. Cell Biosci 2016; 6: 32.
[http://dx.doi.org/10.1186/s13578-016-0100-z] [PMID: 27182434]

[547] Tchou JC, Lin X, Freije D, *et al.* GSTP1 CpG island DNA hypermethylation in hepatocellular carcinomas. Int J Oncol 2000; 16(4): 663-76.
[PMID: 10717233]

[548] Zhong S, Tang MW, Yeo W, Liu C, Lo YM, Johnson PJ. Silencing of GSTP1 gene by CpG island DNA hypermethylation in HBV-associated hepatocellular carcinomas. Clin Cancer Res 2002; 8(4): 1087-92.
[PMID: 11948118]

[549] Wong IH, Lo YM, Yeo W, Lau WY, Johnson PJ. Frequent p15 promoter methylation in tumor and peripheral blood from hepatocellular carcinoma patients. Clin Cancer Res 2000; 6(9): 3516-21.

[PMID: 10999738]

[550] Yeo W, Wong N, Wong WL, Lai PB, Zhong S, Johnson PJ. High frequency of promoter hypermethylation of RASSF1A in tumor and plasma of patients with hepatocellular carcinoma. Liver Int 2005; 25(2): 266-72.
[http://dx.doi.org/10.1111/j.1478-3231.2005.01084.x] [PMID: 15780049]

[551] Hundal R, Shaffer EA. Gallbladder cancer: epidemiology and outcome. Clin Epidemiol 2014; 6: 99-109.
[PMID: 24634588]

[552] Lazcano-Ponce EC, Miquel JF, Muñoz N, *et al.* Epidemiology and molecular pathology of gallbladder cancer. CA Cancer J Clin 2001; 51(6): 349-64.
[http://dx.doi.org/10.3322/canjclin.51.6.349] [PMID: 11760569]

[553] Levy AD, Murakata LA, Rohrmann CA Jr. Gallbladder carcinoma: radiologic-pathologic correlation. Radiographics 2001; 21: 295-314. questionnaire, 549-55
[http://dx.doi.org/10.1148/radiographics.21.2.g01mr16295]

[554] Konstantinidis IT, Deshpande V, Genevay M, *et al.* Trends in presentation and survival for gallbladder cancer during a period of more than 4 decades: a single-institution experience. Arch Surg 2009; 144(5): 441-7.
[http://dx.doi.org/10.1001/archsurg.2009.46] [PMID: 19451486]

[555] Shukla HS. Gallbladder cancer. J Surg Oncol 2006; 93(8): 604-6.
[http://dx.doi.org/10.1002/jso.20524] [PMID: 16724359]

[556] Rana S, Dutta U, Kochhar R, *et al.* Evaluation of CA 242 as a tumor marker in gallbladder cancer. J Gastrointest Cancer 2012; 43(2): 267-71.
[http://dx.doi.org/10.1007/s12029-011-9288-7] [PMID: 21573879]

[557] Zhu AX, Hong TS, Hezel AF, Kooby DA. Current management of gallbladder carcinoma. Oncologist 2010; 15(2): 168-81.
[http://dx.doi.org/10.1634/theoncologist.2009-0302] [PMID: 20147507]

[558] Furuse J. Targeted therapy for biliary-tract cancer. Lancet Oncol 2010; 11(1): 5-6.
[http://dx.doi.org/10.1016/S1470-2045(09)70370-5] [PMID: 20129124]

[559] Kanthan R, Senger JL, Ahmed S, Kanthan SC. Gallbladder Cancer in the 21st Century. J Oncol 2015; 2015: 967472.

[560] Patt YZ, Hassan MM, Lozano RD, *et al.* Phase II trial of cisplatin, interferon alpha-2b, doxorubicin, and 5-fluorouracil for biliary tract cancer. Clin Cancer Res 2001; 7(11): 3375-80.
[PMID: 11705850]

[561] Hidalgo M, Castellano D, Paz-Ares L, *et al.* Phase I-II study of gemcitabine and fluorouracil as a continuous infusion in patients with pancreatic cancer. J Clin Oncol 1999; 17(2): 585-92.
[http://dx.doi.org/10.1200/JCO.1999.17.2.585] [PMID: 10080603]

[562] Hess V, Salzberg M, Borner M, *et al.* Combining capecitabine and gemcitabine in patients with advanced pancreatic carcinoma: a phase I/II trial. J Clin Oncol 2003; 21(1): 66-8.
[http://dx.doi.org/10.1200/JCO.2003.04.029] [PMID: 12506172]

[563] Sharma A, Dwary AD, Mohanti BK, *et al.* Best supportive care compared with chemotherapy for unresectable gall bladder cancer: a randomized controlled study. J Clin Oncol 2010; 28(30): 4581-6.
[http://dx.doi.org/10.1200/JCO.2010.29.3605] [PMID: 20855823]

[564] Subbiah IM, Subbiah V, Tsimberidou AM, *et al.* Targeted therapy of advanced gallbladder cancer and cholangiocarcinoma with aggressive biology: eliciting early response signals from phase 1 trials. Oncotarget 2013; 4(1): 156-65.
[http://dx.doi.org/10.18632/oncotarget.832] [PMID: 23391555]

[565] Thomas MB. Targeted therapies for cancer of the gallbladder. Curr Opin Gastroenterol 2008; 24(3):

372-6.
[http://dx.doi.org/10.1097/MOG.0b013e3282f6a7df] [PMID: 18408467]

[566] Bizama C, García P, Espinoza JA, *et al.* Targeting specific molecular pathways holds promise for advanced gallbladder cancer therapy. Cancer Treat Rev 2015; 41(3): 222-34.
[http://dx.doi.org/10.1016/j.ctrv.2015.01.003] [PMID: 25639632]

[567] Yang G, Zhang L, Li R, Wang L. The role of microRNAs in gallbladder cancer. Mol Clin Oncol 2016; 5(1): 7-13.
[http://dx.doi.org/10.3892/mco.2016.905] [PMID: 27330755]

[568] Srivastava K, Srivastava A, Mittal B. Common genetic variants in pre-microRNAs and risk of gallbladder cancer in North Indian population. J Hum Genet 2010; 55(8): 495-9.
[http://dx.doi.org/10.1038/jhg.2010.54] [PMID: 20520619]

[569] Kono H, Nakamura M, Ohtsuka T, *et al.* High expression of microRNA-155 is associated with the aggressive malignant behavior of gallbladder carcinoma. Oncol Rep 2013; 30(1): 17-24.
[http://dx.doi.org/10.3892/or.2013.2443] [PMID: 23660842]

[570] Peng HH, Zhang YD, Gong LS, Liu WD, Zhang Y. Increased expression of microRNA-335 predicts a favorable prognosis in primary gallbladder carcinoma. Onco Targets Ther 2013; 6: 1625-30.
[PMID: 24250228]

[571] Li Z, Yu X, Shen J, Law PT, Chan MT, Wu WK. MicroRNA expression and its implications for diagnosis and therapy of gallbladder cancer. Oncotarget 2015; 6(16): 13914-21.
[http://dx.doi.org/10.18632/oncotarget.4227] [PMID: 26040010]

[572] Qiu Y, Luo X, Kan T, *et al.* TGF-β upregulates miR-182 expression to promote gallbladder cancer metastasis by targeting CADM1. Mol Biosyst 2014; 10(3): 679-85.
[http://dx.doi.org/10.1039/c3mb70479c] [PMID: 24445397]

[573] Kitamura T, Connolly K, Ruffino L, *et al.* The therapeutic effect of histone deacetylase inhibitor PCI-24781 on gallbladder carcinoma in BK5.erbB2 mice. J Hepatol 2012; 57(1): 84-91.
[http://dx.doi.org/10.1016/j.jhep.2012.01.018] [PMID: 22326466]

[574] Sekine S, Shimada Y, Nagata T, *et al.* Role of aquaporin-5 in gallbladder carcinoma. Eur Surg Res 2013; 51(3-4): 108-17.
[http://dx.doi.org/10.1159/000355675] [PMID: 24217644]

[575] Chang Y, Liu C, Yang J, *et al.* MiR-20a triggers metastasis of gallbladder carcinoma. J Hepatol 2013; 59(3): 518-27.
[http://dx.doi.org/10.1016/j.jhep.2013.04.034] [PMID: 23665284]

[576] Zhou H, Guo W, Zhao Y, *et al.* MicroRNA-26a acts as a tumor suppressor inhibiting gallbladder cancer cell proliferation by directly targeting HMGA2. Int J Oncol 2014; 44(6): 2050-8.
[http://dx.doi.org/10.3892/ijo.2014.2360] [PMID: 24682444]

[577] Zhou H, Guo W, Zhao Y, *et al.* MicroRNA-135a acts as a putative tumor suppressor by directly targeting very low density lipoprotein receptor in human gallbladder cancer. Cancer Sci 2014; 105(8): 956-65.
[http://dx.doi.org/10.1111/cas.12463] [PMID: 24903309]

[578] Letelier P, García P, Leal P, *et al.* miR-1 and miR-145 act as tumor suppressor microRNAs in gallbladder cancer. Int J Clin Exp Pathol 2014; 7(5): 1849-67.
[PMID: 24966896]

[579] Ma MZ, Chu BF, Zhang Y, *et al.* Long non-coding RNA CCAT1 promotes gallbladder cancer development *via* negative modulation of miRNA-218-5p. Cell Death Dis 2015; 6: e1583.
[http://dx.doi.org/10.1038/cddis.2014.541] [PMID: 25569100]

[580] Jin K, Xiang Y, Tang J, *et al.* miR-34 is associated with poor prognosis of patients with gallbladder cancer through regulating telomere length in tumor stem cells. Tumour Biol 2014; 35(2): 1503-10.
[http://dx.doi.org/10.1007/s13277-013-1207-z] [PMID: 24078448]

[581] Ma MZ, Li CX, Zhang Y, *et al.* Long non-coding RNA HOTAIR, a c-Myc activated driver of malignancy, negatively regulates miRNA-130a in gallbladder cancer. Mol Cancer 2014; 13: 156. [http://dx.doi.org/10.1186/1476-4598-13-156] [PMID: 24953832]

[582] Cai J, Xu L, Cai Z, Wang J, Zhou B, Hu H. MicroRNA-146b-5p inhibits the growth of gallbladder carcinoma by targeting epidermal growth factor receptor. Mol Med Rep 2015; 12(1): 1549-55. [http://dx.doi.org/10.3892/mmr.2015.3461] [PMID: 25760482]

[583] Li G, Pu Y. MicroRNA signatures in total peripheral blood of gallbladder cancer patients. Tumour Biol 2015; 36(9): 6985-90. [http://dx.doi.org/10.1007/s13277-015-3412-4] [PMID: 25861754]

[584] Appierto V, Di Cosimo S, Reduzzi C, Pala V, Cappelletti V, Daidone MG. How to study and overcome tumor heterogeneity with circulating biomarkers: The breast cancer case. Semin Cancer Biol 2017; 44: 106-16. [http://dx.doi.org/10.1016/j.semcancer.2017.04.007] [PMID: 28442298]

[585] Dayde D, Tanaka I, Jain R, Tai MC, Taguchi A. Predictive and Prognostic Molecular Biomarkers for Response to Neoadjuvant Chemoradiation in Rectal Cancer. Int J Mol Sci 2017; 18(3): 18. [http://dx.doi.org/10.3390/ijms18030573] [PMID: 28272347]

[586] Pérez-Callejo D, Romero A, Provencio M, Torrente M. Liquid biopsy based biomarkers in non-small cell lung cancer for diagnosis and treatment monitoring. Transl Lung Cancer Res 2016; 5(5): 455-65. [http://dx.doi.org/10.21037/tlcr.2016.10.07] [PMID: 27826527]

[587] Siravegna G, Marsoni S, Siena S, Bardelli A. Integrating liquid biopsies into the management of cancer. Nat Rev Clin Oncol 2017; 14(9): 531-48. [http://dx.doi.org/10.1038/nrclinonc.2017.14] [PMID: 28252003]

[588] Schwarzenbach H, Nishida N, Calin GA, Pantel K. Clinical relevance of circulating cell-free microRNAs in cancer. Nat Rev Clin Oncol 2014; 11(3): 145-56. [http://dx.doi.org/10.1038/nrclinonc.2014.5] [PMID: 24492836]

[589] Lan H, Lu H, Wang X, Jin H. MicroRNAs as potential biomarkers in cancer: opportunities and challenges. Biomed Res Int 2015; 2015: 125094. [http://dx.doi.org/10.1155/2015/125094]

CHAPTER 2

Perspectives in Breast Cancer Treatment and Principal New Agents

Andrea Nicolini[1,#,*], **Paola Ferrari**[1] and **Pier Mario Biava**[2]

[1] *Department of Oncology, Transplantations and New Technologies in Medicine, University of Pisa, Italy*

[2] *Scientific Institute of Research and Care Multimedica, Milan, Italy*

Abstract: In the last decades, the constant and relevant advances in genetics and molecular biology profoundly changed our knowledge on etiopathology, origin and progression of breast cancer disease. The different spread of program screenings, environmental and lifestyles changes have variously affected the epidemiology of disease in the different geographic areas. In the meanwhile, many new drugs have been experimented and entered into clinical practice. Thus, all the main aspects of the disease are continuously involved and need a regular updating. This chapter faces up with this difficult task and all the issues are discussed taking into account the most important scientific contributions in the field. The first part of the chapter is devoted to epidemiology, etiopathology and origin of the disease; thereafter it focuses on medical treatment and presents the current common lines of therapy in the different settings with particular reference to the more recent target therapies. As to this, the currently proposed treatments to overcome the occurrence of "de novo" or acquired resistance are also described. The last section discusses further innovative biological approaches and suggests a strategy for a more successful antitumoral immune manipulation.

Keywords: Breast cancer, Cytotoxic therapy, Endocrine therapy, Epidemiology, Etiopathology, Molecular profiling, New agents, Target therapy.

EPIDEMIOLOGY

Incidence

In the overall, in 2008 1,384,000 females were diagnosed with breast cancer [1]. This corresponds to a standardised rate of 42.3 new cases per 100,000 population [2]. "Breast cancer has the highest incidence of any cancer amongst females in

* **Corresponding Author Andrea Nicolini:** Department of Oncology, Transplantations and New Technologies in Medicine, University of Pisa, Italy; Tel/Fax: +39050992141; E-mail: andrea.nicolini@med.unipi.it

Retired

most regions of the World, with the exception of several countries in Eastern and Western Africa, Central and South America and Southern Asia, where cervical cancer is more common. Also the incidence of lung cancer is higher in a few countries including China and Vietnam" [1]. The highest incidence rates of breast cancer are "recorded in Western Europe, Australia/New Zealand and Northern Europe, whilst the lowest rates are in Eastern Africa, Middle Africa and Melanesia [1]. In the overall, 89% of breast cancers are diagnosed from the age of 40 onwards, with a relevant difference between more developed (95%) and less developed (84%) countries" [1].

"Much of the variation between countries in stage at diagnosis can be directly attributed to public awareness as well as the availability and utilisation of organised breast cancer screening [3 - 4]. Another major factor is the availability and quality of health care services; even when women in less developed countries present early, they may still receive delayed or inadequate treatment" [3 - 4].

ETIOPATHOLOGY AND ORIGIN

Breast Cancer Classification

Histopathology

Breast cancer pathologic classification is based upon the new WHO 2012 classification [5]. Adenocarcinoma represents more than 95% of breast cancers and origins from the terminal duct lobular unit (TDLU). Cancers originating from the ducts are known as ductal carcinomas, while those originating from the lobules are known as lobular carcinomas. However, it has been found that these two subtypes mainly differ regarding E-cadherin expression [6]. Invasive or infiltrating ductal carcinoma (IDC) is the most common form of invasive breast cancer and refers to malignant abnormal proliferation of neoplastic cells penetrating through the duct wall into stroma.

Ductal carcinoma *in situ* (DCIS) is a neoplastic proliferation of epithelial cells limited to the ducts or lobules, with cellular and nuclear atypia, potential malignant capacity, and tendency to develop subsequent invasive breast cancer. The spread of DCIS throughout the ducts into the lobular acini producing extensive lesions is termed lobular cancerization. DCIS is considered a precursor lesion for the subsequent development of invasive carcinoma. In the last decades, a great increase in the detection of DCIS has been observed, mainly due to the widespread use of screening mammography and increasing awareness of breast cancer in the general population. Death due to DCIS is extremely rare; however, in some instances it can occur, either because of undetected invasive component or due to recurrence of invasive lesion after the treatment.

Lobular carcinoma *in situ* (LCIS) is the intralobular proliferation of small, fairly uniform, and loosely cohesive cells with or without pagetoid involvement of terminal ducts. LCIS "represents a risk factor and a nonobligatory precursor for the subsequent development of invasive carcinoma" [7]. "It is usually found incidentally in breast specimen or biopsy performed for other reasons. It is multicentric in about 70% of cases and bilateral in approximately 30%–40% of cases" [8]. "Microscopically, LCIS generally leaves the lobular architecture intact; it is composed by a relatively uniform population of round, small-t--medium-sized cells with normochromatic nuclei filling the distended lobules in a noncohesive pattern. Pleomorphism, mitosis, and necrosis are absent or rarely present" [9 - 10]. LCIS is characterized by two main immunohistochemical markers: the lack of E-cadherin and β-catenin expressions and the positivity for high molecular weight (HMW) keratin [11] Invasive carcinoma "develops in 25%–35% (about 1% per year) of patients observed for more than 20 years" [6]. The relative risk of subsequent development of invasive carcinoma among patients with LCIS ranges from 6.9 to 12 times than that in women without LCIS. The current recommended management for LCIS is, therefore, "a lifelong follow-up with or without tamoxifen treatment" [8].

Invasive ductal carcinomas (IDCs) "are breast cancers having malignant ductal proliferation along with stromal invasion in the presence or absence of DCIS. IDC is classified into many histological subtypes according to a wide range of criteria, including cell type, amount, type and location of secretion, architectural features, immunohistochemical profile" [8].

Approximately 40%-75% of IDCs lack sufficient morphological features to be classified as specific histological types and are generally designated as IDC not otherwise specified (NOS) or "no special type" (NST) [12].

Invasive lobular carcinoma (ILC) is the second most common histologic subtype of breast cancer and represents approximately 10% of all breast cancers [13]. The "classical" ILC type is formed by small noncohesive cells infiltrating stroma in a single-file pattern [14]; this phenotype is due to the deregulation of cell–cell adhesion properties, primarily due to the lack of E-cadherin expression; approximately 90% of ILCs lack E-cadherin protein expression. A number of variants have been described, based on architectural (alveolar, solid and trabecular) or cytological features (pleomorphic, apocrine, histiocytoid, signet ring); these subtypes are called 'mixed-nonclassic' ILC [15]. Molecular analysis of ILC shows that these tumors have distinct genomic profiles compared to IDCs [16 - 17]. "Compared with grade-matched IDC, ILC is associated with a higher nodal stage, higher absolute number of positive nodes and higher ratio of positive nodes. These differences are maintained in the different size subgroups" [18].

ILC incidence has increased in the past two decades, mainly in more than 50 years old women [19 - 20], and this increase has been reported to be due to the frequent use of hormone replacement therapies [21]. Although ILC usually displays features associated with good prognosis (low-intermediate grade, low Ki-67, ER positivity, HER2 negativity), a few studies suggest that the overall outcome is worse as compared to stage-matched IDC [14 - 15]. Gastrointestinal, peritoneal and ovarian metastases are more frequent [14]. However, consistent with many studies, patients with ILC have a favorable prognosis compared to patients with IDC [22 - 24], but others have found a similar [25 - 27] or a worse prognosis [28]. In particular, Colleoni *et al.* reported the outcome of 7372 women with luminal BC, whose 5707 (76.2%) with IDC, 851 (11.4%) with ILC and the remaining with mixed tumors or other histological subtypes. After a median follow-up of 5.8 years, a worse DFS and distant metastasis-free survival for patients with ILC were found. Also, significantly poorer overall survival (OS) occurred in patients with ILC [28]. In the study by Iorfida *et al.*, patients with classic ILC showed longer disease-free survival (DFS) and breast-cancer specific survival (BCSS) than patients with non-classic ILC. OS was worse for patients with the solid and the mixed non-classic variant [15].

A poorer prognosis for pleomorphic invasive lobular carcinoma (PILC) has been reported in several studies [29 - 31]. Pleomorphic lobular carcinoma is now recognized as "a rare and aggressive variant of lobular carcinoma, accounting for <1% of all breast cancers" [32]. Generally, it affects postmenopausal women and shows hormone receptor expression and negative staining for E-cadherin. Higher-grade cytological features, lymphovascular invasion, and advanced stage at presentation are common. HER2 is overexpressed in up to 30% of the cases. Prognosis is poorer than that of classic lobular carcinoma, and the outcome of pleomorphic lobular cancer patients is similar to that of patients with high-grade, invasive ductal carcinoma [33 - 34].

Tubular carcinoma represents less than 2% of invasive breast cancers [35]; it is frequently detected by screening mammography. Tubular carcinomas "are nearly always hormone receptor-positive and well-differentiated, with low proliferation. HER2 is generally neither overexpressed nor amplified" [36]. When compared with invasive carcinoma of no special type, "tubular carcinoma is more likely to be diagnosed at older age and be smaller in size. Nodal involvement is reported in 4%–17% of cases. The prognosis is very good, with 5-year DFS and OS of 94% and 88%, respectively" [37].

Invasive cribriform carcinoma "represents 0.1%–0.6% of breast cancers and the median age at diagnosis is 54–63 years" [38]. This lesion is "generally estrogen receptor-positive (ER+), low grade, and low proliferating. Axillary lymph-nodal metastases are reported in approximately 10% of the cases. ICC has an excellent prognosis, mainly in its pure form" [39].

Pure mucinous carcinoma accounts for 1%–4% of all breast cancers; the median age at diagnosis is 71 years [40]. These lesions are commonly well-differentiated, hormone receptor (HR)+ and HER2− and the axillary lymph nodes are not involved. Prognosis is favorable, with a 5-year survival rate of 94% and a 10-year survival rate of 89% [40]. Pure mucinous carcinoma produces abundant extracellular and/or intracellular mucin, and the mixed form also contains common infiltrating ductal carcinoma [41]. Pure mucinous carcinomas "are generally defined as containing more than 90% of mucin, and mixed mucinous carcinomas are those containing 50%–90% of mucin. The presence of less than 50% of mucin identifies ductal carcinoma with a mucinous component" [42].

Neuroendocrine carcinomas are characterized by "a diffuse expression of neuroendocrine markers (synaptophysin, chromogranin) with morphologic neuroendocrine features. Three main histologic types exist: neuroendocrine tumor, well-differentiated neuroendocrine carcinoma, poorly differentiated/small cell invasive breast carcinoma with neuroendocrine differentiation" [5, 43]. These tumors "are generally low-grade ER+, and PR+, and HER2−, with favorable prognosis. The high-grade small-cell variant is described among the poor prognostic group" [44].

Medullary carcinoma is less than 2% of breast carcinoma and often involves younger women [45] and BRCA1 mutated patients [46]. It is composed of "poorly differentiated cells, with a prominent lymphocytic infiltrate, both within the tumor and at the periphery" [47]. Most of medullary cancer are triple-negative at immunohistochemistry, aneuploid and highly proliferative [48 - 49]. Despite its unfavorable characteristics, also including a more advanced stage and larger tumor size, medullary breast cancer shows similar prognosis in both BCSS and OS compared with IDC after adjusting and matching the confounding factors [50]. Among TNBC patients, medullary breast cancer patients have similar survival compared with IDC patients. Therefore, it has been hypothesized that molecular subtype may be a principal confounder for metastatic breast cancer prognosis [50].

Adenoid cystic carcinoma of the breast accounts for less than 1% of breast cancers. The median age at diagnosis of 60 years [51]. Nodal involvement is reported in 0%–8% of the cases [52]. These tumors are generally HR− and HER2− [53]. The prognosis is generally good, with 10 year OS exceeding 90% [54].

Juvenile carcinoma, or secretory breast carcinoma represents the 0.1%–0.2% of all breast cancers and the median age at diagnosis is 25–40 years [55]. Secretory breast cancer is usually triple negative with low Ki67 expression. These tumors are generally associated with a good prognosis [56].

"Apocrine carcinoma accounts for 0.3%–4% of breast cancers, mostly because of the lack of uniform criteria for its diagnosis. Pure apocrine carcinomas are generally ER−, PR− and androgen receptor (AR) positive" [57]. HER2 is overexpressed in up to 54% of the cases [58]. While the prognosis of apocrine carcinoma has been reported to be the same as other types of breast carcinoma, in a large retrospective analysis pure apocrine carcinoma (ER−, PR−, AR+) showed a worse DFS compared with apocrine-like carcinoma (ER or PR+, AR−), whose outcome was similar to infiltrating ductal carcinoma [59]. However, among "triple-negative breast cancers, the prognosis of the apocrine histotype was intermediate between the histotypes with the best outcome (like medullary carcinomas) and the ones determining the worst outcome (like metaplastic carcinomas)" [32].

Invasive micropapillary carcinoma is characterized by extensive lymphatic vessel invasion [45,60]; it shows a nodal metastases involvement that has been reported in up to 70% of the cases, and high rate of recurrences in skin and in the chest wall [61 - 62]. Median age at diagnosis is 52.5 years. The expression of HR is reported in about 60% of the cases. A positive HER2 status is present in up to 50% of the cases, and the prognosis of this subtype is poor, with a 10-year OS of 48% [63].

Metaplastic cancers include mixed tumors, namely squamous cell carcinoma, spindle cell carcinoma, metaplastic carcinoma with mesenchymal differentiation, and mixed metaplastic carcinoma. Two subtypes, low grade adenosquamous carcinoma and fibromatosis-like metaplastic carcinoma, have more favorable prognosis [45]. The incidence ranges between 0.2%–0.6%, and the median age at presentation is 47–61 years [64 - 65]. These tumors are generally poorly differentiated and triple-negative, with high Ki67 and p53 positivity. Local or distant relapse occurs in >50% of the cases [66]. The prognosis is poorer than that of triple-negative invasive ductal carcinoma [65,67]. In a study conducted on squamous-cell subtype, the median OS was 37 months [68].

Clinical Symptoms

Although the use of screening mammograms has increased the number of asymptomatic breast cancers, still many breast cancers present with locoregional sign/symptoms. The most common sign is a palpable breast mass, more often painless, hard, with irregular margins; other symptoms/signs include breast pain, nipple retraction, secretion, skin nodules/ulceration, presence of axillary nodes/mass, signs on inflammatory breast cancer (erithema and edema of a third or more of the skin of the breast). Symptoms of metastatic disease are related to the location and the extension of the secondary lesion(s). The most common sites of metastatic spread are bone, liver, lung, lymph-nodes, brain; symptoms include bone pain, cough and other respiratory symptoms, abdominal pain fatigue, neurological symptoms [69].

Stage

Breast cancer staging is usually described using the TNM system. This system is based on the size of the tumor (T), the lymph node involvement (N- or N+), and the presence of metastases (M0 or M1) [70].

Trends in Breast Cancer Incidence Varies in Different Countries

"Overall incidence rates were generally stable in the European countries, but were increasing gradually amongst females aged 50– 69 years old, except for Sweden where there was a significant decrease within this age group from 2002 onwards. Incidence had also stabilised in the United States after a sharp decline between 1999 and 2003, whilst decreases were recorded for females in Canada and Australia (significant for the 50–69 age group only)" [71].

"Population-based screening using mammography (especially amongst women aged 50–69 years)" [72], "prevalence of known risk factors such as obesity, earlier puberty, excessive alcohol consumption, reduced physical activity and having children (if any) at an older age maybe account for an increased rate of breast cancer incidence" [71,73-74]. Conversely, significant recent decreases in the incidence rate within the 50–69 age group in a few developed countries can be attributed to a decrease of the use of hormone replacement therapy [75 - 76].

Two Principal Distinct Pathways Each With a Different Age-Specific Incidence Rate Curve and Their Association with Histopathological and Molecular Subtypes

Breast cancer molecular subtypes derive from two main cell types [77 - 84]: basal-like breast cancer from the basal/myoepithelial cells, and non-basal-like cancer (*i.e.*, HER2-enriched, luminal B, and luminal A) from a more luminal-like cell district.

Lilienfeld [85] and de Waard [86 - 88] introduced the concept that breast cancers develop by two distinct pathways (rather than one), each with a different age-specific incidence rate curve. The first pathway results in mainly premenopausal tumors with peak occurrence early in life, similar to estrogen receptor (ER)–negative cancers in the general US population. The second pathway results in predominantly postmenopausal cancers with peak incidence later in life, similar to late-onset ER-positive cancers. ER-negative rates rise rapidly early in life then flatten or fall soon after menopause whereas ER-positive rates rise continuously irrespective of menopause, albeit more slowly after age 50 years" [89]. Accordingly, recent studies in the United States have reported on ER-positive cancers increase over the long term, whereas ER-negative breast cancers have decreased [90 - 92]. Trends in other countries are similar [93]; this finding supports the hypothesis that different risk factors account for different ER-positive and ER-negative trends during the time [92, 94].

"Incidence rates of infiltrating duct, tubular and lobular carcinomas rise rapidly until age 50 years, then increase more slowly, similar to rates for breast cancer overall. Rates for medullary and inflammatory breast carcinomas increase rapidly until age 50 years, then flatten or fall, similar to ER-negative rates. Finally, rates for papillary and mucinous carcinomas increase steadily with age, similar to ER-positive rates and much like cancers at many other organ sites such as colorectal cancer" [95]. Breast cancers that occur early or late are likely to be most represented by one etiological type, but both types remain during the entire life, with substantial mixing during the intermediate interval time, when many cancers develop, although the proportion of each etiological type within a class of breast cancers depends on its definition; however, the peak ages remain close to 50 and 70 years.

Recently, using the National Cancer Institute's Surveillance, Epidemiology, and End Results (SEER) Residual Tissue Repository, "the age-specific incidence rates for the molecular subtypes with immune-histochemical staining of breast cancer tissue microarrays were estimated and population data were input. Luminal A (defined as ER positive and HER2 negative) incidence rates rose continuously, although more slowly after age 50 years (similar to ER-positive cancers). Basal-like (defined as ER negative and HER2 negative) rates increased rapidly early in life then flattened or fell (similar to ER negative cancer)" [96 - 97]. These patterns are confirmed by successive studies from the California Cancer Registry [98 - 101] and emerging data from SEER's large-scale population-based database [102]. The intrinsic gene set classification algorithm was also applied to approximately 2000 breast cancer cases reported by METABRIC Group [103 - 105]. Luminal A and luminal B cases showed prevalent late modes near age 70 years and minor modes near age 50 years, such as ER-positive cancers in SEER. Basal-like cancers showed an early-onset mode near to age 50 years, similar to ER-negative cancers in SEER. The HER2-enriched age distribution is placed between the luminal and basal-like cancers.

Risk Factors

The American Cancer Society divides breast cancer risk factors into two categories based on whether they are related to lifestyle. Among "factors unrelated to lifestyle, seven breast cancer susceptibility genes or gene sets with high penetrance (relative risk > 5) have been identified; their germline mutations are associated with increased breast cancer risk" [106]. These genes are: "DNA damage repair genes BRCA1/2, cell–cell adhesion gene CDH1, tumor suppressor phosphatase and tensin homolog PTEN, serine threonine kinase SKT11, tumor suppressor TP53, and DNA mismatch repair genes MLH1, MSH2/6, and PMS2. These genes are found to be drivers in only 5%–10% of all breast cancer cases, leaving the majority of breast cancer cases attributable to a complex summation of genetic and epigenetic alterations" [106] on which environmental exposures are thought to operate [107].

Diet and Lifestyles

Alcohol is a well-known risk factor for breast cancer [108 - 109]; 10g alcohol consumed per day (defined as a unit of alcohol by the World Health Organization) increase 7-10% the risk of breast cancer [110 - 113]. The underlying mechanism is, at least in part, sex-hormone driven [114 - 115]. In fact, several studies have demonstrated that female sex-hormone blood concentrations in pre- and postmenopausal women increase concomitant with alcohol consumption [116 - 120]. Moreover, alcohol is more strongly associated to hormone-dependent breast cancers than hormone-independent subtypes [121]. Increased aromatization, insufficient estrogen metabolism in the liver or stimulation of adrenal steroidogenesis are potential mechanisms of increased sex-hormone concentrations by alcohol [119, 122 - 123].

A dose-response relationship exists between alcohol consumption (including light consumption) and the risk of breast cancer. In 2012, there have been 144,000 breast cancer cases and 38,000 breast cancer deaths attributed to alcohol; 18.8% of these cases and 17.5% of these deaths were in light alcohol consumers [124].

Overweight and obesity have been linked to the development of cancer [125 - 127]. The rising of breast cancer incidence in western countries (over 30% in the past 25 years), while related in part to changes in reproductive patterns and improved detection methods, likely also reflects the increased obesity and physical inactivity [128]. In fact, "lifestyle modifications, including maintaining or decreasing to body mass index (BMI) < 25, moderate physical activity, and limiting alcohol to <3 drinks per week, can cumulatively reduce breast cancer risk by >30%" [129].

Obesity positively correlates more with hormone receptor–positive than hormone receptor–negative cancers [130 - 132], although likely obesity also increases the risk of basal-like, triple-negative, and inflammatory breast cancers [133 - 136].

A different behavior seems to be in pre- and postmenopausal women. In premenopausal women, most studies report a null or inverse association of obesity with breast cancer risk [137 - 138]. This is observed mainly in ER/PR positive tumors, while risk for ER/PR negative cancers is increased [139]. Conversely, in postmenopausal women, most studies show a direct link of obesity with breast cancer risk [127, 140]. Regarding ER/PR positive (luminal) breast cancer, "a 33% risk increase in postmenopausal women is estimated per 5 kg per m2 BMI increment" [141]. Concerning molecular subtypes other than luminal breast tumors, results are controversial and vary from no association [142 - 143] to a reduced risk with increased BMI [144], to a positive association reported between increasing BMI and the incidence of triple-negative tumors [145 - 146].

The main hypothesis of an increased risk of breast cancer due to obesity involves peripheral production of sex hormones by fat tissue, that can synthesize estradiol from androgenic precursors through increased aromatase enzyme activity [140].

Obesity negatively influences breast cancer outcome, independently of menopausal status. It has been observed that "obese women with breast cancer usually have larger tumors, advanced disease stage at diagnosis, higher rates of metastasis, and higher rates of distant recurrence" [147 - 151]. "Obese women with breast cancer have increased chance of initial or acquired chemotherapy resistance, higher all-cause mortality and higher breast cancer-related mortality at any age compared to normal weight women with breast cancer" [152].

The poorer survival in obese breast cancer patients compared to non-obese patients has been explained either due to the biology of the disease or because the treatment is less effective in patients who are obese [153]. The "adverse effect of obesity on breast cancer prognosis has been thought to be mainly mediated by higher estrogen levels due to greater aromatase activity in the excess adipose tissue. If this were so, it could be hypothesized that an adverse prognostic effect of obesity might not be seen in premenopausal women due to the already relatively high levels of ovarian estrogen production.

As obesity is a significant predictor of outcome on both estrogen receptor–negative and –positive tumors, other factors apart from estrogen may be responsible for the effect of BMI on prognosis. These mechanisms are still unclear. Increased mitotic count and higher Ki-67 expression in tumors from obese premenopausal women were found" [154 - 155]. Obesity is associated with hyperandrogenism, hyperinsulinemia, and lower serum hormone-binding globulin

[156]. Hyperinsulinaemia is associated with worse outcomes in breast cancer patients [157 - 158]. Visceral adipose tissue directly correlates with estradiol and inversely with sex hormone binding globulin (SHBG) [159]; moreover, it is associated with increased insulin and insulin-like growth factors [160 - 161], which can promote a hormonal environment favoring proliferation of cancer mammary cells [160, 162]. Adipose tissue inflammation often occurs in obesity and type 2 diabetes; proinflammatory cytokines and prostaglandin E2 produced by the infiltrating macrophages with chemokine monocyte chemoattractant protein-1 can stimulate tumor cell proliferation and invasion and promote tumor-related angiogenesis. So, it has been postulated that chronic adipose tissue inflammation, rather than body mass index-defined obesity per se, can account for a high risk of postmenopausal estrogen-dependent breast cancer. In addition, in premenopausal women, "obesity and type 2 diabetes, or perhaps the associated inflammation, promote estrogen-independent, notably triple-negative, breast cancer development, invasion and metastasis by mechanisms that may involve macrophage-secreted cytokines, adipokines and insulin" [163].

Also "treatment-related factors may contribute to worse outcome in obese breast cancer patients, as they may receive less chemotherapy due to comorbidities" [164 - 165].

Radiations

Patients exposed to therapeutic chest radiation, particularly before age 20, may have an increased risk of developing breast cancer [166].

Hormones

Approximately 70% of breast cancer express estrogen receptors (ERs): the relationship between ER expression and cancer initiation is not clear, as ER positive cells are non-proliferating in normal breast epithelium, that is found adjacent to proliferating cells, suggesting a paracrine effect. The progesterone receptor (PR) is an estrogen-regulated gene and is variably expressed in ER positive tumors [167].

Current users of exogenous hormones have a slight risk of developing breast cancer [168]. Estrogen only hormone replacement therapy (HRT) for postmenopausal women is associated with a small but significant risk of invasive breast cancer [169 - 170]. A small decline in breast cancer incidence has been reported in several countries after 2002, when the use of postmenopausal HRT started to decline. This reduction involves mainly invasive lobular carcinoma, the type of breast cancer that is more frequently associated with HRT [54, 56].

Regarding endogenous hormones, breast cancer risk has been correlated with earlier age of menarche, later age of menopause, nulliparity, late age of first birth, all of them determining a greater cumulative number of ovarian cycles. These observations are consistent with the correlation between estrogen levels and breast cancer risk [171].

Increased endogenous estrogen and testosterone concentrations are positively correlated with an increased risk of postmenopausal breast cancer, mainly regarding ER+/PR+ tumors. "The relationship between circulating estrogen levels in postmenopausal women and the risk of breast cancer is well established, with an approximately 2-fold higher risk among women in the top 20-25% (*versus* bottom 20-25%) of levels" [172 - 174]. The postmenopausal disappearance of ovarian function results in withdrawal of ovarian estrogen and progesterone. This causes homeostasis imbalances resulting in a total decrease of energy consumption. This, together with a decrease in physical activity and high fat diet, significantly account for obesity in postmenopausal women, that may contribute to breast cancer development through several different mechanisms [175].

In premenopausal women with high estradiol values significantly increased risks of breast cancer was found; the association was stronger for ER+/PR+ tumors, which also were positively correlated with higher levels of total and free testosterone and androstenedione [176].

In spite of the well-known protective effect of full-term pregnancy for breast cancer [177 - 179], it has been suggested "that parity is associated with an early increase in risk, followed by long-term protection [180 - 188]. Alternatively, the early risk and late protection for parity could reflect another qualitative (crossover or reversing) age interaction for breast cancer" [189], where parity increases risk of early-onset subtypes and reduces risk of late-onset ones [23]. Namely, parity and multiple live births decrease the risk for ER-positive and luminal A breast cancer subtypes and increase that of ER-negative tumors [133 - 134, 190-192]. Patients with early age at first birth and lack of breast feeding seem to be at high risk for basal-like (triple-negative and BRCA1) cancers [131, 191, 193-197], mainly among some specific ethnic groups.

Also, insulin-like growth factor-1 (IGF-1), a mitogenic growth factor, and prolactin have been implicated in breast carcinogenesis, and their levels have been correlated with an increased breast cancer risk [198 - 199].

Genetic Factors

"Approximately 5–10% of the total breast cancer burden follows a Mendelian inheritance pattern (autosomal dominant) and is characterized as hereditary. An additional 15–20% of breast cancer cases are familial, referred to women who have two or more first- or second-degree relatives with the disease. Among hereditary breast cancer cases, at least 30% are attributed to germline mutations in the high penetration genes BRCA1 and BRCA2" [200].

BRCA 1 and BRCA2

Breast cancer susceptibility genes *BRCA1* and *BRCA2* have been identified in the mid-1990s. These genes account for less than 20% of familial clustering of breast cancer.

BRCA1 and *BRCA2* are two different tumor suppressor genes involved in response to cellular stress through the activation of DNA repair processes [200 - 204]. Cells lacking BRCA1 or BRCA2 protein are deficient in repairing DNA double-strand breaks.

BRCA1 is implicated in DNA damage repair, cell cycle regulation, transcriptional regulation, chromatin remodelling, while BRCA2 is involved primarily in DNA recombination and repair [205]. Tumors that arise in heterozygote carriers of germline mutations usually have an inactivated wild type allele.

Inherited breast and ovarian cancers occur in patients with mutations in the *BRCA1* and *BRCA2* genes and about 30–40% of sporadic tumors occur in those with loss of BRCA1 expression [206 - 207]. The lifetime risk of breast and ovarian cancer in BRCA1 mutation carriers is 60% and 59% respectively, while in BRCA2 mutation carriers it is 55% and 16.5% respectively [208].

For carriers of BRCA mutations, the risk of both breast and ovarian cancer increases about 10–15% in each decade following the age of 40 years [205]. Moreover, in BRCA1-mutation carriers ovarian cancer occurs earlier than BRCA2-mutation carriers and sporadic cases [209].

BRCA1

In about 80% of BRCA1 mutation carriers breast cancer is diagnosed prior to menopause [210 - 213]. The *BRCA1* gene is located in human chromosome 17q21 [17]. The BRCA1 is a multi-functional protein that interacts with different protein partners in various compartments to play key roles in apoptosis, genetic instability, and also tumorigenesis [214 - 216].

In cancer patients, BRCA1 mutations are mainly in three domains named N-terminal RING domain encoded by exons 2–7 (amino acids 1-109), coding regions of exons 11–13, and BRCA1 C-terminus encoded by exons 16–24 (amino acids 1650–1863) or BRCT domain [217]. These three domains interact with different partner proteins and define BRCA1 subcellular localization.

The RING domain of BRCA1 interacts with another RING domain containing protein BARD1 (BRCA1 associated RING domain protein 1) [218]. BARD1 mutations prevail in hereditary breast and ovarian cancers [219 - 221].

Exons 11–13 contain most clinically relevant mutations important for the tumor suppressor function of BRCA1 [222]. This region interacts with several proteins involved in transcription, DNA repair, and cell cycle progression and the interacting partners include retinoblastoma protein (Rb), c-Myc, RAD50, and RAD51 [215, 223]. Exon 11 has two nuclear localization sequences (NLS) and encodes almost 60% of the BRCA1 protein [221, 224 - 225]. The NLS sequences are placed in between amino acids 501–507 and 607–614 and promote the interaction with importin-alpha, which mediates BRCA1 transport from the cytosol to the nucleus [226]. NLSs mutations induce cytosolic localization of BRCA1 and subsequent increase in unrepaired mutations and chromosomal abnormalities in malignancies.

The C-terminal (BRCT) domain accounts for interactions with substrates of DNA damage-activated kinases [227]. This region interacts with p53, ER-a, c-Myc, STAT1, CtIP, and ZBRK1 transcription regulators as well as with RNA polymerase II, so playing a critical role in transcriptional regulation [228 - 235]. BRCA1 cancer-associated point mutations impede its interaction with RNA polymerase II, thus indicating that BRCA1 takes part of core transcriptional machinery [236]. Recently it has been reported that BRCA1 autoregulates its own transcription to maintain genome integrity in response to genotoxic stress [237 - 238].

BRCA1 mutation prevails in basal-like subtype [239 - 241] and nearly 70% of BRCA1-mutated breast cancers express basal cytokeratins [240] and are triple negative (*i.e.* ER, PR and HER2 negative) and p53 positive [210, 242 - 244].

On the other side, tumors with functional BRCA1 are mostly of the luminal type, they are less aggressive, responsive to endocrine therapies, and with improved survival [245 - 246].

BRCA2

The *BRCA2* gene contains 27 exons with eight internally repeated sequence called BRC motif, which is considered to be the major domain to interact with RAD51 [247 - 250]. Nuclear localization signals in human BRCA2 have been identified

[251] and colocalize with BRCA1 in subnuclear foci in somatic cells [252]. Like BRCA1, BRCA2 is an important transcriptional co-regulator [253]. BRCA2 interacts with SMAD3 to produce a complex that co-activates Smad3-dependent transcriptional activation of plasminogen activator inhibitor-1 (PAI-1) and collaborates with histone acetyltransferases in androgen receptor-mediated transcription [254 - 255].

The BRCA2 C-terminal domain has DNA binding property [256] and many tumor derived mutations occur in the BRCA2 C-terminal domain [257]. The BRCA2, like BRCA1, co-localizes with RAD51 during meiosis on chromosome axes [252]. The co-localization of BRCA2 with RAD51 suggests the involvement of BRCA2 in the repair of DNA damage by homologous recombination (HR) pathway [257]. All these findings suggest tumor suppressor role of BRCA2 by maintaining genome stability. BRCA2- associated breast tumors are mainly (75%) ER positive and p53 negative, like to breast cancer in the general population [210, 242 - 244].

Germline Mutations other than BRCA1/2

Rare germline mutations in several genes other than BRCA1/2, such as *TP53*, *PTEN* and *ATM* have been shown to confer an increased breast cancer risk. It is likely that most of the genetic susceptibility is a result of multiple low-penetrance alleles that co-exist in highly penetrant combinations in a polygenic model [258].

The evolution of sequencing technologies enabled parallel testing of multiple genes, leading to concomitant analysis of breast cancer predisposition genes with either high or intermediate penetration. The identification of the breast cancer hereditary syndromes that have emerged is of major importance, so that genetic counseling, testing, screening and prevention strategies can be offered to affected individuals.

According to National Comprehensive Cancer Network (NCCN] guidelines, "gene panels are aimed at individuals who are tested negative for BRCA1/2 and for those whose family pedigree is suggestive of more than one hereditary syndrome" [204]. "Penetrance of a cancer predisposition gene is a definition referred to its relative risk of causing a particular type of cancer.

High penetrant genes are associated with a cancer relative risk higher than 5, as opposed to low penetrant genes that display a relative risk of around 1.5 Intermediate-penetrant genes exhibit relative cancer risks from 1.5 to 5" [259].

High Penetrant Genes (Low Frequency)

TP53 (Li–Fraumeni syndrome). Thirty to 50% of breast cancers harbor somatic mutations in *p53* [260]. *TP53*, also named "the guardian of the genome", is the most frequently mutated gene in human cancers. It plays a central role in sensing genotoxic and non-genotoxic stresses and transducing an antiproliferative effect (cell cycle arrest or apoptosis) in response [261 - 262]. Inheritance of a *p53* mutant allele causes the Li-Fraumeni familial syndrome, with multiple early-onset cancers, including breast cancer.

PTEN (Cowden/PTEN hamartoma tumor syndrome). Dysfunctional PTEN impedes to activate cell cycle arrest and apoptosis, which accounts for promoting cell survival [263 - 264]. Germline PTEN mutations cause a hereditary cancer predisposition syndrome (Cowden's syndrome), with an elevate incidence of breast, uterine, thyroid and skin cancers [265].

SKT-11 (Peutz–Jeghers syndrome). The SKT-11 gene is a tumor suppressor gene, participating in membrane bonding and apoptosis [266] and it is a negative regulator of the mTOR pathway [267]. Germline mutations in SKT-11 account for the Peutz–Jeghers syndrome (PJS), a rare autosomal dominant disorder with multiple gastrointestinal hamartomatous polyps and mucocutaneous pigmentations of the lips, buccal mucosa and digits. These individuals are at high risk for colorectal, breast, small bowel, pancreatic, gastric and ovarian cancer [268 - 270].

CDH1 – hereditary diffuse gastric Ca. The E-Cadherin gene is a calcium dependent cell–cell adhesion molecule expressed in junctions between epithelial cells [271]. CDH1 germline mutations have been associated with hereditary diffuse gastric cancer (HDGC). "Approximately 30% of families with HDGC due to CDH1 mutations also include women with lobular breast cancer (LBC). It has been estimated that the cumulative risk of LBC in female CDH1 mutation carriers from HDGC families reaches 39% at 80 years old" [272].

Moderate Penetrant Genes

Checkpoint kinase 2 (CHEK2). CHEK2 gene "encodes for a serine threonine kinase, which is activated in response to DNA double strand breaks (DSBs), the most lethal form of DNA damage, contributing to signal transduction to downstream repair proteins" [273]. Check2 mutation 110delC has been shown to increase breast cancer risk by 2–3-fold [274]. "This mutation is particularly frequent in Northern European population, where it carries a lifetime risk of breast cancer as high as 37%" [275]. This mutation is associated with male breast cancer. The majority of CHEK2-associated breast cancer is ER positive and therefore, might be subjective to chemoprevention with tamoxifen [276].

Ataxia-telangiectasia (ATM). ATM is "a multifunctional gene that plays a pivotal role in double-strand break (DSB) repair and in cell cycle progression. Homozygous ATM mutation carriers suffer from ataxia telangiectasia (AT), a disorder characterized by cerebral ataxia, immunodeficiency and increased risk of certain malignancies, including breast cancer" [277].

PALB2 and BRIP1. PALB2 and BRIP1 are two genes whose biallelic mutations are associated with Fanconi anemia (FA), an autosomal recessive disorder characterized by developmental abnormalities, bone marrow failure and increased risk of malignancies [278].

PALB2 is a 'binding partner and localizer of BRCA2', contributing to DNA repair mechanism HR and tumor suppression [279]. Classification of PALB2 as a breast cancer susceptibility gene was based on data showing that 1.08% of individuals with hereditary breast cancer negative for BRCA1/2 harbor a monoallelic mutation in PALB2, which confers a 2-fold high risk [280]. BRIP1 encodes a protein that was identified as the binding partner of BRCA1. Truncating mutations of the gene have been detected in BRCA1/2 negative breast cancer families and the relative breast cancer risk has been estimated to be approximately 2-fold [281]. BRIP1 missense mutations have been found in high risk Jewish women who are BRCA1/2 negative [282].

RAD51C. RAD51C is an integral component of FA/BRCA pathway and plays an important role in DSB repair through HR. Homozygous mutations in RAD51C are associated with a FA phenotype, whereas heterozygous mutations have been identified in breast and ovarian cancer families [283 - 285].

RAD51D. RAD51D is a co-factor of DSB repair mechanism through HR. A recent study that analyzed 841 BRCA negative patients either from breast cancer/ovarian cancer families or with early onset breast cancer/ovarian cancer identified three RAD51D mutations in four breast cancer/ovarian cancer families with only one ovarian cancer case [286 - 288].

MRE11, RAD50, NBS1. The MRN complex, formed by of MRE11, RAD50 and NBS1 proteins is involved in DSB repair and maintenance of genomic integrity [289 - 293]. In an attempt to assess the role of rare MRN variants as intermediate-risk breast cancer susceptibility alleles, a recent study has performed a mutation analysis of MRN genes in a large number of early-onset breast cancer cases, confirming that rare missense substitutions and less frequently, protein-truncating variants of MRN genes confer a 2–3-fold increased risk of breast cancer [294].

FANCM. FANCM is a FA gene that plays a crucial role in DNA repair process [295]. In a recent study, the c.5101C > T FANCM nonsense mutation has been associated with breast cancer risk in a Finnish population. The highest frequency of the mutation was observed in triple- negative breast cancer patients [296].

Breast Cancer Genome

The first report of a comprehensive sequence analysis of 21,000 of the best studied and annotated human genes, including data from breast and colon cancer samples was published in 2006. This screen confirmed cancer genes known to be associated with breast cancer (*e.g.*, *p53* and *BRCA1*), as well as genes not previously characterized as associated with cancer. This research suggested that an individual breast cancer harbored an average of about 90 mutant genes, of which a small minority are expected to contribute to the neoplastic process. The functionally important mutations were often different among tumors; however, when the genes were analyzed by functional groups, mutations in signal transduction pathway and transcription factor genes were found in nearly all breast tumor samples [297]. Overall, many different genetic and epigenetic alterations have been found to be involved in the development of breast cancer.

Genetic Changes

They mainly include somatic mutations and copy number aberrations.

Somatic Mutations

In a recent study, the genomes of 100 tumors for somatic copy number changes and mutation in the coding exons of 21,416 protein coding genes were examined and 7,421 somatic point-mutations were identified. The number of somatic mutations was different in different tumors [298].

Somatic mutations were observed with regard to genes well known to regulate in breast cancer apoptosis, cell cycle and transcription as BRCA1, RB1, TP53, PTEN, AKT1, CDH1, GATA3, and PIK3CA, or signal transduction as APC, ARID1A, ARID2, ASXL1, BAP1, KRAS, MAP2K4, MLL2, MLL3, NF1, SETD2, SF3B1, SMAD4, and STK11. Moreover, new genes as ARID1B, CASP8, MAP3K13, MAP2K13, NCOR1, SMARCD1, and CDKN1B, not previously considered, were found mutated [298].

In the studied tumors, mutations in at least 40 cancer genes with 73 different combinations of mutated cancer genes were found, to underline the genetic diversity and the complexity of classifying these tumors based on genetic subtype.

Copy Number Aberrations (CNAs)

In about 1,000 human breast tumors with long-term clinical follow-up deletions in the PPP2R2A, MTAP, and MAP2K4 genes and somatic CNAs were found in about 40% of genes, likely as a result of either cis- or trans-acting CNAs. The subtypes without or with less CNAs had a better outcome than those with multiple CNAs [37, 299].

Epigenetic Changes

Various epigenetic alterations, such as miRNA expression, DNA methylation and histone modifications, can affect gene expression.

miRNA Expression

The Cancer Genome Atlas Network identified the top 3,662 variably expressed genes and seven different miRNA subtypes in 525 tumors and 22 tumor-adjacent normal tissues [300]. These miRNA subtypes are correlated with the primary mRNA subtypes. Interestingly, miRNA groups 4 and 5 showed high overlapping with the basal-like mRNA subtype and included many TP53 mutations. The remaining miRNA groups [37, 299, 301-303] were formed by a mixture of luminal A, luminal B, and HER2-enriched subtypes, and poorly correlated with the PAM50-defined subtypes.

DNA Methylation and Histone Modification

The Cancer Genome Atlas Network defined five distinct DNA methylation groups (group 1-5) based on methylation arrays of 802 breast tumors [300]. Group 3 showed a hypermethylation phenotype and significantly corresponded to luminal-B mRNA subtype, while group 5 with the lowest levels of DNA methylation corresponded to basal-like mRNA subtypes.

The methylation status of 20-30% of genes differed in a highly metastatic breast cancer cell line, MDA-MB-468LN compared to that found in a less metastatic cell line, MDA-MB-468GFB [304].

Besides, consistent with a study conducted on 880 human breast carcinomas, changes in histone acetylation and methylation patterns can be an early sign of breast cancer [305]. Moreover, moderate to low levels of lysine acetylation and lysine and arginine methylation occurred in basal carcinomas (triple negative) and HER2 positive subtypes, that is in cancers with worse prognosis [305].

Genome-Wide Association Studies and Effective Therapeutic Genetic Treatment

It is evident that if one considers all different potential germline mutations, somatic mutations, copy number aberrations, and methylation changes in each tumor, it is very hard task to define cancer sub-types based on genetic characteristics and to tailor a single effective therapeutic genetic treatment to an individual patient.

THE CANCER STEM CELL (CSC) THEORY

The cancer stem cell (CSC) theory states that breast cancer, such as many other cancers origin by a subpopulation of cells with stem cell properties. These cells may mediate metastasis and account for treatment resistance as they are resistant to chemotherapy and radiotherapy [306].

In 2003 Al-Hajj *et al.* [307] first reported that a subpopulation within samples taken from primary breast tumors produce palpable tumors in non-obese diabetic severe combined immunodeficient (NOD/SCID) immunocompromised mice. All tumorigenic cells expressed CD44, alone or with ESA (epithelial specific antigen, EpCAM) but did not express CD24.

Some different hypotheses have been formulated to explain the CSCs origin [308]. One theory proposes that "CSCs result from the deregulation of dormant normal stem cell self-renewal and differentiation pathways, resulting in cancer cells with both self-renewal and differentiation capabilities. Evidence for this theory arises from the similarities between normal stem cells and CSCs and that normal stem cells are highly susceptible to mutations and oncogenic transformation due to their long lifespans" [309 - 311]. Al-Hajj *et al.* believed that "breast CSCs(BCSCs) are likely to originate from basal mammary stem/progenitor cells due to the similarities between the cell surface profile of basal cells and the putative BCSCs they discovered" [307].

CSCs may also derive from progenitor cells. According to this hypothesis, through the mutation occurred in progenitor cells during the process of differentiation, they can obtain the ability of self-renewal, and then form cancer stem cells [312].

Another theory suggests that "CSCs develop from differentiated cells through epithelial-mesenchymal transition (EMT). Cells that have undergone EMT are susceptible to transformation and have many characteristics and behaviors similar to those of normal and neoplastic stem cells" [313 - 314]. "CD44+/CD24-/low cells from normal and carcinoma human breast tissues express many genes associated with cells that have undergone EMT. Furthermore, cells that have undergone an EMT formed 10-fold more mammospheres than did corresponding control cells, suggesting an increase in tumorigenicity in EMT-induced cells" [314].

A combination of epidermal growth factor receptor (EGFR) pathway activation and loss of INK4A tumor-suppressor function induces a high-grade glioblastoma multiforme phenotype from differentiated astrocytes [315]. Different types of CSC may derive from different origin and lead to the formation of heterogeneous tumor. Liu *et al.* compared gene signatures between BCSCs and their putative normal mammary cells of origin, showing the possible origin of four kinds of BCSCs. Luminal A CSCs were most relevant to luminal progenitor cell or mature luminal cell; luminal B and HER2 CSCs were most relevant to bipotent-enriched progenitor cell; basal-like CSCs were most relevant to bipotent-enriched progenitor cell or mature luminal cell [316].

Progenitor Cells with Stem Cell-Like Properties as the Major Origin of Breast Cancer and Heterogeneous Breast Cancer Subtypes

In hematopoietic malignancies committed progenitor cells may serve as cells of origin for malignancy [317]. Similar results have been shown for breast cancers [318 - 320] as was demonstrated by Molyneux *et al.* in 2010 [312]. Transformed committed progenitor cells may also de-differentiate and acquire stem cell-like properties. A model for committed progenitors assuming stem cell properties during regeneration/wound-healing responses has been reported for prostate testis and pancreas [321 - 323]. Stem cells could be the cell of origin of cancer as they are self-renewing, long-lived and hence likely able to accumulate mutations over time [301, 324 - 325]. Tumors could arise from committed progenitor cells, with enough residual multipotent potential to give rise tumor cellular heterogeneity, comprised of lineage-committed tumor cells and small proportions of tumor-initiating cells. Stem cells may acquire the first genetic aberration and give rise to progeny that is predisposed to further mutations, triggering the onset of the disease [326].

Following this hypothesis, factors which determine the cell fate decisions of the normal progenitor cell are likely to be key determinants of tumor phenotype, response to therapy and molecular subtype. Thus, transcription factors that specify normal mammary cell fate, such as Elf5 or Gata3, also have an important effect on tumor progression [326 - 330] and therapeutic resistance [327].

Additional Sources of Breast Cancer Heterogeneity

In addition to the cell of origin, many other factors contribute to the molecular tumor heterogeneity [37, 331 - 333]. As already mentioned, transcription factors account for differential signatures [326 - 327, 334-336]. Clonal evolution of tumors could result in tumor heterogeneity [337] also through in loss or in gain mutations that confer survival advantages [324]. In the first (MMTV-wnt-1) of three mouse models of mammary cancer (MMTV-wnt-1, MMTV-neu and P53+/-) it was found that both mammary stem and progenitor cells served at cellular level as the targets of wnt-1-induced tumorigenesis. Moreover, the luminal epithelial marker CD61/beta-integrin defined a highly tumorigenic CSC population compared with the CD61- subset. CD61 expression also defined a CSC subset in 50%of P53-/- derived tumors. Unlike this, no CSC was identified in the MMTV-neu/erbB2 model. These findings suggest the evolution of different CSCs sustaining specific mammary tumors [338].

The tumor cell microenvironment also can account for tumor heterogeneity. In fact, tumor cells directly in contact with blood vessels are different from those in an anaerobic environment far from the vasculature [339].

The suggestion that mammary stem cells are located in a 'suprabasal' niche within the mammary gland [340 - 344], imply that they respond to extracellular cues that regulate their fate [345]. There are examples in multiple tissues suggesting that the niche provides signals to maintain the self-renewal capacity of stem cells and upon exiting of this niche, stem cells undergo differentiation [346]. The existence of a stem cell niche [347], and the effect of hormones and paracrine factors on mammary stem cells during mammary gland development [348 - 351], may similarly affect cancer stem cells resident in breast tumors. Furthermore, tumors contain more than tumor cells; for example, endothelial cells, immune-infiltrated populations and fibroblasts. Each cell type plays a relevant role during the initiation and progression of the disease [352 - 354]. Epigenetics play a strong role during development [355] including mammary development [356], thus they define different lineages and specify epithelial functions. Therefore, it is reasonable to think that epigenetic remodeling is a mechanism to define committed decisions [357 - 359]. Together with genomic instability, the underlying epigenetic mechanisms of normal mammary morphogenesis must be considered as a key factor that contributes to tumor heterogeneity [360]; following this, many types of cancer retain a hierarchical organization.

Breast CSCs

Breast CSCs are present in mesenchymal-like (EMT) and epithelial-like (mesenchymal-epithelial transition (MET)) states. The gene-expression profiles of mesenchymal-like and epithelial-like breast CSCs look like those of basal and luminal stem cells of the normal breast. Breast CSCs, due to their plasticity, can transit between EMT- and MET-like states; moreover, these cells can spread to distant tissues and grow at metastatic sites.

The CD24-/CD44+ and ALDH+ Breast CSC Populations

In human breast cancer, the cell surface marker profile CD24-/CD44+ first permitted to identify tumor-initiating cells [307]. Recently, it has been reported that normal and malignant breast stem cells expressing the enzyme aldehyde dehydrogenase (ALDH), are also enriched for tumor-initiating characteristics [311]. In primary breast xenografts, CD24-/CD44+ and ALDH+ identified similar, but non-identical cell populations, each able to initiate tumors in NOD/SCID mice [311]. Tumor cells that simultaneously expressed both CD24-/CD44+ and ALDH+ showed to be the most tumorigenic. In fact, in NOD/SCID mice they gave rise to tumors from as few as 20 cells [311].

However, it remains unclear whether tumors contain multiple types of CSCs and whether CSC markers identify distinct CSC populations. According to Liu *et al.*, breast CSCs are present in distinct EMT and MET states with different CSC markers. Authors propose that reversible transitions between mesenchymal-like and epithelial-like stem cell states in a process regulated by the tumor microenvironment are necessary for these tumor cells to invade and form metastasis at distant sites.

In primary breast cancer, CD24-/CD44+ cells are located primarily at the tumor-invasive edge adjacent to the tumor stroma, ALDH1+ cells are located more centrally, consistent with studies showing that ALDH+ CSCs are generated by hypoxia [361], in tumor internal zones.

To characterize the breast CSC populations, Liu *et al.* isolated CD24-/CD44+ or ALDH-expressing cell populations from a total of 45 primary human breast cancers and performed gene-expression analyses using Affymetrix microarrays. A large number of genes have reciprocal expression patterns within CD24-/CD44+ and ALDH+ populations.

A set of genes expressed in CD24-/CD44+ and ALDH+ populations displayed overlap with those from tumorsphere populations generated from unsorted cells. This suggests that despite significant differences between cells characterized by CD24-/CD44+ or ALDH expression, both of these populations share characteristics with tumorsphere- forming cells, a property of stemness.

These results suggest that these CSC markers identify two stem cell populations in human breast cancers.

Gene-Expression Patterns of the ALDH+ and CD24-/CD44+ Breast CSCs

When a comparison of expression of EMT- and MET-associated genes in sorted CD24-/CD44+ and ALDH+ tumor cells isolated from primary human breast cancers was performed, it was found that EMT-associated genes, including vimentin, zinc finger E-box-binding homeobox 1 (ZEB1), ZEB2, b-catenin (CTNNB1), and matrix metalloproteinase-1 (MMP9), were significantly enriched in the CD24-/CD44+ populations, and was correspondingly decreased in the ALDH+ population. Conversely, genes associated with the alternative epithelial-like state, such as cadherin, occludin, claudins, and desmoplakin, were elevated in the ALDH+ population and correspondingly diminished in the CD24-/CD44+ cell population. These results were confirmed by quantitative RT-PCR. The proliferation marker Ki67 was preferentially expressed in ALDH+ compared with CD24-/CD44+ cells [311]. These studies suggest that in primary human breast cancers, the CD24-/CD44+markers identify mesenchymal-like CSCs, with high invasive capacity and low proliferative potential, whereas ALDH expression identifies epithelial-like proliferative CSCs.

Breast CSCs Plasticity

The molecular pathways linking stem-cell like characteristics with EMT are by far to be defined. However, many findings show that induction of EMT in differentiated breast epithelial tumor cells are enough to origin cancer stem cells prone to metastasize.

Transition between epithelial-like and mesenchymal-like states is a relevant property mainly regulated by the tumor microenvironment. The cross-talk between cancer cells, the surrounding stroma and infiltrating immune cells occurs through secretion of matrix metalloproteinases (MMPs), extracellular matrix (ECM) elements, growth factors and cytokines. This allows localized tissue remodeling to induce tumor cell motility, EMT and metastasization. The tumor stroma in the large majority is constituted by cancer associated fibroblasts (CAFs). CAFs can originate from resident fibroblasts, epithelial and endothelial cells and also from mesenchymal progenitors coming from bone marrow or adipose tissue through circulation. The origin of CAFs from resident epithelial cells by the EMT process involves the formation of motile cells from non-motile epithelial cells and loss of adhesion marker E-cadherin. IL-6 generated in the microenvironment and several transcription factors TGF-beta driven are implicated in this process and in E-cadherin repression, such as Snail, Slug, dEF1 (ZEB1), S1P1 (ZEB2) and bHLH factors (twist and E12/E42). It has been reported that beta-catenin following an autocrine action of EGFR could be mediator of E-cadherin dysfunction and promote EMT [362]. EMT likely comprehends multi-step cellular changes regulated in differential and sequential ways and CAFs take part of the tumor stroma with supportive functions [363]. EMT in breast cancer cell lines *in vitro* can be induced by added cytokines (TNF-alpha) and growth factors (TGF-beta, PDGF, HGF and FGF) through their cell surface receptors. In this way, extracellular effectors such as RAS and other small GTPases, SRC, NF-kB, SMADs, beta-catenin and integrins are activated. Alterations in constituents of the ECM and in response to ECM remodeling by secreted MMP-5 can also promote EMT. This underlines the relevance of tumor microenvironment in initiating EMT [362]. Endothelial to mesenchymal transition (EndMT) is characterized by the loss of endothelial markers like CD31 and gain of mesenchymal markers like fibroblast specific protein-1 (FSP-1) and SMA [364]. Although the molecular mechanisms of EndMT in tumors is not yet fully understood, this process seems to involve TGFbeta1 and Notch signaling pathways among others [365]. Moreover, an increased expression of Snail and Slug in the vasculature of breast has been found [366], thus suggesting that the same factors can induce EndMT and EMT. Inhibition of TGF-beta signaling and stimulation of BMPs induce the reverted process that is mesenchymal to epithelial transition (MET). Another mechanism of EMT reversal involves expression of GATA-3, a transcription factor regulating T lymphocyte differentiation and maturation. Breast CSC signal transduction pathways involving Wnt and NFkB induce EMT, whereas HER2 induces MET [367].

In addition, microRNA circuits within breast CSCs may further regulate EMT/MET transitions by coordinately regulating multiple pathways [361, 368]. Overall, these findings support a model in which mesenchymal-like breast CSCs mediate tumor invasion into the blood, where they survive due to anoikis resistance and form micrometastasis in distant organs. The tumor microenvironment at these sites induces a MET that is required for breast CSC self-renewal. This generates additional breast CSCs, as well as differentiated cells that recapitulate the cellular hierarchy at metastatic sites. The plasticity of CSCs that allows them to transition between mesenchymal- like and epithelial-like states may play a critical role in the ability of these cells to form metastasis at distant sites. Many recent studies showing the importance of cellular plasticity in metastasis support such a model [369]. This is confirmed by the observations that in patients with metastatic breast cancer, circulating tumor cells and bone marrow micrometastasis are highly enriched for cells expressing the EMT breast CSC marker CD24-/CD44+ [370]. Furthermore, these cells do not express the proliferation marker Ki67, consistent with their quiescence. Conversely, macrometastasis express the stem cell markers ALDH and CD24-/CD44+ in proportions similar to those found in primary tumors. It is currently thought that targeting CSC markers allows to decrease cancer recurrences and overcome cancer drug resistance, and that this ultimately will improve cancer treatment [371 - 373].

MOLECULAR PROFILING IN BREAST CANCER

Gene expression microarray studies permit to distinguish between different molecular tumor classes based on contemporaneous expression analyses of thousands of genes. Perou *et al.* [331] using this technique and a subset cluster analysis, could define a dendogram with two main groups that clinically corresponded to ER-positive and ER-negative. The former, called luminal group, had relatively high expression of many ER-responsive genes, luminal cytokeratins and other luminal associated markers. The latter, called ER-negative group, was further divided into basal-like, ErbB2-positive and normal-like subclasses (Figs. **1**, **2**).

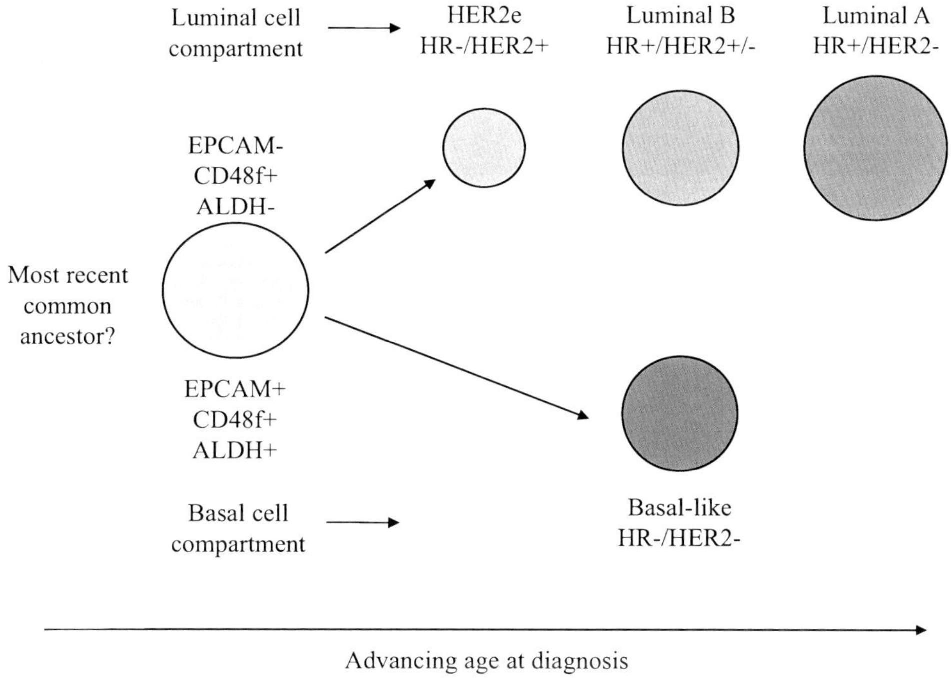

Fig. (1). Breast cancer may be viewed as consisting of two main cell types of origin, *i.e.* luminal *vs.* basal/myoepithelial. HER2-enriched (HER2e) and luminal intrinsic molecular subtypes are initiated from the luminal cell compartment, while basal-like breast cancers (ER-/HER2-) are initiated within the basal/myoepithelial compartment.

Luminal Subtype

About 75% of ER+ and/or PR+ breast cancers are included in this subtype. Currently, most studies support the concept of two luminal-like subclasses, luminal A and luminal B, as suggested by the work of Hu *et al.* [333, 374 - 378].

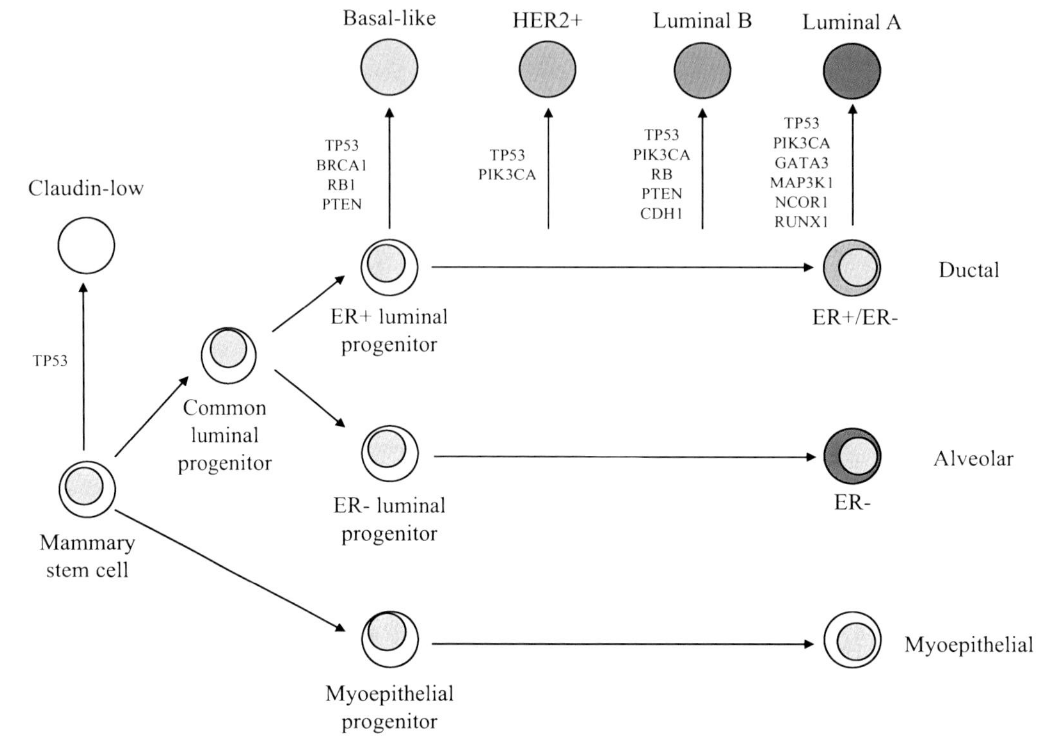

Fig. (2). The five major breast cancer types are linked to a normal epithelial counterpart based on gene expression profiling. The HER2+ subtype could originate through the amplification of the HER2 locus in a luminal target cell that is either ER+ or ER-. Commonly mutated genes in the different subtypes also are shown.

Luminal-A

These tumors include 50%-60% of all breast cancers. Usually they are HER2 negative, have low histological grade, low degree of nuclear pleomorphism, and lower levels of proliferation related genes, such as Ki67 (proliferating cell nuclear antigen) index by immunohistochemistry and include special histological types with good prognosis (*i.e.*, tubular, invasive cribriform, mucinous and lobular) [379]. Bone is the commonest site of recurrence, whereas liver, lung and central nervous system metastatic involvement occurs in less than 10% of patients. Hormonal therapy is the main treatment of these patients [380 - 381].

Luminal-B

Luminal-B tumors comprehend 15%-20% of breast cancers having a more aggressive phenotype, higher histological grade, proliferative index and a worse prognosis [382]. In spite of the different methodological problems and those potentially deriving from tumor heterogeneity [383], Cheang *et al.* [384] suggested the Ki67 index with cut-off point 14% as a potential proliferation marker to distinguish luminal-B tumors from luminal-A. Immunohistochemically, luminal-B subtype can be ER-positive, HER2-negative and Ki 67 high or ER and HER-2 positive, but up to 6% of them are negative for both ER and HER2 [385]. Overall survival in untreated luminal-B breast cancers is as in the basal-like and HER2-positive subtypes which are commonly considered as high-risk tumors [374]. Luminal-B tumors respond less than luminal A subtype to endocrine therapy and less than HER2 positive and basal-like subtypes to preoperative chemotherapy with anthracyclines or taxanes [386 - 389].

HER2-Positive

Human epidermal growth factor receptor-2 is a member of the family of four membrane tyrosine kinases. The phosphorylated tyrosine residues activate downstream pathways and crosstalk with other membrane signaling pathways. Many genes implicated in cell proliferation, survival, differentiation, angiogenesis, invasion and metastasis are governed by transcription factors stimulated by the pathway [390 - 393]. About 15-20% of breast cancers are HER2 positive. These tumors have poor prognosis, are highly proliferative, 75% have a high histological and nuclear grade and more than 40% have p53 mutations [394]. About half HER2+ breast cancers are ER+, but commonly with lower ER expression. These tumors have increased sensitivity to doxorubicin, relative resistance to endocrine therapy and tend to spread to the brain and visceral organs [395 - 396].

Triple Negative Breast Cancer (TNBC)

The basal-like subtype ranges from 8% to 37% of all breast cancers [397]. Basal-like cancers have high histological and nuclear grade, conspicuous lymphocytic infiltrate and medullary features with high mitotic and proliferative indices. Most of them are infiltrating ductal tumors with solid growth pattern, clinically aggressive and with high tendency to metastasize to the brain and lung [398]. Moreover, they express high levels of basal myoepithelial markers, such as CK5, CK 14, CK 17 and laminin, and do not express ER, PR and HER2, therefore also called triple-negative. However, the terms triple-negative and basal-like are not synonymous and there is approximately 20%-30% discordance when different studies are considered [399].

Recently, basing on gene expression profiles of 587 TNBC cases, six different subtypes termed basal-like1 (BL1), basal-like2 (BL2), immunomodulatory (IM), mesenchymal (M), mesenchymalstem-like (MSL), and luminal androgen receptor (LAR) were defined (Fig. **3**) [400]. Microarray and immunohistochemical analyses found that basal-like subtype is about three quarters of BRCA1-related breast cancers, and BRCA1 carriers often express triple-negative phenotype [400].

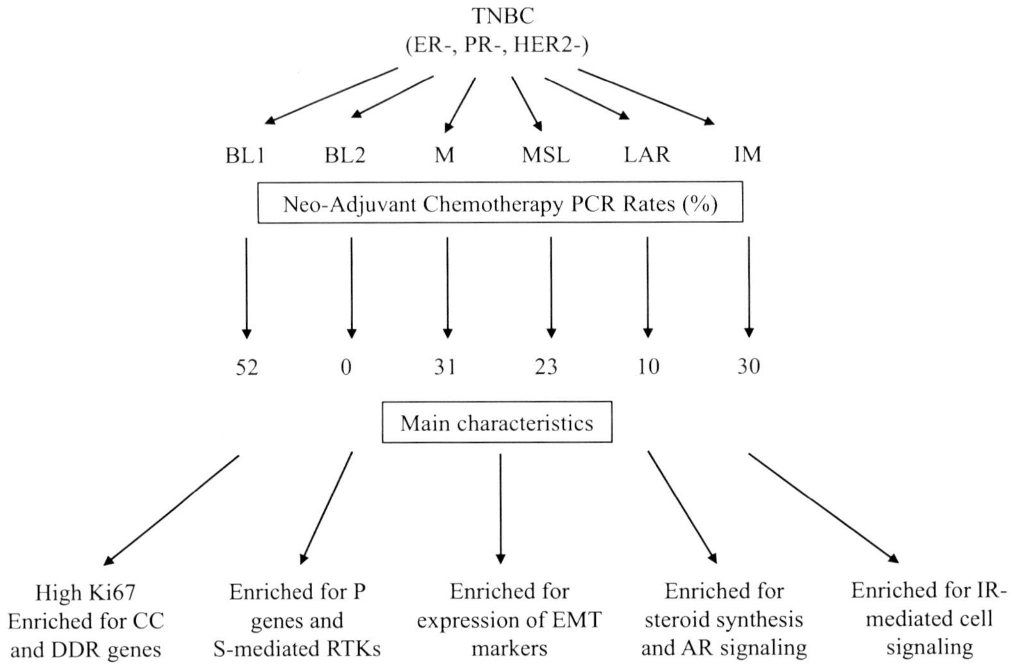

Fig. (3). Common current subclassification of triple negative breast cancer (TNBC) based on meta-analysis of gene expression profiling data from 21 breast cancer dataset (Ref n. 631). TNBC: triple negative breast cancer; BL1: basal-like 1; BL2: basal-like 2; M: mesenchymal-like; MSL: mesenchymal/stem-like; LAR: luminal androgen receptor expression; IM: immunomodulatory; PCR: pathological complete response; CC: cell cycle; DDR: DNA damage response; P: proliferation; S: survival; RTK: receptor tyrosine kinase; EMT: epithelial to mesenchymal transition; AR: androgen receptor; IR: immune response (also see text).

Normal Breast-Like

These tumors comprehend about 5%-10% of all breast carcinomas, and have a prognosis intermediate between luminal and basal-like cancers. They are ER, PR and HER2 negative, therefore they as well can be classified as triple-negative. However, it has been reported that they could be a technical artifact from high contamination with normal tissue during the microarrays [401].

STANDARD THERAPY FOR BREAST CANCER

The main scientific societies, as National Comprehensive Cancer Network (NCCN) and European Society for Medical Oncology (ESMO) elaborate and regularly update guidelines for breast cancer treatment [402 - 403]. Here the current "state of the art" in the different settings is summarized.

Local Treatment

Radiotherapy and Surgery

Lumpectomy followed or not by whole breast irradiation or total mastectomy with or without sentinel node biopsy are the main options in DCIS and LCIS treatment. In invasive breast cancer, lumpectomy and total mastectomy are the principal surgical options. In the first case, whole breast irradiation is mandatory; axillary, supraclavicular, infraclavicular and internal mammary regions irradiation must be added in N+ tumors. In the second case, radiation therapy to chest wall, supracalvicular, infraclavicular and internal mammary nodes must be performed if tumor is greater than 5 cm or margins are positive, or in N+ tumors (>3 lymph-nodes involved).

In tumors with clinically positive axillary lymph-nodes, axillary level I and II dissection must be performed; otherwise, sentinel node mapping and excision is the usual procedure, followed by axillary dissection if sentinel node is positive or not identified. Recent guidelines, however, permit to avoid axillary dissection in the case of T1/T2 tumors with 1 or 2 positive sentinel lymph-nodes, breast conserving surgery, whole breast radiotherapy planned, no prior neoadjuvant therapy [402].

Systemic Treatment

Most breast cancer patients, in addition to local radio and/or surgical therapy need systemic medical treatment. This systemic treatment is administered as neoadjuvant, adjuvant or salvage therapy.

Neoadjuvant Therapy

This therapy is indicated to allow a potentially radical resection or a breast conserving surgery; pathologic complete response (pCR) as surrogate marker of OS is one additional more recent aim to be validated. Locally advanced or biologically aggressive disease is the main indication for neoadjuvant therapy in breast cancer. Therefore, the clinical staging IIIA or IIIB with tumors involving the chest wall, skin or both, or inflammatory breast cancer, are good candidates for neoadjuvant therapy [404]. Also, it must be considered for women with clinical stage IIA and IIB tumors who wish to have breast-conserving surgery. A few different studies in the early 2000s have reported that neoadjuvant chemotherapy successfully decreased local and regional recurrences even in large T3 and T4 tumors [405 - 406].

The significant favorable value of pCR following neoadjuvant chemotherapy on the clinical outcome was first reported in NSABBP-B18 and NSABBP-B27 trials [407 - 408] and successively confirmed in cohort studies and meta-analyses [409 - 410]. In 2012, pertuzumab for the first time got an accelerated approval in the neoadjuvant setting for HER2 positive breast cancer patients due to a significant improvement in pCR. In spite of this, a formal validation of pCR as a surrogate marker of survival in breast cancer is still lacking because of emerging of some limitations in different studies. The main of them is a methodological problem that is the heterogeneity of patient populations (different molecular subtype and stage), the different pCR definitions adopted, the different therapeutic regimens administered and whether additional therapies after surgery are given or not. It is recommended all these aspects should be taken into account in trying to validate the surrogacy of pCR in a meta-analysis of published studies. Accordingly, pCR was found to be associated with improved DFS in luminal B/HER2 negative, HER2 positive/non luminal and triple negative breast cancers, but not in luminal A or luminal B/HER2 positive ones [411] and in some trials where intensified/dose dense chemotherapy was compared with standard dose regimens [412]. Moreover, for an individual patient, the delay of operation could be potentially harmful, as some tumor cells might become resistant during neoadjuvant treatment and metastasize [413]. Besides, the EORTC10902 trial randomized 698 early stage breast cancer patients. PFS, OS, or local recurrence rate were not different when comparing the same regimen given pre-operatively and post-operatively [414]. In a meta-analysis, no differences with regard to death, disease progression or distant disease recurrence were found between neoadjuvant and adjuvant treatment. However, "the rate of local recurrence was higher in the neoadjuvant group; this was mainly in trials where surgery was avoided in cases of clinical complete response" [415].

The main cytotoxic regimens evaluated in neoadjuvant settings include AC followed by docetaxel or paclitaxel, epirubicin/paclitaxel-CMF, and a dose-dense sequence of epirubicin and paclitaxel. Overall, pCR is between 15% and 20% [407 - 408, 416].

HR Positive, HER2 Negative Breast Cancer

"Despite the lower response rate of hormone receptor positive breast cancer patients to chemotherapy, compared to patients with other subtypes of cancer, chemotherapy is still the main neoadjuvant systemic therapy that has been widely studied and used" [417 - 418].

"A broad use of endocrine therapy in neoadjuvant setting is limited due to the slow response rate of tumors, requiring long duration of therapy and risking the benefit of early surgical intervention" [419]. Unlike in premenopausal patients, where there are not many available data on the use of neoadjuvant endocrine treatment, in postmenopausal women the clinical response rate was significantly higher in patients receiving AIs than in those treated with tamoxifen; overall, the pCR was less than 10% [420]. Commonly, the patient's wish and the clinical scenario are used to select who among HR positive, HER2 negative breast cancer patients must receive neoadjuvant chemotherapy.

HR Negative HER2 Positive Patients

Two phase II trials, Neosphere and Tryphaena, tested the role of trastuzumab [421 - 422] and showed a significant improvement in pCR achievement. "The addition of trastuzumab to chemotherapy produced different outcomes according to HR status, with pCR rates 1.5 and 2-fold lower in ER positive than in ER negative cases" [423]. Moreover, data from Neo-Sphere, NeoALTTO, GeparQuinto, ACOSOG Z1041, CALGB 40601, NOAH and TBCRC006 trials further confirmed these findings [421, 424 - 441].

The GeparQuinto trial showed the superiority, in terms of pCRs achieved, of trastuzumab plus chemotherapy compared with lapatinib plus chemotherapy [425]. The neo-ALTTO trial [424], the CHER-LOB study [430], the NSABP B-41 trial [431] showed a significant improvement by using double targeting therapy with lapatinib and trastuzumab to achieve breast conserving surgery and/or pCR.

Pertuzumab is a recombinant humanized monoclonal antibody inhibiting the ligand-dependent dimerization of HER2 and its downstream signaling. Pertuzumab and trastuzumab bind different epitopes of HER2 receptor, therefore they have a complementary mechanism of action. The US FDA approved the use of pertuzumab associated with trastuzumab and docetaxel as primary therapy for HER2 positive breast cancers [432].

HR Positive HER2 Positive Patients

In a pooled analysis of the German neoadjuvant studies it was found that pCR may be a suitable surrogate endpoint for HER-2 positive/HR negative but not for HER-2 positive/HR positive breast cancer patients [433]. Thus, patients with HER-2+ HR+ tumors not showing a pCR do not necessarily have a poor prognosis, as these patients can take benefit from adjuvant hormonal treatment which can improve clinical outcome. It could be reasonable to submit these patients to hormonal therapy combined with HER-2 blocking agents, thus avoiding chemotherapy-related toxicity [433 - 434]. The HER-2 dual blockade without concomitant chemotherapy was evaluated in two phase II neoadjuvant trials [421, 429]. In TBCRC006, "49% of pCR/downstaging was reported with trastuzumab–lapatinib without chemotherapy, 54% in patients with HR positive tumors receiving also letrozole, while this percentage was 40% in HR negative group. Moreover, the efficacy of trastuzumab and pertuzumab combination, without chemotherapy, was evaluated in one arm of the NeoSphere trial, reporting 17% of pCR, but in the subset of patients with HR positive tumors, not receiving any endocrine treatment, this percentage was only 6%. Alterations of the PI3K and ER pathway genes have been correlated with poor outcome in terms of RFS and lower pathological response rate in patients with HR positive tumors (while not in HR negative tumors) receiving neoadjuvant chemotherapy and trastuzumab, suggesting that cross-talks between the two pathways are bidirectional and may influence chemotherapy and trastuzumab efficacy" [435].

Triple Negative Breast Cancer Patients

The role of neoadjuvant therapy in TNBC subtype cancers is somewhat controversial, because not always higher pCR rates correspond to a more prolonged PFS and/or OS [407, 436].

Adjuvant Treatment

In the 2011 and 2013 St. Gallen International Breast Cancer Conferences panel members agreed that the different subtypes can be defined only by genetic array testing, although immunohistochemistry can be an acceptable approximation [437 - 438].

The strongest prognostic factors for disease recurrence and death are patient age, comorbidity, tumor size, tumor grade, number of involved lymph-nodes and HER2 status. A validated model [439] is available to estimate 10-year DFS and OS [440]. Recently, biologic characteristics of tumors have gained importance in adjuvant treatment decision making. Oncotype DX, PAM50, Mammaprint are comprehensive genomic profiling tools used for prognosis. They provide sub-typing of breast cancers and can predict their response to adjuvant therapy [441 - 446].

HR Positive Breast Cancer

In ER-positive disease breast cancer, current guidelines recommend to consider chemotherapy in a) node positive disease and b) node negative disease with tumor > 10 mm or tumor > 5 mm and HER2 positive or high grade or unfavorable features [402].

In 2001, the Early Breast Cancer Trialists' Collaborative Group (EBCTCG) reported the collective data of the randomized trials in early breast cancer adjuvant systemic chemotherapy from 1985 to 2000. This report showed the long-term benefits of adjuvant endocrine therapy with 57% (age < 50 years) and 45% (age 50-69 years) reduction of overall mortality in 15 years in hormone receptor positive breast cancer patients receiving post-operative adjuvant chemotherapy followed by 5 years tamoxifen [447]. Subsequently, after the initial report, EBCTCG reported the 10-year follow-up results concerning differences in efficacy between different chemotherapy regimens; taxane plus anthracycline based or high cumulative dosage anthracycline based regimens had an absolute gain of about 33% in breast cancer mortality [448].

NSABP B-16 trial compared adjuvant AC (doxorubicin-cyclophosphamide) given for 4 cycles plus tamoxifen *versus* tamoxifen alone. A 15% proportional reduction in average annual hazard, relapse, and death at 10 years, and a 25% relative risk reduction were observed compared to tamoxifen only [449].

HR Negative Breast Cancer

CMF is an old chemotherapy regimen first introduced by Bonadonna *et al.* [450]. "Based on an indirect comparison, the degree of risk reduction in overall breast cancer related morbidity and mortality of anthracycline based regimen was much greater than that of CMF, which was 10%" [451]. An EBCTCG meta-analysis found "a reduction of 6.2% in absolute breast cancer related mortality at 10 years follow up, when using CMF adjuvant therapy compared to no adjuvant chemotherapy" [452]. "A meta-analytical comparison of CMF *versus* anthracycline-based chemotherapy regimen in the adjuvant setting showed that the standard 4 cycles of AC and the standard CMF were equivalent (RR 0.98, SE 0.05, 2p=0.67), but anthracycline-based regimens with substantially higher cumulative dosage than the initial standard regimen of 4 AC cycles (*e.g.*, CAF or CEF) were significantly superior to the standard CMF" [448].

The second important step in breast cancer adjuvant chemotherapy was the introduction of taxanes. The BIG 02-98 trial [453] and the BCIRG001 study [454] showed superiority of docetaxel over doxorubicin. The Intergroup 9344 (INT 9344) trial [455] and the NSABP B-28 study compared 4 cycles of AC *versus* 4 AC plus 4 paclitaxel (T). Both trials showed that "adding T resulted in relative DFS improvement of 17%, but lesser degree of improvement in OS (7%)" [456]. Based on the result from the EBCTCG meta-analysis, "the overall absolute reduction of recurrence by additional taxane to the anthracycline regimen was 2.8% and the reduction of mortality with recurrence was 1.3%. However, this improvement was diluted when a very well dosed anthracycline-based regimen was used. The improvement in clinical outcome was sustained over a period of 5 years" [457]. The CALGB 9741 study compared 4 arms with different dosing schedules of AC-T [457]. SWOG S0221 compared AC+G (filgrastim) *versus* dose dense AC in different combinations with dose dense paclitaxel for 6 cycles or weekly paclitaxel for 3 months [458]. GEICAM/2003-02 study compared FAC with FAC followed by weekly paclitaxel for 8 weeks in node negative high-risk patients. Additional paclitaxel improved progression-free survival (PFS) at 63.3-months follow-up [459].

A trial compared the combination cyclophosphamide-docetaxel (TC) with AC in stage I to III breast cancer. At a median follow up of 7 years, overall DFS and OS were significantly improved with TC [460]. Based on current guidelines, preferred regimens include AC followed by paclitaxel and the TC combination. In general, anthracycline-containing regimens are preferred in node positive patients [402].

Adjuvant Endocrine Therapy

Three main categories of drugs are used in adjuvant setting: selective estrogen receptor modulators (SERMs), aromatase inhibitors (AIs) and luteinizing hormone releasing hormone (LHRH) agonists.

The main drug of SERMs is tamoxifen, which competitively binds to estrogen receptors, interfere with DNA synthesis and inhibit G0-G1 cell cycle progression [461]. AIs (exemestane, anastrozole and letrozole) inhibit conversion at peripheral tissues of androstenedione and other hormones to estradiol (E2), that is the main source of estrogen in postmenopausal women [462]. LHRH agonists inhibit ovaries which in premenopausal patients are the main source of E2.

Premenopausals

In premenopausal women, tamoxifen is the drug of choice; "adjuvant tamoxifen decreases the annual odds of recurrence by 39% and the annual odds of death by 31%, irrespective of the use of chemotherapy, age, menopausal status and nodal status" [463]. Tamoxifen for 5 years has been the standard of care for years. After 5 years of tamoxifen, for women postmenopausal at that time, current guidelines recommend to consider an AI for further 5 years [464 - 465]. In the ATLAS trial 80,000 women were randomized to continue tamoxifen for 10 years *versus* stopping at 5 years. The 10 years treatment arm had a 4% improvement in breast cancer related mortality [466]. Basing upon the result of this study, for patients who are premenopausal after 5 years of tamoxifen, it is recommended to consider tamoxifen for additional 5 years [402]. During endocrine therapy, a patient's menstrual status can change from premenopausal to a menopausal state. Therefore, the use of different endocrine adjuvant therapies in sequence could be considered accordingly.

The role of ovarian ablation or suppression with LHRH agonists in premenopausal hormone receptor positive breast cancer patients is still partially definite [467 - 468]. Two meta-analyses did not show a significant reduction in recurrence or death [469 - 470].

Recent data support the use of exemestane plus an LHRH agonist in premenopausal women. In two phase III trials this combination significantly reduced recurrences as compared to tamoxifen plus LHRH agonist [471].

Postmenopausal

AIs are the treatment of choice, based on the improved efficacy compared to tamoxifen. The ATAC trial demonstrated that anastrozole is superior to tamoxifen with prolonged DFS, significantly reduced distant metastases and contralateral breast cancers [472]. The MA-17 trial evaluated additional 5 years of letrozole or placebo in post-menopausal patients who had completed 5 years of adjuvant tamoxifen [473]. In the BIG-98 trial 3 groups were enrolled: one receiving letrozole for 5 years, another receiving tamoxifen for 5 years, and the last one receiving both treatments in sequence. In both trials, the letrozole arms gave better results [474]. Several trials (ITA, IES, ABCSG8, ARNO95) have studied the use of tamoxifen for 2-3 years followed by an AI, all demonstrating a superiority of the sequential treatment *versus* tamoxifen alone for 5 years [475 - 477]. Other studies (MA-17, ABCSG-6) evaluated an extended therapy (tamoxifen for 5 years followed by an AI) and showed a significant superiority for the extended therapy in comparison with 5 years of tamoxifen alone [478 - 480]. Basing on these results, AI is the gold standard adjuvant endocrine therapy for post-menopausal women, and current guidelines recommend the following options: a) an AI for 5 years; b) tamoxifen for 2-3 years followed by an AI for 2-3 years to complete 5 years of therapy; c) tamoxifen for 2-3 years followed by an AI for 5 years; d) tamoxifen for 4-6 years followed by an AI for 5 years; e) for patients with contraindications to AIs, tamoxifen for 5-10 years.

Adjuvant Treatment in HER2 Positive Breast Cancer

HER2+ HR+ and HER2- HR- tumors are two different subtypes, showing different prognosis in absence of anti HER-2 therapy [331, 333].

In HER2+ HR+ tumor, recurrences occur continuously even after more than 10–15 years of follow up. Conversely, HER2- HR- tumors relapse usually within the first 5 years. As to the sites of relapse, bone and soft tissue are more common in ER+ disease, while visceral involvement is more often observed in ER- patients. Also, sensitivity to chemotherapy is quite different between the two subsets [380, 463, 481].

Moreover, emerging data suggest "the involvement of the bidirectional cross-talk between the HER-2 and HR pathways not only in endocrine resistance but also in resistance to HER-2 blocking agents" [482].

HER2 Positive HR Negative Patients

Trastuzumab is a monoclonal antibody that binds to the juxtamembrane domain of HER2 receptor [483]. It induces the activity of p21 or p27, which then cause transcription inhibition and promotes antibody dependent cell – mediated cytotoxicity (ADCC) [484].

The BCIRG 006 trial enrolled early stage HER2+ breast cancer patients and compared three arms: AC-T (adriamycin, cyclophosphamide, and paclitaxel), AC-TH (adriamycin, cyclophosphamide, paclitaxel and trastuzumab), and TCH (docetaxel, cyclophosphamide, and trastuzumab). After 65 months as median follow-up, DFS was 75%, 84%, and 81% in the 3 arms respectively [485]. NCCTG (North Central Cancer Treatment Group) N9831, NSABP (National Surgical Adjuvant Breast and Bowel Project) B-31, FinHER (Finland Herceptin), HERA, NOAH (Neo-adjuvant Herceptin), FNCLCC-PACS (Federation Nationale des Centres de Lutte Contre le Cancer-Programmes d'Actions Concertees Sein) 04, BCIRG (Breast Cancer International Research Group) 006 trials reported that trastuzumab plus chemotherapy gave better clinical outcome than conventional cytotoxic adjuvant therapy. These trials included HER2+/HR+ patients, with up 51% to 72.2% of ER+ patients [486 - 490]. The HERA trial compared 2 years *versus* 1 year of adjuvant trastuzumab for patients with HER2+ breast cancer. No difference in PFS was shown between two groups [487, 491], therefore, one year of adjuvant trastuzumab is currently the standard of care for early stage HER2+ breast cancers. Also in N- small tumors a DFS advantage was observed [492].

Lapatinib inhibits HER2 and HER1 (=EGFR). It binds to the intracellular domain of HER2 protein, thus the efficacy is preserved for the truncated form of HER2 protein.

The results of the ongoing ALTTO trial will provide data on dual HER2 blockade with lapatinib plus trastuzumab; another ongoing study is evaluating pertuzumab and trastuzumab combination with standard chemotherapy regimens [493].

HER2 Positive HR Positive Patients

HR+/HER2+ tumors have lower grade and more common bone involvement; HER2 affects HR positive disease by higher grade, younger age and increased frequency of visceral metastases. Also, triple positive breast cancers occur more often in younger patients [494].

HR positivity also suggests lower sensitivity to anti-HER-2 therapies [495 - 496]. In fact, cross-talk between HER2 and HR pathways can account for resistance to trastuzumab and lapatinib [497].

A retrospective study conducted in patients with HER2+HR+ breast cancers who had received adjuvant chemotherapy followed or not by hormonal therapy and trastuzumab reported "higher DFS and OS in patients who had received chemotherapy, trastuzumab and endocrine therapy than those observed in patients treated with chemotherapy and trastuzumab without subsequent endocrine therapy. At multivariate analysis, administration of endocrine therapy in addition to chemotherapy and trastuzumab resulted the only independent prognostic factor for DFS, with a trend in OS, confirming that endocrine treatment confers benefit when added to chemotherapy and trastuzumab in patients with HR positive/HER-2 positive early breast cancer" [498].

Triple Negative Breast Cancer

A large meta-analysis by EBCTCG demonstrated benefit of adjuvant chemotherapy in TNBC [499]. So, chemotherapy is the mainstay of adjuvant therapy for TNBC, and these patients respond well to both anthracycline or taxane-based regimens [387, 500]. A retrospective analysis in 6,444 patients enrolled in three large CALGB trials confirmed the benefit of adjuvant chemotherapy, with 55% relative reduction and 28% absolute reduction of recurrent risk [457].

Systemic Therapy in Metastatic Disease

Patients must be initially stratified according to whether bone metastases are present, then by hormone receptor status and HER2 expression. In patients with bone disease, bisphosphonates or denosumab therapy must be considered.

Cytotoxic Therapy

Women with symptomatic disease, life-threatening visceral recurrence, or refractory to endocrine treatments, should receive chemotherapy. Combination therapy generally provides higher rates of objective response, but is associated with increased toxicity and a little survival benefit. Furthermore, sequential single agents have less toxicity [501-502].

Active drugs in metastatic breast cancer include anthracyclines (doxorubicin, epirubicin, liposomal doxorubicin), taxanes (paclitaxel, docetaxel nab-paclitaxel), anti-metabolite (capecitabine, gemcitabine), vinorelbine, eribulin, cyclophosph-amide, carboplatin, cisplatin, ixabepilone [500]. The regimens are those just discussed in the adjuvant setting. Regimens including anthracyclines and/or taxanes are more common as first and second line, while vinorelbine-capecitabine and eribulin are following available lines.

Endocrine Therapy

Patients with ER and/or PR positive disease are candidates for endocrine therapy; this must be chosen according to menopausal status and previous treatments. A trial with endocrine therapy is recommended also in ER/PR negative patients, especially if disease is localized to bone or soft tissue or with visceral asymptomatic disease, irrespective of HER2 status [402].

In premenopausal women who are within 1 year of adjuvant antiestrogen exposure, the preferred therapy is ovarian suppression plus AIs. In premenopausal patients without previous exposure to an antiestrogen, initial treatment is tamoxifen alone or in combination with ovarian suppression [503].

For postmenopausal women, treatment options include AIs, SERMs (tamoxifen), ER downregulators (fulvestrant), progestins. As first line therapy, AIs appear to have slightly better outcome compared with other endocrine therapies, although modest [504-506]. Fulvestrant has also demonstrated an efficacy similar to anastrozole in patients progressed on tamoxifen treatment [507-508]. Efficacies of fulvestrant (given at dose 250 mg monthly) and exemestane after progression on non-steroidal AIs are comparable [509]. Fulvestrant high dose (500 mg monthly) has shown a better PFS and OS as compared with low dose [510].

Activation of the mammalian target of rapamycin (mTOR) signaling transduction pathway is a mechanism of resistance to endocrine therapy. In a phase II study, in women with ER positive HER2 negative metastatic breast cancer, the tamoxifen plus everolimus (an mTOR inhibitor) combination was superior to tamoxifen alone in terms of clinical benefit and time to progression (TTP) [511]. In the BOLERO-2 trial, conducted in patients with ER positive HER2 negative breast cancer progressed on non-steroidal AI therapy, the combination of exemestane plus everolimus significantly prolonged PFS *versus* exemestane plus placebo [512-513].

HER2 Targeted Therapy

Anti HER-2 agents are effective in patients with HER-2-positive disease, independently on HR status [514]. Nevertheless, there were some limitations in the analysis, "such as the simplistic definition of HR status (positive *vs*. negative), which does not take into account the degree of HR expression, and the lack of centralized evaluation" [515-516]. Recent data seem to confirm the hypothesis that "some heterogeneity exists among HER-2 positive subsets, mostly related to HR expression" [517].

HER2 Positive HR Negative Breast Cancer

Patients with HER2 positive metastatic breast cancer may benefit from HER2 targeted therapy. In HER2 positive HR negative patients, trastuzumab in combination with chemotherapy (taxanes, carboplatin, vinorelbin, capecitabine) has been the principal option in first line therapy for several years [468, 518]. In a phase III study, the efficacy of pertuzumab plus trastuzumab and docetaxel was compared with trastuzumab-docetaxel combination as first line treatment. There was a significant improvement in PFS and OS with a 34% reduction in the risk of death [519-520]. So, this combination is recommended as preferred option in this setting [402]. Continuation of HER2 blockade for patients progressing on first line trastuzumab containing regimen is recommended [521]. Currently, the association trastuzumab + pertuzumab is under evaluation in combination with several drugs in metastatic patients and in neo-adjuvant setting.

Antibody-drug conjugates (ADCs) represent a novel perspective in breast cancer treatment. Ado-trastuzumab emtansine (T-DM1) is an antibody-drug conjugate between trastuzumab and the microtubule inhibitory agent DM1 (a maytansine derivative). In patients previously treated with trastuzumab and a taxane, T-DM1 therapy resulted in a significant improvement in PFS and OS as compared with lapatinib-capecitabine combination [522]. So, T-DM1 is recommended as preferred option for HER2 positive metastatic breast cancer patients who have previously received a trastuzumab-based regimen [402]. Other regimen in this setting include pertuzumab plus trastuzumab combination with or without a cytotoxic agent (vinorelbine or a taxane) [523-524] and capecitabine plus lapatinib [525-526].

HER2 Positive HR Positive Breast Cancer

HER2+/HR+ patients benefit from endocrine therapy less than HER2-/HR+ patients [527-529]. Experimental studies point out that cross-talks between HER2 and ER pathways may be involved in resistance to endocrine therapy [530-531]. Trastuzumab in association with tamoxifen or fulvestrant may restore tumor sensitivity to these hormonal agents [532-533]. Therefore, concomitant inhibition of both HER2 and ER pathways is thought to be more effective than ER inhibition alone.

Current guidelines on systemic therapy for HER-2 positive/HR positive advanced breast cancer patients consider "endocrine treatment with either trastuzumab or lapatinib or endocrine therapy alone as an acceptable first-line treatment.

Endocrine therapy alone is included as an option because the trials of endocrine therapy with or without HER2-targeted therapy did not demonstrate a survival advantage" [534-536]. Nevertheless, prospective observational study on patients with HER-2 positive advanced breast cancer, including 530 patients with HER-2 positive/HR positive tumors, showed that, "with or without chemotherapy, outcomes were more favorable for the HR positive subset, since dual targeting of HR and HER-2 pathways was associated with more prolonged PFS and OS compared with HER-2 based therapy alone" [537].

HER2 Positive ER Positive PR Positive/Negative Breast Cancer

Tumors expressing both ER and PR have a more favorable prognosis [517, 538-539]. ER positive/PR negative tumors are a different breast cancer subset, more aggressive and with poor outcome [539] and higher levels of EGFR and HER-2 [517]. On this basis, absence of PR expression is considered an unfavorable prognostic factor even in the HER2+ patients [540]. A California Cancer Registry trial showed that the survival of ER+/PR+ /HER2+ subtype was higher than that of ER+/PR-/HER2+ subtype in all stages [541].

Angiogenesis Inhibitors

Bevacizumab, a humanized monoclonal antibody against the vascular endothelial growth factor (VEGF), has been evaluated in a series of trials. When bevacizumab was added to paclitaxel as first line therapy, a superiority in terms of PFS was observed [542]; similar results were obtained adding bevacizumab to docetaxel [543], capecitabine, anthracyclines [544-545]. No improvement in OS or in quality of life was observed [546]. Bevacizumab is currently used in first line chemotherapy, in association with taxanes.

NEW EXPERIMENTAL ANTICANCER AGENTS IN THE DIFFERENT CLINICAL SETTINGS: RATIONALE, MECHANISM OF ACTION, EXPERIMENTAL AND CLINICAL DATA

PI3K/Akt/mTOR Inhibition

The PI3K/Akt/mTOR pathway regulates multiple crucial biological processes as cell growth and proliferation, apoptosis, cell-cycle, DNA repair, angiogenesis and glucose metabolism [547-548].

Growth factors, as EGF bind to the tyrosine kinase receptor, with subsequent PI3K phosphorylation and pathway activation. The tumor suppressor PTEN gene inhibits PI3K, so hindering the Akt activation. In breast cancer, PI3K/Akt/mTOR pathway is often activated, due either to somatic mutations in *PTEN, PI3CA* or

AKT genes or amplification/mutation of the tyrosine kinase receptor [549]. "Amplification of upstream receptors such as HER2, loss of negative regulators such as PTEN, amplification of downstream targets such as protein kinase B (PKB or Akt) and activating mutations or genetic amplification of the alpha catalytic subunit of PI3K have all been described in breast cancer" [550].

The activation of this pathway has been highlighted as frequently responsible of resistance to endocrine therapy [548]. In fact, in HR positive breast cancer there is also a crosstalk between ER and PI3K signal transduction, that is partially responsible for the development of hormone resistance [550-551]. Therefore, there are ongoing studies combining endocrine therapy with selective PI3K or dual PI3K/mTOR inhibitors (NCT01610284.) and evaluating whether dual inhibition of IGF1R and mTOR [552] blocks compensatory feed-back pathway activation and improve the long-term outcomes of patients with HR positive HER2 negative MBC (NCT01234857).

HER2 Negative Breast Cancer

PI3K Inhibitors

They are mainly used in HR positive HER 2 negative breast cancer.

Selective PI3K inhibitors specifically inhibit individual isoforms. The main target is PI3K-alpha, as this gene is frequently mutated [547]. Class I pan-PI3K inhibitors block all isoforms of class I PI3K (alpha-delta).

Wortmannin is a fungal metabolite and a pan-specific irreversible PI3K inhibitor, targeting the p110 subunit. It inhibits cell growth in different cancer cell lines [553] and it has been found to synergize with chemotherapy [554]. LY294002 acts as wortmannin and in experimental studies it has been found to increase cytotoxic therapy in different tumors [554-555].

PX-866 is a synthetic derivative of wortmannin and is a potent inhibitor of PI3K [556]. In phase I studies this drug was well tolerated and increased stable disease rate in patients with advanced solid tumors [557]. Its combination with docetaxel has bee reported to be feasible in patients with solid tumors [558].

BKM120, or Buparlisib is an oral selective inhibitor of pan-class I PI3K [559]. Buparlisib alone and with endocrine therapy has shown antitumoral activity in ER+ breast cancer cell lines and xenografts. A phase Ib study evaluated buparlisib plus letrozole in 51 patients with metastatic ER+ breast cancer refractory to endocrine therapy. The clinical benefit rate at the MTD was 31%, including two objective responses. Moreover, the combination was safe and showed clinical activity independent on PIK3CA mutation status [560]. BKM120 is under evaluation in in combination with chemotherapy or endocrine therapy in many different clinical trials in breast cancer.

Belle 2, Belle 3 and Belle 4 are trials where BKM120 has been investigated in combination with fulvestrant or paclitaxel (see Table **1**).

Table 1. Principal trials on pan-PI3K/Akt or Akt inhibitors in HR+/HER2- or TPN breast cancer.

Trial	**Drug**	**Phase**	**Patients and setting**	**Endpoints**	**ClinicalTrials.gov identifier**
BKM 120 + fulvestrant *vs.* placebo + fulvestrant (BELLE 2)	Pan-PI3K inhibitor	III	Postmenopausal, HR+ HER2- advanced or MBC patients refractory to AI	Primary: PFS Secondary: OS, ORR, CBR, safety	NCT01610284
BKM 120 + fulvestrant *vs.* placebo + fulvestrant (BELLE 3)	Pan-PI3K inhibitor	III	Postmenopausal, HR+ HER2- advanced or MBC patients refractory to AI+mTOR inhibitor	Primary: PFS Secondary: OS, ORR, CBR, safety	NCT 01633060
BKM 120 + paclitaxel (BELLE 4)	Pan-PI3K inhibitor	II	HER2-, advanced or MBC	Primary: PFS Secondary: OS, ORR, CBR, time to response, duration of response, safety	NCT 01572727
GDC 0941/GDC0980 + fulvestrant (Fergie)	Class I selective PI3K inhibitor	II	Patients with advanced or MBC who have previously received an AI	Primary: PFS Secondary: ORR, CBR, duration of response	NCT 01437566
BYL719 monotherapy (PIKNIC)	P110a inhibitor	II	Advanced or MBC progressing after first line therapy. Patients with advanced hormone receptor positive tumors must have an alteration of the PI3K pathway.	Primary: ORR Secondary: CBR, PFS, safety	NCT02506556
BYL719 or BKM120 + letrozole	P110a inhibitor Pan-PI3K inhibitor	II	Neoadjuvant therapy in postmenopausal women with HR+ HER2- BC	Primary: PCR Secondary: ORR, safety, rate of breast conserving surgery	NCT01923168

(Table 1) contd.....

Trial	**Drug**	**Phase**	**Patients and setting**	**Endpoints**	**ClinicalTrials.gov identifier**
MK 2206	Akt inhibitor	II	Advanced or MBC with PIK3CA or AKT mutation, and/or PTEN loss/PTEN mutation.	Primary: ORR Secondary: PFS	NCT01277757
MK 2206	Akt inhibitor	II	Neoadjuvant treatment with anastrozole (postmenopausal patients) or anastrozole and goserelin (premenopausal patients) in clinical stage 2 or 3 PIK3CA mutant ER+, HER2- BC	Primary: PCR rate Secondary: clinical RR, safety	NCT01776008
GDC 0068 (Ipasertib) + paclitaxel	Akt inhibitor	II	Neoadjuvant treatment in TNBC	Primary: PCR rate Secondary: ORR, breast conserving surgery rate, safety	NCT02301988
GDC 0068	Akt inhibitor	II	Front line treatment in TN MBC	Primary: PFS Secondary: OS, ORR, duration of response	NCT02162719

BC: breast cancer; TN: triple negative; MBC: metastatic breast cancer; AI: aromatase inhibitor; DFS: disease-free survival; PFS: progression-free survival; OS: overall survival; RR: response rate; ORR: overall response rate; CBR: clinical benefit rate; PCR: pathological complete response.

SAR245408 (XL147), a pan-PI3K inhibitor is currently under investigation in phase I studies in solid tumors. A phase I/II study (NCT01082068) where this agent is used in association with letrozole is undergoing.

GDC-0941 (pictilisib), a potent, selective class I PI3K inhibitor and BYL719, a selective class 1-alpha inhibitor, are currently evaluated in clinical studies. GDC-0980 is being evaluated as alternative to GDC-0941 with fulvestrant (Table **1**).

Akt Inhibitors

Perifosine is being evaluated in a phase II trial in metastatic breast cancer (NCT00054145).

MK-2206 is currently under evaluation in advanced breast cancer with PIK3CA mutation, or Akt mutation, or PTEN loss/mutation and as neoadjuvant therapy with anastrozole (Table **1**).

GDC0068 is under evaluation with GDC-0973 (MEK inhibitor) in solid tumors (NCT01562275).

mTOR Inhibitors

Everolimus

As the PI3K-Akt-mTOR pathway is involved in tumor escape from hormone dependence, the mTOR inhibitor everolimus has been studied in combination with hormone therapy (AIs, tamoxifen or fulvestrant) [464, 513]. The principal trials are shown in Table **2**.

Table 2. Principal trials on mTOR inhibitor (everolimus) in HR+/HER2- breast cancer.

Trial	**Phase**	**Patients and setting**	**Results/Endpoints**	**ClinicalTrials.gov identifier**	**Reference**
Everolimus + Exemestane (BOLERO 2)	III	Postmenopausal women with HR+ advanced BC with recurrence/progression during or after NSAIs	PFS 7.8 *versus* 3.2 months (p<0.0001) OS 31.0 *vs*. 26.6 months (p ns)	NCT00863655	[512]
Everolimus + Tamoxifen (TAMRAD)	II	Postmenopausal women with HR+, HER2-, AI-resistant MBC	TTP 8.6 *vs*. 4.5 months (p 0.0021) CBR 61% *vs*. 42%(p 0.045)	NCT01298713	[511]
Everolimus + Fulvestrant	II	Postmenopausal, HR+ HER2- MBC patients after failure of AI therapy	TTP: 7.4 months	NCT00570921	[561]
Adjuvant everolimus + HT *vs*. HT (SWOG/NSABP S1207)	III	HR+, HER- breast cancer patients after standard adjuvant or neoadjuvant CT	Primary: DFS Secondary: OS, safety	NCT01674140	
Adjuvant Everolimus + HT for 2-3 years (UNICANCER)	III	HR+, HER- breast cancer patients after 2-3 years adjuvant HT	Primary: DFS Secondary: OS, safety	NCT01805271	
Everolimus + letrozole (BOLERO 4)	II	Postmenopausal, HR+ HER2- advanced or MBC first line	Primary: PFS Secondary: ORR, safety	NCT01698918	

(Table 2) contd.....

Trial	Phase	Patients and setting	Results/Endpoints	ClinicalTrials.gov identifier	Reference
Everolimus + exemestane *vs.* everolimus *vs.* capecitabine (BOLERO 6)	II	Postmenopausal, HR+ HER2- advanced or MBC patients refractory to anastrozole or letrozole	Primary: PFS Secondary: ORR, safety	NCT01783444	

BC: breast cancer; MBC: metastatic breast cancer; HT: hormonetherapy; CT: chemotherapy; AI: aromatase inhibitor; NSAI: not steroidal AI; CBR: clinical benefit rate; DFS: disease-free survival; PFS: progression-free survival; TTP: time to progression; OS: overall survival; ORR: overall response rate.

At present, exemestane in association with everolimus is an important option in clinical practice as first and second line therapy for metastatic breast cancer with resistance to a previous endocrine treatment [402].

Data from BOLERO-2 and TAMRAD support the use of combination of everolimus with endocrine therapy in patients who develop acquired resistance. In fact, PI3K/AKkt/mTOR signaling is often activated after long term estrogen deprivation. Besides, the BOLERO-4 study is evaluating the everolimus-letrozole combination as first line treatment in metastatic breast cancer patients with endocrine naïve disease (Table **2**).

The ongoing phase II study PrE0102 (NCT01797120) compares the combination of fulvestrant plus everolimus with fulvestrant plus placebo in metastatic breast cancer patients resistant to AIs. This combination has shown efficacy and tolerable toxicity in heavily pretreated metastatic ER+ breast cancer patients after AI failure [561].

As to the adjuvant and neoadjuvant setting, two phase III studies, the SWOG-NSABP S1207 and the UNICANCER will evaluate everolimus in pre- and post-menopausal HR+ early breast cancer patients (Table **2**). In ER+ patients, a phase II study was conducted to establish whether neoadjuvant everolimus plus letrozole for operable breast cancer patients would increase the clinical response. Patients who received everolimus plus letrozole showed higher response than patients receiving letrozole plus placebo. In the everolimus plus letrozole arm also a decrease of Ki67 expression in surgical specimens was found. This decrease demonstrated to be a prognostic factor for a favorable long-term outcome [562] and suggests a downregulation of the proliferation molecular pathways [563].

Dual PI3K-mTOR Inhibitors

A phase I trial of BGT226, an oral dual PI3K/mTOR inhibitor was conducted in patients with advanced solid tumors. Nausea (68%), diarrhea (61%), vomiting

(49%) and fatigue (19%) were the most common side effects [564].

BEZ235 and XL765 are currently under evaluation in combination with hormone or chemotherapy in phase I and II trials (NCT01300962, NCT01495247, NCT01082068).

Dual mTorc1/2 Inhibitors

The dual mTORC1/2 mTOR kinase inhibitor AZD8055 was compared with everolimus in ER+ endocrine resistant BC cells. *In vitro* models of tamoxifen (TamR) or estrogen deprivation resistance (MCF7-X) were treated with everolimus or AZD8055 alone or combined with anti-hormone fulvestrant. Everolimus inhibited mTORC1 but not mTORC2/Akt signaling, unlike AZD8055, which quickly inhibited both molecular pathways. Thus, AZD8055 was much better inhibitor of TamR and MCF7-X. In TamR or MCF7-X cells treated with AZD8055 plus fulvestrant better control of resistant growth was obtained rather than when they were used alone. AZD8055 with tamoxifen or estrogen deprivation also effectively inhibited endocrine responsive MCF-7 cells. AZD8055 inhibited ER ser167 phosphorylation in TamR and MCF7-X; however, it did affect neither ER ser118 nor expression of several ER-regulated genes, suggesting that its impact was largely ER-independent [565].

The interaction of nimbolide, a IGF1 inhibitor, and a PI3K inhibitor (LY294002) was investigated in MCF-7 and MDA-MB-231 cells. In nimbolide treated breast cancer cells a significant decrease of the protein expression of IGF signaling molecules and cyclins occurred. As a result, nimbolide decreased the proliferation of breast cancer cells. A cell cycle arrest at the G0/G1 phase and accumulation of apoptotic cells were observed [566].

SGK1 can promote proliferation and survival of cancer cells independently of Akt. So, potent and selective SGK1 inhibitors can be given in tumors displaying elevated SGK1 activity. These tumors in which NDRG1 phosphorylation is not suppressed by Akt inhibitors likely are those more resistant to Akt inhibitors. These tumors could be more efficaciously treated with mTOR inhibitors able to decrease SGK1 activity. Additionally, as Akt and SGK1 are related protein kinases, it has been proposed to evaluate the efficacy of treatment combining a dual Akt/SGK inhibitor on suppressing growth of tumors showing high SGK1 activity. As SGK1 expression is also induced by many steroid hormones [567] authors caution on their frequent use in cancer patients due to risk of inducing resistance to Akt inhibitors and promoting cancer proliferation

In the study by Fu *et al.* [568], the mammalian target of rapamycin (mTOR), protein kinase B (Akt), or mitogen-activated protein kinase kinase (MEK)

inhibitors alone or combined, improved the efficacy of conventional endocrine therapy but the addition of a single agent Akt inhibitor to fulvestrant showed superior efficacy in overcoming endocrine resistance and induced apoptosis and tumor regression.

Finally, recently laboratory and clinical findings have shown that PI3K inhibitors fail to suppress RB phosphorylation during treatment with single agent PI3K inhibitor. In this further study the combination of PI3K and CDK4/6 inhibitors allowed to overcome intrinsic and adaptive resistance and led to tumor regression [569].

HER2 Positive Breast Cancer

The HER superfamily is composed of tyrosine kinase receptors (TKRs) involved in the regulation of proliferation and survival of epithelial cells. The family includes four receptors: HER1 (epidermal growth factor receptor (EGFR)), HER2 (neu, C-erbB2), HER3 and HER4. HER2 is considered "an orphan receptor as it has no known ligand. The other three HER receptors have known ligands and form either homodimers or heterodimers upon ligand binding. HER2 can heterodimerise with any of the other receptors and is considered the preferred dimerisation partner. Dimerisation results in the autophosphorylation of tyrosine residues within the cytoplasmic domain of the receptors and initiates signal transduction *via* the PI3K/Akt and RAS/MAPK pathways" (570-571). "Although HER2-overexpression identifies patients who are likely to respond to therapy with trastuzumab, not all patients benefit from treatment. Approximately 15% of patients relapse after therapy, indicating the presence of de novo or acquired resistance" [572]. Several resistance mechanisms have been proposed, including 1) altered receptor–antibody interaction, 2) increased downstream signalling from other members of the HER family or other receptors [573], 3) constitutive activation of downstream factors [574-575]. Defective apoptosis pathways also can account for trastuzumab resistance. Immunologic factors, *i.e.* stromal tumor infiltrating lymphocytes [576], defective Fc receptors [577] interfering with normal immune responses to trastuzumab have also been hypothesised.

Experimental studies suggest that inhibitors of PI3K/Akt/mTOR pathway can synergize with trastuzumab in resistant cells [578].

In 2 phase Ib studies patients with trastuzumab-resistant metastatic breast cancer received everolimus with paclitaxel and trastuzumab or everolimus with vinorelbin and trastuzumab. In these studies, the association increased the overall response rate (ORR) [579-580].

In a phase II study, evaluating everolimus with paclitaxel and trastuzumab in 55 advanced breast cancer patients, the overall response rate (ORR) was 21.8%, the clinical benefit rate was 36.4% and the median PFS was 5.5 months. The median estimated overall survival was 18.1 months [581].

The BOLERO-3 is a phase III trial evaluating the addition of vinorelbine to everolimus plus trastuzumab in patients already treated with a taxane and with cancers resistant to trastuzumab. Principal data are shown in Table **3**.

A significant improved PFS in patients with HR- cancers, but not with HR+ cancers occurred. It is thought that cross-talk between the ER and the HER2 pathways allows ER to provide an escape mechanism when HER2 but not ER are inhibited [578]. Therefore, in HR+ and HER2+ patients, ER inhibition may be critical to improve outcome. Survival data from this study are still immature [582]. It should be noted that "the everolimus dose used in the BOLERO-2 study was 10 mg daily, which is the recommended dose in combination with endocrine therapy; however, the BOLERO-3 used a dose of 5 mg daily, which was based on the results of the phase 1b study by Jerusalem and colleagues" [580].

The BOLERO-1 trial (NCT00876395) and a phase II study (NCT00912340) evaluate everolimus in combination or compared to trastuzumab (Table **3**).

Table 3. Principal trials on PI3K or mTOR or dual PI3K/mTOR inhibitors in HER2+ breast cancer.

Trial	Drug	Phase	Patients and/or setting	Results/Endpoints	ClinicalTrials.gov identifier	Reference
Everolimus + Trastuzumab and Vinorelbine *vs.* Trastuzumab + Vinorelbine (BOLERO-3)	mTOR inhibitor	III	HER2+ advanced or MBC resistant to trastuzumab and pre-treated with a taxane	PFS: 7.0 *vs.* 5.78 months (p=0.0067)	NCT01007942	[582]
Everolimus plus Trastuzumab + Paclitaxel *vs.* Trastuzumab + Paclitaxel (BOLERO-1)	mTOR inhibitor	III	HER2+ advanced or MBC, 1st line	PFS: 14.95 *vs.* 14.49 months in the whole population (p=0.1166) PFS: 20.27 *vs.* 13.08 months (p=0.0049) in HR- patients	NCT00876395	[581]
Everolimus ± Trastuzumab	mTOR inhibitor	II	Hormone refractory MBC	Primary: ORR Secondary: CBR, PFS	NCT00912340	

(Table 3) contd.....

Trial	Drug	Phase	Patients and/or setting	Results/Endpoints	ClinicalTrials.gov identifier	Reference
Everolimus+trastuzumab + vinorelbine	mTOR inhibitor	II	BC brain metastases	Primary: response rate Secondary: extracranial response, toxicity	NCT01305941	
Temsirolimus	mTOR inhibitor	II	Advanced or MBC	Primary: CBR	NCT00376688	
Ridaforolimus + Trastuzumab	mTOR inhibitor	II	HER2+ Trastuzumab refractory MBC	PR 15%, SD 41%, CBR 34.3% PFS 5.4, OS 17.7 months	NCT00736970	[585]
BKM120 + trastuzumab	PanPI3K inhibitor	I/II	HER2+, trastuzumab resistant MBC	Primary: ORR, dose-limiting toxicity Secondary: CBR, PFS	NCT01132664	
BKM120 + trastuzumab + paclitaxel (NeoPHOEBE)	PanPI3K inhibitor	II	Neo adjuvant treatment of HER2+, PIK3CA wild-type and PIK3CA mutant primary BC	Primary: PCR rate Secondary: ORR, breast conserving surgery rate, safety	NCT01816594	
BEZ235 + Trastuzumab	Dual PI3K/mTOR inhibitor	I/II	HER2+, trastuzumab resistant MBC	Primary: PFS, dose limiting toxicity Secondary: ORR, CBR, safety	NCT01471847	

BC: breast cancer; MBC: metastatic breast cancer; PFS: progression-free survival; OS: overall survival; ORR: overall response rate; CBR: clinical benefit rate; PCR: pathological complete response; SD: stable disease.

Temsirolimus

As preclinical data suggest a synergy of HER2 inhibition and mTOR inhibition in breast and lung cancer models, a phase I study was conducted to evaluate a combination of temsirolimus and neratinib (see below). This combination demonstrated antitumor activity in multiple tumor types [583] and is under evaluation in HER2+ or triple negative MBC (NCT01111825 Phase I/II).

Ridaforolimus

Ridaforolimus (MK-8669) is a novel agent mTOR inhibitor, with significant effects on metabolism and cell growth [584]. The results of a study of ridaforolimus in ER-positive and HER2-positive breast cancer are available [585] (Table **3**).

As mTOR inhibition can activate compensatory insulin-like growth factor receptor (IGFR) signaling, a combination of ridaforolimus and dalotuzumab (anti-IGF1R antibody) has been evaluated in a phase I study in 17 patients with advanced cancer. mTOR-related stomatitis and asthenia were the main dose-limiting toxicities. There were 6 partial responses (3 in patients with breast cancer); in 10 of 23 patients with breast cancer and in 6 of 11 patients with ER+/high-proliferative breast cancer an antitumor activity occurred [549].

Ridaforolimus may overcome trastuzumab resistance by inhibiting PI3K signaling. A combination of ridaforolimus plus trastuzumab demonstrated antitumor activity in trastuzumab resistant HER2+ MBC [585]. A phase II study is ongoing (NCT00736970) (Table **3**).

BKM120

BKM120 is under evaluation in combination with lapatinib in HER2 positive, trastuzumab resistant MBC (NCT01589861); further two trials use a combination of BKM120 plus trastuzumab in HER2+ breast cancer that has failed to respond to trastuzumab or BKM120 with trastuzumab and paclitaxel in HER2+ primary breast cancer prior to surgery (NCT01132664 and NCT01816594) (Table **3**).

MK2206

A phase I trial was conducted on the Akt inhibitor MK-2206 in combination with trastuzumab, in 31 HER2 positive, treatment refractory breast and gastric cancer patients. There were one CR in a patient with metastatic breast cancer, one partial response and 5 stable diseases for at least 4 months. Moreover, trastuzumab administration did not affect the pharmacokinetics of MK-2206 [586].

As there is preclinical synergism between taxanes and MK-2206, a combination of MK-2206 and paclitaxel was evaluated in a phase I study in 22 patients with advanced cancer, including metastatic breast cancer. MK-2206 was well tolerated with paclitaxel and a preliminary antitumor activity was documented [587].

MK-2206 is currently evaluated in phase I trials with lapatinib (NCT01245205 and NCT01281163).

BEZ235

It is under evaluation with trastuzumab in a Phase Ib/II study in patients who failed prior to trastuzumab (NCT01471847).

Biomarkers of Response to PI3K-Akt-mTOR Inhibitors

The Cancer Genome Atlas Network described the most common mutations related to PI3K-Akt-mTOR pathway characterizing breast tumors by subtype [300]. Luminal/ER+ tumors show the highest rate of PIK3CA mutations, mainly the luminal A subtype. HER2+ tumors also have a high rate of PIK3CA mutations. PTEN loss is observed less commonly. PIK3CA, PTEN and AKT1 mutations seem to be mutually exclusive [588]. Phosphorylated Akt, S6 and 4EBP1 are markers of activation of the PI3K pathway [300]. Elevated levels of pS6K suggest worse prognosis, and high levels of pS6K and pAkt predict a good response to rapamycin or rapamycin analogs in breast cancer cells lines and other tumors [589-592].

Inositol polyphosphate 4 phosphatase (INPP4B) deletion may occur with PTEN loss and it correlates with poor prognosis. Tumors without INPP4B may also be candidates to be targeted with PI3k inhibitors [593], while cells with PIK3CA mutations or PTEN loss seem to be more sensitive to PI3K/Akt inhibitors [516, 594-596].

Ribosomal S6 kinases RPS6KA2 (RSK3) and RPS6KA6 (RSK4), NOTCH, c-myc activation and mutations of the mitogen-activated protein kinase pathway can account for resistance to PI3K inhibitors. Therefore, PI3K inhibitors given with MEK or RSK inhibitors can overcome this resistance [597-603]. KRAS mutations bypassing the PI3K pathway also can account for resistance to PI3K/mTOR inhibitors [597, 601-602]. This occurs more often in basal-like/TNBCs, where the combination of MEK inhibitors and PI3K inhibitors is more effective than in other breast cancer subtypes [603].

Data from Clinical Studies

In metastatic breast cancer patients treated with inhibitors of the PI3K/Akt/mTOR pathway combined with endocrine therapy, anti-HER2 therapy or chemotherapy, PIK3CA mutations correlated with improved TTP compared with patients with wild type tumors [604], while PTEN status had no correlation with clinical outcome. In the TAMRAD trial, PI3K mutations, PTEN and pAkt did not affect the response to everolimus. Everolimus was more effective in patients with low PI3K expression or with low LKB1, a known mTOR suppressor, and high phospho-4E binding protein, which is downstream of mTOR (Figs. **4-5**). These findings point out that the patients who mostly benefit from everolimus are those in whom mTOR is activated independently of PI3K [605]. However, in the overall these findings also arise the question on largely incomplete knowledge of the molecular network sustaining tumor growth.

Novel Anti-HER2 Strategies

Experimental evidence suggests that co-inhibition of HER2, other members of the HER family and/or the downstream pathway by trastuzumab/lapatinib combination with other targeted therapies might prevent or at least delay time to resistance [606-607]. However, so far no biomarkers but HER2 have been validated for selection of patients who benefit from therapies targeting the HER2 pathway.

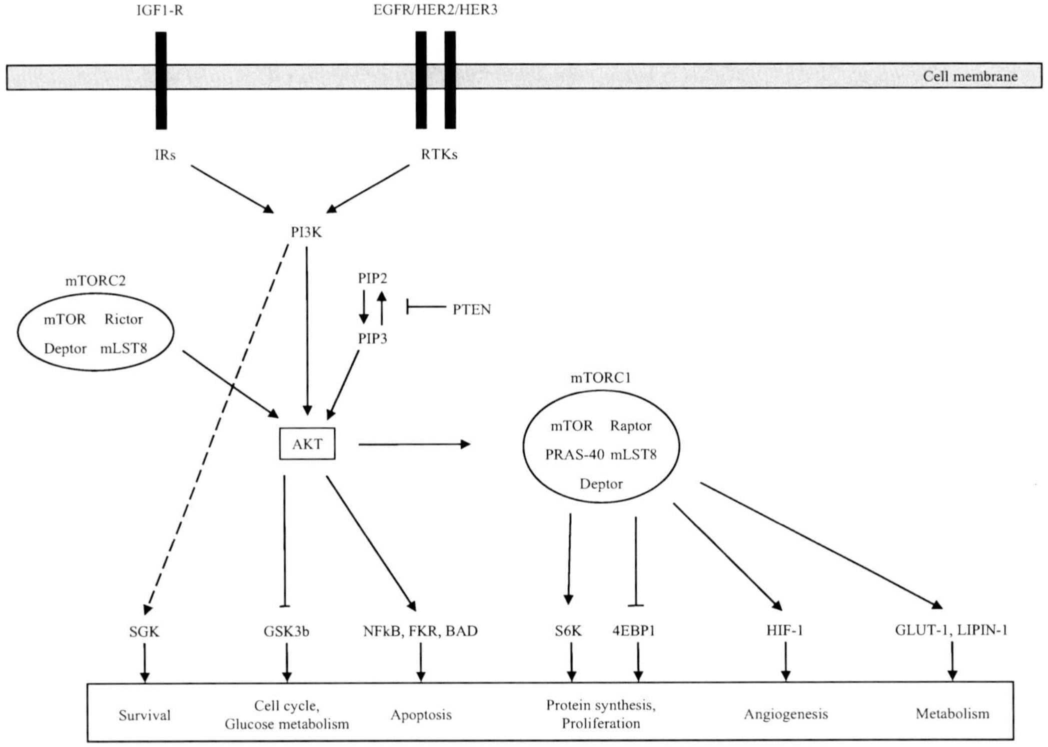

Fig. (4). The PI3K-Akt-mTOR signalling pathway. RTKs: receptors with protein tyrosine kinase activity; PIP2: phosphatidylinositol-4,5 biphosphate; PIP3: phosphatidylinositol-3,4,5 triphosphate; 4EBP1: eukaryotic initiation factor 4E binding protein 1; Glut1: glucose transporter 1; HIF-1: hypoxia inducible factor 1; IRS: insulin receptor substrate; mTORC1/2: mammalian target of rapamycin complex 1/2; PI3K: phosphatidylinositol 3 kinase; PTEN: phosphatase and tensin homolog deleted on chromosome 10; S6K1: S6 kinase 1; GSK3: glycogen synthase kinase-3; FKR: forkhead family transcriptor factors (also see text).

Ongoing trials are combining anti HER-2 agents with drugs blocking other signaling pathways. A possible approach is the simultaneous inhibition of HER2 and insulin growth factor receptor (IGFR-1). IGFR-1 inhibition has been shown to restore sensitivity to trastuzumab in animal models [608]. Besides, concomitant inhibition of HER-2 and SRC showed to have a central role in multiple trastuzumab-resistance mechanisms [609].

Afatinib

Afatinib is an oral TK inhibitor that irreversibly inhibits HER-1, 2 and 4 [610]. A phase II study investigated afatinib activity in 41 advanced or metastatic HER2+ breast cancer patients progressing after trastuzumab. Four patients (10% of 41 treated) showed PR. Fifteen patients (37%) had SD and 19 (46%) obtained clinical benefit [611]. Afatinib is under evaluation in phase I trials in combination with trastuzumab in patients with HER2-positive advanced breast cancer (NCT00950742; NCT01649271). A phase III study of vinorelbine plus trastuzumab or afatinib in metastatic patients who progressed to a chemotherapy regimen combined with trastuzumab has completed the enrollment and results are available (LUX-Breast 1) (Table **4**) [612]. Another phase II trial is also evaluating afatinib with or without vinorelbine in patients with inflammatory or metastatic breast cancer (NCT01325428) (Table **4**).

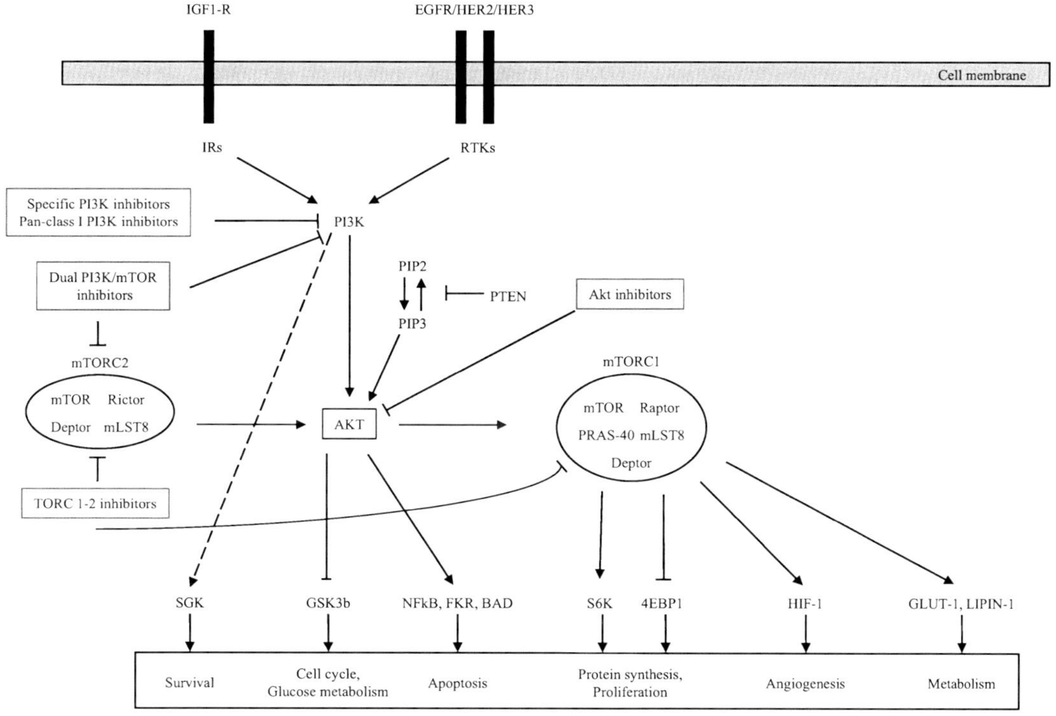

Fig. (5). Classes of PI3K-Akt-mTOR signalling pathway inhibitors in breast cancer (also see text).

While phase I and II studies have demonstrated promising activity in HER2-positive breast cancer, in phase III studies an unfavorable risk-benefit analysis was reported. So, the successful development of afatinib in breast cancer will thus depend on aggressively preventing and managing its associated toxicities [613].

Neratinib

Neratinib is an irreversible inhibitor of HER-1, -2 and -4 [614] that is currently in phase III of clinical development for the treatment of breast cancer (Table **4**).

Table 4. Principal trials on anti-EGFR therapy.

Trial	Drug	Phase	Patients and setting	Results/Endponts	ClinicalTrials.gov identifier	Reference
Pertuzumab + Trastuzumab + Docetaxel (CLEOPATRA)	Anti-human epidermal growth factor receptor 2 (HER2) humanized monoclonal antibody	III	Previously untreated HER2+ MBC	PFS: 18.5 *vs.* 12.4 months (p<0.001) OS 56.5 *vs.* 40.8 months (p<0.001)	NCT00567190	[519] [520]
Trastuzumab-Emtansine (T-DM1) *vs.* capecitabine + lapatinib (EMILIA)	Antibody-drug conjugate between trastuzumab and the microtubule inhibitory agent DM1	III	HER2+ locally advanced MBC who have received prior Trastuzumab-based therapy	PFS: 9.6 *vs.* 6.4 months (p<0.001) OS 30.9 *vs.*. 25.1 months (p<0.001)	NCT00829166	[522]
Afatinib + vinorelbine *vs.* Trastuzumab + vinorelbin (LUX-Breast1)	HER1, 2, 4 inhibitor	III	HER2+ MBC progressing after CT + trastuzumab	PFS: 5.5 *vs.* 5.6 months (p ns)	NCT01125566	[612]
Afatinib alone and in combination with weekly paclitaxel or weekly vinorelbine (LUX-Breast2)	HER1, 2, 4 inhibitor	II	HER2+ MBC patients who failed HER2-targeted treatment in the neoadjuvant or adjuvant setting	Primary: ORR Secondary: duration of response, PFS, safety	NCT01271725	
Afatinib + Letrozole	HER1, 2, 4 inhibitor	II	Hormone refractory MBC	Primary: % of progression free pts Secondary: PFS, OS, ORR	NCT00708214	
Afatinib	HER1, 2, 4 inhibitor	II	HER2+ MBC patients after failure of trastuzumab containing regimens.	Primary: ORR Secondary: PFS, OS, duration of response	NCT00431067	

(Table 4) contd.....

Trial	Drug	Phase	Patients and setting	Results/Endponts	ClinicalTrials.gov identifier	Reference
Afatinib + Trastuzumab(DAFNE)	HER1, 2, 4 inhibitor	II	Neoadjuvant treatment for patients with locally advanced or operable BC receiving taxane-anthracycline containing CT	PCR rate: 49.2% Clinical responses: 96.3%; breast conserving surgery in 59.4%	NCT01594177	
Afatinib *vs.* afatinib plus vinorelbine	HER1, 2, 4 inhibitor	II	HER2+ locally advanced or metastatic inflammatory BC	Primary: CBR Secondary: ORR, duration of response, PFS, safety	NCT01325428	
Adjuvant Neratinib (ExteNET)	HER1, 2, 4 inhibitor	III	HER-2+ BC patients after adjuvant treatment with trastuzumab.	Primary: DFS Secondary: time to distant recurrence, distant disease-free survival, OS, safety	NCT00878709	[618]
Neratinib	HER1, 2, 4 inhibitor	II	Advanced HER2+ BC patients	Primary: PFS Secondary: safety, ORR, CBR, duration of response	NCT00300781	
Neratinib + Paclitaxel *vs.* Trastuzumab + Paclitaxel (NEFERTT)	HER1, 2, 4 inhibitor	II	First-line, HER2+ locally advanced or MBC patients	Primary: PFS Secondary: OS, ORR, CBR, duration of response, safety	NCT00915018	
Neratinib + Capecitabine *vs.* Lapatinib + Capecitabine (NALA)	HER1, 2, 4 inhibitor	III	Patients with HER2+ MBC who have received 2 or more prior HER2 directed regimens in the metastatic setting	Primary: PFS, OS Secondary: ORR, CBR, DOR, safety	NCT01808573	
Neratinib + trastuzumab + paclitaxel	HER1, 2, 4 inhibitor	II	Neoadjuvant treatment in locally advanced HER2+ BC patients	Primary: PCR Secondary: recurrence free interval, OS, safety	NCT01008150	

BC: breast cancer; MBC: metastatic breast cancer; CT: chemotherapy; DFS: disease-free survival; PFS: progression-free survival; OS: overall survival; ORR: overall response rate; CBR: clinical benefit rate; PCR: pathological complete response; DOR: duration of response.

In a phase II trial evaluating 136 HER-2-positive advanced BC patients [615], median PFS were 22.3 and 39.6 weeks and ORR were 24% and 56%, respectively in pretreated and in trastuzumab-naïve patients.

In a phase I/II study, safety, clinical activity, and pharmacokinetics of neratinib plus vinorelbine combination were evaluated in patients with HER2-positive MBC; the primary efficacy end point was objective response (OR). The most common toxicities were diarrhea, neutropenia, and nausea. The OR rate was 41% in lapatinib naïve and 8% in lapatinib treated patients [616].

A phase III trial is ongoing to assess the benefit and risk of neratinib plus paclitaxel combination in the first-line setting of MBC [617] (Table **4**).

In the adjuvant setting, the results of the ExteNET trial (neratinib after trastuzumab-based adjuvant therapy in patients with HER2+ breast cancer) have been recently published. Between 2009 and 2011, 2840 patients were enrolled. At 2-year follow-up, 70 invasive disease-free survival events had occurred in patients who had received neratinib compared with 109 events in those who were given placebo (p=0·0091). The 2-year invasive disease-free survival rate was 93.9% in the neratinib group and 91.6% in the placebo group [618].

Dual anti-HER therapies are more efficacious than single agents. In a phase I study, neratinib was combined with trastuzumab and paclitaxel in 21 metastatic HER2+ patients. Objective responses (CR and PR), occurred in 8 patients (38%), with a clinical benefit (CR + PR+ SD) ≥24 weeks in 11 patients (52%). Median TTP was 3.7 months [619].

A phase I/II trial evaluated the safety and efficacy of neratinib plus capecitabine in patients with HER2-positive metastatic breast cancer. "Patients with trastuzumab-pretreated HER2-positive metastatic breast cancer received neratinib plus capecitabine at the MTD. The primary end point was objective response rate (ORR). The most common drug-related adverse events were diarrhea and palmar-plantar erythrodysesthesia syndrome. The ORR was 64% (n = 39 of 61) in patients with no prior lapatinib exposure and 57% (n = 4 of 7) in patients previously treated with lapatinib. Median PFS was 40.3 and 35.9 weeks, respectively" [620].

Phase II trials with neratinib monotherapy in patients with HER-2 positive metastatic brain tumors are ongoing (NCT01494662). Neratinib is also under evaluation with trastuzumab plus paclitaxel in MBC (NCT01423123) and with temsirolimus (NCT01111825).

MM-111

MM-111 is a bi-specific monoclonal antibody that reversibly targets the HER-2 and -3 heterodimer [621]. A phase I-II study is evaluating its efficacy as single agent in HER2+ advanced breast cancer patients who progressed after trastuzumab or lapatinib [622]. A phase I trial is studying MM-111 plus trastuzumab in HER2-positive, heregulin-positive, advanced and refractory breast cancer (NCT01097460).

Pertuzumab-TDM-1 Combination

Pertuzumab and TDM-1 have been administered in combination or in sequence in HER2 positive breast cancer cell lines. A pre-treatment with pertuzumab markedly reduced the efficacy of TDM-1 [623].

A phase IIa study evaluated the safety and efficacy of TDM-1 and pertuzumab, administered at full doses, in 64 patients with HER2-positive locally advanced or metastatic breast cancer. The ORR was 41% overall, and 57% in first-line patients. "Median, progression-free survival was 6.6 and 7.7 months, respectively. Exploratory biomarker analyses demonstrated that patients with above-median tumor HER2 mRNA levels had a numerically higher ORR than patients with below-median levels (44% v 33%, respectively)" [624].

Cycline-Dependent Kinases

Cell cycle dysregulation is one of the principal features of cancer. Cyclin dependent kinases (CDKs) regulate cell cycle. CDK4 and CDK6 are responsible for the transition from G1 phase to S phase through the phosphorylation state of pRB (phosphorylated retinoblastoma). When pRB is hyperphosphorylated, the G1/S transition is possible, and the cell cycle is upregulated [625].

Selective CDK4/6 inhibitors can arrest the cells at this G1/S phase checkpoint, and at least three CDK4/6 inhibitors are currently being evaluated; LY2835219, LEE011 (ribocilclib) and PD0332991 (palbociclib) [626-628]. Many different trials are ongoing (Table **5**).

Table 5. Principal trials on CDK4/6 inhibitors.

Trial	Phase	Patients and setting	Results/Endponts	ClinicalTrials.gov identifier
Ribociclib + letrozole (Monaleesa2)	III	Postmenopausal HR+ HER2- MBC, 1st line HT	Primary: PFS Secondary: OS, ORR, CBR, safety	NCT01958021
Ribociclib +Fulvestrant (Monaleesa-3)	III	Postmenopausal HR+ HER2- advanced BC patients 1st or 2nd line HT	Primary: PFS Secondary: OR, ORR, safety	NCT02422615

(Table 5) contd.....

Trial	Phase	Patients and setting	Results/Endponts	ClinicalTrials.gov identifier
Ribociclib or Placebo + Tamoxifen and Goserelin or a NSAI and Goserelin (Monaleesa-7)	III	Premenopausal HR+ HER2-, advanced BC patients	Primary: PFS Secondary: OS, CBR, safety, TTR, DOR	NCT02278120
Ribociclib (LEE011) + letrozole *vs*. letrozole (Monaleesa)	II	Neoadjuvant treatment of primary BC	Primary: cell cycle response rate Secondary: safety	NCT01919229
Palbociclib + letrozole *vs*. Letrozole (Paloma-1)	II	HR+, HER2-, advanced BC patients, 1st line	Median PFS 20.2 *vs*. 10.2 months (p=0.0004)	NCT00721409
Palbociclib + Letrozole *vs*. Placebo + Letrozole (Paloma2)	III	Postmenopausal ER+ HER2- BC patients, 1st line	Primary: PFS Secondary: OS, ORR, DOR	NCT01740427
Palbociclib + Fulvestrant *vs*. Fulvestrant alone (Paloma-3)	III	HR+, HER2 negative metastatic BC patients whose disease has progressed after prior endocrine therapy	PFS: 9.2 *vs*. 3.8 months (p>0.001) CBR 34.0% *vs*. 19.0% (p<0.001)	NCT01942135
Palbociclib + Letrozole *vs*. Placebo +Letrozole (Paloma-4)	III	Previously untreated postmenopausal ER+ HER2 advanced BC patients	Primary: PFS Secondary: OS, ORR, DOR	NCT02297438
Palbociclib + Exemestane *vs*. Capecitabine (PEARL)	III	Postmenopausal HR+/HER2- MBC patients whose disease was refractory to previous NSAI	Primary: PFS Secondary: OS, ORR, CBR, DOR	NCT02028507
Palbociclib + Trastuzumab ± Letrozole (PATRICIA)	II	Post-menopausal HER2+ locally advanced or metastatic BC patients who have received chemotherapy and trastuzumab for metastatic disease	Primary: PFS Secondary: ORR, CBR, safety	NCT02448420
Palbociclib + Fulvestrant or Letrozole (PARSIFAL)	II	ER+, HER2- locally advanced or metastatic BC patients	Primary: PFS Secondary: OS, ORR, CBR, safety	NCT02491983
Palbociclib monotherapy *vs*. Palbociclib + endocrine therapy) on which the patient has progressed (TREND)	II	Postmenopausal ER+, HER2- advanced BC patients	Primary: ORR, CBR Secondary: PFS, OS, TTP, DOR	

(Table 5) contd.....

Trial	Phase	Patients and setting	Results/Endponts	ClinicalTrials.gov identifier
Palbociclib + standard HT (Penelope-B)	III	HR+ HER2- BC with residual disease after neoadjuvant CT	Primary: DFS Secondary: OS, safety	NCT01864746
Palbociclib + Letrozole (PALLET)	II	Neoadjuvant therapy in postmenopausal ER+ primary BC patients	Primary: Ki-67, CCR Secondary: PCR, safety	NCT02296801
Palbociclib + Letrozole *vs*. 3 FEC-3 Docetaxel CT (NeoPAL)	II	Neoadjuvant treatment of postmenopausal stage II-IIIA PAM 50 ROR-defined low or intermediate risk luminal BC patients	Primary: number of patients with residual cancer burden Secondary: clinical response, safety	NCT02400567
Palbociclib + standard adjuvant endocrine therapy *vs*. standard adjuvant endocrine therapy (PALLAS)	III	HR+ HER2- early breast cancer	Primary: DFS Secondary: safety	NCT02513394
Abemaciclib + fulvestrant *vs*. fulvestrant (MONARCH-2)	III	HR+, HER2- locally advanced or metastatic BC, 1st line or after TAM or AI failure	Primary: PFS Secondary: OS, ORR, CBR, DOR	NCT02107703
Abemaciclib +NSAI *vs*. NSAI (MONARCH 3)	III	Postmenopausal HR+, HER2- locoregionally recurrent or metastatic BC, 1st line	Primary: PFS Secondary: OS, ORR, CBR, DOR	NCT02246621
Abemaciclib + Anastrozole *vs*. Abemaciclib *vs*. Anastrozole (neoMONARCH)	II	Neoadjuvant therapy in postmenopausal HR+, HER2- BC patients	Primary: Ki-67 Secondary: CCR, clinical OR, safety	NCT02441946

BC: breast cancer; MBC: metastatic breast cancer; HT: hormonetherapy; CT: chemotherapy; AI: aromatase inhibitor; NSAI: not steroidal AI; DFS: disease-free survival; PFS: progression-free survival; OS: overall survival; OR: overall response; ORR: overall response rate; CBR: clinical benefit rate; PCR: pathological complete response; CCR: clinical complete response; DOR: duration of response; TTP: time to progression; TTR: time to response.

Potential Targets in TNBC

BL1 and BL2 subtypes, which overexpress cell cycle and DNA damage response genes, are sensitive to cisplatin; M and MSL subtypes, with high expression of genes involved in EMT and growth pathways, respond to PI3K/mTOR inhibitors and abl/src inhibitors. The LAR subtype is sensitive to the AR antagonist bicalutamide. The IM subtype is sustained by genes involved in immune cell processes. Its gene expression profile is similar to that of medullary breast cancer, that has a favorable prognosis despite its high grade [629-630] (Fig. **6**).

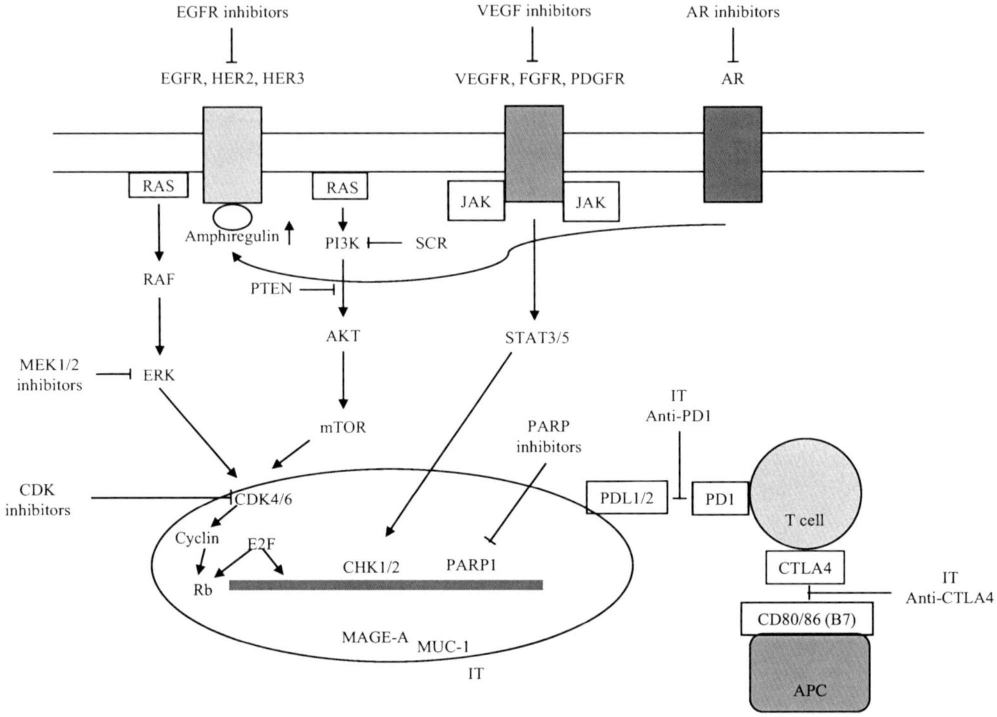

Fig. (6). Principal targets to inhibit relevant molecular pathways in TNBC. EGFR inhibitors: cetuximab, panitumumab, gefitinib, erlotinib, lapatinib, afatinib; VEGFR inhibitors: bevacizumab, sunitinib; MEK1/2 inhibitors: trametinib, cobimetinib; CDK inhibitors: dinaciclib, LEE001; AR inhibitors: enzalutamide; PARP inhibitors: veliparib, olaparib; IT (immunotherapy): nivolumab, pembrolizumab, ipilimumab (also see text).

Targeting the VEGF Pathway

Aberrant VEGF pathway activation shown in basal-like BC has led to investigations targeting VEGF and its downstream receptors. Bevacizumab, the anti-VEGF antibody, has been shown to increase DFS when combined with paclitaxel in patients with TNBC, compared to paclitaxel alone. However, the OS was not affected [532]. Other small-molecule multikinase inhibitors targeting VEGFR and other receptor tyrosine kinases, such as sunitinib demonstrated a limited activity [631-634].

CALGB 40603 trial (NCT00861705) showed that addition of both carboplatin and bevacizumab to neoadjuvant chemotherapy (paclitaxel followed by doxorubicin and cyclophosphamide) significantly increased pCR rate in stage II to

III TNBC patients [635]; however, data on PFS and OS are still expected [630].

Targeting the EGFR Pathway

As EGFR is upregulated in most basal like BCs, it represents a potential therapeutic target. However, patients with HER2- tumors and overexpression of EGFR did not benefit from the addition of lapatinib [636].

Poor results were found in two independent phase II clinical trials evaluating the efficacy of cetuximab combined with either carboplatin or cisplatin [637-638].

MEK Inhibitors

"Constitutive activity of pathways downstream from EGFR may be one explanation for lack of response to EGFR-targeted therapies. A possible target being explored is the mitogen activated protein kinase/extracellular signal-regulated kinase (MEK), a signaling pathway with a central role in promoting tumor initiation and progression. Activation of this pathway has been shown to be associated with an increased risk of metastasis in patients with breast cancer" [639-640]. In a study, basal-like BC appeared to be mostly sensitive to MEK inhibitors [639]. This study also showed a potential negative feedback loop between the MEK cascade and thePI3K-PTEN-Akt signaling suggesting concurrent treatment with both MEK and PI3K inhibitors [639].

A randomized phase II trial is evaluating efficacy and safety of the MEK inhibitor cobimetinib combined with paclitaxel as first line therapy of patients with metastatic or locally advanced, TNBC (NCT02322814).

Targeting Androgen Receptor (AR)

Up to one third of TNBC express the androgen receptor (AR) which is localized in the nucleus, which is indicative of transcriptionally active receptors. AR is mostly present in the "luminal AR (LAR)" molecular subtype of TNBC. Also, other non-LAR molecular subtypes of TNBC are dependent on AR protein likely by upregulation of EGFR ligand amphiregulin (AREG). AR inhibition significantly decreases proliferation, anchorage-independent growth, migration, and invasion and increases apoptosis in TNBC lines (SUM159PT, HCC1806, BT549, and MDA-MB-231). *In vivo*, enzalutamide significantly decreases viability of SUM159PT and HCC1806 xenografts. Therefore, enzalutamide is a promising targeted therapy for different molecular subtypes of AR+ TNBC [641].

Single agent enzalutamide is currently under evaluation in a phase II study in advanced, AR+, TNBC patients (NCT01889238).

Poly ADP Ribose Polymerase (PARP) Inhibitors

The principal ongoing trials on PARP inhibitors are shown in Table **6**.

Table 6. Principal trials on PARP inhibitors.

Trial	**Phase**	**Patients**	**Results/Endponts**	**ClinicalTrials.gov identifier**
Olaparib *vs.* physician choice (OlympiAD)	III	MBC patients with germline BRCA 1/2 mutations.	Primary: PFS Secondary: OS, ORR, safety	NCT02000622
Olaparib *vs.* Placebo (OlympiA)	III	Adjuvant treatment in gBRCA1/2 mutated high risk HER2- primary BC patients after definitive local treatment and neoadjuvant or adjuvant chemotherapy	Primary: DFS Secondary: OS, safety	NCT02032823
Talazoparib *vs.* physician's choice (EMBRACA)	III	Advanced and/or metastatic BC patients with BRCA mutation who have received no more than 2 prior CT regimens for metastatic disease	Primary: PFS Secondary: OS, ORR, DOR	NCT01945775
Talazoparib (ABRAZO)	II	Locally advanced and/or metastatic BC patients with BRCA mutation	Primary: ORR Secondary: OS, PFS, CBR, DOR	NCT02034916
Talazoparib (TBB)	II	BRCA1 and BRCA2 Wild-Type Patients with either advanced TPN BC and Homologous Recombination deficiency or advanced HER2-BC with germline or somatic mutation in Homologous Recombination pathway genes	Primary: ORR Secondary: PFS, CBR	NCT02401347
Carboplatin + Paclitaxel ± Veliparib	III	HER2- metastatic or locally advanced unresectable BRCA-associated BC	Primary: PFS Secondary: OS, CBR, ORR, DOR	NCT02163694
Veliparib + Carboplatin + Standard Neoadjuvant CT or Carboplatin + Standard Neoadjuvant CT *vs.* Standard Neoadjuvant CT	III	Neoadjuvant therapy in early stage TNBC	Primary: Rate of eligibility for breast conservation after therapy Secondary: DFS, OS, ORR, PCR	NCT02032277

(Table 6) contd.....

Trial	Phase	Patients	Results/Endponts	ClinicalTrials.gov identifier
Niraparib *vs.* physician choice (BRAVO)	III	Previously treated, HER2-, germline BRCA mutation-positive advanced/metastatic BC patients)	Primary: PFS Secondary: OS	NCT01905592
Rucaparib (RUBY)	II	BRCAness genomic signature MBC	Primary: CBR Secondary: PFS, OS, safety	NCT02505048

BC: breast cancer; BC: TN: triple negative; MBC: metastatic breast cancer; CT: chemotherapy; DFS: disease-free survival; PFS: progression-free survival; OS: overall survival; ORR: overall response rate; CBR: clinical benefit rate; PCR: pathological complete response; DOR: duration of response.

Poly(ADP-Ribose) polymerases (PARPs) are important components of base excision repair pathway for single strand DNA breaks; so, patients with BRCA1 and 2 mutations are susceptive of DNA breakage due to defective DNA repair machinery. Currently, "synthetic lethality is the hypothesized mechanism of action for PARP inhibitors in tumors with HR deficiency, as it is thought that the simultaneous blockage of both pathways prevents the tumor cells from repairing DNA damage" [642]. "Gene microarray expression profiling has shown considerable similarities between BRCA1-mutated tumors and basal tumors" [241]. This shared phenotype has been termed "BRCAness" [643]. What is unknown is whether the basal phenotype depends on the BRCA loss or if the BRCA loss is a result of the basal phenotype [644].

Basal-like cancers with deficient BRCA1 pathway may respond to PARP inhibitors [629, 645-651]. Clinical trials using PARP inhibitors, such as olaparib and BSI-201, showed clinical efficacy in BRCA1/2 -associated breast, ovarian, and prostate cancers, as well as sporadic basal-like breast cancers [645, 652-654].

Immunotherapy

Although breast cancer historically is not considered strongly immunogenic cancer a few experimental and clinical findings suggest a relevant role of immunotherapy also in this disease [655]. Recently, tumor-infiltrating lymphocytes (TILs) have acquired relevance both as prognostic biomarkers and also as predictive factors of the efficacy of conventional chemotherapies [656-657]. Increased TILs, consistent with gene expression profiling and immunohistochemistry (IHC) staining, are associated with better prognosis [658-660]. In the neoadjuvant GeparSixto trial (NCT01426880), patients with high levels of TILs had a greater pCR rate than the patients with low levels of TILs. mRNA expression levels of 12 immune genes were measured in 481 tumors. They included immune activating (CXCL9, CCL5, CD8A, CD80,

CXCL13, IGKC, CD21) and presumed immunosuppressive factors (IDO1, PD-1, PD-L1, CTLA4, FOXP3). An immunosuppressive marker, particularly PD-1 and PD-L1 showed a positive correlation with the evaluated immune-activating markers, stromal TILs and treatment response. In order to explain the unexpected concomitant positive correlation of immunoactivating and immunosuppressive factors, authors suggest "a feedback activation of immunosuppressive pathways as part of the immune reaction" and report on "concordant results by other studies" [661]. In TNBC, TILs more often and at higher level than in the other molecular subtypes have been observed likely indicating a stronger immune response against a more immunogenic phenotype. Moreover, the existence of an immunomodulatory subtype of TNBC with high expression of genes involved in antigen processing and T cell functions suggests immunotherapy as a promising target therapy against TNBCs [622]. As main advantages of immunotherapies they are usually less toxic and patients may also less likely to develop resistance due to the ability of the immune system to contemporaneously target many different antigens and to adapt to mutagenic cancer cells. Promising immunotherapies for breast cancer include therapeutic vaccines, checkpoint inhibitors/immune modulators, adoptive cell therapy and adjuvant immunotherapies.

Therapeutic Vaccines

The main aim is to promote an immune response against tumor-specific or tumor associated antigens (TAAs) therefore also attacking cancer cells bearing these antigens. In TNBC TAAs not shared by normal cells offer an attractive target for this kind of therapy. Cancer testis antigens such as melanoma-associated antigens (MAGE-A) A3 and A4 and cancer /testis antigen 1 (CTAGIA/NY-ESO-1) are not present in somatic cells but are expressed in a substantial proportion of TNBC (652-653). Another potential target is the glycosylated form of mucin 1 (MUC-1). MUC-1 is a cell surface-associated antigen that is aberrantly glycosylated by epithelial cancer cells. It derives glycopeptide epitopes able to bind cell surface major histocompatibility complex (MHC) molecules. In such a way, they can be recognized by CTLs and can be bound by antibodies accounting for an antibody-dependent cell-mediated cytotoxicity (ADCC). MUC-1 is overexpressed in 90% of breast cancers including TNBC. Due to these characteristics, it is considered an ideal cancer vaccine target and recently an immune-recognition of tumor associated mucin MUC-1 by a fully synthetic aberrantly glycosylated MUC-1 tripartite vaccine has been reported [662]. A few phase I/II trials are ongoing [663].

Checkpoint Inhibitors/Immune Modulators

This approach target molecules serving as checks and balances in the regulation of immune responses. These treatments block inhibitory molecules such as cytotoxic T lymphocyte antigen 4 (CTLA-4), B and T lymphocyte attenuator (BTLA), PD-1 and its ligands PD-L1, PD-L2, and T cell immunoglobulin and mucin domain-containing protein 3 (TIM-3), lymphocyte-activation gene 3 (LAG-3); or alternatively activate stimulatory molecules such as tumor necrosis factor (TNF) receptor superfamily, member 4 (TNFRSF4/OX-40/CD134) and TNF receptor superfamily, member 9 (TNFRSF9/CD137) in order to increase anti-cancer immune responses. A PD-L1 overexpression correlating with higher rate of TILs was observed in 20% of TNBC patients [664]. A retrospective analysis confirmed this and showed that higher levels of PD-L1 correlated with increased TILs and were associated with improved RFS [665]. Further studies have reported that higher levels of PD-L1 correlated with higher TILs and better prognosis in neo-adjuvant treatment of TNBC [666-667]. However, whether or not PD-L1 expression predicts response to anti-PD-1/PD-L1 immuno- therapy is to be confirmed. Many phase I, II and two phase III trials are ongoing [663].

Adoptive Cell Therapy

Adoptive T cell transfer is another modality of immunotherapy for breast cancer. In this modality, T cells, removed and genetically modified or treated with chemicals are re-introduced into the patient with the goal of improving anti-cancer response of the immune system. Some trials using adoptive T cell transfer techniques are currently ongoing [663].

Adjuvant Immunotherapies

Adjuvants are substances that are either used alone or combined with other immunotherapies to increase the immune response. Some adjuvant immunotherapies use ligands—molecules that bind to proteins such as receptors—to help control the immune response. These ligands can be either stimulating (agonists) or blocking (antagonists). Some clinical trials are planned or ongoing [663].

INNOVATIVE BIOLOGICAL APPROACHES: TUMOR CELL REPROGRAMMING AND CONVENTIONAL ANTIESTROGEN THERAPY PLUS IMMUNE STIMULATION; PLANNING OF A NEW ANTITUMORAL TREATMENT

Reprogramming and Differentiation Treatment of Cancer Cells

The term "cancer cell reprogramming" is recently used in cancer diseases "to define any kind of intervention aimed at transforming cancer cells into terminally differentiated cells". In similar manner act the differentiation treatments, which have the scope "to induce terminal differentiation of cancer cells, forcing cancer stem cells to become transit amplifying cells that can then be treated by other therapies such as chemotherapy or radiation" [668]. As described in the cited article, one of the first applications of differentiation therapy is the removal of the block to differentiation in human myeloid leukemia [669]. This block of differentiation allows myeloid leukaemic cells to continue to proliferate and/or prevents the terminal differentiation and apoptosis as it happens in normal white blood cells. In chronic myeloid leukemia, the bcr-abl (t9/22) translocation causes a fusion product that is an activated tyrosine kinase in myelocytes. Imatinib mesylate blocks the binding of ATP to the activated tyrosine kinase, prevents phosphorylation, and allows the leukemic cells to differentiate and undergo apoptosis. In acute promyelocytic leukemia, fusion of the retinoic acid receptor-alpha with the gene coding for promyelocytic protein, the PML-RAR alpha (t15:17) translocation, accounts for a fusion product blocking the activity of the promyelocytic protein. This protein is required for the formation of the granules of promyelocytes. Retinoic acids (RA) bind to the retinoic acid receptor (RAR alpha) component of the fusion product, resulting in degradation of the fusion protein by ubiquitinization. This allows normal PML to participate in granule formation and differentiation of the promyelocytes. In cases of RA resistance, the association with vitamin D3 can induce differentiation of leukemic cell lines [670]. Finally, retinoid and their receptors play an important role in controlling the progression of many other tumors [671].

The Role of the Embryonic Microenvironment in Controlling Tumor Growth

There are series of researches demonstrating that the factors present in the embryo during organogenesis are able to prevent the cancer cell induction [672-673]. Exposure of developing embryos to chemical carcinogens during organogenesis may lead to malformations, but not to cancer. Thus, "the embryonic microenvironment is able to correct for the mutations or epigenetic alterations induced by carcinogens and prevent cancer formation" [674]. As stated above, the "blastocyst environment also controls growth of transplanted malignant cells, whereas malignant cells placed in other sites of the developing embryo, such as the perivitelline space, are not controlled" [675]. On the other hand, more differentiated tissues of the embryo may also regulate growth of cancer cells that are normally found in that tissue. Many kinds of cancers including leukemia, melanoma, hepatocellular carcinoma and breast cancer converted to normal developing tissue when placed in an appropriate embryonic microenvironment [676-677]. Malignant melanoma cells placed into the extra-embryonal membrane of the Zebrafish differentiated into normal neural crest like cells [678]. Thus, "different embryonic microenvironments may have different differentiation potentials related to normal cellular differentiation. For example, the placenta may also regulate transplanted leukemia cells" [679]. Many other experiments demonstrated similar results [680-681] so that it has been concluded that "embryo-mother cross-talk is very important in determining the arrest of tumor growth, because both maternal (decidua) and embryonic tissues contain substances with anti-cancer properties" [682]. These promising results open new perspectives not only in cancer biology comprehension, but also in conceiving new different therapeutic strategies. In fact, in the last years many studies have clarified that "embryonic cells share fundamental features with tumor cells" [683], such as embryonic proteins (AFP, or ABC transporters), common molecular pathways (*i.e.* beta-catenin/TCF/WNT Notch, BMP and Hedgehog signals) [684-686], a metabolism based on glycolytic way to produce energy [687]. In addition, EMT observed in cancer tissue [314, 688] may also be conceived as an embryonic program of multiplication and development [689].

To understand how the substances taken from developing embryo or other biological substances are able to induce differentiation of cancer cells, important results on reprogramming and differentiation treatment are here recorded.

The Use of Stem Cell Differentiation Stage Factors (SCDFs)

One of the most important approach in reprogramming cancer cells is the use of stem cell differentiation stage factors (SCDSFs) taken from zebrafish embryos at different stages of development. These factors are first of all (99%) proteins with low molecular weight and (1%) nucleic acids with regulatory properties [690]. The use of this network of proteins and nucleic acids demonstrated "a significant inhibition of the proliferation of different human tumor lines *in vitro* (glioblastoma multiforme, melanoma, hepatocellular carcinoma, breast carcinoma, kidney adenocarcinoma, colon adenocarcinoma, acute lymphoblastic leukemia)" [691]. The molecules that play a key role in the regulation process of the cell cycle are, for example, p53 [692] and pRb [693]: that means that SCDSFs are able to modulate the regulation of transcriptional and post-translational processes. In addition, the treatment of colon cancer with SCDSFs induces caspase 3 activation with a consequent induction of an apoptotic pathway. There is also "a concurrent significant normalization of the ratio of e-cadherin to beta catenin, with an increase in e-cadherin levels" [694]. Finally, SCDSFs induce "an almost complete growth-inhibition of cell proliferation of CaCO2 colon cancer cell line after the concurrent treatment with 5 Fluorouracil" [695], suggesting that "zebrafish embryo factors improve chemotherapy efficacy with 5 Fluorouracil. Subcutaneous injection of SCDFs with primary Lewis Lung Carcinoma cells into C57BL/6 female syngeneic mice produces a highly significant difference ($P<0.001$) between treated and control mice both in terms of primary tumor growth and survival compared to untreated controls" [696]. A randomized clinical trial, conducted between January 1th, 2001 and April 30th, 2004 on 179 patients with intermediate-advanced hepatocellular carcinoma unresponsive to conventional treatment (transplantation, resection, ablation therapy or chemoembolization) have been treated with a product developed following the aforementioned *in vitro* and *in vivo* studies. This product, administered sublingually at a dose of 30 drops to the patients three times a day, demonstrated "a regression in 19.8% of the patients (2.4% of complete regression, and 17.4% of partial regression) and a non-progression in 16% with over 60% survival rate after 40 months in the patients who responded to treatment, compared to 10% in the remaining patients. The performance status improved in 82.6% of the treated patients, also in those with advanced stages of the disease" [697].

Another clinical trial was conducted on 50 patients with "advanced" stage HCC from 2005 to 2010. In 13,1% of the patients with "advanced" stage there was demonstrated a complete response with disappearance of the tumoral lesions and normalization of AlFP levels. No side effects occurred [698].

Other Reprogramming Treatments

Other important reprogramming treatments are here recorded in synthetic way. For example, "extracts taken from prophase amphibian oocytes of axolotl (AOE) and xenopus (XOE) induce re-expression of some tumor suppressor genes. RARB, CST6, CCND2, GAS2 and CDKN2A are silenced or expressed at very low levels in MCF-7 and HCC 1954 cell lines representing luminal and basal breast cancer phenotypes" [699]. Extracts of AOE and XOE prophase oocytes induce re-expression of these genes in both breast cancer cell lines.

Also, human umbilical cord matrix stem cells (hUCMSCs) are unique stem cells derived from Warton's jelly (termed Warton's jelly stromal cells). Other reprogramming treatments regard metastatic growth of MDA-231 human breast carcinoma cells in a mouse xenograft experimental model treated with un-engineered (naïve) human umbilical cord matrix stem cells (hUCMSC) [700]. The inhibited growth of cancer cells appears to be due to blocking of ERK-1/2 and PI3K/Akt signaling with activation of intrinsic apoptosis signals.

In prostate cancer, "a subpopulation of cancer stem-like cells that are highly tumorigenic expresses some essential stem cell associated transcription factors (Oct3/4, Sox2, Nanog, Klf4 and c-myc)" [701]. Due to their importance in prostate cancer growth, these cancer stem cells seem a likely target for novel strategies for prostate cancer therapy. One of these novel strategies is related to the enzyme telomerase. "New gene constructs to reprogram telomerase have been engineered and validated: 1) small interfering RNA against wild type mouse telomerase RNA (••MTer-siRNA); 2) mutant-template mouse telomerase RNA

(MT-mTer) which encodes incorrect mouse telomeric repeats.

Lentiviral delivery of ••-mTer-siRNA to mouse prostate cancer cells caused growth inhibition and rapid apoptosis in cancer progenitor cells isolated from human prostatectomy specimens" [702]. In addition, we record here that "the transfection of human PC-3 and LNCaP prostate carcinoma cells with small interfering double-stranded RNA (siRNA) oligonucleotides against the MAT-8 gene produced a specific down-regulation of MAT-8 expression and a significant decrease in cellular proliferation of both cancer cell lines" [703].

Other examples of reprogramming cancer cells are the epigenetic reprogramming of breast cancer. "Site specific DNA methylation and prolonged repression of the tumor suppressor Maspin and the oncogene Sox2 can be obtained in breast cancer cells by Zinc-finger ATFs targeting DNA methyltransferase-3a to the promoters of these genes" [704].

Other way to reprogram cancer cells consists of the "stimulation of immunity to cancer by reprogramming tumor-associated dendritic cells. The endocytic activity and immune activation of ovarian cancer-associated dendritic cells (DC) may be selectively increased by treatment with the immunostimulatory miRNA miR-155. Nanoparticles carrying oligonucleotide duplexes mimicking the bulged structure of endogenous pre-miRNA strongly enhanced miR-155 activity. Tumor infiltrating DCs were transformed from immunosuppressive to highly immunostimulatory cells, activating strong antitumor responses" [705].

Also, the transfer of chromosome 3p fragments in a novel epithelial ovarian cancer cell line model (OV-90) induce tumor suppression. Reprogramming of the transcriptome appears to be a consequence of the chromosome 3 transfer [706].

Conventional Antiestrogen Therapy Plus Immune Stimulation

Recently, we have reported on promising results and the potential rationale [548] of a new schedule of conventional antiestrogen therapy combined with an immune stimulation. They both are here briefly summarized.

In 24 with higher levels of hormone dependency (55% of ER+ PR+) among 31 studied endocrine dependent metastatic breast cancer patients, a longer than 5 years OS and a delayed median PFS of 45 months were found compared with 7 remaining with lower hormone dependency (20% of ER+ PR+) in whom OS was shorter than 5 years and median PFS was 20 months. Our explanation of these findings was that "a direct antagonism with estrogen at the promoting regions of DNA by antiestrogen likely accounts for downregulation of the genes favoring proliferation, survival and angiogenesis. It derives a stable or decreased tumor burden and a quiescent state of tumoral cells (G0-G1 phase). Under this particular condition, a lower genomic instability and a decreased total protein synthesis with lowering of circulating growth factors, surface receptors, cytokines and key regulatory enzymes occurs. This likely accounts for a downregulation .... thus delaying acquired hormone resistance" [548]. Furthermore, it is likely that "in any hormone dependent breast cancer patient, a portion only of tumoral burden is under cytostatic effect of hormone treatment. The portion of tumor burden not well controlled by hormone treatment continues to work and to foster with antiestrogens" [548]. It is noteworthy that in this pilot study 16% of patients survived more than 10 years in complete remission, compared with 1.6% of long term survivors so far reported in the same population of MBC [707].

A New Direction for Antitumoral Treatment

It is clear that cancer is a complex disease, that is sustained by many alterations present at genetic (mutations, amplifications, deletions, *etc.*) and epigenetic (hypo and hyper methylation, noncoding RNAs) levels. These alterations foster a great deal of pathological interconnected molecular pathways, that morphologically account for a moreless undifferentiated phenotype and functionally for an endless proliferation. Recently Hanahan and Weinberg [708] as "an organizing principle for rationalizing the complexity of neoplastic disease" have defined "eight biological capabilities acquired during the multistep development of human tumors" that they called "hallmarks of cancer". These "hallmarks" include: sustaining proliferative signaling, evading growth suppressors, resisting cell deaths, enabling replicative immortality, inducing angiogenesis, activating invasion and metastasis, reprogramming of energy and metabolism, evading immune destruction.

In this milestone article, the authors analyze each hallmark and report on the principal scientific literature data supporting the key role of these eight hallmarks in the progression of cancer. After having reminded their previous metaphor [708], according to which "numerous signaling molecules affecting cancer cells operate as nodes and branches of elaborated integrated circuits that are reprogrammed derivatives of the circuits operating in normal cells", they show "a circuitry that is aligned with individual hallmarks of cancer". They state that it mimics "the intracellular integrated circuit" that "can be segmented into distinct subcircuits, each of which is specialized to support a discrete cell-biological property in normal cells and is reprogrammed in order to implement a hallmark capability in cancer cells".

Indeed, the illustrative representation of the pathological mechanisms fostering the progression of cancer disease is consistently supported by many data from experimental and clinical research. Nevertheless, it is not exhaustive and does not recapitulate all the findings and the experience that has been acquired in the field. In fact, although for any cancer the same eight hallmarks are activated, the entire circuitry, the circuits and subcircuits are likely different as they reflect different genetic and epigenetic alterations and different contribution from tumor microenvironment. This is supported by the finding that any cancer has its own genetic signature and biological aggressiveness. Besides favored by the genetic instability and biological plasticity, the portrait of the entire activated circuitry, circuits and subcircuits significantly changes during the disease, as documented by temporal heterogeneity of the tumor and the arising of resistance to treatments. This makes very difficult to eradicate advanced cancer disease. However, there are findings suggesting that immune tolerance is impaired when cancer cells are in resting state or in response to conventional therapy mainly in the condition of minimal residual cancer disease.

Consistent with this are data from some clinical studies conducted by us and other authors. In the just above reported pilot study in endocrine dependent MBC, in spite of the advanced disease and large tumor burden, more than doubled OS than that expected occurred. According to our interpretative model, in these patients due to antiestrogens the hormone dependent cancer cells are in a resting state (G0-G1 state) that blocks tumor growth. This and the following downregulation of the mechanisms sustaining the immune evasion permitted an active immune stimulation with interferon-beta-interleukin-2 sequence [548]. In another pilot study conducted by us in gastrointestinal cancers with high risk of relapse, again almost a doubled 5 years OS than that expected occurred. In this study [709], starting from the first year after conventional adjuvant chemotherapy and up to the fifth year, patients received 2-3 cycles of additional adjuvant chemotherapy by infusional 5-FU plus leucovorin. We hypothesized that in chemosensitive patients 5-FU-leucovorin regimen mimics at least in part the mechanism of action of antiestrogens. Therefore, in this population of patients with minimal residual disease, residual cancer cells more times and at regular intervals likely were forced to remain in a resting state. This can have blocked tumor growth and favored the recovery of an immunosurveillance. Similarly, in another experimental study [710] a prolonged median OS occurred in metastatic breast cancer patients on clinical benefit following conventional therapy who as maintenance treatment had received retinoids and interleukin 2 in addition to tamoxifen.

Also, it is well known that about 10%-30% of colorectal cancer patients recurring at liver after primary surgery are definitely cured following conventional adjuvant chemotherapy that commonly includes infusional 5-FU plus leucovorin and/or radical resection of liver metastases [711-714]. It is noteworthy that only about

20% of colorectal cancer patients are chemosensitive to adjuvant chemotherapy with infusional 5-FU-leucovorin regimen [715]. This last and the previous example of gastrointestinal cancers at high risk of relapse submitted to additional adjuvant chemotherapy underline the relevance of an undetectable tumor burden.

Most known molecular pathways sustaining tumor growth take part of autocrine or paracrine loops involving cancer cell [716].

When cancer cell is in a resting state (G0-G1 state), all its activities including the synthesis of relevant proteins acting as enzymes or growth factors and the entire circuitry fostering tumor growth are downregulated. An undetectable tumor burden is likely another condition favoring the downregulation of the hallmarks including impairment of immune tolerance. Therefore, in a resting state of cancer cells or in the case of an undetectable tumor burden, an induced or active immune manipulation or a spontaneous recovery of immune system are more likely to be obtained.

Most used conventional antiblastics are toxic for cancer cell and aim at disrupting DNA or cytoplasmic organization while the more recently introduced biological drugs usually target a specific or few molecular pathways within the same subcircuit. Instead, a decrease of tumor burden possibly to minimal residual disease or a resting state of cancer cells should be addressed by reductive surgical interventions and/or a synergic pharmacological combination. In the case, such therapeutic interventions are successful, attempts should be made to contemporaneously stimulate the immune system. In particular, these attempts could be orchestrated with the concomitant use of drugs capable to inhibit the well-known mechanisms of the suppression of cell mediated immunity and drugs capable to stimulate the immune system.

CONCLUSIONS AND PERSPECTIVES

Breast cancer has an huge social relevance being among women worldwide the most commonly diagnosed cancer and the second leading cause of cancer-related death. In 2012 there were about 1,68 million new diagnosed worldwide and 520,000 deaths [128]. Since many decades continuous efforts have been made by the scientific communities to improve the comprehension of the disease and the clinical outcome. As to the former, a tremendous progress has been attained by researches in genetics and molecular biology. As to the latter, the spread of the screening programs have significantly increased the rate of early detected cancers and more efficacious and better tailored treatments have significantly improved the prognosis of some subsets in advanced disease. In spite of some undeniable results, metastatic breast cancer remains a main hurdle to overcome due to the common occurrence of resistance to the different available drugs. This chapter has examined the main aspects of the disease on the light of the recent advances. In particular, the principal current treatments in the different settings have been described. Moreover, the main options to overcome drug resistance have been reported. Cancer cell reprogramming and an innovative schedule of hormone immunotherapy based on a new therapeutic strategy joining tumor growth with immune evasion are new perspectives of cancer treatment that are worthwhile to be intensively investigated. In conclusion, it is clear that the understanding and therapy of cancer is a work in progress that needs to be regularly up-dated. This process mainly involves breast cancer which is the leading cancer among women.

CONFLICT OF INTEREST

The authors declare no conflict of interest, financial or otherwise.

ACKNOWLEDGEMENTS

Declared none.

REFERENCES

[1] Ferlay J, Shin H, Bray F, Forman D, Mathers C, Parkin D. GLOBOCAN 2008: cancer incidence and mortality worldwide.IARC CancerBase No 10 (version 20). Lyon: IARC 2010.

[2] Ahmad O, Boschi-Pinto C, Lopez A, Murray C, Lozano R, Inoue M. Age standardization of rates: a new WHO standard. GPE Discussion Paper Series No. 31. Geneva: World Health Organization; 2001.

[3] Hortobagyi GN, de la Garza Salazar J, Pritchard K, *et al.* The global breast cancer burden: variations in epidemiology and survival. Clin Breast Cancer 2005; 6(5): 391-401. [http://dx.doi.org/10.3816/CBC.2005.n.043] [PMID: 16381622]

[4] Parkin DM, Fernández LM. Use of statistics to assess the global burden of breast cancer. Breast J 2006; 12(1) (Suppl. 1): S70-80. [http://dx.doi.org/10.1111/j.1075-122X.2006.00205.x] [PMID: 16430400]

[5] Agarwal G, Ramakant P, Forgach ER, *et al.* Breast cancer care in developing countries. World J Surg 2009; 33(10): 2069-76. [http://dx.doi.org/10.1007/s00268-009-0150-z] [PMID: 19653033]

[6] Coughlin SS, Ekwueme DU. Breast cancer as a global health concern. Cancer Epidemiol 2009; 33(5): 315-8. [http://dx.doi.org/10.1016/j.canep.2009.10.003] [PMID: 19896917]

[7] Shin HR, Joubert C, Boniol M, *et al.* Recent trends and patterns in breast cancer incidence among Eastern and Southeastern Asian women. Cancer Causes Control 2010; 21(11): 1777-85. [http://dx.doi.org/10.1007/s10552-010-9604-8] [PMID: 20559704]

[8] Giles GG, Amos A. Evaluation of the organised mammographic screening programme in Australia. Ann Oncol 2003; 14(8): 1209-11. [http://dx.doi.org/10.1093/annonc/mdg326] [PMID: 12881380]

[9] Kumle M. Declining breast cancer incidence and decreased HRT use. Lancet 2008; 372(9639): 608-10. [http://dx.doi.org/10.1016/S0140-6736(08)61255-6] [PMID: 18722851]

[10] Rossouw JE, Anderson GL, Prentice RL, *et al.* Risks and benefits of estrogen plus progestin in healthy postmenopausal women: principal results From the Women's Health Initiative randomized controlled

trial. JAMA 2002; 288(3): 321-33.
[http://dx.doi.org/10.1001/jama.288.3.321] [PMID: 12117397]

[11] Prat A, Adamo B, Fan C, *et al.* Gene expression analyses across six cancer types identify basal-like breast cancer as a unique molecular entity. Sci Rep 2013; 3: 1-12.
[http://dx.doi.org/10.1038/srep03544]

[12] Anderson WF, Matsuno R. Breast cancer heterogeneity: a mixture of at least two main types? J Natl Cancer Inst 2006; 98(14): 948-51.
[http://dx.doi.org/10.1093/jnci/djj295] [PMID: 16849671]

[13] Li JJ, Li SA, Mohla S, Rochefort H, Maudelonde T. Hormonal Carcinogenesis V Advances in Experimental Medicine and Biology. New York: Springer 2008; Vol. 617.
[http://dx.doi.org/10.1007/978-0-387-69080-3]

[14] Ellis MJ, Ding L, Shen D, *et al.* Whole-genome analysis informs breast cancer response to aromatase inhibition. Nature 2012; 486(7403): 353-60.
[PMID: 22722193]

[15] Ithimakin S, Day KC, Malik F, *et al.* HER2 drives luminal breast cancer stem cells in the absence of HER2 amplification: implications for efficacy of adjuvant trastuzumab. Cancer Res 2013; 73(5): 1635-46.
[http://dx.doi.org/10.1158/0008-5472.CAN-12-3349] [PMID: 23442322]

[16] Korkaya H, Wicha MS. HER2 and breast cancer stem cells: more than meets the eye. Cancer Res 2013; 73(12): 3489-93.
[http://dx.doi.org/10.1158/0008-5472.CAN-13-0260] [PMID: 23740771]

[17] Korkaya H, Wicha MS. Breast cancer stem cells: we've got them surrounded. Clin Cancer Res 2013; 19(3): 511-3.
[http://dx.doi.org/10.1158/1078-0432.CCR-12-3450] [PMID: 23251003]

[18] Prat A, Perou CM. Mammary development meets cancer genomics. Nat Med 2009; 15(8): 842-4.
[http://dx.doi.org/10.1038/nm0809-842] [PMID: 19661985]

[19] Lilienfeld AM, Johnson EA. The age distribution in female breast and genital cancers. Cancer 1955; 8(5): 875-82.
[http://dx.doi.org/10.1002/1097-0142(1955)8:5<875::AID-CNCR2820080504>3.0.CO;2-Z] [PMID: 13261037]

[20] de Waard F. Premenopausal and postmenopausal breast cancer: one disease or two? J Natl Cancer Inst 1979; 63(3): 549-52.
[http://dx.doi.org/10.1093/jnci/63.3.549] [PMID: 381750]

[21] De Waard F, Baanders-Vanhalewijn EA, Huizinga J. The bimodal age distribution of patients with mammary carcinoma; evidence for the existence of 2 types of human breast cancer. Cancer 1964; 17(2): 141-51.
[http://dx.doi.org/10.1002/1097-0142(196402)17:2<141::AID-CNCR2820170202>3.0.CO;2-Z] [PMID: 14123673]

[22] de LAIVE J, Baanders-Vanhalewijn EA, Baanders-van Halewijn EA. On the bimodal age distribution of mammary carcinoma. Br J Cancer 1960; 14(3): 437-48.
[http://dx.doi.org/10.1038/bjc.1960.47] [PMID: 13782349]

[23] Anderson WF, Rosenberg PS, Menashe I, Mitani A, Pfeiffer RM. Age-related crossover in breast cancer incidence rates between black and white ethnic groups. J Natl Cancer Inst 2008; 100(24): 1804-14.
[http://dx.doi.org/10.1093/jnci/djn411] [PMID: 19066264]

[24] DeSantis C, Howlader N, Cronin KA, Jemal A. Breast cancer incidence rates in U.S. women are no longer declining. Cancer Epidemiol Biomarkers Prev 2011; 20(5): 733-9.
[http://dx.doi.org/10.1158/1055-9965.EPI-11-0061] [PMID: 21357727]

[25] Anderson WF, Katki HA, Rosenberg PS. Breast cancer incidence in the United States: current and future trends. J Natl Cancer Inst 2011; 103(18): 1397-402.
[http://dx.doi.org/10.1093/jnci/djr257] [PMID: 21753181]

[26] DeSantis C, Ma J, Bryan L, Jemal A. Breast cancer statistics, 2013. CA Cancer J Clin 2014; 64(1): 52-62.
[http://dx.doi.org/10.3322/caac.21203] [PMID: 24114568]

[27] Anderson WF, Rosenberg PS, Petito L, *et al.* Divergent estrogen receptor-positive and -negative breast cancer trends and etiologic heterogeneity in Denmark. Int J Cancer 2013; 133(9): 2201-6.
[http://dx.doi.org/10.1002/ijc.28222] [PMID: 23616071]

[28] Toriola AT, Colditz GA. Trends in breast cancer incidence and mortality in the United States: implications for prevention. Breast Cancer Res Treat 2013; 138(3): 665-73.
[http://dx.doi.org/10.1007/s10549-013-2500-7] [PMID: 23546552]

[29] Knudson AG. Two genetic hits (more or less) to cancer. Nat Rev Cancer 2001; 1(2): 157-62.
[http://dx.doi.org/10.1038/35101031] [PMID: 11905807]

[30] Gierach GL, Burke A, Anderson WF. Epidemiology of triple negative breast cancers. Breast Dis 2010; 32(1-2): 5-24.
[http://dx.doi.org/10.3233/BD-2010-0319] [PMID: 21965309]

[31] Anderson WF, Luo S, Chatterjee N, *et al.* Human epidermal growth factor receptor-2 and estrogen receptor expression, a demonstration project using the residual tissue repository of the Surveillance, Epidemiology, and End Results (SEER) program. Breast Cancer Res Treat 2009; 113(1): 189-96.
[http://dx.doi.org/10.1007/s10549-008-9918-3] [PMID: 18256926]

[32] Telli ML, Chang ET, Kurian AW, *et al.* Asian ethnicity and breast cancer subtypes: a study from the California Cancer Registry. Breast Cancer Res Treat 2011; 127(2): 471-8.
[http://dx.doi.org/10.1007/s10549-010-1173-8] [PMID: 20957431]

[33] Kurian AW, Fish K, Shema SJ, Clarke CA. Lifetime risks of specific breast cancer subtypes among women in four racial/ethnic groups. Breast Cancer Res 2010; 12(6): R99.
[http://dx.doi.org/10.1186/bcr2780] [PMID: 21092082]

[34] Keegan TH, DeRouen MC, Press DJ, Kurian AW, Clarke CA. Occurrence of breast cancer subtypes in adolescent and young adult women. Breast Cancer Res 2012; 14(2): R55.
[http://dx.doi.org/10.1186/bcr3156] [PMID: 22452927]

[35] Clarke CA, Keegan TH, Yang J, *et al.* Age-specific incidence of breast cancer subtypes: understanding the black-white crossover. J Natl Cancer Inst 2012; 104(14): 1094-101.
[http://dx.doi.org/10.1093/jnci/djs264] [PMID: 22773826]

[36] 2012.www.seer.cancer.gov

[37] Curtis C, Shah SP, Chin SF, *et al.* The genomic and transcriptomic architecture of 2,000 breast tumours reveals novel subgroups. Nature 2012; 486(7403): 346-52.
[PMID: 22522925]

[38] Parker JS, Mullins M, Cheang MC, *et al.* Supervised risk predictor of breast cancer based on intrinsic subtypes. J Clin Oncol 2009; 27(8): 1160-7.
[http://dx.doi.org/10.1200/JCO.2008.18.1370] [PMID: 19204204]

[39] Dawson SJ, Rueda OM, Aparicio S, Caldas C. A new genome-driven integrated classification of breast cancer and its implications. EMBO J 2013; 32(5): 617-28.
[http://dx.doi.org/10.1038/emboj.2013.19] [PMID: 23395906]

[40] Lakhani S, Ellis IO, Schnitt SJ, Eds. WHO classification of tumour of the breast. 4th ed., Lyon: IARC 2012.

[41] Vinay K, Abul KA, Jon CA, Nelson F. Robbins and Cotran Pathologic Basis of Disease. 2010.

[42] Fattaneh AT, Peter D. WHO Classification Pathology and Genetics of Tumours of the Breast and Female Genital Organs. Lyon, France: IARC Press 2003.

[43] Rosai J. Rosai and Ackerman's Surgical Pathology. 10th ed., Lyon, France: Elsevier 2011.

[44] Hanby AM, Hughes TA. *In situ* and invasive lobular neoplasia of the breast. Histopathology 2008; 52(1): 58-66.
[http://dx.doi.org/10.1111/j.1365-2559.2007.02891.x] [PMID: 18171417]

[45] Schnitt SJ, Morrow M. Lobular carcinoma *in situ*: current concepts and controversies. Semin Diagn Pathol 1999; 16(3): 209-23.
[PMID: 10490198]

[46] Mastracci TL, Tjan S, Bane AL, O'Malley FP, Andrulis IL. E-cadherin alterations in atypical lobular hyperplasia and lobular carcinoma *in situ* of the breast. Mod Pathol 2005; 18(6): 741-51.
[http://dx.doi.org/10.1038/modpathol.3800362] [PMID: 15696125]

[47] Farid M. Essentials of Diagnostic Breast Pathology, Practical Approach. 1st ed., Berlin: Springer 2007.

[48] Li CI, Anderson BO, Daling JR, Moe RE. Trends in incidence rates of invasive lobular and ductal breast carcinoma. JAMA 2003; 289(11): 1421-4.
[http://dx.doi.org/10.1001/jama.289.11.1421] [PMID: 12636465]

[49] Rakha EA, Ellis IO. Lobular breast carcinoma and its variants. Semin Diagn Pathol 2010; 27(1): 49-61.
[http://dx.doi.org/10.1053/j.semdp.2009.12.009] [PMID: 20306830]

[50] Iorfida M, Maiorano E, Orvieto E, *et al.* Invasive lobular breast cancer: subtypes and outcome. Breast Cancer Res Treat 2012; 133(2): 713-23.
[http://dx.doi.org/10.1007/s10549-012-2002-z] [PMID: 22399188]

[51] Ciriello G, Gatza ML, Beck AH, *et al.* Comprehensive molecular portraits of invasive lobular breast cancer. Cell 2015; 163(2): 506-19.
[http://dx.doi.org/10.1016/j.cell.2015.09.033] [PMID: 26451490]

[52] Desmedt C, Zoppoli G, Gundem G, *et al.* Genomic characterization of primary invasive lobular breast cancer. J Clin Oncol 2016; 34(16): 1872-81.
[http://dx.doi.org/10.1200/JCO.2015.64.0334] [PMID: 26926684]

[53] Fernández B, Paish EC, Green AR, *et al.* Lymph-node metastases in invasive lobular carcinoma are different from those in ductal carcinoma of the breast. J Clin Pathol 2011; 64(11): 995-1000.
[http://dx.doi.org/10.1136/jclinpath-2011-200151] [PMID: 21712309]

[54] Li CI, Weiss NS, Stanford JL, Daling JR. Hormone replacement therapy in relation to risk of lobular and ductal breast carcinoma in middle-aged women. Cancer 2000; 88(11): 2570-7.
[http://dx.doi.org/10.1002/1097-0142(20000601)88:11<2570::AID-CNCR20>3.0.CO;2-O] [PMID: 10861435]

[55] Talman ML, Jensen MB, Rank F. Invasive lobular breast cancer. Prognostic significance of histological malignancy grading. Acta Oncol 2007; 46(6): 803-9.
[http://dx.doi.org/10.1080/02841860601137397] [PMID: 17653904]

[56] Suhrke P, Mæhlen J, Zahl PH. Hormone therapy use and breast cancer incidence by histological subtypes in Sweden and Norway. Breast J 2012; 18(6): 549-56.
[http://dx.doi.org/10.1111/tbj.12001] [PMID: 23002918]

[57] Toikkanen S, Pylkkänen L, Joensuu H. Invasive lobular carcinoma of the breast has better short- and long-term survival than invasive ductal carcinoma. Br J Cancer 1997; 76(9): 1234-40.
[http://dx.doi.org/10.1038/bjc.1997.540] [PMID: 9365176]

[58] Li CI, Moe RE, Daling JR. Risk of mortality by histologic type of breast cancer among women aged 50 to 79 years. Arch Intern Med 2003; 163(18): 2149-53.

[http://dx.doi.org/10.1001/archinte.163.18.2149] [PMID: 14557212]

[59] Wasif N, Maggard MA, Ko CY, Giuliano AE. Invasive lobular *vs*. ductal breast cancer: a stage-matched comparison of outcomes. Ann Surg Oncol 2010; 17(7): 1862-9. [http://dx.doi.org/10.1245/s10434-010-0953-z] [PMID: 20162457]

[60] Lee JH, Park S, Park HS, Park BW. Clinicopathological features of infiltrating lobular carcinomas comparing with infiltrating ductal carcinomas: a case control study. World J Surg Oncol 2010; 8: 34. [http://dx.doi.org/10.1186/1477-7819-8-34] [PMID: 20423478]

[61] Santiago RJ, Harris EE, Qin L, Hwang WT, Solin LJ. Similar long-term results of breast-conservation treatment for Stage I and II invasive lobular carcinoma compared with invasive ductal carcinoma of the breast: The University of Pennsylvania experience. Cancer 2005; 103(12): 2447-54. [http://dx.doi.org/10.1002/cncr.21071] [PMID: 15887223]

[62] Viale G, Rotmensz N, Maisonneuve P, *et al.* Lack of prognostic significance of "classic" lobular breast carcinoma: a matched, single institution series. Breast Cancer Res Treat 2009; 117(1): 211-4. [http://dx.doi.org/10.1007/s10549-008-0112-4] [PMID: 18629634]

[63] Colleoni M, Rotmensz N, Maisonneuve P, *et al.* Outcome of special types of luminal breast cancer. Ann Oncol 2012; 23(6): 1428-36. [http://dx.doi.org/10.1093/annonc/mdr461] [PMID: 22039080]

[64] Eusebi V, Magalhaes F, Azzopardi JG. Pleomorphic lobular carcinoma of the breast: an aggressive tumor showing apocrine differentiation. Hum Pathol 1992; 23(6): 655-62. [http://dx.doi.org/10.1016/0046-8177(92)90321-S] [PMID: 1592388]

[65] Weidner N, Semple JP. Pleomorphic variant of invasive lobular carcinoma of the breast. Hum Pathol 1992; 23(10): 1167-71. [http://dx.doi.org/10.1016/0046-8177(92)90035-2] [PMID: 1398644]

[66] Bentz JS, Yassa N, Clayton F. Pleomorphic lobular carcinoma of the breast: clinicopathologic features of 12 cases. Mod Pathol 1998; 11(9): 814-22. [PMID: 9758360]

[67] Montagna E, Maisonneuve P, Rotmensz N, *et al.* Heterogeneity of triple-negative breast cancer: histologic subtyping to inform the outcome. Clin Breast Cancer 2013; 13(1): 31-9. [http://dx.doi.org/10.1016/j.clbc.2012.09.002] [PMID: 23098574]

[68] Rosen PP. Rosen's Breast Pathology. Philadelphia, PA: Lippincott Williams & Wilkins 2001.

[69] Jacobs M, Fan F, Tawfik O. Clinicopathologic and biomarker analysis of invasive pleomorphic lobular carcinoma as compared with invasive classic lobular carcinoma: an experience in our institution and review of the literature. Ann Diagn Pathol 2012; 16(3): 185-9. [http://dx.doi.org/10.1016/j.anndiagpath.2011.10.001] [PMID: 22197540]

[70] Diab SG, Clark GM, Osborne CK, Libby A, Allred DC, Elledge RM. Tumor characteristics and clinical outcome of tubular and mucinous breast carcinomas. J Clin Oncol 1999; 17(5): 1442-8. [http://dx.doi.org/10.1200/JCO.1999.17.5.1442] [PMID: 10334529]

[71] Oakley GJ III, Tubbs RR, Crowe J, *et al.* HER-2 amplification in tubular carcinoma of the breast. Am J Clin Pathol 2006; 126(1): 55-8. [http://dx.doi.org/10.1309/E0YEKHBP3YYQYUBD] [PMID: 16753605]

[72] Memis A, Ozdemir N, Parildar M, Ustun EE, Erhan Y. Mucinous (colloid) breast cancer: mammographic and US features with histologic correlation. Eur J Radiol 2000; 35(1): 39-43. [http://dx.doi.org/10.1016/S0720-048X(99)00124-2] [PMID: 10930764]

[73] Louwman MW, Vriezen M, van Beek MW, *et al.* Uncommon breast tumors in perspective: incidence, treatment and survival in the Netherlands. Int J Cancer 2007; 121(1): 127-35. [http://dx.doi.org/10.1002/ijc.22625] [PMID: 17330844]

[74] Venable JG, Schwartz AM, Silverberg SG. Infiltrating cribriform carcinoma of the breast: a distinctive

clinicopathologic entity. Hum Pathol 1990; 21(3): 333-8. [http://dx.doi.org/10.1016/0046-8177(90)90235-W] [PMID: 2312110]

[75] Di Saverio S, Gutierrez J, Avisar E. A retrospective review with long term follow up of 11,400 cases of pure mucinous breast carcinoma. Breast Cancer Res Treat 2008; 111(3): 541-7. [http://dx.doi.org/10.1007/s10549-007-9809-z] [PMID: 18026874]

[76] Clayton F. Pure mucinous carcinomas of breast: morphologic features and prognostic correlates. Hum Pathol 1986; 17(1): 34-8. [http://dx.doi.org/10.1016/S0046-8177(86)80152-6] [PMID: 3002950]

[77] Tan PH, Tse GM, Bay BH. Mucinous breast lesions: diagnostic challenges. J Clin Pathol 2008; 61(1): 11-9. [http://dx.doi.org/10.1136/jcp.2006.046227] [PMID: 17873114]

[78] Righi L, Sapino A, Marchiò C, Papotti M, Bussolati G. Neuroendocrine differentiation in breast cancer: established facts and unresolved problems. Semin Diagn Pathol 2010; 27(1): 69-76. [http://dx.doi.org/10.1053/j.semdp.2009.12.003] [PMID: 20306832]

[79] Sapino A, Papotti M, Righi L, Cassoni P, Chiusa L, Bussolati G. Clinical significance of neuroendocrine carcinoma of the breast. Ann Oncol 2001; 12(S2) (Suppl. 2): S115-7. [http://dx.doi.org/10.1093/annonc/12.suppl_2.S115] [PMID: 11762336]

[80] Tavassoli FA, Devilee P. World Health Organization Classification of Tumors: Pathology and Genetics of Tumors of the Breast and Female Genital Organs. Lyon, France: IARC Press 2003.

[81] Work ME, Andrulis IL, John EM, *et al.* Risk factors for uncommon histologic subtypes of breast cancer using centralized pathology review in the Breast Cancer Family Registry. Breast Cancer Res Treat 2012; 134(3): 1209-20. [http://dx.doi.org/10.1007/s10549-012-2056-y] [PMID: 22527103]

[82] Ridolfi RL, Rosen PP, Port A, Kinne D, Miké V. Medullary carcinoma of the breast: a clinicopathologic study with 10 year follow-up. Cancer 1977; 40(4): 1365-85. [http://dx.doi.org/10.1002/1097-0142(197710)40:4<1365::AID-CNCR2820400402>3.0.CO;2-N] [PMID: 907958]

[83] Vu-Nishino H, Tavassoli FA, Ahrens WA, Haffty BG. Clinicopathologic features and long-term outcome of patients with medullary breast carcinoma managed with breast-conserving therapy (BCT). Int J Radiat Oncol Biol Phys 2005; 62(4): 1040-7. [http://dx.doi.org/10.1016/j.ijrobp.2005.01.008] [PMID: 15990007]

[84] Vincent-Salomon A, Gruel N, Lucchesi C, *et al.* Identification of typical medullary breast carcinoma as a genomic sub-group of basal-like carcinomas, a heterogeneous new molecular entity. Breast Cancer Res 2007; 9(2): R24. [http://dx.doi.org/10.1186/bcr1666] [PMID: 17417968]

[85] Wang XX, Jiang YZ, Liu XY, Li JJ, Song CG, Shao ZM. Difference in characteristics and outcomes between medullary breast carcinoma and invasive ductal carcinoma: a population based study from SEER 18 database. Oncotarget 2016; 7(16): 22665-73. [http://dx.doi.org/10.18632/oncotarget.8142] [PMID: 27009810]

[86] Vranic S, Bender R, Palazzo J, Gatalica Z. A review of adenoid cystic carcinoma of the breast with emphasis on its molecular and genetic characteristics. Hum Pathol 2013; 44(3): 301-9. [http://dx.doi.org/10.1016/j.humpath.2012.01.002] [PMID: 22520948]

[87] Arpino G, Clark GM, Mohsin S, Bardou VJ, Elledge RM. Adenoid cystic carcinoma of the breast: molecular markers, treatment, and clinical outcome. Cancer 2002; 94(8): 2119-27. [http://dx.doi.org/10.1002/cncr.10455] [PMID: 12001107]

[88] Cadoo KA, McArdle O, O'Shea AM, Power CP, Hennessy BT. Management of unusual histological types of breast cancer. Oncologist 2012; 17(9): 1135-45. [http://dx.doi.org/10.1634/theoncologist.2012-0134] [PMID: 22826373]

[89] Ghabach B, Anderson WF, Curtis RE, Huycke MM, Lavigne JA, Dores GM. Adenoid cystic carcinoma of the breast in the United States (1977 to 2006): a population-based cohort study. Breast Cancer Res 2010; 12(4): R54.
[http://dx.doi.org/10.1186/bcr2613] [PMID: 20653964]

[90] Ozguroglu M, Tascilar K, Ilvan S, Soybir G, Celik V. Secretory carcinoma of the breast. Case report and review of the literature. Oncology 2005; 68(2-3): 263-8.
[http://dx.doi.org/10.1159/000086782] [PMID: 16015043]

[91] Herz H, Cooke B, Goldstein D. Metastatic secretory breast cancer. Non-responsiveness to chemotherapy: case report and review of the literature. Ann Oncol 2000; 11(10): 1343-7.
[http://dx.doi.org/10.1023/A:1008387800525] [PMID: 11106125]

[92] Japaze H, Emina J, Diaz C, *et al.* 'Pure' invasive apocrine carcinoma of the breast: a new clinicopathological entity? Breast 2005; 14(1): 3-10.
[http://dx.doi.org/10.1016/j.breast.2004.06.003] [PMID: 15695074]

[93] Vranic S, Tawfik O, Palazzo J, *et al.* EGFR and HER-2/neu expression in invasive apocrine carcinoma of the breast. Mod Pathol 2010; 23(5): 644-53.
[http://dx.doi.org/10.1038/modpathol.2010.50] [PMID: 20208479]

[94] Dellapasqua S, Maisonneuve P, Viale G, *et al.* Immunohistochemically defined subtypes and outcome of apocrine breast cancer. Clin Breast Cancer 2013; 13(2): 95-102.
[http://dx.doi.org/10.1016/j.clbc.2012.11.004] [PMID: 23245877]

[95] Pettinato G, Manivel CJ, Panico L, Sparano L, Petrella G. Invasive micropapillary carcinoma of the breast: clinicopathologic study of 62 cases of a poorly recognized variant with highly aggressive behavior. Am J Clin Pathol 2004; 121(6): 857-66.
[http://dx.doi.org/10.1309/XTJ7VHB49UD78X60] [PMID: 15198358]

[96] Zekioglu O, Erhan Y, Ciris M, Bayramoglu H, Ozdemir N. Invasive micropapillary carcinoma of the breast: high incidence of lymph node metastasis with extranodal extension and its immunohistochemical profile compared with invasive ductal carcinoma. Histopathology 2004; 44(1): 18-23.
[http://dx.doi.org/10.1111/j.1365-2559.2004.01757.x] [PMID: 14717664]

[97] Walsh MM, Bleiweiss IJ. Invasive micropapillary carcinoma of the breast: eighty cases of an underrecognized entity. Hum Pathol 2001; 32(6): 583-9.
[http://dx.doi.org/10.1053/hupa.2001.24988] [PMID: 11431712]

[98] Chen L, Fan Y, Lang RG, *et al.* Breast carcinoma with micropapillary features: clinicopathologic study and long-term follow-up of 100 cases. Int J Surg Pathol 2008; 16(2): 155-63.
[http://dx.doi.org/10.1177/1066896907307047] [PMID: 18417672]

[99] Pezzi CM, Patel-Parekh L, Cole K, Franko J, Klimberg VS, Bland K. Characteristics and treatment of metaplastic breast cancer: analysis of 892 cases from the National Cancer Data Base. Ann Surg Oncol 2007; 14(1): 166-73.
[http://dx.doi.org/10.1245/s10434-006-9124-7] [PMID: 17066230]

[100] Jung SY, Kim HY, Nam BH, *et al.* Worse prognosis of metaplastic breast cancer patients than other patients with triple-negative breast cancer. Breast Cancer Res Treat 2010; 120(3): 627-37.
[http://dx.doi.org/10.1007/s10549-010-0780-8] [PMID: 20143153]

[101] Yerushalmi R, Hayes MM, Gelmon KA. Breast carcinoma--rare types: review of the literature. Ann Oncol 2009; 20(11): 1763-70.
[http://dx.doi.org/10.1093/annonc/mdp245] [PMID: 19602565]

[102] Tse GM, Tan PH, Putti TC, Lui PC, Chaiwun B, Law BK. Metaplastic carcinoma of the breast: a clinicopathological review. J Clin Pathol 2006; 59(10): 1079-83.
[http://dx.doi.org/10.1136/jcp.2005.030536] [PMID: 16467167]

[103] Hennessy BT, Krishnamurthy S, Giordano S, *et al.* Squamous cell carcinoma of the breast. J Clin

Oncol 2005; 23(31): 7827-35.
[http://dx.doi.org/10.1200/JCO.2004.00.9589] [PMID: 16258085]

[104] Burstein HJ, Harris JR, Morrow M. Malignant tumors of the breast.Cancer, principles and practice of oncology. Philadelphia, PA: Lippincott Williams and Wilkins 2011.

[105] Edge SB, Byr DR, Compton CC, Eds. American Joint Committee on cancer (AJCC) Cancer Staging manual. 7th ed., New York: Springer 2009.

[106] Cancer Facts and Figures 2012. Boston, MA, USA: American Cancer Society 2012.

[107] Economopoulou P, Dimitriadis G, Psyrri A. Beyond BRCA: new hereditary breast cancer susceptibility genes. Cancer Treat Rev 2015; 41(1): 1-8.
[http://dx.doi.org/10.1016/j.ctrv.2014.10.008] [PMID: 25467110]

[108] Singletary KW, Gapstur SM. Alcohol and breast cancer: review of epidemiologic and experimental evidence and potential mechanisms. JAMA 2001; 286(17): 2143-51.
[http://dx.doi.org/10.1001/jama.286.17.2143] [PMID: 11694156]

[109] Boffetta P, Hashibe M. Alcohol and cancer. Lancet Oncol 2006; 7(2): 149-56.
[http://dx.doi.org/10.1016/S1470-2045(06)70577-0] [PMID: 16455479]

[110] Dumitrescu RG, Shields PG. The etiology of alcohol-induced breast cancer. Alcohol 2005; 35(3): 213-25.
[http://dx.doi.org/10.1016/j.alcohol.2005.04.005] [PMID: 16054983]

[111] Smith-Warner SA, Spiegelman D, Yaun SS, *et al.* Alcohol and breast cancer in women: a pooled analysis of cohort studies. JAMA 1998; 279(7): 535-40.
[http://dx.doi.org/10.1001/jama.279.7.535] [PMID: 9480365]

[112] Benzon Larsen S, Vogel U, Christensen J, *et al.* Interaction between ADH1C Arg(272)Gln and alcohol intake in relation to breast cancer risk suggests that ethanol is the causal factor in alcohol related breast cancer. Cancer Lett 2010; 295(2): 191-7.
[http://dx.doi.org/10.1016/j.canlet.2010.02.023] [PMID: 20350778]

[113] Tjonneland A. Lifestyle Risk Factors for Breast Cancer - Diet, Cancer and Health, a Prospective Cohort Study in Denmark. Copenhagen: University of Copenhagen 2008.

[114] Dumitrescu RG, Cotarla I. Understanding breast cancer risk -- where do we stand in 2005? J Cell Mol Med 2005; 9(1): 208-21.
[http://dx.doi.org/10.1111/j.1582-4934.2005.tb00350.x] [PMID: 15784178]

[115] Mitrunen K, Hirvonen A. Molecular epidemiology of sporadic breast cancer. The role of polymorphic genes involved in oestrogen biosynthesis and metabolism. Mutat Res 2003; 544(1): 9-41.
[http://dx.doi.org/10.1016/S1383-5742(03)00016-4] [PMID: 12888106]

[116] Rinaldi S, Peeters PH, Bezemer ID, *et al.* Relationship of alcohol intake and sex steroid concentrations in blood in pre- and post-menopausal women: the European Prospective Investigation into Cancer and Nutrition. Cancer Causes Control 2006; 17(8): 1033-43.
[http://dx.doi.org/10.1007/s10552-006-0041-7] [PMID: 16933054]

[117] Muti P, Trevisan M, Micheli A, *et al.* Alcohol consumption and total estradiol in premenopausal women. Cancer Epidemiol Biomarkers Prev 1998; 7(3): 189-93.
[PMID: 9521430]

[118] Coutelle C, Höhn B, Benesova M, *et al.* Risk factors in alcohol associated breast cancer: alcohol dehydrogenase polymorphism and estrogens. Int J Oncol 2004; 25(4): 1127-32.
[PMID: 15375565]

[119] Dorgan JF, Baer DJ, Albert PS, *et al.* Serum hormones and the alcohol-breast cancer association in postmenopausal women. J Natl Cancer Inst 2001; 93(9): 710-5.
[http://dx.doi.org/10.1093/jnci/93.9.710] [PMID: 11333294]

[120] Reichman ME, Judd JT, Longcope C, *et al.* Effects of alcohol consumption on plasma and urinary

hormone concentrations in premenopausal women. J Natl Cancer Inst 1993; 85(9): 722-7. [http://dx.doi.org/10.1093/jnci/85.9.722] [PMID: 8478958]

[121] Li CI, Chlebowski RT, Freiberg M, *et al.* Alcohol consumption and risk of postmenopausal breast cancer by subtype: the women's health initiative observational study. J Natl Cancer Inst 2010; 102(18): 1422-31.
[http://dx.doi.org/10.1093/jnci/djq316] [PMID: 20733117]

[122] Purohit V. Can alcohol promote aromatization of androgens to estrogens? A review. Alcohol 2000; 22(3): 123-7.
[http://dx.doi.org/10.1016/S0741-8329(00)00124-5] [PMID: 11163119]

[123] Etique N, Chardard D, Chesnel A, Merlin JL, Flament S, Grillier-Vuissoz I. Ethanol stimulates proliferation, ERalpha and aromatase expression in MCF-7 human breast cancer cells. Int J Mol Med 2004; 13(1): 149-55.
[PMID: 14654987]

[124] Shield KD, Soerjomataram I, Rehm J. Alcohol Use and Breast Cancer: A Critical Review. Alcohol Clin Exp Res 2016; 40(6): 1166-81.
[http://dx.doi.org/10.1111/acer.13071] [PMID: 27130687]

[125] Vucenik I, Stains JP. Obesity and cancer risk: evidence, mechanisms, and recommendations. Ann N Y Acad Sci 2012; 1271: 37-43.
[http://dx.doi.org/10.1111/j.1749-6632.2012.06750.x] [PMID: 23050962]

[126] Obesity and Overweight. Geneva, Switzerland: World Health Organization 2011.

[127] Wiseman M. The second World Cancer Research Fund/American Institute for Cancer Research expert report. Food, nutrition, physical activity, and the prevention of cancer: a global perspective. Proc Nutr Soc 2008; 67(3): 253-6.
[http://dx.doi.org/10.1017/S002966510800712X] [PMID: 18452640]

[128] Ferlay J, Soerjomataram I, Dikshit R, *et al.* Cancer incidence and mortality worldwide: sources, methods and major patterns in GLOBOCAN 2012. Int J Cancer 2015; 136(5): E359-86.
[http://dx.doi.org/10.1002/ijc.29210] [PMID: 25220842]

[129] Howell A, Anderson AS, Clarke RB, *et al.* Risk determination and prevention of breast cancer. Breast Cancer Res 2014; 16(5): 446.
[http://dx.doi.org/10.1186/s13058-014-0446-2] [PMID: 25467785]

[130] Wu Q, Ishikawa T, Sirianni R, *et al.* 27-Hydroxycholesterol promotes cell-autonomous, ER-positive breast cancer growth. Cell Reports 2013; 5(3): 637-45.
[http://dx.doi.org/10.1016/j.celrep.2013.10.006] [PMID: 24210818]

[131] Nelson ER, Wardell SE, Jasper JS, *et al.* 27-Hydroxycholesterol links hypercholesterolemia and breast cancer pathophysiology. Science 2013; 342(6162): 1094-8.
[http://dx.doi.org/10.1126/science.1241908] [PMID: 24288332]

[132] Althuis MD, Fergenbaum JH, Garcia-Closas M, Brinton LA, Madigan MP, Sherman ME. Etiology of hormone receptor-defined breast cancer: a systematic review of the literature. Cancer Epidemiol Biomarkers Prev 2004; 13(10): 1558-68.
[PMID: 15466970]

[133] Millikan RC, Newman B, Tse CK, *et al.* Epidemiology of basal-like breast cancer. Breast Cancer Res Treat 2008; 109(1): 123-39.
[http://dx.doi.org/10.1007/s10549-007-9632-6] [PMID: 17578664]

[134] Yang XR, Chang-Claude J, Goode EL, *et al.* Associations of breast cancer risk factors with tumor subtypes: a pooled analysis from the Breast Cancer Association Consortium studies. J Natl Cancer Inst 2011; 103(3): 250-63.
[http://dx.doi.org/10.1093/jnci/djq526] [PMID: 21191117]

[135] Chang S, Buzdar AU, Hursting SD. Inflammatory breast cancer and body mass index. J Clin Oncol

1998; 16(12): 3731-5.
[http://dx.doi.org/10.1200/JCO.1998.16.12.3731] [PMID: 9850015]

[136] Schairer C, Li Y, Frawley P, *et al.* Risk factors for inflammatory breast cancer and other invasive breast cancers. J Natl Cancer Inst 2013; 105(18): 1373-84.
[http://dx.doi.org/10.1093/jnci/djt206] [PMID: 24046390]

[137] Cheraghi Z, Poorolajal J, Hashem T, Esmailnasab N, Doosti Irani A. Effect of body mass index on breast cancer during premenopausal and postmenopausal periods: a meta-analysis. PLoS One 2012; 7(12): e51446.
[http://dx.doi.org/10.1371/journal.pone.0051446] [PMID: 23236502]

[138] Anderson GL, Neuhouser ML. Obesity and the risk for premenopausal and postmenopausal breast cancer. Cancer Prev Res (Phila) 2012; 5(4): 515-21.
[http://dx.doi.org/10.1158/1940-6207.CAPR-12-0091] [PMID: 22392012]

[139] Renehan AG, Zwahlen M, Egger M. Adiposity and cancer risk: new mechanistic insights from epidemiology. Nat Rev Cancer 2015; 15(8): 484-98.
[http://dx.doi.org/10.1038/nrc3967] [PMID: 26205341]

[140] Renehan AG, Tyson M, Egger M, Heller RF, Zwahlen M. Body-mass index and incidence of cancer: a systematic review and meta-analysis of prospective observational studies. Lancet 2008; 371(9612): 569-78.
[http://dx.doi.org/10.1016/S0140-6736(08)60269-X] [PMID: 18280327]

[141] Suzuki R, Orsini N, Saji S, Key TJ, Wolk A. Body weight and incidence of breast cancer defined by estrogen and progesterone receptor status--a meta-analysis. Int J Cancer 2009; 124(3): 698-712.
[http://dx.doi.org/10.1002/ijc.23943] [PMID: 18988226]

[142] Ahn J, Schatzkin A, Lacey JV Jr, *et al.* Adiposity, adult weight change, and postmenopausal breast cancer risk. Arch Intern Med 2007; 167(19): 2091-102.
[http://dx.doi.org/10.1001/archinte.167.19.2091] [PMID: 17954804]

[143] Feigelson HS, Patel AV, Teras LR, Gansler T, Thun MJ, Calle EE. Adult weight gain and histopathologic characteristics of breast cancer among postmenopausal women. Cancer 2006; 107(1): 12-21.
[http://dx.doi.org/10.1002/cncr.21965] [PMID: 16718671]

[144] Setiawan VW, Yang HP, Pike MC, *et al.* Type I and II endometrial cancers: have they different risk factors? J Clin Oncol 2013; 31(20): 2607-18.
[http://dx.doi.org/10.1200/JCO.2012.48.2596] [PMID: 23733771]

[145] Ritte R, Lukanova A, Berrino F, *et al.* Adiposity, hormone replacement therapy use and breast cancer risk by age and hormone receptor status: a large prospective cohort study. Breast Cancer Res 2012; 14(3): R76.
[http://dx.doi.org/10.1186/bcr3186] [PMID: 22583394]

[146] Phipps AI, Chlebowski RT, Prentice R, *et al.* Body size, physical activity, and risk of triple-negative and estrogen receptor-positive breast cancer. Cancer Epidemiol Biomarkers Prev 2011; 20(3): 454-63.
[http://dx.doi.org/10.1158/1055-9965.EPI-10-0974] [PMID: 21364029]

[147] Biglia N, Peano E, Sgandurra P, *et al.* Body mass index (BMI) and breast cancer: impact on tumor histopathologic features, cancer subtypes and recurrence rate in pre and postmenopausal women. Gynecol Endocrinol 2013; 29(3): 263-7.
[http://dx.doi.org/10.3109/09513590.2012.736559] [PMID: 23174088]

[148] Loi S, Milne RL, Friedlander ML, *et al.* Obesity and outcomes in premenopausal and postmenopausal breast cancer. Cancer Epidemiol Biomarkers Prev 2005; 14(7): 1686-91.
[http://dx.doi.org/10.1158/1055-9965.EPI-05-0042] [PMID: 16030102]

[149] Carmichael AR. Obesity and prognosis of breast cancer. Obes Rev 2006; 7(4): 333-40.
[http://dx.doi.org/10.1111/j.1467-789X.2006.00261.x] [PMID: 17038127]

[150] Protani M, Coory M, Martin JH. Effect of obesity on survival of women with breast cancer: systematic review and meta-analysis. Breast Cancer Res Treat 2010; 123(3): 627-35. [http://dx.doi.org/10.1007/s10549-010-0990-0] [PMID: 20571870]

[151] Ewertz M, Jensen MB, Gunnarsdóttir KA, *et al.* Effect of obesity on prognosis after early-stage breast cancer. J Clin Oncol 2011; 29(1): 25-31. [http://dx.doi.org/10.1200/JCO.2010.29.7614] [PMID: 21115856]

[152] Bradshaw PT, Ibrahim JG, Stevens J, *et al.* Postdiagnosis change in bodyweight and survival after breast cancer diagnosis. Epidemiology 2012; 23(2): 320-7. [http://dx.doi.org/10.1097/EDE.0b013e31824596a1] [PMID: 22317813]

[153] Rose DP, Vona-Davis L. Influence of obesity on breast cancer receptor status and prognosis. Expert Rev Anticancer Ther 2009; 9(8): 1091-101. [http://dx.doi.org/10.1586/era.09.71] [PMID: 19671029]

[154] McTiernan A, Rajan KB, Tworoger SS, *et al.* Adiposity and sex hormones in postmenopausal breast cancer survivors. J Clin Oncol 2003; 21(10): 1961-6. [http://dx.doi.org/10.1200/JCO.2003.07.057] [PMID: 12743149]

[155] Daling JR, Malone KE, Doody DR, Johnson LG, Gralow JR, Porter PL. Relation of body mass index to tumor markers and survival among young women with invasive ductal breast carcinoma. Cancer 2001; 92(4): 720-9. [http://dx.doi.org/10.1002/1097-0142(20010815)92:4<720::AID-CNCR1375>3.0.CO;2-T] [PMID: 11550140]

[156] Ivandić A, Prpić-Krizevac I, Sucić M, Jurić M. Hyperinsulinemia and sex hormones in healthy premenopausal women: relative contribution of obesity, obesity type, and duration of obesity. Metabolism 1998; 47(1): 13-9. [http://dx.doi.org/10.1016/S0026-0495(98)90186-X] [PMID: 9440471]

[157] Goodwin PJ, Ennis M, Pritchard KI, *et al.* Insulin-like growth factor binding proteins 1 and 3 and breast cancer outcomes. Breast Cancer Res Treat 2002; 74(1): 65-76. [http://dx.doi.org/10.1023/A:1016075709022] [PMID: 12150454]

[158] Goodwin PJ, Ennis M, Pritchard KI, *et al.* Fasting insulin and outcome in early-stage breast cancer: results of a prospective cohort study. J Clin Oncol 2002; 20(1): 42-51. [http://dx.doi.org/10.1200/JCO.2002.20.1.42] [PMID: 11773152]

[159] Kaaks R, McTiernan A. Obesity and Sex Hormones.Cancer Prevention and Management Through Exercise and Weight Control. CRC Press 2005; pp. 289-300. [http://dx.doi.org/10.1201/9781420026641.sec5]

[160] Yu H, Rohan T. Role of the insulin-like growth factor family in cancer development and progression. J Natl Cancer Inst 2000; 92(18): 1472-89. [http://dx.doi.org/10.1093/jnci/92.18.1472] [PMID: 10995803]

[161] Zimmet PZ. Hyperinsulinemia--how innocent a bystander? Diabetes Care 1993; 16(S3) (Suppl. 3): 56-70. [http://dx.doi.org/10.2337/diacare.16.3.56] [PMID: 8299479]

[162] Kaaks R. Nutrition, hormones, and breast cancer: is insulin the missing link? Cancer Causes Control 1996; 7(6): 605-25. [http://dx.doi.org/10.1007/BF00051703] [PMID: 8932921]

[163] Rose DP, Gracheck PJ, Vona-Davis L. The Interactions of Obesity, Inflammation and Insulin Resistance in Breast Cancer. Cancers (Basel) 2015; 7(4): 2147-68. [http://dx.doi.org/10.3390/cancers7040883] [PMID: 26516917]

[164] Madarnas Y, Sawka CA, Franssen E, Bjarnason GA. Are medical oncologists biased in their treatment of the large woman with breast cancer? Breast Cancer Res Treat 2001; 66(2): 123-33. [http://dx.doi.org/10.1023/A:1010635328299] [PMID: 11437098]

[165] Griggs JJ, Lyman GH. Variation in adjuvant breast cancer chemotherapy dosing in overweight and obese women; Abstract 145. 26th San Antonio Breast Cancer Symposium. 2003; San Antonio (TX, USA). 2003.

[166] Ronckers CM, Erdmann CA, Land CE. Radiation and breast cancer: a review of current evidence. Breast Cancer Res 2005; 7(1): 21-32.
[http://dx.doi.org/10.1186/bcr970] [PMID: 15642178]

[167] Abeloff MD, Wolff AC, Weber BL, Zaks TZ, Sacchini V, McCormick B. Cancer of the breast. 2008.
[http://dx.doi.org/10.1016/B978-0-443-06694-8.50099-3]

[168] Breast cancer and hormone replacement therapy: collaborative reanalysis of data from 51 epidemiological studies of 52,705 women with breast cancer and 108,411 women without breast cancer. Lancet 1997; 350(9084): 1047-59.
[http://dx.doi.org/10.1016/S0140-6736(97)08233-0] [PMID: 10213546]

[169] Reeves GK, Beral V, Green J, Gathani T, Bull D. Hormonal therapy for menopause and breast-cancer risk by histological type: a cohort study and meta-analysis. Lancet Oncol 2006; 7(11): 910-8.
[http://dx.doi.org/10.1016/S1470-2045(06)70911-1] [PMID: 17081916]

[170] Ravdin PM, Cronin KA, Howlader N, *et al.* The decrease in breast-cancer incidence in 2003 in the United States. N Engl J Med 2007; 356(16): 1670-4.
[http://dx.doi.org/10.1056/NEJMsr070105] [PMID: 17442911]

[171] Key T, Appleby P, Barnes I, Reeves G. Endogenous sex hormones and breast cancer in postmenopausal women: reanalysis of nine prospective studies. J Natl Cancer Inst 2002; 94(8): 606-16.
[http://dx.doi.org/10.1093/jnci/94.8.606] [PMID: 11959894]

[172] Brown SB, Hankinson SE. Endogenous estrogens and the risk of breast, endometrial, and ovarian cancers. Steroids. 2015; 99 (Pt A): 8-10.
[http://dx.doi.org/10.1016/j.steroids.2014.12.013]

[173] Missmer SA, Eliassen AH, Barbieri RL, Hankinson SE. Endogenous estrogen, androgen, and progesterone concentrations and breast cancer risk among postmenopausal women. J Natl Cancer Inst 2004; 96(24): 1856-65.
[http://dx.doi.org/10.1093/jnci/djh336] [PMID: 15601642]

[174] Zeleniuch-Jacquotte A, Shore RE, Koenig KL, *et al.* Postmenopausal levels of oestrogen, androgen, and SHBG and breast cancer: long-term results of a prospective study. Br J Cancer 2004; 90(1): 153-9.
[http://dx.doi.org/10.1038/sj.bjc.6601517] [PMID: 14710223]

[175] Boonyaratanakornkit V, Pateetin P. The role of ovarian sex steroids in metabolic homeostasis, obesity, and postmenopausal breast cancer: molecular mechanisms and therapeutic implications. Biomed Res Int 2015; 2015: 140-6.
[http://dx.doi.org/10.1155/2015/140196]

[176] Eliassen AH, Missmer SA, Tworoger SS, *et al.* Endogenous steroid hormone concentrations and risk of breast cancer among premenopausal women. J Natl Cancer Inst 2006; 98(19): 1406-15.
[http://dx.doi.org/10.1093/jnci/djj376] [PMID: 17018787]

[177] Pike MC, Krailo MD, Henderson BE, Casagrande JT, Hoel DG. 'Hormonal' risk factors, 'breast tissue age' and the age-incidence of breast cancer. Nature 1983; 303(5920): 767-70.
[http://dx.doi.org/10.1038/303767a0] [PMID: 6866078]

[178] Lane-Claypon JE. A further report on cancer of the breast with special reference to its associated antecedent conditions. Rep Public Health Med Subj (Lond) 1926; 2.

[179] MacMahon B, Cole P, Lin TM, *et al.* Age at first birth and breast cancer risk. Bull World Health Organ 1970; 43(2): 209-21.
[PMID: 5312521]

[180] Woods KL, Smith SR, Morrison JM. Parity and breast cancer: evidence of a dual effect. BMJ 1980; 281(6237): 419-21.
[http://dx.doi.org/10.1136/bmj.281.6237.419] [PMID: 7427299]

[181] Hsieh C, Pavia M, Lambe M, *et al.* Dual effect of parity on breast cancer risk. Eur J Cancer 1994; 30A(7): 969-73.
[http://dx.doi.org/10.1016/0959-8049(94)90125-2] [PMID: 7946593]

[182] Lambe M, Hsieh C, Trichopoulos D, Ekbom A, Pavia M, Adami HO. Transient increase in the risk of breast cancer after giving birth. N Engl J Med 1994; 331(1): 5-9.
[http://dx.doi.org/10.1056/NEJM199407073310102] [PMID: 8202106]

[183] Albrektsen G, Heuch I, Kvåle G. The short-term and long-term effect of a pregnancy on breast cancer risk: a prospective study of 802,457 parous Norwegian women. Br J Cancer 1995; 72(2): 480-4.
[http://dx.doi.org/10.1038/bjc.1995.359] [PMID: 7640236]

[184] Pathak DR. Dual effect of first full term pregnancy on breast cancer risk: empirical evidence and postulated underlying biology. Cancer Causes Control 2002; 13(4): 295-8.
[http://dx.doi.org/10.1023/A:1015282916368] [PMID: 12074498]

[185] Melbye M, Wohlfahrt J, Olsen JH, *et al.* Induced abortion and the risk of breast cancer. N Engl J Med 1997; 336 (2): 81v85.

[186] Lyons TR, Schedin PJ, Borges VF. Pregnancy and breast cancer: when they collide. J Mammary Gland Biol Neoplasia 2009; 14(2): 87-98.
[http://dx.doi.org/10.1007/s10911-009-9119-7] [PMID: 19381788]

[187] Tiede B, Kang Y. From milk to malignancy: the role of mammary stem cells in development, pregnancy and breast cancer. Cell Res 2011; 21(2): 245-57.
[http://dx.doi.org/10.1038/cr.2011.11] [PMID: 21243011]

[188] Kobayashi S, Sugiura H, Ando Y, *et al.* Reproductive history and breast cancer risk. Breast Cancer 2012; 19(4): 302-8.
[http://dx.doi.org/10.1007/s12282-012-0384-8] [PMID: 22711317]

[189] Anderson WF, Jatoi I, Sherman ME. Qualitative age interactions in breast cancer studies: mind the gap. J Clin Oncol 2009; 27(32): 5308-11.
[http://dx.doi.org/10.1200/JCO.2009.22.9450] [PMID: 19826117]

[190] Colditz GA, Rosner BA, Chen WY, Holmes MD, Hankinson SE. Risk factors for breast cancer according to estrogen and progesterone receptor status. J Natl Cancer Inst 2004; 96(3): 218-28.
[http://dx.doi.org/10.1093/jnci/djh025] [PMID: 14759989]

[191] Shinde SS, Forman MR, Kuerer HM, *et al.* Higher parity and shorter breastfeeding duration: association with triple-negative phenotype of breast cancer. Cancer 2010; 116(21): 4933-43.
[http://dx.doi.org/10.1002/cncr.25443] [PMID: 20665494]

[192] Phipps AI, Chlebowski RT, Prentice R, *et al.* Reproductive history and oral contraceptive use in relation to risk of triple-negative breast cancer. J Natl Cancer Inst 2011; 103(6): 470-7.
[http://dx.doi.org/10.1093/jnci/djr030] [PMID: 21346227]

[193] Breast cancer and breastfeeding: collaborative reanalysis of individual data from 47 epidemiological studies in 30 countries, including 50302 women with breast cancer and 96973 women without the disease. Lancet 2002; 360(9328): 187-95.
[http://dx.doi.org/10.1016/S0140-6736(02)09454-0] [PMID: 12133652]

[194] Phipps AI, Malone KE, Porter PL, Daling JR, Li CI. Reproductive and hormonal risk factors for postmenopausal luminal, HER-2-overexpressing, and triple-negative breast cancer. Cancer 2008; 113(7): 1521-6.
[http://dx.doi.org/10.1002/cncr.23786] [PMID: 18726992]

[195] Gaudet MM, Press MF, Haile RW, *et al.* Risk factors by molecular subtypes of breast cancer across a

population-based study of women 56 years or younger. Breast Cancer Res Treat 2011; 130(2): 587-97. [http://dx.doi.org/10.1007/s10549-011-1616-x] [PMID: 21667121]

[196] Kotsopoulos J, Lubinski J, Salmena L, *et al.* Breastfeeding and the risk of breast cancer in BRCA1 and BRCA2 mutation carriers. Breast Cancer Res 2012; 14(2): R42. [http://dx.doi.org/10.1186/bcr3138] [PMID: 22405187]

[197] Li CI, Beaber EF, Tang MT, Porter PL, Daling JR, Malone KE. Reproductive factors and risk of estrogen receptor positive, triple-negative, and HER2-neu overexpressing breast cancer among women 20-44 years of age. Breast Cancer Res Treat 2013; 137(2): 579-87. [http://dx.doi.org/10.1007/s10549-012-2365-1] [PMID: 23224237]

[198] Hankinson SE, Willett WC, Colditz GA, *et al.* Circulating concentrations of insulin-like growth factor-I and risk of breast cancer. Lancet 1998; 351(9113): 1393-6. [http://dx.doi.org/10.1016/S0140-6736(97)10384-1] [PMID: 9593409]

[199] Tworoger SS, Eliassen AH, Sluss P, Hankinson SE. A prospective study of plasma prolactin concentrations and risk of premenopausal and postmenopausal breast cancer. J Clin Oncol 2007; 25(12): 1482-8. [http://dx.doi.org/10.1200/JCO.2006.07.6356] [PMID: 17372279]

[200] Siegel R, Naishadham D, Jemal A. Cancer statistics, 2013. CA Cancer J Clin 2013; 63(1): 11-30. [http://dx.doi.org/10.3322/caac.21166] [PMID: 23335087]

[201] Carroll JC, Cremin C, Allanson J, *et al.* Hereditary breast and ovarian cancers. Can Fam Physician 2008; 54(12): 1691-2. [PMID: 19074707]

[202] Lalloo F, Evans DG. Familial breast cancer. Clin Genet 2012; 82(2): 105-14. [http://dx.doi.org/10.1111/j.1399-0004.2012.01859.x] [PMID: 22356477]

[203] Khatcheressian JL, Hurley P, Bantug E, *et al.* Breast cancer follow-up and management after primary treatment: American Society of Clinical Oncology clinical practice guideline update. J Clin Oncol 2013; 31(7): 961-5. [http://dx.doi.org/10.1200/JCO.2012.45.9859] [PMID: 23129741]

[204] https://www.nccn.org/professionals/physician_gls/pdf/genetics_screening.pdf

[205] King MC, Marks JH, Mandell JB. Breast and ovarian cancer risks due to inherited mutations in BRCA1 and BRCA2. Science 2003; 302(5645): 643-6. [http://dx.doi.org/10.1126/science.1088759] [PMID: 14576434]

[206] Tavormina PL, Shiang R, Thompson LM, *et al.* Thanatophoric dysplasia (types I and II) caused by distinct mutations in fibroblast growth factor receptor 3. Nat Genet 1995; 9(3): 321-8. [http://dx.doi.org/10.1038/ng0395-321] [PMID: 7773297]

[207] Miyakis S, Sourvinos G, Spandidos DA. Differential expression and mutation of the ras family genes in human breast cancer. Biochem Biophys Res Commun 1998; 251(2): 609-12. [http://dx.doi.org/10.1006/bbrc.1998.9527] [PMID: 9792821]

[208] Mavaddat N, Peock S, Frost D, *et al.* Cancer risks for BRCA1 and BRCA2 mutation carriers: results from prospective analysis of EMBRACE. J Natl Cancer Inst 2013; 105(11): 812-22. [http://dx.doi.org/10.1093/jnci/djt095] [PMID: 23628597]

[209] Boyd J, Sonoda Y, Federici MG, *et al.* Clinicopathologic features of BRCA-linked and sporadic ovarian cancer. JAMA 2000; 283(17): 2260-5. [http://dx.doi.org/10.1001/jama.283.17.2260] [PMID: 10807385]

[210] Veronesi A, de Giacomi C, Magri MD, *et al.* Familial breast cancer: characteristics and outcome of BRCA 1-2 positive and negative cases. BMC Cancer 2005; 5: 70. [http://dx.doi.org/10.1186/1471-2407-5-70] [PMID: 15996267]

[211] Atchley DP, Albarracin CT, Lopez A, *et al.* Clinical and pathologic characteristics of patients with

BRCA-positive and BRCA-negative breast cancer. J Clin Oncol 2008; 26(26): 4282-8. [http://dx.doi.org/10.1200/JCO.2008.16.6231] [PMID: 18779615]

[212] Tung N, Wang Y, Collins LC, *et al.* Estrogen receptor positive breast cancers in BRCA1 mutation carriers: clinical risk factors and pathologic features. Breast Cancer Res 2010; 12(1): R12. [http://dx.doi.org/10.1186/bcr2478] [PMID: 20149218]

[213] Tung N, Miron A, Schnitt SJ, *et al.* Prevalence and predictors of loss of wild type BRCA1 in estrogen receptor positive and negative BRCA1-associated breast cancers. Breast Cancer Res 2010; 12(6): R95. [http://dx.doi.org/10.1186/bcr2776] [PMID: 21080930]

[214] Deng CX, Scott F. Role of the tumor suppressor gene Brca1 in genetic stability and mammary gland tumor formation. Oncogene 2000; 19(8): 1059-64. [http://dx.doi.org/10.1038/sj.onc.1203269] [PMID: 10713690]

[215] Deng CX, Brodie SG. Roles of BRCA1 and its interacting proteins. BioEssays 2000; 22(8): 728-37. [http://dx.doi.org/10.1002/1521-1878(200008)22:8<728::AID-BIES6>3.0.CO;2-B] [PMID: 10918303]

[216] Bhattacharyya A, Ear US, Koller BH, Weichselbaum RR, Bishop DK. The breast cancer susceptibility gene BRCA1 is required for subnuclear assembly of Rad51 and survival following treatment with the DNA cross-linking agent cisplatin. J Biol Chem 2000; 275(31): 23899-903. [http://dx.doi.org/10.1074/jbc.C000276200] [PMID: 10843985]

[217] Clark SL, Rodriguez AM, Snyder RR, Hankins GD, Boehning D. Structure-Function Of The Tumor Suppressor BRCA1. Comput Struct Biotechnol J 2012; 1(1): e201204005. [http://dx.doi.org/10.5936/csbj.201204005] [PMID: 22737296]

[218] Wu LC, Wang ZW, Tsan JT, *et al.* Identification of a RING protein that can interact *in vivo* with the BRCA1 gene product. Nat Genet 1996; 14(4): 430-40. [http://dx.doi.org/10.1038/ng1296-430] [PMID: 8944023]

[219] Ghimenti C, Sensi E, Presciuttini S, *et al.* Germline mutations of the BRCA1-associated ring domain (BARD1) gene in breast and breast/ovarian families negative for BRCA1 and BRCA2 alterations. Genes Chromosomes Cancer 2002; 33(3): 235-42. [http://dx.doi.org/10.1002/gcc.1223] [PMID: 11807980]

[220] Walsh T, Casadei S, Lee MK, *et al.* Mutations in 12 genes for inherited ovarian, fallopian tube, and peritoneal carcinoma identified by massively parallel sequencing. Proc Natl Acad Sci USA 2011; 108(44): 18032-7. [http://dx.doi.org/10.1073/pnas.1115052108] [PMID: 22006311]

[221] Li ML, Greenberg RA. Links between genome integrity and BRCA1 tumor suppression. Trends Biochem Sci 2012; 37(10): 418-24. [http://dx.doi.org/10.1016/j.tibs.2012.06.007] [PMID: 22836122]

[222] Al-Mulla F, Bland JM, Serratt D, Miller J, Chu C, Taylor GT. Age-dependent penetrance of different germline mutations in the BRCA1 gene. J Clin Pathol 2009; 62(4): 350-6. [http://dx.doi.org/10.1136/jcp.2008.062646] [PMID: 19329713]

[223] Scully R, Chen J, Plug A, *et al.* Association of BRCA1 with Rad51 in mitotic and meiotic cells. Cell 1997; 88(2): 265-75. [http://dx.doi.org/10.1016/S0092-8674(00)81847-4] [PMID: 9008167]

[224] Thakur S, Zhang HB, Peng Y, *et al.* Localization of BRCA1 and a splice variant identifies the nuclear localization signal. Mol Cell Biol 1997; 17(1): 444-52. [http://dx.doi.org/10.1128/MCB.17.1.444] [PMID: 8972225]

[225] Wilson CA, Payton MN, Elliott GS, *et al.* Differential subcellular localization, expression and biological toxicity of BRCA1 and the splice variant BRCA1-delta11b. Oncogene 1997; 14(1): 1-16. [http://dx.doi.org/10.1038/sj.onc.1200924] [PMID: 9010228]

[226] Chen CF, Li S, Chen Y, Chen PL, Sharp ZD, Lee WH. The nuclear localization sequences of the

BRCA1 protein interact with the importin-alpha subunit of the nuclear transport signal receptor. J Biol Chem 1996; 271(51): 32863-8.
[http://dx.doi.org/10.1074/jbc.271.51.32863] [PMID: 8955125]

[227] Mohammad DH, Yaffe MB. 14-3-3 proteins, FHA domains and BRCT domains in the DNA damage response. DNA Repair (Amst) 2009; 8(9): 1009-17.
[http://dx.doi.org/10.1016/j.dnarep.2009.04.004] [PMID: 19481982]

[228] Wang Q, Zhang H, Kajino K, Greene MI. BRCA1 binds c-Myc and inhibits its transcriptional and transforming activity in cells. Oncogene 1998; 17(15): 1939-48.
[http://dx.doi.org/10.1038/sj.onc.1202403] [PMID: 9788437]

[229] Zhang H, Somasundaram K, Peng Y, *et al.* BRCA1 physically associates with p53 and stimulates its transcriptional activity. Oncogene 1998; 16(13): 1713-21.
[http://dx.doi.org/10.1038/sj.onc.1201932] [PMID: 9582019]

[230] Zheng L, Annab LA, Afshari CA, Lee WH, Boyer TG. BRCA1 mediates ligand-independent transcriptional repression of the estrogen receptor. Proc Natl Acad Sci USA 2001; 98(17): 9587-92.
[http://dx.doi.org/10.1073/pnas.171174298] [PMID: 11493692]

[231] Ouchi T, Lee SW, Ouchi M, Aaronson SA, Horvath CM. Collaboration of signal transducer and activator of transcription 1 (STAT1) and BRCA1 in differential regulation of IFN-gamma target genes. Proc Natl Acad Sci USA 2000; 97(10): 5208-13.
[http://dx.doi.org/10.1073/pnas.080469697] [PMID: 10792030]

[232] Wong AK, Ormonde PA, Pero R, *et al.* Characterization of a carboxy-terminal BRCA1 interacting protein. Oncogene 1998; 17(18): 2279-85.
[http://dx.doi.org/10.1038/sj.onc.1202150] [PMID: 9811458]

[233] Yu X, Wu LC, Bowcock AM, Aronheim A, Baer R. The C-terminal (BRCT) domains of BRCA1 interact *in vivo* with CtIP, a protein implicated in the CtBP pathway of transcriptional repression. J Biol Chem 1998; 273(39): 25388-92.
[http://dx.doi.org/10.1074/jbc.273.39.25388] [PMID: 9738006]

[234] Zheng L, Pan H, Li S, *et al.* Sequence-specific transcriptional corepressor function for BRCA1 through a novel zinc finger protein, ZBRK1. Mol Cell 2000; 6(4): 757-68.
[http://dx.doi.org/10.1016/S1097-2765(00)00075-7] [PMID: 11090615]

[235] Anderson SF, Schlegel BP, Nakajima T, Wolpin ES, Parvin JD. BRCA1 protein is linked to the RNA polymerase II holoenzyme complex *via* RNA helicase A. Nat Genet 1998; 19(3): 254-6.
[http://dx.doi.org/10.1038/930] [PMID: 9662397]

[236] Scully R, Anderson SF, Chao DM, *et al.* BRCA1 is a component of the RNA polymerase II holoenzyme. Proc Natl Acad Sci USA 1997; 94(11): 5605-10.
[http://dx.doi.org/10.1073/pnas.94.11.5605] [PMID: 9159119]

[237] De Siervi A, De Luca P, Byun JS, *et al.* Transcriptional autoregulation by BRCA1. Cancer Res 2010; 70(2): 532-42.
[http://dx.doi.org/10.1158/0008-5472.CAN-09-1477] [PMID: 20068145]

[238] Di LJ, Fernandez AG, De Siervi A, Longo DL, Gardner K. Transcriptional regulation of BRCA1 expression by a metabolic switch. Nat Struct Mol Biol 2010; 17(12): 1406-13.
[http://dx.doi.org/10.1038/nsmb.1941] [PMID: 21102443]

[239] Foulkes WD. Re: Potential for bias in studies on efficacy of prophylactic surgery for BRCA1 and BRCA2 mutation carriers. J Natl Cancer Inst 2003; 95(17): 1344.
[http://dx.doi.org/10.1093/jnci/djg053] [PMID: 12953090]

[240] Lakhani SR, Reis-Filho JS, Fulford L, *et al.* Prediction of BRCA1 status in patients with breast cancer using estrogen receptor and basal phenotype. Clin Cancer Res 2005; 11(14): 5175-80.
[http://dx.doi.org/10.1158/1078-0432.CCR-04-2424] [PMID: 16033833]

[241] Sorlie T, Tibshirani R, Parker J, *et al.* Repeated observation of breast tumor subtypes in independent

gene expression data sets. Proc Natl Acad Sci USA 2003; 100(14): 8418-23. [http://dx.doi.org/10.1073/pnas.0932692100] [PMID: 12829800]

[242] Jóhannsson OT, Idvall I, Anderson C, *et al.* Tumour biological features of BRCA1-induced breast and ovarian cancer. Eur J Cancer 1997; 33(3): 362-71. [http://dx.doi.org/10.1016/S0959-8049(97)89007-7] [PMID: 9155518]

[243] Loman N, Johannsson O, Bendahl PO, Borg A, Fernö M, Olsson H. Steroid receptors in hereditary breast carcinomas associated with BRCA1 or BRCA2 mutations or unknown susceptibility genes. Cancer 1998; 83(2): 310-9. [http://dx.doi.org/10.1002/(SICI)1097-0142(19980715)83:2<310::AID-CNCR15>3.0.CO;2-W] [PMID: 9669814]

[244] Musolino A, Bella MA, Bortesi B, *et al.* BRCA mutations, molecular markers, and clinical variables in early-onset breast cancer: a population-based study. Breast 2007; 16(3): 280-92. [http://dx.doi.org/10.1016/j.breast.2006.12.003] [PMID: 17257844]

[245] Huber LJ, Yang TW, Sarkisian CJ, Master SR, Deng CX, Chodosh LA. Impaired DNA damage response in cells expressing an exon 11-deleted murine Brca1 variant that localizes to nuclear foci. Mol Cell Biol 2001; 21(12): 4005-15. [http://dx.doi.org/10.1128/MCB.21.12.4005-4015.2001] [PMID: 11359908]

[246] Catteau A, Harris WH, Xu CF, Solomon E. Methylation of the BRCA1 promoter region in sporadic breast and ovarian cancer: correlation with disease characteristics. Oncogene 1999; 18(11): 1957-65. [http://dx.doi.org/10.1038/sj.onc.1202509] [PMID: 10208417]

[247] Bork P, Blomberg N, Nilges M. Internal repeats in the BRCA2 protein sequence. Nat Genet 1996; 13(1): 22-3. [http://dx.doi.org/10.1038/ng0596-22] [PMID: 8673099]

[248] Chen PL, Chen CF, Chen Y, Xiao J, Sharp ZD, Lee WH. The BRC repeats in BRCA2 are critical for RAD51 binding and resistance to methyl methanesulfonate treatment. Proc Natl Acad Sci USA 1998; 95(9): 5287-92. [http://dx.doi.org/10.1073/pnas.95.9.5287] [PMID: 9560268]

[249] Wong AK, Pero R, Ormonde PA, Tavtigian SV, Bartel PL. RAD51 interacts with the evolutionarily conserved BRC motifs in the human breast cancer susceptibility gene brca2. J Biol Chem 1997; 272(51): 31941-4. [http://dx.doi.org/10.1074/jbc.272.51.31941] [PMID: 9405383]

[250] Pellegrini L, Yu DS, Lo T, *et al.* Insights into DNA recombination from the structure of a RAD51-BRCA2 complex. Nature 2002; 420(6913): 287-93. [http://dx.doi.org/10.1038/nature01230] [PMID: 12442171]

[251] Spain BH, Larson CJ, Shihabuddin LS, Gage FH, Verma IM. Truncated BRCA2 is cytoplasmic: implications for cancer-linked mutations. Proc Natl Acad Sci USA 1999; 96(24): 13920-5. [http://dx.doi.org/10.1073/pnas.96.24.13920] [PMID: 10570174]

[252] Chen J, Silver DP, Walpita D, *et al.* Stable interaction between the products of the BRCA1 and BRCA2 tumor suppressor genes in mitotic and meiotic cells. Mol Cell 1998; 2(3): 317-28. [http://dx.doi.org/10.1016/S1097-2765(00)80276-2] [PMID: 9774970]

[253] Welcsh PL, King MC. BRCA1 and BRCA2 and the genetics of breast and ovarian cancer. Hum Mol Genet 2001; 10(7): 705-13. [http://dx.doi.org/10.1093/hmg/10.7.705] [PMID: 11257103]

[254] Preobrazhenska O, Yakymovych M, Kanamoto T, *et al.* BRCA2 and Smad3 synergize in regulation of gene transcription. Oncogene 2002; 21(36): 5660-4. [http://dx.doi.org/10.1038/sj.onc.1205732] [PMID: 12165866]

[255] Shin S, Verma IM. BRCA2 cooperates with histone acetyltransferases in androgen receptor-mediated transcription. Proc Natl Acad Sci USA 2003; 100(12): 7201-6.

[http://dx.doi.org/10.1073/pnas.1132020100] [PMID: 12756300]

[256] Yang H, Jeffrey PD, Miller J, *et al.* BRCA2 function in DNA binding and recombination from a BRCA2-DSS1-ssDNA structure. Science 2002; 297(5588): 1837-48. [http://dx.doi.org/10.1126/science.297.5588.1837] [PMID: 12228710]

[257] Jasin M. Homologous repair of DNA damage and tumorigenesis: the BRCA connection. Oncogene 2002; 21(58): 8981-93. [http://dx.doi.org/10.1038/sj.onc.1206176] [PMID: 12483514]

[258] Antoniou AC, Easton DF. Models of genetic susceptibility to breast cancer. Oncogene 2006; 25(43): 5898-905. [http://dx.doi.org/10.1038/sj.onc.1209879] [PMID: 16998504]

[259] Jorde L, Carey J, Bamshad M, White R, Eds. Medical genetics. 2nd ed., Mosby 2000.

[260] Olivier M, Eeles R, Hollstein M, Khan MA, Harris CC, Hainaut P. The IARC TP53 database: new online mutation analysis and recommendations to users. Hum Mutat 2002; 19(6): 607-14. [http://dx.doi.org/10.1002/humu.10081] [PMID: 12007217]

[261] Lacroix M, Toillon RA, Leclercq G. p53 and breast cancer, an update. Endocr Relat Cancer 2006; 13(2): 293-325. [http://dx.doi.org/10.1677/erc.1.01172] [PMID: 16728565]

[262] Gunther EJ, Moody SE, Belka GK, *et al.* Impact of p53 loss on reversal and recurrence of conditional Wnt-induced tumorigenesis. Genes Dev 2003; 17(4): 488-501. [http://dx.doi.org/10.1101/gad.1051603] [PMID: 12600942]

[263] García JM, Silva J, Peña C, *et al.* Promoter methylation of the PTEN gene is a common molecular change in breast cancer. Genes Chromosomes Cancer 2004; 41(2): 117-24. [http://dx.doi.org/10.1002/gcc.20062] [PMID: 15287024]

[264] Shen WH, Balajee AS, Wang J, *et al.* Essential role for nuclear PTEN in maintaining chromosomal integrity. Cell 2007; 128(1): 157-70. [http://dx.doi.org/10.1016/j.cell.2006.11.042] [PMID: 17218262]

[265] Ngeow J, Sesock K, Eng C. Breast cancer risk and clinical implications for germline PTEN mutation carriers. Breast Cancer Res Treat 2017; 165(1): 1-8. [http://dx.doi.org/10.1007/s10549-015-3665-z] [PMID: 26700035]

[266] Collins SP, Reoma JL, Gamm DM, Uhler MD. LKB1, a novel serine/threonine protein kinase and potential tumour suppressor, is phosphorylated by cAMP-dependent protein kinase (PKA) and prenylated *in vivo*. Biochem J 2000; 345(Pt 3): 673-80. [http://dx.doi.org/10.1042/bj3450673] [PMID: 10642527]

[267] Corradetti MN, Inoki K, Bardeesy N, DePinho RA, Guan KL. Regulation of the TSC pathway by LKB1: evidence of a molecular link between tuberous sclerosis complex and Peutz-Jeghers syndrome. Genes Dev 2004; 18(13): 1533-8. [http://dx.doi.org/10.1101/gad.1199104] [PMID: 15231735]

[268] Volikos E, Robinson J, Aittomäki K, *et al.* LKB1 exonic and whole gene deletions are a common cause of Peutz-Jeghers syndrome. J Med Genet 2006; 43(5): e18. [http://dx.doi.org/10.1136/jmg.2005.039875] [PMID: 16648371]

[269] Lim W, Olschwang S, Keller JJ, *et al.* Relative frequency and morphology of cancers in STK11 mutation carriers. Gastroenterology 2004; 126(7): 1788-94. [http://dx.doi.org/10.1053/j.gastro.2004.03.014] [PMID: 15188174]

[270] Hemminki A. The molecular basis and clinical aspects of Peutz-Jeghers syndrome. Cell Mol Life Sci 1999; 55(5): 735-50. [http://dx.doi.org/10.1007/s000180050329] [PMID: 10379360]

[271] Graziano F, Humar B, Guilford P. The role of the E-cadherin gene (CDH1) in diffuse gastric cancer

susceptibility: from the laboratory to clinical practice. Ann Oncol 2003; 14(12): 1705-13. [http://dx.doi.org/10.1093/annonc/mdg486] [PMID: 14630673]

[272] Pharoah PD, Guilford P, Caldas C. Incidence of gastric cancer and breast cancer in CDH1 (E-cadherin) mutation carriers from hereditary diffuse gastric cancer families. Gastroenterology 2001; 121(6): 1348-53. [http://dx.doi.org/10.1053/gast.2001.29611] [PMID: 11729114]

[273] Stracker TH, Usui T, Petrini JH. Taking the time to make important decisions: the checkpoint effector kinases Chk1 and Chk2 and the DNA damage response. DNA Repair (Amst) 2009; 8(9): 1047-54. [http://dx.doi.org/10.1016/j.dnarep.2009.04.012] [PMID: 19473886]

[274] CHEK2*1100delC and susceptibility to breast cancer: a collaborative analysis involving 10,860 breast cancer cases and 9,065 controls from 10 studies. Am J Hum Genet 2004; 74(6): 1175-82. [http://dx.doi.org/10.1086/421251] [PMID: 15122511]

[275] Gage M, Wattendorf D, Henry LR. Translational advances regarding hereditary breast cancer syndromes. J Surg Oncol 2012; 105(5): 444-51. [http://dx.doi.org/10.1002/jso.21856] [PMID: 22441895]

[276] Cybulski C, Huzarski T, Byrski T, *et al.* Estrogen receptor status in CHEK2-positive breast cancers: implications for chemoprevention. Clin Genet 2009; 75(1): 72-8. [http://dx.doi.org/10.1111/j.1399-0004.2008.01111.x] [PMID: 19021634]

[277] Ahmed M, Rahman N. ATM and breast cancer susceptibility. Oncogene 2006; 25(43): 5906-11. [http://dx.doi.org/10.1038/sj.onc.1209873] [PMID: 16998505]

[278] Alter BP, Greene MH, Velazquez I, Rosenberg PS. Cancer in Fanconi anemia. Blood 2003; 101(5): 2072. [http://dx.doi.org/10.1182/blood-2002-11-3597] [PMID: 12584146]

[279] Xia B, Sheng Q, Nakanishi K, *et al.* Control of BRCA2 cellular and clinical functions by a nuclear partner, PALB2. Mol Cell 2006; 22(6): 719-29. [http://dx.doi.org/10.1016/j.molcel.2006.05.022] [PMID: 16793542]

[280] Rahman N, Seal S, Thompson D, *et al.* PALB2, which encodes a BRCA2-interacting protein, is a breast cancer susceptibility gene. Nat Genet 2007; 39(2): 165-7. [http://dx.doi.org/10.1038/ng1959] [PMID: 17200668]

[281] Seal S, Thompson D, Renwick A, *et al.* Truncating mutations in the Fanconi anemia J gene BRIP1 are low-penetrance breast cancer susceptibility alleles. Nat Genet 2006; 38(11): 1239-41. [http://dx.doi.org/10.1038/ng1902] [PMID: 17033622]

[282] Catucci I, Milgrom R, Kushnir A, *et al.* Germline mutations in BRIP1 and PALB2 in Jewish high cancer risk families. Fam Cancer 2012; 11(3): 483-91. [http://dx.doi.org/10.1007/s10689-012-9540-8] [PMID: 22692731]

[283] Meindl A, Hellebrand H, Wiek C, *et al.* Germline mutations in breast and ovarian cancer pedigrees establish RAD51C as a human cancer susceptibility gene. Nat Genet 2010; 42(5): 410-4. [http://dx.doi.org/10.1038/ng.569] [PMID: 20400964]

[284] Pelttari LM, Heikkinen T, Thompson D, *et al.* RAD51C is a susceptibility gene for ovarian cancer. Hum Mol Genet 2011; 20(16): 3278-88. [http://dx.doi.org/10.1093/hmg/ddr229] [PMID: 21616938]

[285] Osorio A, Endt D, Fernández F, *et al.* Predominance of pathogenic missense variants in the RAD51C gene occurring in breast and ovarian cancer families. Hum Mol Genet 2012; 21(13): 2889-98. [http://dx.doi.org/10.1093/hmg/dds115] [PMID: 22451500]

[286] Loveday C, Turnbull C, Ramsay E, *et al.* Germline mutations in RAD51D confer susceptibility to ovarian cancer. Nat Genet 2011; 43(9): 879-82. [http://dx.doi.org/10.1038/ng.893] [PMID: 21822267]

[287] Osher DJ, De Leeneer K, Michils G, *et al.* Mutation analysis of RAD51D in non-BRCA1/2 ovarian and breast cancer families. Br J Cancer 2012; 106(8): 1460-3. [http://dx.doi.org/10.1038/bjc.2012.87] [PMID: 22415235]

[288] Gutiérrez-Enríquez S, Bonache S, de Garibay GR, *et al.* About 1% of the breast and ovarian Spanish families testing negative for BRCA1 and BRCA2 are carriers of RAD51D pathogenic variants. Int J Cancer 2014; 134(9): 2088-97. [http://dx.doi.org/10.1002/ijc.28540] [PMID: 24130102]

[289] Rupnik A, Grenon M, Lowndes N. The MRN complex. Curr Biol 2008; 18(11): R455-7. [http://dx.doi.org/10.1016/j.cub.2008.03.040] [PMID: 18522810]

[290] Zhang ZH, Yang LS, Huang F, Hao JH, Su PY, Sun YH. Current evidence on the relationship between two polymorphisms in the NBS1 gene and breast cancer risk: a meta-analysis. Asian Pac J Cancer Prev 2012; 13(11): 5375-9. [http://dx.doi.org/10.7314/APJCP.2012.13.11.5375] [PMID: 23317186]

[291] Heikkinen K, Karppinen SM, Soini Y, Mäkinen M, Winqvist R. Mutation screening of Mre11 complex genes: indication of RAD50 involvement in breast and ovarian cancer susceptibility. J Med Genet 2003; 40(12): e131. [http://dx.doi.org/10.1136/jmg.40.12.e131] [PMID: 14684699]

[292] Tommiska J, Seal S, Renwick A, *et al.* Evaluation of RAD50 in familial breast cancer predisposition. Int J Cancer 2006; 118(11): 2911-6. [http://dx.doi.org/10.1002/ijc.21738] [PMID: 16385572]

[293] Bartkova J, Tommiska J, Oplustilova L, *et al.* Aberrations of the MRE11-RAD50-NBS1 DNA damage sensor complex in human breast cancer: MRE11 as a candidate familial cancer-predisposing gene. Mol Oncol 2008; 2(4): 296-316. [http://dx.doi.org/10.1016/j.molonc.2008.09.007] [PMID: 19383352]

[294] Damiola F, Pertesi M, Oliver J, *et al.* Rare key functional domain missense substitutions in MRE11A, RAD50, and NBN contribute to breast cancer susceptibility: results from a Breast Cancer Family Registry case-control mutation-screening study. Breast Cancer Res 2014; 16(3): R58. [http://dx.doi.org/10.1186/bcr3669] [PMID: 24894818]

[295] Wang W. Emergence of a DNA-damage response network consisting of Fanconi anaemia and BRCA proteins. Nat Rev Genet 2007; 8(10): 735-48. [http://dx.doi.org/10.1038/nrg2159] [PMID: 17768402]

[296] Kiiski JI, Pelttari LM, Khan S, *et al.* Exome sequencing identifies FANCM as a susceptibility gene for triple-negative breast cancer. Proc Natl Acad Sci USA 2014; 111(42): 15172-7. [http://dx.doi.org/10.1073/pnas.1407909111] [PMID: 25288723]

[297] Sjöblom T, Jones S, Wood LD, *et al.* The consensus coding sequences of human breast and colorectal cancers. Science 2006; 314(5797): 268-74. [http://dx.doi.org/10.1126/science.1133427] [PMID: 16959974]

[298] Stephens PJ, Tarpey PS, Davies H, *et al.* The landscape of cancer genes and mutational processes in breast cancer. Nature 2012; 486(7403): 400-4. [PMID: 22722201]

[299] Stranger BE, Forrest MS, Clark AG, *et al.* Genome-wide associations of gene expression variation in humans. PLoS Genet 2005; 1(6): e78. [http://dx.doi.org/10.1371/journal.pgen.0010078] [PMID: 16362079]

[300] Comprehensive molecular portraits of human breast tumours. Nature 2012; 490(7418): 61-70. [http://dx.doi.org/10.1038/nature11412] [PMID: 23000897]

[301] Polyak K. Heterogeneity in breast cancer. J Clin Invest 2011; 121(10): 3786-8. [http://dx.doi.org/10.1172/JCI60534] [PMID: 21965334]

[302] Viale G. The current state of breast cancer classification. Ann Oncol 2012; 23 (Suppl. 10): x207-10. [http://dx.doi.org/10.1093/annonc/mds326] [PMID: 22987963]

[303] Hsiao YH, Chou MC, Fowler C, Mason JT, Man YG. Breast cancer heterogeneity: mechanisms, proofs, and implications. J Cancer 2010; 1: 6-13. [http://dx.doi.org/10.7150/jca.1.6] [PMID: 20842218]

[304] Rodenhiser DI, Andrews J, Kennette W, *et al.* Epigenetic mapping and functional analysis in a breast cancer metastasis model using whole-genome promoter tiling microarrays. Breast Cancer Res 2008; 10(4): R62. [http://dx.doi.org/10.1186/bcr2121] [PMID: 18638373]

[305] Elsheikh SE, Green AR, Rakha EA, *et al.* Global histone modifications in breast cancer correlate with tumor phenotypes, prognostic factors, and patient outcome. Cancer Res 2009; 69(9): 3802-9. [http://dx.doi.org/10.1158/0008-5472.CAN-08-3907] [PMID: 19366799]

[306] Liu S, Cong Y, Wang D, *et al.* Breast cancer stem cells transition between epithelial and mesenchymal states reflective of their normal counterparts. Stem Cell Reports 2013; 2(1): 78-91. [http://dx.doi.org/10.1016/j.stemcr.2013.11.009] [PMID: 24511467]

[307] Al-Hajj M, Wicha MS, Benito-Hernandez A, Morrison SJ, Clarke MF. Prospective identification of tumorigenic breast cancer cells. Proc Natl Acad Sci USA 2003; 100(7): 3983-8. [http://dx.doi.org/10.1073/pnas.0530291100] [PMID: 12629218]

[308] Velasco-Velázquez MA, Homsi N, De La Fuente M, Pestell RG. Breast cancer stem cells. Int J Biochem Cell Biol 2012; 44(4): 573-7. [http://dx.doi.org/10.1016/j.biocel.2011.12.020] [PMID: 22249027]

[309] Hartwig FP, Nedel F, Collares T, Tarquinio SB, Nör JE, Demarco FF. Oncogenic somatic events in tissue-specific stem cells: a role in cancer recurrence? Ageing Res Rev 2014; 13: 100-6. [http://dx.doi.org/10.1016/j.arr.2013.12.004] [PMID: 24374269]

[310] Ponti D, Costa A, Zaffaroni N, *et al.* Isolation and *in vitro* propagation of tumorigenic breast cancer cells with stem/progenitor cell properties. Cancer Res 2005; 65(13): 5506-11. [http://dx.doi.org/10.1158/0008-5472.CAN-05-0626] [PMID: 15994920]

[311] Ginestier C, Hur MH, Charafe-Jauffret E, *et al.* ALDH1 is a marker of normal and malignant human mammary stem cells and a predictor of poor clinical outcome. Cell Stem Cell 2007; 1(5): 555-67. [http://dx.doi.org/10.1016/j.stem.2007.08.014] [PMID: 18371393]

[312] Molyneux G, Geyer FC, Magnay FA, *et al.* BRCA1 basal-like breast cancers originate from luminal epithelial progenitors and not from basal stem cells. Cell Stem Cell 2010; 7(3): 403-17. [http://dx.doi.org/10.1016/j.stem.2010.07.010] [PMID: 20804975]

[313] Morel AP, Lièvre M, Thomas C, Hinkal G, Ansieau S, Puisieux A. Generation of breast cancer stem cells through epithelial-mesenchymal transition. PLoS One 2008; 3(8): e2888. [http://dx.doi.org/10.1371/journal.pone.0002888] [PMID: 18682804]

[314] Mani SA, Guo W, Liao MJ, *et al.* The epithelial-mesenchymal transition generates cells with properties of stem cells. Cell 2008; 133(4): 704-15. [http://dx.doi.org/10.1016/j.cell.2008.03.027] [PMID: 18485877]

[315] Bachoo RM, Maher EA, Ligon KL, *et al.* Epidermal growth factor receptor and Ink4a/Arf: convergent mechanisms governing terminal differentiation and transformation along the neural stem cell to astrocyte axis. Cancer Cell 2002; 1(3): 269-77. [http://dx.doi.org/10.1016/S1535-6108(02)00046-6] [PMID: 12086863]

[316] Liu X, Feng D, Liu D, *et al.* Dissecting the Origin of Breast Cancer Subtype Stem Cell and the Potential Mechanism of Malignant Transformation. PLoS One 2016; 11(10): e0165001. [http://dx.doi.org/10.1371/journal.pone.0165001] [PMID: 27768723]

[317] Jamieson CH, Ailles LE, Dylla SJ, *et al.* Granulocyte-macrophage progenitors as candidate leukemic

stem cells in blast-crisis CML. N Engl J Med 2004; 351(7): 657-67.
[http://dx.doi.org/10.1056/NEJMoa040258] [PMID: 15306667]

[318] Lim E, Vaillant F, Wu D, *et al.* Aberrant luminal progenitors as the candidate target population for basal tumor development in BRCA1 mutation carriers. Nat Med 2009; 15(8): 907-13.
[http://dx.doi.org/10.1038/nm.2000] [PMID: 19648928]

[319] Dontu G, Al-Hajj M, Abdallah WM, Clarke MF, Wicha MS. Stem cells in normal breast development and breast cancer. Cell Prolif 2003; 36(S1) (Suppl. 1): 59-72.
[http://dx.doi.org/10.1046/j.1365-2184.36.s.1.6.x] [PMID: 14521516]

[320] Li Y, Welm B, Podsypanina K, *et al.* Evidence that transgenes encoding components of the Wnt signaling pathway preferentially induce mammary cancers from progenitor cells. Proc Natl Acad Sci USA 2003; 100(26): 15853-8.
[http://dx.doi.org/10.1073/pnas.2136825100] [PMID: 14668450]

[321] Villadsen R, Fridriksdottir AJ, Rønnov-Jessen L, *et al.* Evidence for a stem cell hierarchy in the adult human breast. J Cell Biol 2007; 177(1): 87-101.
[http://dx.doi.org/10.1083/jcb.200611114] [PMID: 17420292]

[322] Barroca V, Lassalle B, Coureuil M, *et al.* Mouse differentiating spermatogonia can generate germinal stem cells *in vivo*. Nat Cell Biol 2009; 11(2): 190-6.
[http://dx.doi.org/10.1038/ncb1826] [PMID: 19098901]

[323] Xu X, D'Hoker J, Stangé G, *et al.* Beta cells can be generated from endogenous progenitors in injured adult mouse pancreas. Cell 2008; 132(2): 197-207.
[http://dx.doi.org/10.1016/j.cell.2007.12.015] [PMID: 18243096]

[324] Magee JA, Piskounova E, Morrison SJ. Cancer stem cells: impact, heterogeneity, and uncertainty. Cancer Cell 2012; 21(3): 283-96.
[http://dx.doi.org/10.1016/j.ccr.2012.03.003] [PMID: 22439924]

[325] Smalley M, Ashworth A. Stem cells and breast cancer: A field in transit. Nat Rev Cancer 2003; 3(11): 832-44.
[http://dx.doi.org/10.1038/nrc1212] [PMID: 14668814]

[326] Asselin-Labat ML, Sutherland KD, Vaillant F, *et al.* Gata-3 negatively regulates the tumor-initiating capacity of mammary luminal progenitor cells and targets the putative tumor suppressor caspase-14. Mol Cell Biol 2011; 31(22): 4609-22.
[http://dx.doi.org/10.1128/MCB.05766-11] [PMID: 21930782]

[327] Kalyuga M, Gallego-Ortega D, Lee HJ, *et al.* ELF5 suppresses estrogen sensitivity and underpins the acquisition of antiestrogen resistance in luminal breast cancer. PLoS Biol 2012; 10(12): e1001461.
[http://dx.doi.org/10.1371/journal.pbio.1001461] [PMID: 23300383]

[328] Chakrabarti R, Hwang J, Andres Blanco M, *et al.* Elf5 inhibits the epithelial-mesenchymal transition in mammary gland development and breast cancer metastasis by transcriptionally repressing Snail2. Nat Cell Biol 2012; 14(11): 1212-22.
[http://dx.doi.org/10.1038/ncb2607] [PMID: 23086238]

[329] Chou J, Lin JH, Brenot A, Kim JW, Provot S, Werb Z. GATA3 suppresses metastasis and modulates the tumour microenvironment by regulating microRNA-29b expression. Nat Cell Biol 2013; 15(2): 201-13.
[http://dx.doi.org/10.1038/ncb2672] [PMID: 23354167]

[330] Yan W, Cao QJ, Arenas RB, Bentley B, Shao R. GATA3 inhibits breast cancer metastasis through the reversal of epithelial-mesenchymal transition. J Biol Chem 2010; 285(18): 14042-51.
[http://dx.doi.org/10.1074/jbc.M110.105262] [PMID: 20189993]

[331] Perou CM, Sørlie T, Eisen MB, *et al.* Molecular portraits of human breast tumours. Nature 2000; 406(6797): 747-52.
[http://dx.doi.org/10.1038/35021093] [PMID: 10963602]

[332] Prat A, Parker JS, Karginova O, *et al.* Phenotypic and molecular characterization of the claudin-low intrinsic subtype of breast cancer. Breast Cancer Res 2010; 12(5): R68. [http://dx.doi.org/10.1186/bcr2635] [PMID: 20813035]

[333] Sørlie T, Perou CM, Tibshirani R, *et al.* Gene expression patterns of breast carcinomas distinguish tumor subclasses with clinical implications. Proc Natl Acad Sci USA 2001; 98(19): 10869-74. [http://dx.doi.org/10.1073/pnas.191367098] [PMID: 11553815]

[334] Gallego-Ortega D, Oakes SR, Lee HJ, Piggin CL, Ormandy CJ. ELF5, normal mammary development and the heterogeneous phenotypes of breast cancer. Breast Cancer Manag 2013; 2: 489-98. [http://dx.doi.org/10.2217/bmt.13.50]

[335] Carroll JS, Liu XS, Brodsky AS, *et al.* Chromosome-wide mapping of estrogen receptor binding reveals long-range regulation requiring the forkhead protein FoxA1. Cell 2005; 122(1): 33-43. [http://dx.doi.org/10.1016/j.cell.2005.05.008] [PMID: 16009131]

[336] Kouros-Mehr H, Slorach EM, Sternlicht MD, Werb Z. GATA-3 maintains the differentiation of the luminal cell fate in the mammary gland. Cell 2006; 127(5): 1041-55. [http://dx.doi.org/10.1016/j.cell.2006.09.048] [PMID: 17129787]

[337] Nowell PC. The clonal evolution of tumor cell populations. Science 1976; 194(4260): 23-8. [http://dx.doi.org/10.1126/science.959840] [PMID: 959840]

[338] Vaillant F, Asselin-Labat ML, Shackleton M, Forrest NC, Lindeman GJ, Visvader JE. The mammary progenitor marker CD61/beta3 integrin identifies cancer stem cells in mouse models of mammary tumorigenesis. Cancer Res 2008; 68(19): 7711-7. [http://dx.doi.org/10.1158/0008-5472.CAN-08-1949] [PMID: 18829523]

[339] Charles N, Ozawa T, Squatrito M, *et al.* Perivascular nitric oxide activates notch signaling and promotes stem-like character in PDGF-induced glioma cells. Cell Stem Cell 2010; 6(2): 141-52. [http://dx.doi.org/10.1016/j.stem.2010.01.001] [PMID: 20144787]

[340] Gudjonsson T, Villadsen R, Nielsen HL, Rønnov-Jessen L, Bissell MJ, Petersen OW. Isolation, immortalization, and characterization of a human breast epithelial cell line with stem cell properties. Genes Dev 2002; 16(6): 693-706. [http://dx.doi.org/10.1101/gad.952602] [PMID: 11914275]

[341] Péchoux C, Gudjonsson T, Ronnov-Jessen L, Bissell MJ, Petersen OW. Human mammary luminal epithelial cells contain progenitors to myoepithelial cells. Dev Biol 1999; 206(1): 88-99. [http://dx.doi.org/10.1006/dbio.1998.9133] [PMID: 9918697]

[342] Stingl J, Eaves CJ, Kuusk U, Emerman JT. Phenotypic and functional characterization *in vitro* of a multipotent epithelial cell present in the normal adult human breast. Differentiation 1998; 63(4): 201-13. [http://dx.doi.org/10.1111/j.1432-0436.1998.00201.x] [PMID: 9745711]

[343] Stingl J, Eaves CJ, Zandieh I, Emerman JT. Characterization of bipotent mammary epithelial progenitor cells in normal adult human breast tissue. Breast Cancer Res Treat 2001; 67(2): 93-109. [http://dx.doi.org/10.1023/A:1010615124301] [PMID: 11519870]

[344] Chepko G, Dickson RB. Ultrastructure of the putative stem cell niche in rat mammary epithelium. Tissue Cell 2003; 35(2): 83-93. [http://dx.doi.org/10.1016/S0040-8166(02)00107-6] [PMID: 12747930]

[345] Chepko G, Smith GH. Mammary epithelial stem cells: our current understanding. J Mammary Gland Biol Neoplasia 1999; 4(1): 35-52. [http://dx.doi.org/10.1023/A:1018752519356] [PMID: 10219905]

[346] Spradling A, Drummond-Barbosa D, Kai T. Stem cells find their niche. Nature 2001; 414(6859): 98-104. [http://dx.doi.org/10.1038/35102160] [PMID: 11689954]

[347] Blanpain C, Horsley V, Fuchs E. Epithelial stem cells: turning over new leaves. Cell 2007; 128(3): 445-58.
[http://dx.doi.org/10.1016/j.cell.2007.01.014] [PMID: 17289566]

[348] Lee HJ, Gallego-Ortega D, Ledger A, *et al.* Progesterone drives mammary secretory differentiation *via* RankL-mediated induction of Elf5 in luminal progenitor cells. Development 2013; 140(7): 1397-401.
[http://dx.doi.org/10.1242/dev.088948] [PMID: 23462470]

[349] Fata JE, Werb Z, Bissell MJ. Regulation of mammary gland branching morphogenesis by the extracellular matrix and its remodeling enzymes. Breast Cancer Res 2004; 6(1): 1-11.
[http://dx.doi.org/10.1186/bcr634] [PMID: 14680479]

[350] Lee HJ, Ormandy CJ. Interplay between progesterone and prolactin in mammary development and implications for breast cancer. Mol Cell Endocrinol 2012; 357(1-2): 101-7.
[http://dx.doi.org/10.1016/j.mce.2011.09.020] [PMID: 21945475]

[351] Sternlicht MD, Kouros-Mehr H, Lu P, Werb Z. Hormonal and local control of mammary branching morphogenesis. Differentiation 2006; 74(7): 365-81.
[http://dx.doi.org/10.1111/j.1432-0436.2006.00105.x] [PMID: 16916375]

[352] Joyce JA, Pollard JW. Microenvironmental regulation of metastasis. Nat Rev Cancer 2009; 9(4): 239-52.
[http://dx.doi.org/10.1038/nrc2618] [PMID: 19279573]

[353] Mantovani A, Allavena P, Sica A, Balkwill F. Cancer-related inflammation. Nature 2008; 454(7203): 436-44.
[http://dx.doi.org/10.1038/nature07205] [PMID: 18650914]

[354] Tlsty TD, Coussens LM. Tumor stroma and regulation of cancer development. Annu Rev Pathol 2006; 1: 119-50.
[http://dx.doi.org/10.1146/annurev.pathol.1.110304.100224] [PMID: 18039110]

[355] Meissner A, Mikkelsen TS, Gu H, *et al.* Genome-scale DNA methylation maps of pluripotent and differentiated cells. Nature 2008; 454(7205): 766-70.
[PMID: 18600261]

[356] Pal B, Bouras T, Shi W, *et al.* Global changes in the mammary epigenome are induced by hormonal cues and coordinated by Ezh2. Cell Reports 2013; 3(2): 411-26.
[http://dx.doi.org/10.1016/j.celrep.2012.12.020] [PMID: 23375371]

[357] Widschwendter M, Fiegl H, Egle D, *et al.* Epigenetic stem cell signature in cancer. Nat Genet 2007; 39(2): 157-8.
[http://dx.doi.org/10.1038/ng1941] [PMID: 17200673]

[358] Schuebel K, Chen W, Baylin SB. CIMPle origin for promoter hypermethylation in colorectal cancer? Nat Genet 2006; 38(7): 738-40.
[http://dx.doi.org/10.1038/ng0706-738] [PMID: 16804535]

[359] Lee TI, Jenner RG, Boyer LA, *et al.* Control of developmental regulators by Polycomb in human embryonic stem cells. Cell 2006; 125(2): 301-13.
[http://dx.doi.org/10.1016/j.cell.2006.02.043] [PMID: 16630818]

[360] Wicha MS, Liu S, Dontu G. Cancer stem cells: an old idea--a paradigm shift. Cancer Res 2006; 66(4): 1883-90; discussion 1895-6.

[361] Conley SJ, Gheordunescu E, Kakarala P, *et al.* Antiangiogenic agents increase breast cancer stem cells *via* the generation of tumor hypoxia. Proc Natl Acad Sci USA 2012; 109(8): 2784-9
[http://dx.doi.org/10.1073/pnas.1018866109] [PMID: 22308314]

[362] Al Saleh S, Sharaf LH, Luqmani YA. Signalling pathways involved in endocrine resistance in breast cancer and associations with epithelial to mesenchymal transition (Review). Int J Oncol 2011; 38(5): 1197-217.

[PMID: 21318221]

[363] Potenta S, Zeisberg E, Kalluri R. The role of endothelial-to-mesenchymal transition in cancer progression. Br J Cancer 2008; 99(9): 1375-9. [http://dx.doi.org/10.1038/sj.bjc.6604662] [PMID: 18797460]

[364] Kalluri R, Weinberg RA. The basics of epithelial-mesenchymal transition. J Clin Invest 2009; 119(6): 1420-8. [http://dx.doi.org/10.1172/JCI39104] [PMID: 19487818]

[365] Lin F, Wang N, Zhang TC. The role of endothelial-mesenchymal transition in development and pathological process. IUBMB Life 2012; 64(9): 717-23. [http://dx.doi.org/10.1002/iub.1059] [PMID: 22730243]

[366] Parker BS, Argani P, Cook BP, *et al.* Alterations in vascular gene expression in invasive breast carcinoma. Cancer Res 2004; 64(21): 7857-66. [http://dx.doi.org/10.1158/0008-5472.CAN-04-1976] [PMID: 15520192]

[367] Liu S, Clouthier SG, Wicha MS. Role of microRNAs in the regulation of breast cancer stem cells. J Mammary Gland Biol Neoplasia 2012; 17(1): 15-21. [http://dx.doi.org/10.1007/s10911-012-9242-8] [PMID: 22331423]

[368] Liu S, Patel SH, Ginestier C, *et al.* MicroRNA93 regulates proliferation and differentiation of normal and malignant breast stem cells. PLoS Genet 2012; 8(6): e1002751. [http://dx.doi.org/10.1371/journal.pgen.1002751] [PMID: 22685420]

[369] Biddle A, Liang X, Gammon L, *et al.* Cancer stem cells in squamous cell carcinoma switch between two distinct phenotypes that are preferentially migratory or proliferative. Cancer Res 2011; 71(15): 5317-26. [http://dx.doi.org/10.1158/0008-5472.CAN-11-1059] [PMID: 21685475]

[370] Sheridan C, Kishimoto H, Fuchs RK, *et al.* CD44+/CD24- breast cancer cells exhibit enhanced invasive properties: an early step necessary for metastasis. Breast Cancer Res 2006; 8(5): R59. [http://dx.doi.org/10.1186/bcr1610] [PMID: 17062128]

[371] Ranji P, Salmani Kesejini T, Saeedikhoo S, Alizadeh AM. Targeting cancer stem cell-specific markers and/or associated signaling pathways for overcoming cancer drug resistance. Tumour Biol 2016; 37(10): 13059-75. [http://dx.doi.org/10.1007/s13277-016-5294-5] [PMID: 27561758]

[372] Visvader JE. Keeping abreast of the mammary epithelial hierarchy and breast tumorigenesis. Genes Dev 2009; 23(22): 2563-77. [http://dx.doi.org/10.1101/gad.1849509] [PMID: 19933147]

[373] Liu S, Dontu G, Mantle ID, *et al.* Hedgehog signaling and Bmi-1 regulate self-renewal of normal and malignant human mammary stem cells. Cancer Res 2006; 66(12): 6063-71. [http://dx.doi.org/10.1158/0008-5472.CAN-06-0054] [PMID: 16778178]

[374] Hu Z, Fan C, Oh DS, *et al.* The molecular portraits of breast tumors are conserved across microarray platforms. BMC Genomics 2006; 7: 96. [http://dx.doi.org/10.1186/1471-2164-7-96] [PMID: 16643655]

[375] van 't Veer LJ, Dai H, van de Vijver MJ, *et al.* Gene expression profiling predicts clinical outcome of breast cancer. Nature 2002; 415(6871): 530-6. [http://dx.doi.org/10.1038/415530a] [PMID: 11823860]

[376] Sotiriou C, Neo SY, McShane LM, *et al.* Breast cancer classification and prognosis based on gene expression profiles from a population-based study. Proc Natl Acad Sci USA 2003; 100(18): 10393-8. [http://dx.doi.org/10.1073/pnas.1732912100] [PMID: 12917485]

[377] Abd El-Rehim DM, Ball G, Pinder SE, *et al.* High-throughput protein expression analysis using tissue microarray technology of a large well-characterised series identifies biologically distinct classes of breast cancer confirming recent cDNA expression analyses. Int J Cancer 2005; 116(3): 340-50.

[http://dx.doi.org/10.1002/ijc.21004] [PMID: 15818618]

[378] Habashy HO, Powe DG, Abdel-Fatah TM, *et al.* A review of the biological and clinical characteristics of luminal-like oestrogen receptor-positive breast cancer. Histopathology 2012; 60(6): 854-63. [http://dx.doi.org/10.1111/j.1365-2559.2011.03912.x] [PMID: 21906125]

[379] Carey LA, Perou CM, Livasy CA, *et al.* Race, breast cancer subtypes, and survival in the Carolina Breast Cancer Study. JAMA 2006; 295(21): 2492-502. [http://dx.doi.org/10.1001/jama.295.21.2492] [PMID: 16757721]

[380] Kennecke H, Yerushalmi R, Woods R, *et al.* Metastatic behavior of breast cancer subtypes. J Clin Oncol 2010; 28(20): 3271-7. [http://dx.doi.org/10.1200/JCO.2009.25.9820] [PMID: 20498394]

[381] Guarneri V, Conte P. Metastatic breast cancer: therapeutic options according to molecular subtypes and prior adjuvant therapy. Oncologist 2009; 14(7): 645-56. [http://dx.doi.org/10.1634/theoncologist.2009-0078] [PMID: 19608638]

[382] Creighton CJ. The molecular profile of luminal B breast cancer. Biologics 2012; 6: 289-97. [http://dx.doi.org/10.2147/BTT.S29923] [PMID: 22956860]

[383] Dowsett M, Nielsen TO, A'Hern R, *et al.* Assessment of Ki67 in breast cancer: recommendations from the International Ki67 in Breast Cancer working group. J Natl Cancer Inst 2011; 103(22): 1656-64. [http://dx.doi.org/10.1093/jnci/djr393] [PMID: 21960707]

[384] Cheang MC, Chia SK, Voduc D, *et al.* Ki67 index, HER2 status, and prognosis of patients with luminal B breast cancer. J Natl Cancer Inst 2009; 101(10): 736-50. [http://dx.doi.org/10.1093/jnci/djp082] [PMID: 19436038]

[385] Nishimura R, Osako T, Okumura Y, Hayashi M, Toyozumi Y, Arima N. Ki-67 as a prognostic marker according to breast cancer subtype and a predictor of recurrence time in primary breast cancer. Exp Ther Med 2010; 1(5): 747-54. [http://dx.doi.org/10.3892/etm.2010.133] [PMID: 22993598]

[386] Paik S, Shak S, Tang G, *et al.* A multigene assay to predict recurrence of tamoxifen-treated, node-negative breast cancer. N Engl J Med 2004; 351(27): 2817-26. [http://dx.doi.org/10.1056/NEJMoa041588] [PMID: 15591335]

[387] Rouzier R, Perou CM, Symmans WF, *et al.* Breast cancer molecular subtypes respond differently to preoperative chemotherapy. Clin Cancer Res 2005; 11(16): 5678-85. [http://dx.doi.org/10.1158/1078-0432.CCR-04-2421] [PMID: 16115903]

[388] Bhargava R, Beriwal S, Dabbs DJ, *et al.* Immunohistochemical surrogate markers of breast cancer molecular classes predicts response to neoadjuvant chemotherapy: a single institutional experience with 359 cases. Cancer 2010; 116(6): 1431-9. [http://dx.doi.org/10.1002/cncr.24876] [PMID: 20131351]

[389] Ignatiadis M, Bedard P. A meta analysis of gene expression profiling studies identifies clinically relevant oncogenic pathways in basal like breast cancer. Cancer Res 2009; 69: 106. [http://dx.doi.org/10.1158/0008-5472.SABCS-09-106]

[390] Barnes CJ, Kumar R. Biology of the epidermal growth factor receptor family. Cancer Treat Res 2004; 119: 1-13. [http://dx.doi.org/10.1007/1-4020-7847-1_1] [PMID: 15164870]

[391] Bazley LA, Gullick WJ. The epidermal growth factor receptor family. Endocr Relat Cancer 2005; 12 (Suppl. 1): S17-27. [http://dx.doi.org/10.1677/erc.1.01032] [PMID: 16113093]

[392] Moasser MM. The oncogene HER2: its signaling and transforming functions and its role in human cancer pathogenesis. Oncogene 2007; 26(45): 6469-87. [http://dx.doi.org/10.1038/sj.onc.1210477] [PMID: 17471238]

[393] Gutierrez C, Schiff R. HER2: biology, detection, and clinical implications. Arch Pathol Lab Med 2011; 135(1): 55-62.
[PMID: 21204711]

[394] Tsutsui S, Ohno S, Murakami S, Kataoka A, Kinoshita J, Hachitanda Y. Prognostic significance of the coexpression of p53 protein and c-erbB2 in breast cancer. Am J Surg 2003; 185(2): 165-7.
[http://dx.doi.org/10.1016/S0002-9610(02)01203-5] [PMID: 12559449]

[395] Ross JS, Fletcher JA, Linette GP, *et al.* The Her-2/neu gene and protein in breast cancer 2003: biomarker and target of therapy. Oncologist 2003; 8(4): 307-25.
[http://dx.doi.org/10.1634/theoncologist.8-4-307] [PMID: 12897328]

[396] Gabos Z, Sinha R, Hanson J, *et al.* Prognostic significance of human epidermal growth factor receptor positivity for the development of brain metastasis after newly diagnosed breast cancer. J Clin Oncol 2006; 24(36): 5658-63.
[http://dx.doi.org/10.1200/JCO.2006.07.0250] [PMID: 17102066]

[397] Rakha EA, Elsheikh SE, Aleskandarany MA, *et al.* Triple-negative breast cancer: distinguishing between basal and nonbasal subtypes. Clin Cancer Res 2009; 15(7): 2302-10.
[http://dx.doi.org/10.1158/1078-0432.CCR-08-2132] [PMID: 19318481]

[398] Heitz F, Harter P, Lueck HJ, *et al.* Triple-negative and HER2-overexpressing breast cancers exhibit an elevated risk and an earlier occurrence of cerebral metastases. Eur J Cancer 2009; 45(16): 2792-8.
[http://dx.doi.org/10.1016/j.ejca.2009.06.027] [PMID: 19643597]

[399] Kreike B, van Kouwenhove M, Horlings H, *et al.* Gene expression profiling and histopathological characterization of triple-negative/basal-like breast carcinomas. Breast Cancer Res 2007; 9(5): R65.
[http://dx.doi.org/10.1186/bcr1771] [PMID: 17910759]

[400] Foulkes WD, Brunet JS, Stefansson IM, *et al.* The prognostic implication of the basal-like (cyclin E high/p27 low/p53+/glomeruloid-microvascular-proliferation+) phenotype of BRCA1-related breast cancer. Cancer Res 2004; 64(3): 830-5.
[http://dx.doi.org/10.1158/0008-5472.CAN-03-2970] [PMID: 14871808]

[401] Weigelt B, Mackay A, A'hern R, *et al.* Breast cancer molecular profiling with single sample predictors: a retrospective analysis. Lancet Oncol 2010; 11(4): 339-49.
[http://dx.doi.org/10.1016/S1470-2045(10)70008-5] [PMID: 20181526]

[402] Guidelines NC. 2016.https://www.nccn.org/professionals/physician_gls/pdf/breast.pdft

[403] http://www.esmo.org/Guidelines/Breast-Cancer

[404] Deo SV, Bhutani M, Shukla NK, Raina V, Rath GK, Purkayasth J. Randomized trial comparing neo-adjuvant *versus* adjuvant chemotherapy in operable locally advanced breast cancer (T4b N0-2 M0). J Surg Oncol 2003; 84(4): 192-7.
[http://dx.doi.org/10.1002/jso.10323] [PMID: 14756429]

[405] Sweeting RS, Klauber-Demore N, Meyers MO, *et al.* Young women with locally advanced breast cancer who achieve breast conservation after neoadjuvant chemotherapy have a low local recurrence rate. Am Surg 2011; 77(7): 850-5.
[PMID: 21944346]

[406] Akay CL, Meric-Bernstam F, Hunt KK, *et al.* Evaluation of the MD Anderson Prognostic Index for local-regional recurrence after breast conserving therapy in patients receiving neoadjuvant chemotherapy. Ann Surg Oncol 2012; 19(3): 901-7.
[http://dx.doi.org/10.1245/s10434-011-2006-7] [PMID: 21861223]

[407] Bear H. Primary chemotherapy for operable breast cancer: the NSABP experience. Breast Cancer Res 2005; 7 (Suppl. 1): S17.
[http://dx.doi.org/10.1186/bcr1221]

[408] Bear HD, Anderson S, Brown A, *et al.* The effect on tumor response of adding sequential preoperative

docetaxel to preoperative doxorubicin and cyclophosphamide: preliminary results from National Surgical Adjuvant Breast and Bowel Project Protocol B-27. J Clin Oncol 2003; 21(22): 4165-74. [http://dx.doi.org/10.1200/JCO.2003.12.005] [PMID: 14559892]

[409] Mieog JS, van der Hage JA, van de Velde CJ. Preoperative chemotherapy for women with operable breast cancer. Cochrane Database Syst Rev 2007; (2): CD005002. [PMID: 17443564]

[410] Kong X, Moran MS, Zhang N, Haffty B, Yang Q. Meta-analysis confirms achieving pathological complete response after neoadjuvant chemotherapy predicts favourable prognosis for breast cancer patients. Eur J Cancer 2011; 47(14): 2084-90. [http://dx.doi.org/10.1016/j.ejca.2011.06.014] [PMID: 21737257]

[411] von Minckwitz G, Untch M, Blohmer JU, *et al.* Definition and impact of pathologic complete response on prognosis after neoadjuvant chemotherapy in various intrinsic breast cancer subtypes. J Clin Oncol 2012; 30(15): 1796-804. [http://dx.doi.org/10.1200/JCO.2011.38.8595] [PMID: 22508812]

[412] Berruti A, Amoroso V, Gallo F, *et al.* Pathologic complete response as a potential surrogate for the clinical outcome in patients with breast cancer after neoadjuvant therapy: a meta-regression of 29 randomized prospective studies. J Clin Oncol 2014; 32(34): 3883-91. [http://dx.doi.org/10.1200/JCO.2014.55.2836] [PMID: 25349292]

[413] Miller E, Lee HJ, Lulla A, Hernandez L, Gokare P, Lim B. Current treatment of early breast cancer: adjuvant and neoadjuvant therapy. F1000 Res 2014; 3: 198. [PMID: 25400908]

[414] van der Hage JA, van de Velde CJ, Julien JP, Tubiana-Hulin M, Vandervelden C, Duchateau L. Preoperative chemotherapy in primary operable breast cancer: results from the European Organization for Research and Treatment of Cancer trial 10902. J Clin Oncol 2001; 19(22): 4224-37. [http://dx.doi.org/10.1200/JCO.2001.19.22.4224] [PMID: 11709566]

[415] Mauri D, Pavlidis N, Ioannidis JP. Neoadjuvant *versus* adjuvant systemic treatment in breast cancer: a meta-analysis. J Natl Cancer Inst 2005; 97(3): 188-94. [http://dx.doi.org/10.1093/jnci/dji021] [PMID: 15687361]

[416] Loibl S, von Minckwitz G, Raab G, *et al.* Surgical procedures after neoadjuvant chemotherapy in operable breast cancer: results of the GEPARDUO trial. Ann Surg Oncol 2006; 13(11): 1434-42. [http://dx.doi.org/10.1245/s10434-006-9011-2] [PMID: 16983592]

[417] Sataloff DM, Mason BA, Prestipino AJ, Seinige UL, Lieber CP, Baloch Z. Pathologic response to induction chemotherapy in locally advanced carcinoma of the breast: a determinant of outcome. J Am Coll Surg 1995; 180(3): 297-306. [PMID: 7874340]

[418] Fisher B, Bryant J, Wolmark N, *et al.* Effect of preoperative chemotherapy on the outcome of women with operable breast cancer. J Clin Oncol 1998; 16(8): 2672-85. [http://dx.doi.org/10.1200/JCO.1998.16.8.2672] [PMID: 9704717]

[419] Krainick-Strobel UE, Lichtenegger W, Wallwiener D, *et al.* Neoadjuvant letrozole in postmenopausal estrogen and/or progesterone receptor positive breast cancer: a phase IIb/III trial to investigate optimal duration of preoperative endocrine therapy. BMC Cancer 2008; 8: 62. [http://dx.doi.org/10.1186/1471-2407-8-62] [PMID: 18302747]

[420] Thomas E, Holmes FA, Smith TL, *et al.* The use of alternate, non-cross-resistant adjuvant chemotherapy on the basis of pathologic response to a neoadjuvant doxorubicin-based regimen in women with operable breast cancer: long-term results from a prospective randomized trial. J Clin Oncol 2004; 22(12): 2294-302. [http://dx.doi.org/10.1200/JCO.2004.05.207] [PMID: 15197190]

[421] Gianni L, Pienkowski T, Im YH, *et al.* Efficacy and safety of neoadjuvant pertuzumab and trastuzumab in women with locally advanced, inflammatory, or early HER2-positive breast cancer

(NeoSphere): a randomised multicentre, open-label, phase 2 trial. Lancet Oncol 2012; 13(1): 25-32. [http://dx.doi.org/10.1016/S1470-2045(11)70336-9] [PMID: 22153890]

[422] Schneeweiss A, Chia S, Hickish T, *et al.* Pertuzumab plus trastuzumab in combination with standard neoadjuvant anthracycline-containing and anthracycline-free chemotherapy regimens in patients with HER2-positive early breast cancer: a randomized phase II cardiac safety study (TRYPHAENA). Ann Oncol 2013; 24(9): 2278-84.
[http://dx.doi.org/10.1093/annonc/mdt182] [PMID: 23704196]

[423] Guarneri V, Broglio K, Kau SW, *et al.* Prognostic value of pathologic complete response after primary chemotherapy in relation to hormone receptor status and other factors. J Clin Oncol 2006; 24(7): 1037-44.
[http://dx.doi.org/10.1200/JCO.2005.02.6914] [PMID: 16505422]

[424] Baselga J, Bradbury I, Eidtmann H, *et al.* Lapatinib with trastuzumab for HER2-positive early breast cancer (NeoALTTO): a randomised, open-label, multicentre, phase 3 trial. Lancet 2012; 379(9816): 633-40.
[http://dx.doi.org/10.1016/S0140-6736(11)61847-3] [PMID: 22257673]

[426] Untch M, Loibl S, Bischoff J, *et al.* Lapatinib *versus* trastuzumab in combination with neoadjuvant anthracycline-taxane-based chemotherapy (GeparQuinto, GBG 44): a randomised phase 3 trial. Lancet Oncol 2012; 13(2): 135-44.
[http://dx.doi.org/10.1016/S1470-2045(11)70397-7] [PMID: 22257523]

[426] Buzdar AU, Suman VJ, Meric-Bernstam F, *et al.* Fluorouracil, epirubicin, and cyclophosphamide (FEC-75) followed by paclitaxel plus trastuzumab *versus* paclitaxel plus trastuzumab followed by FEC-75 plus trastuzumab as neoadjuvant treatment for patients with HER2-positive breast cancer (Z1041): a randomised, controlled, phase 3 trial. Lancet Oncol 2013; 14(13): 1317-25.
[http://dx.doi.org/10.1016/S1470-2045(13)70502-3] [PMID: 24239210]

[427] Carey LA, Berry DA, Cirrincione CT, *et al.* Molecular Heterogeneity and Response to Neoadjuvant Human Epidermal Growth Factor Receptor 2 Targeting in CALGB 40601, a Randomized Phase III Trial of Paclitaxel Plus Trastuzumab With or Without Lapatinib. J Clin Oncol 2016; 34(6): 542-9.
[http://dx.doi.org/10.1200/JCO.2015.62.1268] [PMID: 26527775]

[428] Prat A, Bianchini G, Thomas M, *et al.* Research-based PAM50 subtype predictor identifies higher responses and improved survival outcomes in HER2-positive breast cancer in the NOAH study. Clin Cancer Res 2014; 20(2): 511-21.
[http://dx.doi.org/10.1158/1078-0432.CCR-13-0239] [PMID: 24443618]

[429] Rimawi MF, Mayer IA, Forero A, *et al.* Multicenter phase II study of neoadjuvant lapatinib and trastuzumab with hormonal therapy and without chemotherapy in patients with human epidermal growth factor receptor 2-overexpressing breast cancer: TBCRC 006. J Clin Oncol 2013; 31(14): 1726-31.
[http://dx.doi.org/10.1200/JCO.2012.44.8027] [PMID: 23569315]

[430] Guarneri V, Frassoldati A, Bottini A, *et al.* Preoperative chemotherapy plus trastuzumab, lapatinib, or both in human epidermal growth factor receptor 2-positive operable breast cancer: results of the randomized phase II CHER-LOB study. J Clin Oncol 2012; 30(16): 1989-95.
[http://dx.doi.org/10.1200/JCO.2011.39.0823] [PMID: 22493419]

[431] Robidoux A, Tang G, Rastogi P, *et al.* Lapatinib as a component of neoadjuvant therapy for HER2-positive operable breast cancer (NSABP protocol B-41): an open-label, randomised phase 3 trial. Lancet Oncol 2013; 14(12): 1183-92.
[http://dx.doi.org/10.1016/S1470-2045(13)70411-X] [PMID: 24095300]

[432] Amiri-Kordestani L, Wedam S, Zhang L, *et al.* First FDA approval of neoadjuvant therapy for breast cancer: pertuzumab for the treatment of patients with HER2-positive breast cancer. Clin Cancer Res 2014; 20(21): 5359-64.
[http://dx.doi.org/10.1158/1078-0432.CCR-14-1268] [PMID: 25204553]

[433] Ju NR, Jeffe DB, Keune J, Aft R. Patient and tumor characteristics associated with breast cancer recurrence after complete pathological response to neoadjuvant chemotherapy. Breast Cancer Res Treat 2013; 137(1): 195-201.
[http://dx.doi.org/10.1007/s10549-012-2312-1] [PMID: 23149464]

[434] Tanioka M, Sasaki M, Shimomura A, *et al.* Pathologic complete response after neoadjuvant chemotherapy in HER2-overexpressing breast cancer according to hormonal receptor status. Breast 2014; 23(4): 466-72.
[http://dx.doi.org/10.1016/j.breast.2014.03.008] [PMID: 24742606]

[435] Takada M, Higuchi T, Tozuka K, *et al.* Alterations of the genes involved in the PI3K and estrogen-receptor pathways influence outcome in human epidermal growth factor receptor 2-positive and hormone receptor-positive breast cancer patients treated with trastuzumab-containing neoadjuvant chemotherapy. BMC Cancer 2013; 13: 241.
[http://dx.doi.org/10.1186/1471-2407-13-241] [PMID: 23679233]

[436] Liedtke C, Mazouni C, Hess KR, *et al.* Response to neoadjuvant therapy and long-term survival in patients with triple-negative breast cancer. J Clin Oncol 2008; 26(8): 1275-81.
[http://dx.doi.org/10.1200/JCO.2007.14.4147] [PMID: 18250347]

[437] Goldhirsch A, Wood WC, Coates AS, Gelber RD, Thürlimann B, Senn HJ. Strategies for subtypes--dealing with the diversity of breast cancer: highlights of the St. Gallen International Expert Consensus on the Primary Therapy of Early Breast Cancer 2011. Ann Oncol 2011; 22(8): 1736-47.
[http://dx.doi.org/10.1093/annonc/mdr304] [PMID: 21709140]

[438] Goldhirsch A, Winer EP, Coates AS, *et al.* Personalizing the treatment of women with early breast cancer: highlights of the St Gallen International Expert Consensus on the Primary Therapy of Early Breast Cancer 2013. Ann Oncol 2013; 24(9): 2206-23.
[http://dx.doi.org/10.1093/annonc/mdt303] [PMID: 23917950]

[439] https://www.adjuvantonline.com/

[440] Olivotto IA, Bajdik CD, Ravdin PM, *et al.* Population-based validation of the prognostic model ADJUVANT! for early breast cancer. J Clin Oncol 2005; 23(12): 2716-25.
[http://dx.doi.org/10.1200/JCO.2005.06.178] [PMID: 15837986]

[441] Dowsett M, Cuzick J, Wale C, *et al.* Prediction of risk of distant recurrence using the 21-gene recurrence score in node-negative and node-positive postmenopausal patients with breast cancer treated with anastrozole or tamoxifen: a TransATAC study. J Clin Oncol 2010; 28(11): 1829-34.
[http://dx.doi.org/10.1200/JCO.2009.24.4798] [PMID: 20212256]

[442] Paik S, Tang G, Shak S, *et al.* Gene expression and benefit of chemotherapy in women with node-negative, estrogen receptor-positive breast cancer. J Clin Oncol 2006; 24(23): 3726-34.
[http://dx.doi.org/10.1200/JCO.2005.04.7985] [PMID: 16720680]

[443] Albain KS, Barlow WE, Shak S, *et al.* Prognostic and predictive value of the 21-gene recurrence score assay in postmenopausal women with node-positive, oestrogen-receptor-positive breast cancer on chemotherapy: a retrospective analysis of a randomised trial. Lancet Oncol 2010; 11(1): 55-65.
[http://dx.doi.org/10.1016/S1470-2045(09)70314-6] [PMID: 20005174]

[444] Gonzalez-Angulo A, Barlow E, Gralow J, *et al.* SWOG S1007: A phase III randomized clinical trial of standard adjuvant endocrine therapy with or without chemotherapy in patients with one to three positive nodes, hormone receptor (HR)-positive, and HER2-negative breast cancer with recurrence score (RS) of 25 or less. 2011 ASCO annual meeting: suppl: abstr TPS104 J Clin Oncol. 29.

[445] Bastien RR, Rodríguez-Lescure Á, Ebbert MT, *et al.* PAM50 breast cancer subtyping by RT-qPCR and concordance with standard clinical molecular markers. BMC Med Genomics 2012; 5: 44.
[http://dx.doi.org/10.1186/1755-8794-5-44] [PMID: 23035882]

[446] Dowsett M, Sestak I, Lopez-Knowles E, *et al.* Comparison of PAM50 risk of recurrence score with oncotype DX and IHC4 for predicting risk of distant recurrence after endocrine therapy. J Clin Oncol

2013; 31(22): 2783-90.
[http://dx.doi.org/10.1200/JCO.2012.46.1558] [PMID: 23816962]

[447] Group EB. Effects of chemotherapy and hormonal therapy for early breast cancer on recurrence and 15-year survival: an overview of the randomised trials. Lancet 2005; 365(9472): 1687-717.
[http://dx.doi.org/10.1016/S0140-6736(05)66544-0] [PMID: 15894097]

[448] Peto R, Davies C, Godwin J, *et al.* Comparisons between different polychemotherapy regimens for early breast cancer: meta-analyses of long-term outcome among 100,000 women in 123 randomised trials. Lancet 2012; 379(9814): 432-44.
[http://dx.doi.org/10.1016/S0140-6736(11)61625-5] [PMID: 22152853]

[449] Fisher B, Redmond C, Legault-Poisson S, *et al.* Postoperative chemotherapy and tamoxifen compared with tamoxifen alone in the treatment of positive-node breast cancer patients aged 50 years and older with tumors responsive to tamoxifen: results from the National Surgical Adjuvant Breast and Bowel Project B-16. J Clin Oncol 1990; 8(6): 1005-18.
[http://dx.doi.org/10.1200/JCO.1990.8.6.1005] [PMID: 2189950]

[450] Bonadonna G, Brusamolino E, Valagussa P, *et al.* Combination chemotherapy as an adjuvant treatment in operable breast cancer. N Engl J Med 1976; 294(8): 405-10.
[http://dx.doi.org/10.1056/NEJM197602192940801] [PMID: 1246307]

[451] Dignam JJ, Huang L, Ries L, Reichman M, Mariotto A, Feuer E. Estimating breast cancer-specific and other-cause mortality in clinical trial and population-based cancer registry cohorts. Cancer 2009; 115(22): 5272-83.
[http://dx.doi.org/10.1002/cncr.24617] [PMID: 19670456]

[452] Palmieri C, Jones A. The 2011 EBCTCG polychemotherapy overview. Lancet 2012; 379(9814): 390-2.
[http://dx.doi.org/10.1016/S0140-6736(11)61823-0] [PMID: 22152852]

[453] Francis P, Crown J, Di Leo A, *et al.* Adjuvant chemotherapy with sequential or concurrent anthracycline and docetaxel: Breast International Group 02-98 randomized trial. J Natl Cancer Inst 2008; 100(2): 121-33.
[http://dx.doi.org/10.1093/jnci/djm287] [PMID: 18182617]

[454] Mackey JR, Martin M, Pienkowski T, *et al.* Adjuvant docetaxel, doxorubicin, and cyclophosphamide in node-positive breast cancer: 10-year follow-up of the phase 3 randomised BCIRG 001 trial. Lancet Oncol 2013; 14(1): 72-80.
[http://dx.doi.org/10.1016/S1470-2045(12)70525-9] [PMID: 23246022]

[455] Henderson IC, Berry DA, Demetri GD, *et al.* Improved outcomes from adding sequential Paclitaxel but not from escalating Doxorubicin dose in an adjuvant chemotherapy regimen for patients with node-positive primary breast cancer. J Clin Oncol 2003; 21(6): 976-83.
[http://dx.doi.org/10.1200/JCO.2003.02.063] [PMID: 12637460]

[456] Mamounas EP, Bryant J, Lembersky B, *et al.* Paclitaxel after doxorubicin plus cyclophosphamide as adjuvant chemotherapy for node-positive breast cancer: results from NSABP B-28. J Clin Oncol 2005; 23(16): 3686-96.
[http://dx.doi.org/10.1200/JCO.2005.10.517] [PMID: 15897552]

[457] Citron ML, Berry DA, Cirrincione C, *et al.* Randomized trial of dose-dense *versus* conventionally scheduled and sequential *versus* concurrent combination chemotherapy as postoperative adjuvant treatment of node-positive primary breast cancer: first report of Intergroup Trial C9741/Cancer and Leukemia Group B Trial 9741. J Clin Oncol 2003; 21(8): 1431-9.
[http://dx.doi.org/10.1200/JCO.2003.09.081] [PMID: 12668651]

[458] Budd GT, Barlow WE, Moore HC, *et al.* SWOG S0221: a phase III trial comparing chemotherapy schedules in high-risk early-stage breast cancer. J Clin Oncol 2015; 33(1): 58-64.
[http://dx.doi.org/10.1200/JCO.2014.56.3296] [PMID: 25422488]

[459] Martín M, Ruiz A, Ruiz Borrego M, *et al.* Fluorouracil, doxorubicin, and cyclophosphamide (FAC)

versus FAC followed by weekly paclitaxel as adjuvant therapy for high-risk, node-negative breast cancer: results from the GEICAM/2003-02 study. J Clin Oncol 2013; 31(20): 2593-9. [http://dx.doi.org/10.1200/JCO.2012.46.9841] [PMID: 23733779]

[460] Jones S, Holmes FA, O'Shaughnessy J, *et al.* Docetaxel With Cyclophosphamide Is Associated With an Overall Survival Benefit Compared With Doxorubicin and Cyclophosphamide: 7-Year Follow-Up of US Oncology Research Trial 9735. J Clin Oncol 2009; 27(8): 1177-83. [http://dx.doi.org/10.1200/JCO.2008.18.4028] [PMID: 19204201]

[461] Dutertre M, Smith CL. Molecular mechanisms of selective estrogen receptor modulator (SERM) action. J Pharmacol Exp Ther 2000; 295(2): 431-7. [PMID: 11046073]

[462] Trunet PF, Vreeland F, Royce C, Chaudri HA, Cooper J, Bhatnagar AS. Clinical use of aromatase inhibitors in the treatment of advanced breast cancer. J Steroid Biochem Mol Biol 1997; 61(3-6): 241-5. [http://dx.doi.org/10.1016/S0960-0760(97)80018-0] [PMID: 9365196]

[463] Effects of chemotherapy and hormonal therapy for early breast cancer on recurrence and 15-year survival: an overview of the randomised trials. Lancet 2005; 365(9472): 1687-717. [http://dx.doi.org/10.1016/S0140-6736(05)66544-0] [PMID: 15894097]

[464] Smith IE, Dowsett M, Yap YS, *et al.* Adjuvant aromatase inhibitors for early breast cancer after chemotherapy-induced amenorrhoea: caution and suggested guidelines. J Clin Oncol 2006; 24(16): 2444-7.

[465] Yu B, Douglas N, Ferin MJ, *et al.* Changes in markers of ovarian reserve and endocrine function in young women with breast cancer undergoing adjuvant chemotherapy. Cancer 2010; 116(9): 2099-105. [PMID: 20187091]

[466] Davies C, Pan H, Godwin J, *et al.* Long-term effects of continuing adjuvant tamoxifen to 10 years *versus* stopping at 5 years after diagnosis of oestrogen receptor-positive breast cancer: ATLAS, a randomised trial. Lancet 2013; 381(9869): 805-16. [http://dx.doi.org/10.1016/S0140-6736(12)61963-1] [PMID: 23219286]

[467] Pritchard KI. Ovarian suppression/ablation in premenopausal ER-positive breast cancer patients. Issues and recommendations. Oncology (Williston Park) 2009; 23(1): 27-33. [PMID: 19283918]

[468] Tan SH, Wolff AC. The role of ovarian ablation in the adjuvant therapy of breast cancer. Curr Oncol Rep 2008; 10(1): 27-37. [http://dx.doi.org/10.1007/s11912-008-0006-y] [PMID: 18366958]

[469] Cuzick J, Ambroisine L, Davidson N, *et al.* Use of luteinising-hormone-releasing hormone agonists as adjuvant treatment in premenopausal patients with hormone-receptor-positive breast cancer: a meta-analysis of individual patient data from randomised adjuvant trials. Lancet 2007; 369(9574): 1711-23. [http://dx.doi.org/10.1016/S0140-6736(07)60778-8] [PMID: 17512856]

[470] Puhalla S, Brufsky A, Davidson N. Adjuvant endocrine therapy for premenopausal women with breast cancer. Breast 2009; 18 (Suppl. 3): S122-30. [http://dx.doi.org/10.1016/S0960-9776(09)70286-3] [PMID: 19914530]

[471] Pagani O, Regan MM, Walley BA, *et al.* Adjuvant exemestane with ovarian suppression in premenopausal breast cancer. N Engl J Med 2014; 371(2): 107-18. [http://dx.doi.org/10.1056/NEJMoa1404037] [PMID: 24881463]

[472] Howell A, Cuzick J, Baum M, *et al.* Results of the ATAC (Arimidex, Tamoxifen, Alone or in Combination) trial after completion of 5 years' adjuvant treatment for breast cancer. Lancet 2005; 365(9453): 60-2. [http://dx.doi.org/10.1016/S0140-6736(04)17666-6] [PMID: 15639680]

[473] Goss PE, Ingle JN, Pater JL, *et al.* Late extended adjuvant treatment with letrozole improves outcome

in women with early-stage breast cancer who complete 5 years of tamoxifen. J Clin Oncol 2008; 26(12): 1948-55.
[http://dx.doi.org/10.1200/JCO.2007.11.6798] [PMID: 18332475]

[474] Mouridsen H, Giobbie-Hurder A, Goldhirsch A, *et al.* Letrozole therapy alone or in sequence with tamoxifen in women with breast cancer. N Engl J Med 2009; 361(8): 766-76.
[http://dx.doi.org/10.1056/NEJMoa0810818] [PMID: 19692688]

[475] Boccardo F, Rubagotti A, Puntoni M, *et al.* Switching to anastrozole *versus* continued tamoxifen treatment of early breast cancer: preliminary results of the Italian Tamoxifen Anastrozole Trial. J Clin Oncol 2005; 23(22): 5138-47.
[http://dx.doi.org/10.1200/JCO.2005.04.120] [PMID: 16009955]

[476] Coombes RC, Hall E, Gibson LJ, *et al.* A randomized trial of exemestane after two to three years of tamoxifen therapy in postmenopausal women with primary breast cancer. N Engl J Med 2004; 350(11): 1081-92.
[http://dx.doi.org/10.1056/NEJMoa040331] [PMID: 15014181]

[477] Jakesz R, Jonat W, Gnant M, *et al.* Switching of postmenopausal women with endocrine-responsive early breast cancer to anastrozole after 2 years' adjuvant tamoxifen: combined results of ABCSG trial 8 and ARNO 95 trial. Lancet 2005; 366(9484): 455-62.
[http://dx.doi.org/10.1016/S0140-6736(05)67059-6] [PMID: 16084253]

[478] Goss PE, Ingle JN, Martino S, *et al.* A randomized trial of letrozole in postmenopausal women after five years of tamoxifen therapy for early-stage breast cancer. N Engl J Med 2003; 349(19): 1793-802.
[http://dx.doi.org/10.1056/NEJMoa032312] [PMID: 14551341]

[479] Goss PE, Ingle JN, Martino S, *et al.* Randomized trial of letrozole following tamoxifen as extended adjuvant therapy in receptor-positive breast cancer: updated findings from NCIC CTG MA.17. J Natl Cancer Inst 2005; 97(17): 1262-71.
[http://dx.doi.org/10.1093/jnci/dji250] [PMID: 16145047]

[480] Jakesz R, Greil R, Gnant M, *et al.* Extended adjuvant therapy with anastrozole among postmenopausal breast cancer patients: results from the randomized Austrian Breast and Colorectal Cancer Study Group Trial 6a. J Natl Cancer Inst 2007; 99(24): 1845-53.
[http://dx.doi.org/10.1093/jnci/djm246] [PMID: 18073378]

[481] Hayes DF, Thor AD, Dressler LG, *et al.* HER2 and response to paclitaxel in node-positive breast cancer. N Engl J Med 2007; 357(15): 1496-506.
[http://dx.doi.org/10.1056/NEJMoa071167] [PMID: 17928597]

[482] Arpino G, Gutierrez C, Weiss H, *et al.* Treatment of human epidermal growth factor receptor 2-overexpressing breast cancer xenografts with multiagent HER-targeted therapy. J Natl Cancer Inst 2007; 99(9): 694-705.
[http://dx.doi.org/10.1093/jnci/djk151] [PMID: 17470737]

[483] Vogel C, Cobleigh MA, Tripathy D, *et al.* First-line, single-agent Herceptin(R) (trastuzumab) in metastatic breast cancer. a preliminary report. Eur J Cancer 2001; 37 (Suppl. 1): 25-9.
[http://dx.doi.org/10.1016/S0959-8049(00)00405-6] [PMID: 11342197]

[484] Barok M, Isola J, Pályi-Krekk Z, *et al.* Trastuzumab causes antibody-dependent cellular cytotoxicity-mediated growth inhibition of submacroscopic JIMT-1 breast cancer xenografts despite intrinsic drug resistance. Mol Cancer Ther 2007; 6(7): 2065-72.
[http://dx.doi.org/10.1158/1535-7163.MCT-06-0766] [PMID: 17620435]

[485] Slamon D, Eiermann W, Robert N, *et al.* Adjuvant trastuzumab in HER2-positive breast cancer. N Engl J Med 2011; 365(14): 1273-83.
[http://dx.doi.org/10.1056/NEJMoa0910383] [PMID: 21991949]

[486] Joensuu H, Bono P, Kataja V, *et al.* Fluorouracil, epirubicin, and cyclophosphamide with either docetaxel or vinorelbine, with or without trastuzumab, as adjuvant treatments of breast cancer: final results of the FinHer Trial. J Clin Oncol 2009; 27(34): 5685-92.

[http://dx.doi.org/10.1200/JCO.2008.21.4577] [PMID: 19884557]

[487] Goldhirsch A, Gelber RD, Piccart-Gebhart MJ, *et al.* 2 years *versus* 1 year of adjuvant trastuzumab for HER2-positive breast cancer (HERA): an open-label, randomised controlled trial. Lancet 2013; 382(9897): 1021-8.
[http://dx.doi.org/10.1016/S0140-6736(13)61094-6] [PMID: 23871490]

[488] Perez EA, Suman VJ, Davidson NE, *et al.* HER2 testing by local, central, and reference laboratories in specimens from the North Central Cancer Treatment Group N9831 intergroup adjuvant trial. J Clin Oncol 2006; 24(19): 3032-8.
[http://dx.doi.org/10.1200/JCO.2005.03.4744] [PMID: 16809727]

[489] Spielmann M, Roché H, Delozier T, *et al.* Trastuzumab for patients with axillary-node-positive breast cancer: results of the FNCLCC-PACS 04 trial. J Clin Oncol 2009; 27(36): 6129-34.
[http://dx.doi.org/10.1200/JCO.2009.23.0946] [PMID: 19917839]

[490] Hortobagyi GN, Perez EA. Integration of trastuzumab into adjuvant systemic therapy of breast cancer: ongoing and planned clinical trials. Semin Oncol 2001; 28(5) (Suppl. 16): 41-6.
[http://dx.doi.org/10.1016/S0093-7754(01)90281-7] [PMID: 11706395]

[491] Piccart-Gebhart MJ, Procter M, Leyland-Jones B, *et al.* Trastuzumab after adjuvant chemotherapy in HER2-positive breast cancer. N Engl J Med 2005; 353(16): 1659-72.
[http://dx.doi.org/10.1056/NEJMoa052306] [PMID: 16236737]

[492] Waks AG, Tolaney SM. The evolving understanding of small HER2-positive breast cancers: matching management to outcomes. Future Oncol 2015; 11(24): 3261-71.
[http://dx.doi.org/10.2217/fon.15.240] [PMID: 26634944]

[493] Von Minckwitz G, Baselga J, Bradbury I, *et al.* Adjuvant Pertuzumab and Herceptin IN IniTial TherapY of Breast Cancer: APHINITY (BIG 4–11/BO25126/TOC4939g). Cancer Res 2011; 71(24) OT1-02-04-abs.

[494] Alqaisi A, Chen L, Romond E, *et al.* Impact of estrogen receptor (ER) and human epidermal growth factor receptor-2 (HER2) co-expression on breast cancer disease characteristics: implications for tumor biology and research. Breast Cancer Res Treat 2014; 148(2): 437-44.
[http://dx.doi.org/10.1007/s10549-014-3145-x] [PMID: 25257728]

[495] Lee CH, Chang YC, Chen CS, *et al.* Crosstalk between nicotine and estrogen-induced estrogen receptor activation induces α9-nicotinic acetylcholine receptor expression in human breast cancer cells. Breast Cancer Res Treat 2011; 129(2): 331-45.
[http://dx.doi.org/10.1007/s10549-010-1209-0] [PMID: 20953833]

[496] Collins DC, Cocchiglia S, Tibbitts P, *et al.* Growth factor receptor/steroid receptor cross talk in trastuzumab-treated breast cancer. Oncogene 2015; 34(4): 525-30.
[http://dx.doi.org/10.1038/onc.2013.586] [PMID: 24469058]

[497] Thery JC, Spano JP, Azria D, Raymond E, Penault Llorca F. Resistance to human epidermal growth factor receptor type 2-targeted therapies. Eur J Cancer 2014; 50(5): 892-901.
[http://dx.doi.org/10.1016/j.ejca.2014.01.003] [PMID: 24462377]

[498] Hayashi N, Niikura N, Yamauchi H, Nakamura S, Ueno NT. Adding hormonal therapy to chemotherapy and trastuzumab improves prognosis in patients with hormone receptor-positive and human epidermal growth factor receptor 2-positive primary breast cancer. Breast Cancer Res Treat 2013; 137(2): 523-31.
[http://dx.doi.org/10.1007/s10549-012-2336-6] [PMID: 23184079]

[499] Clarke M, Coates AS, Darby SC, *et al.* Adjuvant chemotherapy in oestrogen-receptor-poor breast cancer: patient-level meta-analysis of randomised trials. Lancet 2008; 371(9606): 29-40.
[http://dx.doi.org/10.1016/S0140-6736(08)60069-0] [PMID: 18177773]

[500] Martín M, Seguí MA, Antón A, *et al.* Adjuvant docetaxel for high-risk, node-negative breast cancer. N Engl J Med 2010; 363(23): 2200-10.

[http://dx.doi.org/10.1056/NEJMoa0910320] [PMID: 21121833]

[501] Carrick S, Parker S, Wilcken N, Ghersi D, Marzo M, Simes J. Single agent *versus* combination chemotherapy for metastatic breast cancer. Cochrane Database Syst Rev 2005; (2): CD003372. [PMID: 15846660]

[502] Gennari A, Stockler M, Puntoni M, *et al.* Duration of chemotherapy for metastatic breast cancer: a systematic review and meta-analysis of randomized clinical trials. J Clin Oncol 2011; 29(16): 2144-9. [http://dx.doi.org/10.1200/JCO.2010.31.5374] [PMID: 21464403]

[503] Klijn JG, Blamey RW, Boccardo F, Tominaga T, Duchateau L, Sylvester R. Combined tamoxifen and luteinizing hormone-releasing hormone (LHRH) agonist *versus* LHRH agonist alone in premenopausal advanced breast cancer: a meta-analysis of four randomized trials. J Clin Oncol 2001; 19(2): 343-53. [http://dx.doi.org/10.1200/JCO.2001.19.2.343] [PMID: 11208825]

[504] Bonneterre J, Thürlimann B, Robertson JF, *et al.* Anastrozole *versus* tamoxifen as first-line therapy for advanced breast cancer in 668 postmenopausal women: results of the Tamoxifen or Arimidex Randomized Group Efficacy and Tolerability study. J Clin Oncol 2000; 18(22): 3748-57. [http://dx.doi.org/10.1200/JCO.2000.18.22.3748] [PMID: 11078487]

[505] Paridaens RJ, Dirix LY, Beex LV, *et al.* Phase III study comparing exemestane with tamoxifen as first-line hormonal treatment of metastatic breast cancer in postmenopausal women: the European Organisation for Research and Treatment of Cancer Breast Cancer Cooperative Group. J Clin Oncol 2008; 26(30): 4883-90.
[http://dx.doi.org/10.1200/JCO.2007.14.4659] [PMID: 18794551]

[506] Gibson L, Lawrence D, Dawson C, Bliss J. Aromatase inhibitors for treatment of advanced breast cancer in postmenopausal women. Cochrane Database Syst Rev 2009; (4): CD003370. [PMID: 19821307]

[507] Osborne CK, Pippen J, Jones SE, *et al.* Double-blind, randomized trial comparing the efficacy and tolerability of fulvestrant *versus* anastrozole in postmenopausal women with advanced breast cancer progressing on prior endocrine therapy: results of a North American trial. J Clin Oncol 2002; 20(16): 3386-95.
[http://dx.doi.org/10.1200/JCO.2002.10.058] [PMID: 12177098]

[508] Robertson JF, Osborne CK, Howell A, *et al.* Fulvestrant *versus* anastrozole for the treatment of advanced breast carcinoma in postmenopausal women: a prospective combined analysis of two multicenter trials. Cancer 2003; 98(2): 229-38.
[http://dx.doi.org/10.1002/cncr.11468] [PMID: 12872340]

[509] Chia S, Gradishar W, Mauriac L, *et al.* Double-blind, randomized placebo controlled trial of fulvestrant compared with exemestane after prior nonsteroidal aromatase inhibitor therapy in postmenopausal women with hormone receptor-positive, advanced breast cancer: results from EFECT. J Clin Oncol 2008; 26(10): 1664-70.
[http://dx.doi.org/10.1200/JCO.2007.13.5822] [PMID: 18316794]

[510] Di Leo A, Jerusalem G, Petruzelka L, *et al.* Results of the CONFIRM phase III trial comparing fulvestrant 250 mg with fulvestrant 500 mg in postmenopausal women with estrogen receptor-positive advanced breast cancer. J Clin Oncol 2010; 28(30): 4594-600.
[http://dx.doi.org/10.1200/JCO.2010.28.8415] [PMID: 20855825]

[511] Bachelot T, Bourgier C, Cropet C, *et al.* Randomized phase II trial of everolimus in combination with tamoxifen in patients with hormone receptor-positive, human epidermal growth factor receptor 2-negative metastatic breast cancer with prior exposure to aromatase inhibitors: a GINECO study. J Clin Oncol 2012; 30(22): 2718-24.
[http://dx.doi.org/10.1200/JCO.2011.39.0708] [PMID: 22565002]

[512] Yardley DA, Noguchi S, Pritchard KI, *et al.* Everolimus plus exemestane in postmenopausal patients with HR(+) breast cancer: BOLERO-2 final progression-free survival analysis. Adv Ther 2013; 30(10): 870-84.

[http://dx.doi.org/10.1007/s12325-013-0060-1] [PMID: 24158787]

[513] Baselga J, Campone M, Piccart M, *et al.* Everolimus in postmenopausal hormone-receptor-positive advanced breast cancer. N Engl J Med 2012; 366(6): 520-9. [http://dx.doi.org/10.1056/NEJMoa1109653] [PMID: 22149876]

[514] Untch M, Gelber RD, Jackisch C, *et al.* Estimating the magnitude of trastuzumab effects within patient subgroups in the HERA trial. Ann Oncol 2008; 19(6): 1090-6. [http://dx.doi.org/10.1093/annonc/mdn005] [PMID: 18296421]

[515] Slamon DJ, Leyland-Jones B, Shak S, *et al.* Use of chemotherapy plus a monoclonal antibody against HER2 for metastatic breast cancer that overexpresses HER2. N Engl J Med 2001; 344(11): 783-92. [http://dx.doi.org/10.1056/NEJM200103153441101] [PMID: 11248153]

[516] Brufsky A, Lembersky B, Schiffman K, Lieberman G, Paton VE. Hormone receptor status does not affect the clinical benefit of trastuzumab therapy for patients with metastatic breast cancer. Clin Breast Cancer 2005; 6(3): 247-52. [http://dx.doi.org/10.3816/CBC.2005.n.027] [PMID: 16137436]

[517] Arpino G, Weiss H, Lee AV, *et al.* Estrogen receptor-positive, progesterone receptor-negative breast cancer: association with growth factor receptor expression and tamoxifen resistance. J Natl Cancer Inst 2005; 97(17): 1254-61. [http://dx.doi.org/10.1093/jnci/dji249] [PMID: 16145046]

[518] Yamamoto D, Iwase S, Kitamura K, Odagiri H, Yamamoto C, Nagumo Y. A phase II study of trastuzumab and capecitabine for patients with HER2-overexpressing metastatic breast cancer: Japan Breast Cancer Research Network (JBCRN) 00 Trial. Cancer Chemother Pharmacol 2008; 61(3): 509-14. [http://dx.doi.org/10.1007/s00280-007-0497-5] [PMID: 17516068]

[519] Baselga J, Cortés J, Kim SB, *et al.* Pertuzumab plus trastuzumab plus docetaxel for metastatic breast cancer. N Engl J Med 2012; 366(2): 109-19. [http://dx.doi.org/10.1056/NEJMoa1113216] [PMID: 22149875]

[520] Swain SM, Baselga J, Kim SB, *et al.* Pertuzumab, trastuzumab, and docetaxel in HER2-positive metastatic breast cancer. N Engl J Med 2015; 372(8): 724-34. [http://dx.doi.org/10.1056/NEJMoa1413513] [PMID: 25693012]

[521] von Minckwitz G, du Bois A, Schmidt M, *et al.* Trastuzumab beyond progression in human epidermal growth factor receptor 2-positive advanced breast cancer: a german breast group 26/breast international group 03-05 study. J Clin Oncol 2009; 27(12): 1999-2006. [http://dx.doi.org/10.1200/JCO.2008.19.6618] [PMID: 19289619]

[522] Verma S, Miles D, Gianni L, *et al.* Trastuzumab emtansine for HER2-positive advanced breast cancer. N Engl J Med 2012; 367(19): 1783-91. [http://dx.doi.org/10.1056/NEJMoa1209124] [PMID: 23020162]

[523] Baselga J, Gelmon KA, Verma S, *et al.* Phase II trial of pertuzumab and trastuzumab in patients with human epidermal growth factor receptor 2-positive metastatic breast cancer that progressed during prior trastuzumab therapy. J Clin Oncol 2010; 28(7): 1138-44. [http://dx.doi.org/10.1200/JCO.2009.24.2024] [PMID: 20124182]

[524] Cortés J, Fumoleau P, Bianchi GV, *et al.* Pertuzumab monotherapy after trastuzumab-based treatment and subsequent reintroduction of trastuzumab: activity and tolerability in patients with advanced human epidermal growth factor receptor 2-positive breast cancer. J Clin Oncol 2012; 30(14): 1594-600. [http://dx.doi.org/10.1200/JCO.2011.37.4207] [PMID: 22393084]

[525] Geyer CE, Forster J, Lindquist D, *et al.* Lapatinib plus capecitabine for HER2-positive advanced breast cancer. N Engl J Med 2006; 355(26): 2733-43. [http://dx.doi.org/10.1056/NEJMoa064320] [PMID: 17192538]

[526] Cameron D, Casey M, Oliva C, Newstat B, Imwalle B, Geyer CE. Lapatinib plus capecitabine in women with HER-2-positive advanced breast cancer: final survival analysis of a phase III randomized trial. Oncologist 2010; 15(9): 924-34. [http://dx.doi.org/10.1634/theoncologist.2009-0181] [PMID: 20736298]

[527] Dowsett M, Allred C, Knox J, *et al.* Relationship between quantitative estrogen and progesterone receptor expression and human epidermal growth factor receptor 2 (HER-2) status with recurrence in the Arimidex, Tamoxifen, Alone or in Combination trial. J Clin Oncol 2008; 26(7): 1059-65. [http://dx.doi.org/10.1200/JCO.2007.12.9437] [PMID: 18227529]

[528] De Laurentiis M, Arpino G, Massarelli E, *et al.* A meta-analysis on the interaction between HER-2 expression and response to endocrine treatment in advanced breast cancer. Clin Cancer Res 2005; 11(13): 4741-8. [http://dx.doi.org/10.1158/1078-0432.CCR-04-2569] [PMID: 16000569]

[529] Lipton A, Ali SM, Leitzel K, *et al.* Elevated serum Her-2/neu level predicts decreased response to hormone therapy in metastatic breast cancer. J Clin Oncol 2002; 20(6): 1467-72. [http://dx.doi.org/10.1200/JCO.2002.20.6.1467] [PMID: 11896093]

[530] Osborne CK, Schiff R. Mechanisms of endocrine resistance in breast cancer. Annu Rev Med 2011; 62: 233-47. [http://dx.doi.org/10.1146/annurev-med-070909-182917] [PMID: 20887199]

[531] Shou J, Massarweh S, Osborne CK, *et al.* Mechanisms of tamoxifen resistance: increased estrogen receptor-HER2/neu cross-talk in ER/HER2-positive breast cancer. J Natl Cancer Inst 2004; 96(12): 926-35. [http://dx.doi.org/10.1093/jnci/djh166] [PMID: 15199112]

[532] Witters LM, Kumar R, Chinchilli VM, Lipton A. Enhanced anti-proliferative activity of the combination of tamoxifen plus HER-2-neu antibody. Breast Cancer Res Treat 1997; 42(1): 1-5. [http://dx.doi.org/10.1023/A:1005798224288] [PMID: 9116313]

[533] Witters L, Engle L, Lipton A. Restoration of estrogen responsiveness by blocking the HER-2/neu pathway. Oncol Rep 2002; 9(6): 1163-6. [PMID: 12375012]

[534] Johnston S, Pippen J Jr, Pivot X, *et al.* Lapatinib combined with letrozole *versus* letrozole and placebo as first-line therapy for postmenopausal hormone receptor-positive metastatic breast cancer. J Clin Oncol 2009; 27(33): 5538-46. [http://dx.doi.org/10.1200/JCO.2009.23.3734] [PMID: 19786658]

[535] Kaufman B, Mackey JR, Clemens MR, *et al.* Trastuzumab plus anastrozole *versus* anastrozole alone for the treatment of postmenopausal women with human epidermal growth factor receptor 2-positive, hormone receptor-positive metastatic breast cancer: results from the randomized phase III TAnDEM study. J Clin Oncol 2009; 27(33): 5529-37. [http://dx.doi.org/10.1200/JCO.2008.20.6847] [PMID: 19786670]

[536] Giordano SH, Temin S, Kirshner JJ, *et al.* Systemic therapy for patients with advanced human epidermal growth factor receptor 2-positive breast cancer: American Society of Clinical Oncology clinical practice guideline. J Clin Oncol 2014; 32(19): 2078-99. [http://dx.doi.org/10.1200/JCO.2013.54.0948] [PMID: 24799465]

[537] Tripathy D, Kaufman PA, Brufsky AM, *et al.* First-line treatment patterns and clinical outcomes in patients with HER2-positive and hormone receptor-positive metastatic breast cancer from registHER. Oncologist 2013; 18(5): 501-10. [http://dx.doi.org/10.1634/theoncologist.2012-0414] [PMID: 23652380]

[538] Stendahl M, Rydén L, Nordenskjöld B, Jönsson PE, Landberg G, Jirström K. High progesterone receptor expression correlates to the effect of adjuvant tamoxifen in premenopausal breast cancer patients. Clin Cancer Res 2006; 12(15): 4614-8. [http://dx.doi.org/10.1158/1078-0432.CCR-06-0248] [PMID: 16899609]

[539] Creighton CJ, Kent Osborne C, van de Vijver MJ, *et al.* Molecular profiles of progesterone receptor loss in human breast tumors. Breast Cancer Res Treat 2009; 114(2): 287-99. [http://dx.doi.org/10.1007/s10549-008-0017-2] [PMID: 18425577]

[540] Cancello G, Maisonneuve P, Rotmensz N, *et al.* Progesterone receptor loss identifies Luminal B breast cancer subgroups at higher risk of relapse. Ann Oncol 2013; 24(3): 661-8. [http://dx.doi.org/10.1093/annonc/mds430] [PMID: 23022996]

[541] Parise CA, Caggiano V. Breast cancer survival defined by the ER/PR/HER2 subtypes and a surrogate classification according to tumor grade and immunohistochemical biomarkers. J Cancer Epidemiol 2014; 2014: 469251.

[542] Miller K, Wang M, Gralow J, *et al.* Paclitaxel plus bevacizumab *versus* paclitaxel alone for metastatic breast cancer. N Engl J Med 2007; 357(26): 2666-76. [http://dx.doi.org/10.1056/NEJMoa072113] [PMID: 18160686]

[543] Miles DW, Chan A, Dirix LY, *et al.* Phase III study of bevacizumab plus docetaxel compared with placebo plus docetaxel for the first-line treatment of human epidermal growth factor receptor 2-negative metastatic breast cancer. J Clin Oncol 2010; 28(20): 3239-47. [http://dx.doi.org/10.1200/JCO.2008.21.6457] [PMID: 20498403]

[544] Miles D, Zielinski C, Martin M, Vrdoljak E, Robert N. Combining capecitabine and bevacizumab in metastatic breast cancer: a comprehensive review. Eur J Cancer 2012; 48(4): 482-91. [http://dx.doi.org/10.1016/j.ejca.2011.12.007] [PMID: 22257791]

[545] Robert NJ, Diéras V, Glaspy J, *et al.* RIBBON-1: randomized, double-blind, placebo-controlled, phase III trial of chemotherapy with or without bevacizumab for first-line treatment of human epidermal growth factor receptor 2-negative, locally recurrent or metastatic breast cancer. J Clin Oncol 2011; 29(10): 1252-60. [http://dx.doi.org/10.1200/JCO.2010.28.0982] [PMID: 21383283]

[546] O'Shaughnessy J, Miles D. A meta-analysis of overall survival data from three randomized trials of bevacizumab (BV) and first-line chemotherapy as treatment for patients with metastatic breast cancer (MBC). 2010 J Clin Oncol 28 (15) abs 1005

[547] Fruman DA, Rommel C. PI3K and cancer: lessons, challenges and opportunities. Nat Rev Drug Discov 2014; 13(2): 140-56. [http://dx.doi.org/10.1038/nrd4204] [PMID: 24481312]

[548] Nicolini A, Ferrari P, Kotlarova L, Rossi G, Biava PM. The PI3K-AKt-mTOR Pathway and New Tools to Prevent Acquired Hormone Resistance in Breast Cancer. Curr Pharm Biotechnol 2015; 16(9): 804-15. [http://dx.doi.org/10.2174/138920101609150715141545] [PMID: 26201609]

[549] Di Cosimo S, Sathyanarayanan S, Bendell JC, *et al.* Combination of the mTOR inhibitor ridaforolimus and the anti-IGF1R monoclonal antibody dalotuzumab: preclinical characterization and phase I clinical trial. Clin Cancer Res 2015; 21(1): 49-59. [http://dx.doi.org/10.1158/1078-0432.CCR-14-0940] [PMID: 25320355]

[550] Fu X, Osborne CK, Schiff R. Biology and therapeutic potential of PI3K signaling in ER+/HER2-negative breast cancer. Breast 2013; 22 (Suppl. 2): S12-8. [http://dx.doi.org/10.1016/j.breast.2013.08.001] [PMID: 24011769]

[551] Paplomata E, Zelnak A, O'Regan R. Everolimus: side effect profile and management of toxicities in breast cancer. Breast Cancer Res Treat 2013; 140(3): 453-62. [http://dx.doi.org/10.1007/s10549-013-2630-y] [PMID: 23907751]

[552] Zhao L, Vogt PK. Class I PI3K in oncogenic cellular transformation. Oncogene 2008; 27(41): 5486-96. [http://dx.doi.org/10.1038/onc.2008.244] [PMID: 18794883]

[553] Schultz RM, Merriman RL, Andis SL, *et al. In vitro* and *in vivo* antitumor activity of the

phosphatidylinositol-3-kinase inhibitor, wortmannin. Anticancer Res 1995; 15(4): 1135-9. [PMID: 7653991]

[554] Tsao MS, Chow S, Hedley DW, Hedley DW. Inhibition of phosphatidylinositide 3-kinase enhances gemcitabine-induced apoptosis in human pancreatic cancer cells. Cancer Res 2000; 60(19): 5451-5. [PMID: 11034087]

[555] McDonald GT, Sullivan R, Paré GC, Graham CH. Inhibition of phosphatidylinositol 3-kinase promotes tumor cell resistance to chemotherapeutic agents *via* a mechanism involving delay in cell cycle progression. Exp Cell Res 2010; 316(19): 3197-206. [http://dx.doi.org/10.1016/j.yexcr.2010.08.007] [PMID: 20736003]

[556] Ihle NT, Williams R, Chow S, *et al.* Molecular pharmacology and antitumor activity of PX-866, a novel inhibitor of phosphoinositide-3-kinase signaling. Mol Cancer Ther 2004; 3(7): 763-72. [PMID: 15252137]

[557] Hong DS, Bowles DW, Falchook GS, *et al.* A multicenter phase I trial of PX-866, an oral irreversible phosphatidylinositol 3-kinase inhibitor, in patients with advanced solid tumors. Clin Cancer Res 2012; 18(15): 4173-82. [http://dx.doi.org/10.1158/1078-0432.CCR-12-0714] [PMID: 22693357]

[558] Bowles DW, Ma WW, Senzer N, *et al.* A multicenter phase 1 study of PX-866 in combination with docetaxel in patients with advanced solid tumours. Br J Cancer 2013; 109(5): 1085-92. [http://dx.doi.org/10.1038/bjc.2013.474] [PMID: 23942080]

[559] Maira SM, Pecchi S, Huang A, *et al.* Identification and characterization of NVP-BKM120, an orally available pan-class I PI3-kinase inhibitor. Mol Cancer Ther 2012; 11(2): 317-28. [http://dx.doi.org/10.1158/1535-7163.MCT-11-0474] [PMID: 22188813]

[560] Mayer IA, Abramson VG, Isakoff SJ, *et al.* Stand up to cancer phase Ib study of pan-phosphoinositid--3-kinase inhibitor buparlisib with letrozole in estrogen receptor-positive/human epidermal growth factor receptor 2-negative metastatic breast cancer. J Clin Oncol 2014; 32(12): 1202-9. [http://dx.doi.org/10.1200/JCO.2013.54.0518] [PMID: 24663045]

[561] Massarweh S, Romond E, Black EP, *et al.* A phase II study of combined fulvestrant and everolimus in patients with metastatic estrogen receptor (ER)-positive breast cancer after aromatase inhibitor (AI) failure. Breast Cancer Res Treat 2014; 143(2): 325-32. [http://dx.doi.org/10.1007/s10549-013-2810-9] [PMID: 24327334]

[562] Dowsett M, Smith IE, Ebbs SR, *et al.* Prognostic value of Ki67 expression after short-term presurgical endocrine therapy for primary breast cancer. J Natl Cancer Inst 2007; 99(2): 167-70. [http://dx.doi.org/10.1093/jnci/djk020] [PMID: 17228000]

[563] Baselga J, Semiglazov V, van Dam P, *et al.* Phase II randomized study of neoadjuvant everolimus plus letrozole compared with placebo plus letrozole in patients with estrogen receptor-positive breast cancer. J Clin Oncol 2009; 27(16): 2630-7. [http://dx.doi.org/10.1200/JCO.2008.18.8391] [PMID: 19380449]

[564] Markman B, Tabernero J, Krop I, *et al.* Phase I safety, pharmacokinetic, and pharmacodynamic study of the oral phosphatidylinositol-3-kinase and mTOR inhibitor BGT226 in patients with advanced solid tumors. Ann Oncol 2012; 23(9): 2399-408. [http://dx.doi.org/10.1093/annonc/mds011] [PMID: 22357447]

[565] Jordan NJ, Dutkowski CM, Barrow D, *et al.* Impact of dual mTORC1/2 mTOR kinase inhibitor AZD8055 on acquired endocrine resistance in breast cancer *in vitro*. Breast Cancer Res 2014; 16(1): R12. [http://dx.doi.org/10.1186/bcr3604] [PMID: 24457069]

[566] Shaw RJ, Bardeesy N, Manning BD, *et al.* The LKB1 tumor suppressor negatively regulates mTOR signaling. Cancer Cell 2004; 6(1): 91-9. [http://dx.doi.org/10.1016/j.ccr.2004.06.007] [PMID: 15261145]

[567] Engelman JA. Targeting PI3K signalling in cancer: opportunities, challenges and limitations. Nat Rev Cancer 2009; 9(8): 550-62.
[http://dx.doi.org/10.1038/nrc2664] [PMID: 19629070]

[568] Fu X, Creighton CJ, Biswal NC, *et al.* Overcoming endocrine resistance due to reduced PTEN levels in estrogen receptor-positive breast cancer by co-targeting mammalian target of rapamycin, protein kinase B, or mitogen-activated protein kinase kinase. Breast Cancer Res 2014; 16(5): 430.
[http://dx.doi.org/10.1186/s13058-014-0430-x] [PMID: 25212826]

[569] O'Reilly KE, Rojo F, She QB, *et al.* mTOR inhibition induces upstream receptor tyrosine kinase signaling and activates Akt. Cancer Res 2006; 66(3): 1500-8.
[http://dx.doi.org/10.1158/0008-5472.CAN-05-2925] [PMID: 16452206]

[570] Browne BC, O'Brien N, Duffy MJ, Crown J, O'Donovan N. HER-2 signaling and inhibition in breast cancer. Curr Cancer Drug Targets 2009; 9(3): 419-38.
[http://dx.doi.org/10.2174/156800909788166484] [PMID: 19442060]

[571] Castaneda CA, Cortes-Funes H, Gomez HL, Ciruelos EM. The phosphatidyl inositol 3-kinase/AKT signaling pathway in breast cancer. Cancer Metastasis Rev 2010; 29(4): 751-9.
[http://dx.doi.org/10.1007/s10555-010-9261-0] [PMID: 20922461]

[572] Nahta R, Esteva FJ. Herceptin: mechanisms of action and resistance. Cancer Lett 2006; 232(2): 123-38.
[http://dx.doi.org/10.1016/j.canlet.2005.01.041] [PMID: 16458110]

[573] Holbro T, Beerli RR, Maurer F, Koziczak M, Barbas CF III, Hynes NE. The ErbB2/ErbB3 heterodimer functions as an oncogenic unit: ErbB2 requires ErbB3 to drive breast tumor cell proliferation. Proc Natl Acad Sci USA 2003; 100(15): 8933-8.
[http://dx.doi.org/10.1073/pnas.1537685100] [PMID: 12853564]

[574] Rexer BN, Arteaga CL. Intrinsic and acquired resistance to HER2-targeted therapies in HER2 gene-amplified breast cancer: mechanisms and clinical implications. Crit Rev Oncog 2012; 17(1): 1-16.
[http://dx.doi.org/10.1615/CritRevOncog.v17.i1.20] [PMID: 22471661]

[575] Huw LY, O'Brien C, Pandita A, *et al.* Acquired PIK3CA amplification causes resistance to selective phosphoinositide 3-kinase inhibitors in breast cancer. Oncogenesis 2013; 2: e83.
[http://dx.doi.org/10.1038/oncsis.2013.46] [PMID: 24366379]

[576] Loi S. Tumor-infiltrating lymphocytes, breast cancer subtypes and therapeutic efficacy. OncoImmunology 2013; 2(7): e24720.
[http://dx.doi.org/10.4161/onci.24720] [PMID: 24073365]

[577] Clynes RA, Towers TL, Presta LG, Ravetch JV. Inhibitory Fc receptors modulate *in vivo* cytotoxicity against tumor targets. Nat Med 2000; 6(4): 443-6.
[http://dx.doi.org/10.1038/74704] [PMID: 10742152]

[578] Nahta R. Pharmacological strategies to overcome HER2 cross-talk and Trastuzumab resistance. Curr Med Chem 2012; 19(7): 1065-75.
[http://dx.doi.org/10.2174/092986712799320691] [PMID: 22229414]

[579] Andre F, Campone M, O'Regan R, *et al.* Phase I study of everolimus plus weekly paclitaxel and trastuzumab in patients with metastatic breast cancer pretreated with trastuzumab. J Clin Oncol 2010; 28(34): 5110-5.
[http://dx.doi.org/10.1200/JCO.2009.27.8549] [PMID: 20975068]

[580] Jerusalem G, Fasolo A, Dieras V, *et al.* Phase I trial of oral mTOR inhibitor everolimus in combination with trastuzumab and vinorelbine in pre-treated patients with HER2-overexpressing metastatic breast cancer. Breast Cancer Res Treat 2011; 125(2): 447-55.
[http://dx.doi.org/10.1007/s10549-010-1260-x] [PMID: 21107682]

[581] Hurvitz SA, Dalenc F, Campone M, *et al.* A phase 2 study of everolimus combined with trastuzumab and paclitaxel in patients with HER2-overexpressing advanced breast cancer that progressed during

prior trastuzumab and taxane therapy. Breast Cancer Res Treat 2013; 141(3): 437-46. [http://dx.doi.org/10.1007/s10549-013-2689-5] [PMID: 24101324]

[582] O'Regan R, Ozguroglu M, Andre F, Toi M, Jerusalem G, Wilks S. Phase III, randomized, double-blind, placebo-controlled multicenter trial of daily everolimus plus weekly trastuzumab and vinorelbine in trastuzumab- resistant, advanced breast cancer (BOLERO-3). J Clin Oncol 2013; 31. [Suppl.].

[583] Gandhi L, Bahleda R, Tolaney SM, *et al.* Phase I study of neratinib in combination with temsirolimus in patients with human epidermal growth factor receptor 2-dependent and other solid tumors. J Clin Oncol 2014; 32(2): 68-75. [http://dx.doi.org/10.1200/JCO.2012.47.2787] [PMID: 24323026]

[584] Rivera VM, Squillace RM, Miller D, *et al.* Ridaforolimus (AP23573; MK-8669), a potent mTOR inhibitor, has broad antitumor activity and can be optimally administered using intermittent dosing regimens. Mol Cancer Ther 2011; 10(6): 1059-71. [http://dx.doi.org/10.1158/1535-7163.MCT-10-0792] [PMID: 21482695]

[585] Seiler M, Ray-Coquard I, Melichar B, *et al.* Oral ridaforolimus plus trastuzumab for patients with HER2+ trastuzumab-refractory metastatic breast cancer. Clin Breast Cancer 2015; 15(1): 60-5. [http://dx.doi.org/10.1016/j.clbc.2014.07.008] [PMID: 25239224]

[586] Hudis C, Swanton C, Janjigian YY, *et al.* A phase 1 study evaluating the combination of an allosteric AKT inhibitor (MK-2206) and trastuzumab in patients with HER2-positive solid tumors. Breast Cancer Res 2013; 15(6): R110. [http://dx.doi.org/10.1186/bcr3577] [PMID: 24252402]

[587] Gonzalez-Angulo AM, Krop I, Akcakanat A, *et al.* SU2C phase Ib study of paclitaxel and MK-2206 in advanced solid tumors and metastatic breast cancer. J Natl Cancer Inst 2015; 107(3): dju493. [http://dx.doi.org/10.1093/jnci/dju493] [PMID: 25688104]

[588] Oda K, Stokoe D, Taketani Y, McCormick F. High frequency of coexistent mutations of PIK3CA and PTEN genes in endometrial carcinoma. Cancer Res 2005; 65(23): 10669-73. [http://dx.doi.org/10.1158/0008-5472.CAN-05-2620] [PMID: 16322209]

[589] Noh WC, Mondesire WH, Peng J, *et al.* Determinants of rapamycin sensitivity in breast cancer cells. Clin Cancer Res 2004; 10(3): 1013-23. [http://dx.doi.org/10.1158/1078-0432.CCR-03-0043] [PMID: 14871980]

[590] Noh WC, Kim YH, Kim MS, *et al.* Activation of the mTOR signaling pathway in breast cancer and its correlation with the clinicopathologic variables. Breast Cancer Res Treat 2008; 110(3): 477-83. [http://dx.doi.org/10.1007/s10549-007-9746-x] [PMID: 17805960]

[591] O'Reilly T, McSheehy PM. Biomarker Development for the Clinical Activity of the mTOR Inhibitor Everolimus (RAD001): Processes, Limitations, and Further Proposals. Transl Oncol 2010; 3(2): 65-79. [http://dx.doi.org/10.1593/tlo.09277] [PMID: 20360931]

[592] Meric-Bernstam F, Akcakanat A, Chen H, *et al.* PIK3CA/PTEN mutations and Akt activation as markers of sensitivity to allosteric mTOR inhibitors. Clin Cancer Res 2012; 18(6): 1777-89. [http://dx.doi.org/10.1158/1078-0432.CCR-11-2123] [PMID: 22422409]

[593] Bertucci MC, Mitchell CA. Phosphoinositide 3-kinase and INPP4B in human breast cancer. Ann N Y Acad Sci 2013; 1280: 1-5. [http://dx.doi.org/10.1111/nyas.12036] [PMID: 23551093]

[594] O'Brien C, Wallin JJ, Sampath D, *et al.* Predictive biomarkers of sensitivity to the phosphatidylinositol 3′ kinase inhibitor GDC-0941 in breast cancer preclinical models. Clin Cancer Res 2010; 16(14): 3670-83. [http://dx.doi.org/10.1158/1078-0432.CCR-09-2828] [PMID: 20453058]

[595] Sangai T, Akcakanat A, Chen H, *et al.* Biomarkers of response to Akt inhibitor MK-2206 in breast cancer. Clin Cancer Res 2012; 18(20): 5816-28.

[http://dx.doi.org/10.1158/1078-0432.CCR-12-1141] [PMID: 22932669]

[596] Janku F, Wheler JJ, Naing A, *et al.* PIK3CA mutation H1047R is associated with response to PI3K/AKT/mTOR signaling pathway inhibitors in early-phase clinical trials. Cancer Res 2013; 73(1): 276-84.
[http://dx.doi.org/10.1158/0008-5472.CAN-12-1726] [PMID: 23066039]

[597] Janku F, Tsimberidou AM, Garrido-Laguna I, *et al.* PIK3CA mutations in patients with advanced cancers treated with PI3K/AKT/mTOR axis inhibitors. Mol Cancer Ther 2011; 10(3): 558-65.
[http://dx.doi.org/10.1158/1535-7163.MCT-10-0994] [PMID: 21216929]

[598] Janku F, Wheler JJ, Westin SN, *et al.* PI3K/AKT/mTOR inhibitors in patients with breast and gynecologic malignancies harboring PIK3CA mutations. J Clin Oncol 2012; 30(8): 777-82.
[http://dx.doi.org/10.1200/JCO.2011.36.1196] [PMID: 22271473]

[599] Muellner MK, Uras IZ, Gapp BV, *et al.* A chemical-genetic screen reveals a mechanism of resistance to PI3K inhibitors in cancer. Nat Chem Biol 2011; 7(11): 787-93.
[http://dx.doi.org/10.1038/nchembio.695] [PMID: 21946274]

[600] Serra V, Eichhorn PJ, García-García C, *et al.* RSK3/4 mediate resistance to PI3K pathway inhibitors in breast cancer. J Clin Invest 2013; 123(6): 2551-63.
[http://dx.doi.org/10.1172/JCI66343] [PMID: 23635776]

[601] Ihle NT, Lemos R Jr, Wipf P, *et al.* Mutations in the phosphatidylinositol-3-kinase pathway predict for antitumor activity of the inhibitor PX-866 whereas oncogenic Ras is a dominant predictor for resistance. Cancer Res 2009; 69(1): 143-50.
[http://dx.doi.org/10.1158/0008-5472.CAN-07-6656] [PMID: 19117997]

[602] Di Nicolantonio F, Arena S, Tabernero J, *et al.* Deregulation of the PI3K and KRAS signaling pathways in human cancer cells determines their response to everolimus. J Clin Invest 2010; 120(8): 2858-66.
[http://dx.doi.org/10.1172/JCI37539] [PMID: 20664172]

[603] Hoeflich KP, O'Brien C, Boyd Z, *et al. In vivo* antitumor activity of MEK and phosphatidylinositol 3-kinase inhibitors in basal-like breast cancer models. Clin Cancer Res 2009; 15(14): 4649-64.
[http://dx.doi.org/10.1158/1078-0432.CCR-09-0317] [PMID: 19567590]

[604] Oliveira M, Navarro A, De Mattos-Arruda L, *et al.* PI3K pathway (PI3KP) dysregulation and response to PAN-PI3K/AKT/mTOR/dual PI3K-mTOR inhibitors (PI3KPI) in metastatic breast cancer (MBC) patients (Pts). J Clin Oncol 2012; Suppl.: 30.

[605] Treilleux I, Arnedos M, Cropet C, *et al.* Predictive markers of everolimus efficacy in hormone receptor positive (HR+) metastatic breast cancer (MBC): final results of the TAMRAD Trial Translational Study. J Clin Oncol 2013; Suppl.: 31.

[606] Baselga J. Treatment of HER2-overexpressing breast cancer. Ann Oncol 2010; 21 (Suppl. 7): vii36-40.
[http://dx.doi.org/10.1093/annonc/mdq421] [PMID: 20943641]

[607] O'Donovan N, Byrne AT, O'Connor AE, McGee S, Gallagher WM, Crown J. Synergistic interaction between trastuzumab and EGFR/HER-2 tyrosine kinase inhibitors in HER-2 positive breast cancer cells. Invest New Drugs 2011; 29(5): 752-9.
[http://dx.doi.org/10.1007/s10637-010-9415-5] [PMID: 20229355]

[608] Nahta R, Yuan LX, Zhang B, Kobayashi R, Esteva FJ. Insulin-like growth factor-I receptor/human epidermal growth factor receptor 2 heterodimerization contributes to trastuzumab resistance of breast cancer cells. Cancer Res 2005; 65(23): 11118-28.
[http://dx.doi.org/10.1158/0008-5472.CAN-04-3841] [PMID: 16322262]

[609] Zhang S, Huang WC, Li P, *et al.* Combating trastuzumab resistance by targeting SRC, a common node downstream of multiple resistance pathways. Nat Med 2011; 17(4): 461-9.
[http://dx.doi.org/10.1038/nm.2309] [PMID: 21399647]

[610] Solca F, Dahl G, Zoephel A, *et al.* Target binding properties and cellular activity of afatinib (BIBW

2992), an irreversible ErbB family blocker. J Pharmacol Exp Ther 2012; 343(2): 342-50. [http://dx.doi.org/10.1124/jpet.112.197756] [PMID: 22888144]

[611] Lin NU, Winer EP, Wheatley D, *et al.* A phase II study of afatinib (BIBW 2992), an irreversible ErbB family blocker, in patients with HER2-positive metastatic breast cancer progressing after trastuzumab. Breast Cancer Res Treat 2012; 133(3): 1057-65. [http://dx.doi.org/10.1007/s10549-012-2003-y] [PMID: 22418700]

[612] Harbeck N, Huang CS, Hurvitz S, *et al.* Afatinib plus vinorelbine *versus* trastuzumab plus vinorelbine in patients with HER2-overexpressing metastatic breast cancer who had progressed on one previous trastuzumab treatment (LUX-Breast 1): an open-label, randomised, phase 3 trial. Lancet Oncol 2016; 17(3): 357-66. [http://dx.doi.org/10.1016/S1470-2045(15)00540-9] [PMID: 26822398]

[613] Hurvitz SA, Shatsky R, Harbeck N. Afatinib in the treatment of breast cancer. Expert Opin Investig Drugs 2014; 23(7): 1039-47. [http://dx.doi.org/10.1517/13543784.2014.924505] [PMID: 24870559]

[614] Rabindran SK, Discafani CM, Rosfjord EC, *et al.* Antitumor activity of HKI-272, an orally active, irreversible inhibitor of the HER-2 tyrosine kinase. Cancer Res 2004; 64(11): 3958-65. [http://dx.doi.org/10.1158/0008-5472.CAN-03-2868] [PMID: 15173008]

[615] Burstein HJ, Sun Y, Dirix LY, *et al.* Neratinib, an irreversible ErbB receptor tyrosine kinase inhibitor, in patients with advanced ErbB2-positive breast cancer. J Clin Oncol 2010; 28(8): 1301-7. [http://dx.doi.org/10.1200/JCO.2009.25.8707] [PMID: 20142587]

[616] Awada A, Dirix L, Manso Sanchez L, *et al.* Safety and efficacy of neratinib (HKI-272) plus vinorelbine in the treatment of patients with ErbB2-positive metastatic breast cancer pretreated with anti-HER2 therapy. Ann Oncol 2013; 24(1): 109-16. [http://dx.doi.org/10.1093/annonc/mds284] [PMID: 22967996]

[617] Chow LW, Xu B, Gupta S, *et al.* Combination neratinib (HKI-272) and paclitaxel therapy in patients with HER2-positive metastatic breast cancer. Br J Cancer 2013; 108(10): 1985-93. [http://dx.doi.org/10.1038/bjc.2013.178] [PMID: 23632474]

[618] Chan A, Delaloge S, Holmes FA, *et al.* Neratinib after trastuzumab-based adjuvant therapy in patients with HER2-positive breast cancer (ExteNET): a multicentre, randomised, double-blind, placebo-controlled, phase 3 trial. Lancet Oncol 2016; 17(3): 367-77. [http://dx.doi.org/10.1016/S1470-2045(15)00551-3] [PMID: 26874901]

[619] Jankowitz RC, Abraham J, Tan AR, *et al.* Safety and efficacy of neratinib in combination with weekly paclitaxel and trastuzumab in women with metastatic HER2□positive breast cancer: an NSABP Foundation Research Program phase I study. Cancer Chemother Pharmacol 2013; 72(6): 1205-12. [http://dx.doi.org/10.1007/s00280-013-2262-2] [PMID: 24077916]

[620] Saura C, Garcia-Saenz JA, Xu B, *et al.* Safety and efficacy of neratinib in combination with capecitabine in patients with metastatic human epidermal growth factor receptor 2-positive breast cancer. J Clin Oncol 2014; 32(32): 3626-33. [http://dx.doi.org/10.1200/JCO.2014.56.3809] [PMID: 25287822]

[621] Higgins MJ, Gabrail NY, Miller K, *et al.* A phase I/II study of MM-111, a novel bispecific antibody that targets the ErB2/ErB3 heterodimer, in combination with trastuzumab in advanced refractory HER2-positive breast cancer. J Clin Oncol 2011; 29 (Suppl.): TPS119. [http://dx.doi.org/10.1200/jco.2011.29.15_suppl.tps119]

[622] Denlinger CS, Beeram M, Tolcher AW, *et al.* A phase I/II and pharmacologic study of MM-111 in patients with advanced, refractory HER2-positive (HER2) cancers. J Clin Oncol 2010; 28 (Suppl.): TPS169. [http://dx.doi.org/10.1200/jco.2010.28.15_suppl.tps169]

[623] Korkola JE, Lui M, Libby T, Heisler L, Feiler H, Gray JW. Detrimental effects of sequential compared to concurrent treatment of pertuzumab plus T-DM1 in HER2+ breast cancer cell lines. SABCS 2014

S6-07.

[624] Miller KD, Diéras V, Harbeck N, *et al.* Phase IIa trial of trastuzumab emtansine with pertuzumab for patients with human epidermal growth factor receptor 2-positive, locally advanced, or metastatic breast cancer. J Clin Oncol 2014; 32(14): 1437-44. [http://dx.doi.org/10.1200/JCO.2013.52.6590] [PMID: 24733796]

[625] Choi YJ, Anders L. Signaling through cyclin D-dependent kinases. Oncogene 2014; 33(15): 1890-903. [http://dx.doi.org/10.1038/onc.2013.137] [PMID: 23644662]

[626] Patnaik A, Rosen LS, Tolaney SM, *et al.* Clinical activity of LY2835219,a novel cell cycle inhibitor selective for CDK4 and CDK6, in patients with metastatic breast cancer. Proceedings American Association for Cancer Research Annual Meeting. [http://dx.doi.org/10.1158/1538-7445.AM2014-CT232]

[627] Sunkyu K, Alice L, Rajiv C, *et al.* LEE011: an orally bioavailable, selective small molecule inhibitor of CDK4/6--reactivating Rb in cancer. 2013 AACR-NCI-EORTC International Conference on Molecular Targets and Cancer Therapeutics. (Abstract B264)

[628] Slamon DJ, Hurvitz SA, Applebaum S, *et al.* Phase I study ofPD 0332991, cyclin-D kinase (CDK) 4/6 inhibitor in combination with letrozole for first-line treatment of patients with ER-positive, HER2-negative breast cancer. J Clin Oncol 2010; 28(15s): abs3060. [http://dx.doi.org/10.1200/jco.2010.28.15_suppl.3060]

[629] Lehmann BD, Bauer JA, Chen X, *et al.* Identification of human triple-negative breast cancer subtypes and preclinical models for selection of targeted therapies. J Clin Invest 2011; 121(7): 2750-67. [http://dx.doi.org/10.1172/JCI45014] [PMID: 21633166]

[630] Kalimutho M, Parsons K, Mittal D, López JA, Srihari S, Khanna KK. Targeted Therapies for Triple-Negative Breast Cancer: Combating a Stubborn Disease. Trends Pharmacol Sci 2015; 36(12): 822-46. [http://dx.doi.org/10.1016/j.tips.2015.08.009] [PMID: 26538316]

[631] Ivy SP, Wick JY, Kaufman BM. An overview of small-molecule inhibitors of VEGFR signaling. Nat Rev Clin Oncol 2009; 6(10): 569-79. [http://dx.doi.org/10.1038/nrclinonc.2009.130] [PMID: 19736552]

[632] Burstein HJ, Elias AD, Rugo HS, *et al.* Phase II study of sunitinib malate, an oral multitargeted tyrosine kinase inhibitor, in patients with metastatic breast cancer previously treated with an anthracycline and a taxane. J Clin Oncol 2008; 26(11): 1810-6. [http://dx.doi.org/10.1200/JCO.2007.14.5375] [PMID: 18347007]

[633] Kozloff M, Chuang E, Starr A, *et al.* An exploratory study of sunitinib plus paclitaxel as first-line treatment for patients with advanced breast cancer. Ann Oncol 2010; 21(7): 1436-41. [http://dx.doi.org/10.1093/annonc/mdp565] [PMID: 20032126]

[634] Moreno-Aspitia A, Morton RF, Hillman DW, *et al.* Phase II trial of sorafenib in patients with metastatic breast cancer previously exposed to anthracyclines or taxanes: North Central Cancer Treatment Group and Mayo Clinic Trial N0336. J Clin Oncol 2009; 27(1): 11-5. [http://dx.doi.org/10.1200/JCO.2007.15.5242] [PMID: 19047293]

[635] Sikov WM, Berry DA, Perou CM, *et al.* Impact of the addition of carboplatin and/or bevacizumab to neoadjuvant once-per-week paclitaxel followed by dose-dense doxorubicin and cyclophosphamide on pathologic complete response rates in stage II to III triple-negative breast cancer: CALGB 40603 (Alliance). J Clin Oncol 2015; 33(1): 13-21. [http://dx.doi.org/10.1200/JCO.2014.57.0572] [PMID: 25092775]

[636] Finn RS, Press MF, Dering J, *et al.* Estrogen receptor, progesterone receptor, human epidermal growth factor receptor 2 (HER2), and epidermal growth factor receptor expression and benefit from lapatinib in a randomized trial of paclitaxel with lapatinib or placebo as first-line treatment in HER2-negative or unknown metastatic breast cancer. J Clin Oncol 2009; 27(24): 3908-15. [http://dx.doi.org/10.1200/JCO.2008.18.1925] [PMID: 19620495]

[637] Carey LA, Rugo HS, Marcom PK, *et al.* TBCRC 001: randomized phase II study of cetuximab in combination with carboplatin in stage IV triple-negative breast cancer. J Clin Oncol 2012; 30(21): 2615-23.
[http://dx.doi.org/10.1200/JCO.2010.34.5579] [PMID: 22665533]

[638] Baselga J, Gómez P, Greil R, *et al.* Randomized phase II study of the anti-epidermal growth factor receptor monoclonal antibody cetuximab with cisplatin *versus* cisplatin alone in patients with metastatic triple-negative breast cancer. J Clin Oncol 2013; 31(20): 2586-92.
[http://dx.doi.org/10.1200/JCO.2012.46.2408] [PMID: 23733761]

[639] Mirzoeva OK, Das D, Heiser LM, *et al.* Basal subtype and MAPK/ERK kinase (MEK)-phosphoinositide 3-kinase feedback signaling determine susceptibility of breast cancer cells to MEK inhibition. Cancer Res 2009; 69(2): 565-72.
[http://dx.doi.org/10.1158/0008-5472.CAN-08-3389] [PMID: 19147570]

[640] Adeyinka A, Nui Y, Cherlet T, Snell L, Watson PH, Murphy LC. Activated mitogen-activated protein kinase expression during human breast tumorigenesis and breast cancer progression. Clin Cancer Res 2002; 8(6): 1747-53.
[PMID: 12060612]

[641] Barton VN, D'Amato NC, Gordon MA, *et al.* Multiple molecular subtypes of triple-negative breast cancer critically rely on androgen receptor and respond to enzalutamide *in vivo*. Mol Cancer Ther 2015; 14(3): 769-78.
[http://dx.doi.org/10.1158/1535-7163.MCT-14-0926] [PMID: 25713333]

[642] Dizdar O, Arslan C, Altundag K. Advances in PARP inhibitors for the treatment of breast cancer. Expert Opin Pharmacother 2015; 16(18): 2751-8.
[http://dx.doi.org/10.1517/14656566.2015.1100168] [PMID: 26485111]

[643] Turner N, Tutt A, Ashworth A. Hallmarks of 'BRCAness' in sporadic cancers. Nat Rev Cancer 2004; 4(10): 814-9.
[http://dx.doi.org/10.1038/nrc1457] [PMID: 15510162]

[644] Ashworth A. A synthetic lethal therapeutic approach: poly(ADP) ribose polymerase inhibitors for the treatment of cancers deficient in DNA double-strand break repair. J Clin Oncol 2008; 26(22): 3785-90.
[http://dx.doi.org/10.1200/JCO.2008.16.0812] [PMID: 18591545]

[645] Fong PC, Boss DS, Yap TA, *et al.* Inhibition of poly(ADP-ribose) polymerase in tumors from BRCA mutation carriers. N Engl J Med 2009; 361(2): 123-34.
[http://dx.doi.org/10.1056/NEJMoa0900212] [PMID: 19553641]

[646] Rugo HS, Olopade O, DeMichele A, *et al.* Veliparib/carboplatin plus standard neoadjuvant therapy for high-risk breast cancer: First efficacy results from the I-SPY2 trial. Cancer Res 2013; 73(24 Suppl): Abs S5-02.

[647] Telli ML, Jensen KC, Kurian AW, *et al.* PrECOG0105: final efficacy results from a phase II study of gemcitabine (G) and carboplatin plus iniparib (BSI-201) as neoadjuvant therapy for triple negative (TN) and BRCA1/2 mutation associated breast cancer. J Clin Oncol 2013; 31(suppl) abs 1003.

[648] Abkevich V, Timms KM, Hennessy BT, *et al.* Patterns of genomic loss of heterozygosity predict homologous recombination repair defects in epithelial ovarian cancer. Br J Cancer 2012; 107(10): 1776-82.
[http://dx.doi.org/10.1038/bjc.2012.451] [PMID: 23047548]

[649] Turner NC, Reis-Filho JS. Basal-like breast cancer and the BRCA1 phenotype. Oncogene 2006; 25(43): 5846-53.
[http://dx.doi.org/10.1038/sj.onc.1209876] [PMID: 16998499]

[650] Turner NC, Reis-Filho JS, Russell AM, *et al.* BRCA1 dysfunction in sporadic basal-like breast cancer. Oncogene 2007; 26(14): 2126-32.
[http://dx.doi.org/10.1038/sj.onc.1210014] [PMID: 17016441]

[651] Lips EH, Mulder L, Oonk A, *et al.* Triple-negative breast cancer: BRCAness and concordance of clinical features with BRCA1-mutation carriers. Br J Cancer 2013; 108(10): 2172-7. [http://dx.doi.org/10.1038/bjc.2013.144] [PMID: 23558900]

[652] Martin SA, Lord CJ, Ashworth A. DNA repair deficiency as a therapeutic target in cancer. Curr Opin Genet Dev 2008; 18(1): 80-6. [http://dx.doi.org/10.1016/j.gde.2008.01.016] [PMID: 18343102]

[653] Anders CK, Winer EP, Ford JM, *et al.* Poly(ADP-Ribose) polymerase inhibition: "targeted" therapy for triple-negative breast cancer. Clin Cancer Res 2010; 16(19): 4702-10. [http://dx.doi.org/10.1158/1078-0432.CCR-10-0939] [PMID: 20858840]

[654] Tutt A, Robson M, Garber JE, *et al.* Oral poly(ADP-ribose) polymerase inhibitor olaparib in patients with BRCA1 or BRCA2 mutations and advanced breast cancer: a proof-of-concept trial. Lancet 2010; 376(9737): 235-44. [http://dx.doi.org/10.1016/S0140-6736(10)60892-6] [PMID: 20609467]

[655] Nicolini A, Carpi A, Ferrari P, Biava PM, Rossi G. Immunotherapy and Hormone-therapy in Metastatic Breast Cancer: A Review and an Update. Curr Drug Targets 2016; 17(10): 1127-39. [http://dx.doi.org/10.2174/1389450117666160201114752] [PMID: 26844558]

[656] Fridman WH, Galon J, Pagès F, Tartour E, Sautès-Fridman C, Kroemer G. Prognostic and predictive impact of intra- and peritumoral immune infiltrates. Cancer Res 2011; 71(17): 5601-5. [http://dx.doi.org/10.1158/0008-5472.CAN-11-1316] [PMID: 21846822]

[657] Matsumoto H, Koo SL, Dent R, Tan PH, Iqbal J. Role of inflammatory infiltrates in triple negative breast cancer. J Clin Pathol 2015; 68(7): 506-10. [http://dx.doi.org/10.1136/jclinpath-2015-202944] [PMID: 25750267]

[658] Finak G, Bertos N, Pepin F, *et al.* Stromal gene expression predicts clinical outcome in breast cancer. Nat Med 2008; 14(5): 518-27. [http://dx.doi.org/10.1038/nm1764] [PMID: 18438415]

[659] Loi S, Michiels S, Salgado R, *et al.* Tumor infiltrating lymphocytes are prognostic in triple negative breast cancer and predictive for trastuzumab benefit in early breast cancer: results from the FinHER trial. Ann Oncol 2014; 25(8): 1544-50. [http://dx.doi.org/10.1093/annonc/mdu112] [PMID: 24608200]

[660] Liu S, Lachapelle J, Leung S, Gao D, Foulkes WD, Nielsen TO. CD8+ lymphocyte infiltration is an independent favorable prognostic indicator in basal-like breast cancer. Breast Cancer Res 2012; 14(2): R48. [http://dx.doi.org/10.1186/bcr3148] [PMID: 22420471]

[661] Denkert C, von Minckwitz G, Brase JC, *et al.* Tumor-infiltrating lymphocytes and response to neoadjuvant chemotherapy with or without carboplatin in human epidermal growth factor receptor 2-positive and triple-negative primary breast cancers. J Clin Oncol 2015; 33(9): 983-91. [http://dx.doi.org/10.1200/JCO.2014.58.1967] [PMID: 25534375]

[662] Lakshminarayanan V, Thompson P, Wolfert MA, *et al.* Immune recognition of tumor-associated mucin MUC1 is achieved by a fully synthetic aberrantly glycosylated MUC1 tripartite vaccine. Proc Natl Acad Sci USA 2012; 109(1): 261-6. [http://dx.doi.org/10.1073/pnas.1115166109] [PMID: 22171012]

[663] www.cancerresearch.org/cancer-immunotherapy/impacting-all-cancer/breast-cancer

[664] Mittendorf EA, Philips AV, Meric-Bernstam F, *et al.* PD-L1 expression in triple-negative breast cancer. Cancer Immunol Res 2014; 2(4): 361-70. [http://dx.doi.org/10.1158/2326-6066.CIR-13-0127] [PMID: 24764583]

[665] Schalper KA, Velcheti V, Carvajal D, *et al.* *In situ* tumor PD-L1 mRNA expression is associated with increased TILs and better outcome in breast carcinomas. Clin Cancer Res 2014; 20(10): 2773-82. [http://dx.doi.org/10.1158/1078-0432.CCR-13-2702] [PMID: 24647569]

[666] Sabatier R, Finetti P, Mamessier E, *et al.* Prognostic and predictive value of PDL1 expression in breast cancer. Oncotarget 2015; 6(7): 5449-64.
[http://dx.doi.org/10.18632/oncotarget.3216] [PMID: 25669979]

[667] Wimberly H, Brown JR, Schalper K, *et al.* PD-L1 Expression Correlates with Tumor-Infiltrating Lymphocytes and Response to Neoadjuvant Chemotherapy in Breast Cancer. Cancer Immunol Res 2015; 3(4): 326-32.
[http://dx.doi.org/10.1158/2326-6066.CIR-14-0133] [PMID: 25527356]

[668] Sell S, Nicolini A, Ferrari P, Biava PM. Cancer: A Problem of Developmental Biology; Scientific Evidence for Reprogramming and Differentiation Therapy. Curr Drug Targets 2016; 17(10): 1103-10.
[http://dx.doi.org/10.2174/1389450116666150907102717] [PMID: 26343109]

[669] Sell S. On the stem cell origin of cancer. Am J Pathol 2010; 176(6): 2584-494.
[http://dx.doi.org/10.2353/ajpath.2010.091064] [PMID: 20431026]

[670] Muto A, Kizaki M, Yamato K, *et al.* 1,25-Dihydroxyvitamin D3 induces differentiation of a retinoic acid-resistant acute promyelocytic leukemia cell line (UF-1) associated with expression of p21(WAF1/CIP1) and p27(KIP1). Blood 1999; 93(7): 2225-33.
[PMID: 10090931]

[671] Sun SY, Lotan R. Retinoids and their receptors in cancer development and chemoprevention. Crit Rev Oncol Hematol 2002; 41(1): 41-55.
[http://dx.doi.org/10.1016/S1040-8428(01)00144-5] [PMID: 11796231]

[672] Einhorn L. Are there factors preventing cancer development during embryonic life? Oncodev Biol Med 1983; 4(3): 219-29.
[PMID: 6298752]

[673] Pierce GB. The cancer cell and its control by the embryo. Rous-Whipple Award lecture. Am J Pathol 1983; 113(1): 117-24.
[PMID: 6312802]

[674] Brent RL. Radiation teratogenesis. Teratology 1980; 21(3): 281-98.
[http://dx.doi.org/10.1002/tera.1420210304] [PMID: 7006137]

[675] Pierce GB, Lewis SH, Miller GJ, Moritz E, Miller P. Tumorigenicity of embryonal carcinoma as an assay to study control of malignancy by the murine blastocyst. Proc Natl Acad Sci USA 1979; 76(12): 6649-51.
[http://dx.doi.org/10.1073/pnas.76.12.6649] [PMID: 293749]

[676] Webb CG, Gootwine E, Sachs L. Developmental potential of myeloid leukemia cells injected into midgestation embryos. Dev Biol 1984; 101(1): 221-4.
[http://dx.doi.org/10.1016/0012-1606(84)90132-5] [PMID: 6229433]

[677] Hendrix MJ, Seftor EA, Seftor RE, Kasemeier-Kulesa J, Kulesa PM, Postovit LM. Reprogramming metastatic tumour cells with embryonic microenvironments. Nat Rev Cancer 2007; 7(4): 246-55.
[http://dx.doi.org/10.1038/nrc2108] [PMID: 17384580]

[678] Kulesa PM, Kasemeier-Kulesa JC, Teddy JM, *et al.* Reprogramming metastatic melanoma cells to assume a neural crest cell-like phenotype in an embryonic microenvironment. Proc Natl Acad Sci USA 2006; 103(10): 3752-7.
[http://dx.doi.org/10.1073/pnas.0506977103] [PMID: 16505384]

[679] Gootwine E, Webb CG, Sachs L. Participation of myeloid leukaemic cells injected into embryos in haematopoietic differentiation in adult mice. Nature 1982; 299(5878): 63-5.
[http://dx.doi.org/10.1038/299063a0] [PMID: 6955597]

[680] Biava PM, Fiorito A, Negro C, Mariani M. Effects of treatment with embryonic and uterine tissue homogenates on Lewis lung carcinoma development. Cancer Lett 1988; 41(3): 265-70.
[http://dx.doi.org/10.1016/0304-3835(88)90287-X] [PMID: 3409205]

[681] Biava PM, Bonsignorio D. Hoxa. Life-Protecting Factor (LPF): an anti-cancer low molecular weight fraction isolated from pregnant uterine mucosa during embryo organogenesis. J Tumor Marker Oncol 2000; 15: 223-33.

[682] Biava PM, Bonsignorio D, Hoxha M, *et al.* Mother-embryo cross-talk: the anti-cancer substances produced by mother and embryo during cell differentiation. A review of experimental data. J Tumor Marker Oncol 2002; 17: 55-8.

[683] Lee JT, Herlyn M. Embryogenesis meets tumorigenesis. Nat Med 2006; 12(8): 882-4. [http://dx.doi.org/10.1038/nm0806-882] [PMID: 16892028]

[684] Katoh M, Katoh M. Transcriptional regulation of WNT2B based on the balance of Hedgehog, Notch, BMP and WNT signals. Int J Oncol 2009; 34(5): 1411-5. [PMID: 19360354]

[685] Katoh M, Katoh M. Transcriptional mechanisms of WNT5A based on NF-kappaB, Hedgehog, TGFbeta, and Notch signaling cascades. Int J Mol Med 2009; 23(6): 763-9. [http://dx.doi.org/10.3892/ijmm_00000190] [PMID: 19424602]

[686] Peifer M, Polakis P. Wnt signaling in oncogenesis and embryogenesis--a look outside the nucleus. Science 2000; 287(5458): 1606-9. [http://dx.doi.org/10.1126/science.287.5458.1606] [PMID: 10733430]

[687] Christofk HR, Vander Heiden MG, Harris MH, *et al.* The M2 splice isoform of pyruvate kinase is important for cancer metabolism and tumour growth. Nature 2008; 452(7184): 230-3. [http://dx.doi.org/10.1038/nature06734] [PMID: 18337823]

[688] Carro MS, Lim WK, Alvarez MJ, *et al.* The transcriptional network for mesenchymal transformation of brain tumours. Nature 2010; 463(7279): 318-25. [http://dx.doi.org/10.1038/nature08712] [PMID: 20032975]

[689] Borczuk AC, Gorenstein L, Walter KL, Assaad AA, Wang L, Powell CA. Non-small-cell lung cancer molecular signatures recapitulate lung developmental pathways. Am J Pathol 2003; 163(5): 1949-60. [http://dx.doi.org/10.1016/S0002-9440(10)63553-5] [PMID: 14578194]

[690] Biava PM, Canaider S, Facchin F, *et al.* Stem Cell Differentiation Stage Factors from Zebrafish Embryo: A Novel Strategy to Modulate the Fate of Normal and Pathological Human (Stem) Cells. Curr Pharm Biotechnol 2015; 16(9): 782-92. [http://dx.doi.org/10.2174/138920101666150629102825] [PMID: 26201607]

[691] Biava PM, Bonsignorio D, Hoxa M. Cell proliferation curves of different human tumor lines after *in vitro* treatment with Zebrafish embryonic extracts. J Tumor Marker Oncol 2001; 16: 195-202.

[692] Biava PM, Carluccio A. Activation of anti-oncogene p53 produced by embryonic extracts *in vitro* tumor cells. J Tumor Marker Oncol 1977; 12: 9-15.

[693] Biava PM, Bonsignorio D, Hoxa M, *et al.* Post-traslational modification of the retinoblastoma protein (pRb) induced by *in vitro* administration of Zebrafish embryonic extracts on human kidney adenocarcinomacell line. J Tumor Marker Oncol 2002; 17: 59-64.

[694] Cucina A, Biava PM, D'Anselmi F, *et al.* Zebrafish embryo proteins induce apoptosis in human colon cancer cells (Caco2). Apoptosis 2006; 11(9): 1617-28. [http://dx.doi.org/10.1007/s10495-006-8895-4] [PMID: 16820966]

[695] D'Anselmi F, Cucina A, Biava PM, *et al.* Zebrafish stem cell differentiation stage factors suppress Bcl-xL release and enhance 5-Fu-mediated apoptosis in colon cancer cells. Curr Pharm Biotechnol 2011; 12(2): 261-7. [http://dx.doi.org/10.2174/138920111794295864] [PMID: 21043999]

[696] Biava PM, Nicolini A, Ferrari P, Carpi A, Sell S. A systemic approach to cancer treatment: tumor cell reprogramming focused on endocrine-related cancers. Curr Med Chem 2014; 21(9): 1072-81. [http://dx.doi.org/10.2174/0929867321666131201143124] [PMID: 24304275]

[697] Livraghi T, Meloni F, Frosi A, *et al.* Treatment with stem cell differentiation stage factors of intermediate-advanced hepatocellular carcinoma: an open randomized clinical trial. Oncol Res 2005; 15(7-8): 399-408.
[PMID: 16491958]

[698] Livraghi T, Ceriani R, Palmisano A, *et al.* Complete response in 5 out of 38 patients with advanced hepatocellular carcinoma treated with stem cell differentiation stage factors: case reports from a single centre. Curr Pharm Biotechnol 2011; 12(2): 254-60.
[http://dx.doi.org/10.2174/138920111794295855] [PMID: 21044000]

[699] Allegrucci C, Rushton MD, Dixon JE, *et al.* Epigenetic reprogramming of breast cancer cells with oocyte extracts. Mol Cancer 2011; 10(1): 7.
[PMID: 21232089]

[700] Rachakatla RS, Troyer D. Wharton's jelly stromal cells as potential delivery vehicles for cancer therapeutics. Future Oncol 2009; 5(8): 1237-44.
[http://dx.doi.org/10.2217/fon.09.99] [PMID: 19852738]

[701] Xu T, He K, Wang L, Goldkorn A. Prostate tumor cells with cancer progenitor properties have high telomerase activity and are rapidly killed by telomerase interference. Prostate 2011; 71(13): 1390-400.
[http://dx.doi.org/10.1002/pros.21355] [PMID: 21321978]

[702] Xu Y, He K, Goldkorn A. Telomerase targeted therapy in cancer and cancer stem cells. Clin Adv Hematol Oncol 2011; 9(6): 442-55.
[PMID: 21841744]

[703] Grzmil M, Voigt S, Thelen P, Hemmerlein B, Helmke K, Burfeind P. Up-regulated expression of the MAT-8 gene in prostate cancer and its siRNA-mediated inhibition of expression induces a decrease in proliferation of human prostate carcinoma cells. Int J Oncol 2004; 24(1): 97-105.
[PMID: 14654946]

[704] Stolzenburg S, Rots MG, Beltran AS, *et al.* Targeted silencing of the oncogenic transcription factor SOX2 in breast cancer. Nucleic Acids Res 2012; 40(14): 6725-40.
[http://dx.doi.org/10.1093/nar/gks360] [PMID: 22561374]

[705] Cubillos-Ruiz JR, Baird JR, Tesone AJ, *et al.* Reprogramming tumor-associated dendritic cells *in vivo* using miRNA mimetics triggers protective immunity against ovarian cancer. Cancer Res 2012; 72(7): 1683-93.
[http://dx.doi.org/10.1158/0008-5472.CAN-11-3160] [PMID: 22307839]

[706] Quinn MC, Filali-Mouhim A, Provencher DM, Mes-Masson AM, Tonin PN. Reprogramming of the transcriptome in a novel chromosome 3 transfer tumor suppressor ovarian cancer cell line model affected molecular networks that are characteristic of ovarian cancer. Mol Carcinog 2009; 48(7): 648-61.
[http://dx.doi.org/10.1002/mc.20511] [PMID: 19123201]

[707] Greenberg PA, Hortobagyi GN, Smith TL, Ziegler LD, Frye DK, Buzdar AU. Long-term follow-up of patients with complete remission following combination chemotherapy for metastatic breast cancer. J Clin Oncol 1996; 14(8): 2197-205.
[http://dx.doi.org/10.1200/JCO.1996.14.8.2197] [PMID: 8708708]

[708] Hanahan D, Weinberg RA. Hallmarks of cancer: the next generation. Cell 2011; 144(5): 646-74.
[http://dx.doi.org/10.1016/j.cell.2011.02.013] [PMID: 21376230]

[709] Nicolini A, Conte M, Rossi G, Ferrari P, Carpi A, Miccoli P. A new pharmacological approach to gastrointestinal cancer at high risk of relapse based on maintenance of the cytostatic effect. Tumour Biol 2010; 31(5): 523-32.
[http://dx.doi.org/10.1007/s13277-010-0065-1] [PMID: 20589491]

[710] Recchia F, Frati L, Rea S, Torchio P, Sica G. Minimal residual disease in metastatic breast cancer: treatment with IFN-beta, retinoids, and tamoxifen. J Interferon Cytokine Res 1998; 18(1): 41-7.

[http://dx.doi.org/10.1089/jir.1998.18.41] [PMID: 9475666]

[711] Fong Y, Fortner J, Sun RL, Brennan MF, Blumgart LH. Clinical score for predicting recurrence after hepatic resection for metastatic colorectal cancer: analysis of 1001 consecutive cases. Ann Surg 1999; 230(3): 309-18.
[http://dx.doi.org/10.1097/00000658-199909000-00004] [PMID: 10493478]

[712] Scheele J, Stangl R, Altendorf-Hofmann A. Hepatic metastases from colorectal carcinoma: impact of surgical resection on the natural history. Br J Surg 1990; 77(11): 1241-6.
[http://dx.doi.org/10.1002/bjs.1800771115] [PMID: 2253003]

[713] Portier G, Elias D, Bouche O, *et al.* Multicenter randomized trial of adjuvant fluorouracil and folinic acid compared with surgery alone after resection of colorectal liver metastases: FFCD ACHBTH AURC 9002 trial. J Clin Oncol 2006; 24(31): 4976-82.
[http://dx.doi.org/10.1200/JCO.2006.06.8353] [PMID: 17075115]

[714] Liu W, Zhou JG, Sun Y, Zhang L, Xing BC. The role of neoadjuvant chemotherapy for resectable colorectal liver metastases: a systematic review and meta-analysis. Oncotarget 2016; 7(24): 37277-87.
[http://dx.doi.org/10.18632/oncotarget.8671] [PMID: 27074564]

[715] Nicolini A, Carpi A, Ferrari P, Sagripanti A, Anselmi L. A multistep therapy with subcutaneous low dose recombinant interleukin-2, 5-fluorouracil and leucovorin prolongs the response of metastatic colorectal cancer patients: a pilot study. Biomed Pharmacother 1998; 52(7-8): 311-6.
[http://dx.doi.org/10.1016/S0753-3322(98)80027-9] [PMID: 9809175]

[716] Nicolini A, Carpi A, Rossi G. Cytokines in breast cancer. Cytokine Growth Factor Rev 2006; 17(5): 325-37.
[http://dx.doi.org/10.1016/j.cytogfr.2006.07.002] [PMID: 16931107]

CHAPTER 3

Foods and Nutrients as Adjuvants in Cancer Chemo-prevention and Treatment

Natalia G. Vallianou[1,*], **Eleni Geladari**[2], **Angelos Evangelopoulos**[3] and **Christos Kazazis**[4]

[1] *Consultant in Internal Medicine, Evangelismos General Hospital, Athens, Greece*

[2] *Internist, Evangelismos General Hospital, Athens, Greece*

[3] *Biologist, Evangelismos General Hospital, Athens, Greece*

[4] *Honorary Lecturer at Leicester University, Leicester, UK*

Abstract: Foods and nutrients may be useful sources in preventing and treating cancer. Substances like resveratrol, curcumin, pomegranate and honey have turned out to be much promising agents for combating tumor cells, especially for overcoming chemo-resistance. Compounds like resveratrol, which is abundant in red wine, red grapes, blueberries, peanuts and pistachios seem to induce p53-dependent apoptosis. Identification of a receptor of resveratrol in tumor cells, supports the potential of this substance as an anti-cancer agent. This receptor could serve in studies of future resveratrol analogues. Resveratrol is documented to surpass chemo-resistance through inhibition of NF- κB as well as the STAT3 pathway. Resveratrol has shown much promise in preclinical trials and because of its good safety profile it may be an ideal chemo-preventive and chemotherapeutic agent. Curcumin, a yellow component which belongs to the superfamily of polyphenols, is the most potent substance of turmeric, an Indian spice derived from the Curcuma longa plant. Curcumin and its analogues deploy their anti-tumor effects by inhibiting the STAT3 and the NF-κB, which play pivotal-roles in the evolution of tumor cells. Furthermore, it has been demonstrated to inhibit focal adhesion kinase (FAK) phosphorylation and enhance the expressions of several extracellular matrix components, which play a pivotal role in invasion and metastasis. Curcumin has been shown to enhance cell adhesion ability, through induction of extra-cellular matrix substance collagen, fibronectin, and laminin. Taken together, it is suggested that curcumin suppresses FAK activity by means of inhibition of its phosphorylation sites and also induces extra-cellular matrix components to enhance cell adhesion ability, thus, preventing detachment of cancer cells and cell migration. Honey comprises of vitamins, including ascorbic acid, pantothenic acid, niacin and riboflavin and minerals like calcium, copper, iron, magnesium, manganese, and phosphorus. Also, honey is suggested to exhibit anti-cancerous features, while its anti-proliferative and apoptotic effects vary according to the "floral origin and the phenolic content". Finally, the role of pomegranate juice and/or pomegranate extracts in preven-

* **Corresponding author Natalia G. Vallianou:** Consultant in Internal Medicine, Evangelismos General Hospital, Athens, Greece; Tel: +302294092359; E-mail: natalia.vallianou@hotmail.com

Atta-ur-Rahman (Ed.)

ting and treating various tumors, especially prostate cancer will be discussed in this review. The pomegranate fruit is derived from the tree Punica granatum and is cultivated in Mediterranean countries, Russia, India, China, Japan, and the United States. The anti-oxidant properties of pomegranates have been demonstrated to be better compared to red wine or green tea, two dietary compounds which have been promising in preclinical prostate cancer cell lines for reducing metastasis, mainly through inhibiting matrix substances. The roles of resveratrol, pomegranate, honey as well as curcumin as weapons in our battle against cancer will be further elucidated in this review. Many more studies are needed in order to determine the exact dosage of each of the above-mentioned compounds for the prevention and treatment of various types of cancers, respectively. In this view, bioavailability of those substances has to be improved, whilst the maximum dosage that does not cause any serious adverse effects in humans has to be established. In particular, future studies should focus on ameliorating formulas of the above-mentioned compounds, by improving their absorption and kinetics as well as by minimizing adverse side effects, which are estimated to be often in doses that are really helpful for humans in the clinical setting.

Keywords: Apoptosis, Cancer, Curcumin, Honey, Molecular pathways, Pomegranate, Resveratrol.

INTRODUCTION

Foods and nutrients may be useful sources in preventing and treating cancer. Substances like resveratrol, curcumin, honey and pomegranate have turned out to be much promising agents for combating tumor cells, especially for overcoming chemo-resistance. In this chapter, the anti-cancer potential of resveratrol, curcumin, honey and pomegranate will be further discussed. Many studies have documented the anti-cancer effects of the above-mentioned compounds. However, the main problems that remain to be resolved are the development of formulas with better bioavailability and less adverse side effects, as it is commonplace that these substances exert anti-cancer properties both *in vitro* and *in vivo* in experimental models, but these properties are achieved with much heavier dosges when it comes to human beings.

Resveratrol

Resveratrol is a stilbene substance, belonging to the superfamily of phytoalexins, which are compounds synthesized by plants when stress occurs, usually an infection. It is abundant in red wine, red grapes, blueberries, peanuts and pistachios [1 - 3]. Resveratrol (RSV) also possesses anti-oxidant and anti-cancerous properties. The anti-cancerous mechanisms of action of RSV are not well-understood, but it has been suggested from several studies, that they are the result of resveratrol's action in inducing apoptosis [4 - 12] (Fig. **1**).

Resveratrol and p53

Resveratrol induces p53-dependent apoptosis [4 - 12]. p53 is a suppressive oncogene, which exists at low concentrations in normal cells, whilst if there is DNA damage, its concentrations increase because a post-translational change that makes this protein more stable [12 - 19]. p53 is involved in apoptosis and DNA repair [19, 20].

"Acetylation" is documented to augment DNA binding of p53 and has been found to be crucial for enrollment of the protein to its "co-activators"–like "CREB-binding protein/p300" [18]. Those "co-activators" possess "histone acetyl-transferase" properties. "Acetylation" of the p53 seems to be related with CREB-binding protein/p300 (PCAF) and several other p53 co-activator" [19]. It has been suggested that resveratrol induces acetylation of p53. Phosphorylation of specific p53 residues is considered necessary for the activation of certain promoter regions [20].

"Resveratrol-induced apoptosis factor (PCAF)", also a p53 co-activator, and CREB-binding protein/p300 are also co-activators for various transcription factors besides p53. Resveratrol is capable of phosphorylating p53 at N-terminal and C-terminal serine residues in many human tumor cell lines while also of promoting "acetylation" of p53. "Activation" of "extracellularly-regulated kinases" 1 and 2 (ERK1/2) has been implicated in resveratrol-associated "serine phosphorylation of p53 in tumor cells" [4 - 12]. It has been documented that RSV induces serine residue "phosphorylation of a mutant p53" in prostate tumor cells, a phenomenon which induces apoptosis [6, 7].

Resveratrol and COX-2 Cyclooxygenase

Resveratrol and COX-2 Cyclooxygenase (COX), the enzyme involved in prostaglandin synthesis, is secreted by increased production of various inflammatory biomarkers. Constitutive expression of COX-2 has been related to "cancer cell growth and metastasis," increased "cellular adhesion" as well as the "inhibition of apoptosis" [21]. COX-2 is found in the endoplasmic reticulum, Golgi system, and the nuclear "envelope" [22, 23]. COX-1 and COX-2 are located in the nuclear envelope, as well as in the endoplasmic reticulum of "prostaglandin -2 releasing cells". Data have documented that prostaglandin-2 biosynthesis, COXs, and prostaglandin-2 are located in the perinuclear area. Resveratrol induces accumulation of COX-2 in the nuclear region in various cancer cells, such as human breast tumor cells, head and neck tumor cell lines, glioma cells, as well as in ovarian and prostate tumor cell lines [24]. Those data implicate that inducible COX-2 may play a major role in "p53-dependent apoptosis" of tumor cells. Researchers have also documented that COX-2 can be pro-apoptotic, while

Hinz *et al.* have demonstrated that COX-2 inhibitors may be deleterious for certain cancers due to the pro-apoptotic action of this protein [25, 26]. Nevertheless, inhibiting COX-2 pharmacologically has turned out to provide conflicting data [27, 28]. Other data have suggested that over-expression of COX-2 could produce an anti- proliferative result, which could be attributed to p53 and p21 [29, 30]. Other studies support the notion that "constitutive COX-2" expression induces tumor growth exerting anti-apoptotic effect, whereas "inducible COX-2", *e.g.* by resveratrol and located mainly to the nucleus is pro-apoptotic, by means of phosphorylation of the serine residue of p53 [31]. This mechanism could turn out to be unique in treating many tumors. Anti-tumor agents could possibly be utilized in the clinical setting, to affect the "constitutively and inducible expressed" properties of COX-2, especially for combating cancers in which inhibition of COX-2 pharmacologically, induces cell cycle arrest, rather than apoptosis [32 - 34].

Independence of COX-2 Enzymatic Activity

RSV possesses COX-2 features induced by carcinogens in human breast, and esophageal epithelial cells in rodents [35 - 37]. Besides, RSV has been found to inhibit COX-2 activity [38]. Zahner *et al.* have documented that over-expression of COX-2 may increase p53 gene expression and that this feature of COX-2 is totally independent of the enzymatic activity of COX-2 resulting in development of PGE2 [24]. In addition, the COX-2 specific inhibitor NS398, impedes Ser-15 phosphorylation of p53 and p53-dependent apoptosis in cancer cell lines treated with RSV [35 - 37]. NS398 does not affect ERK1/2 activation in RSV-treated cells [38, 39]. These studies support the notion that NS398 inhibits RSV-induced COX-2 in the nuclei, where COX-2 forms compositions with SUMO-1, pERK1/2, and p53. In other words, COX-2 can lose its ability to form complexes with other nucleoproteins and thus cannot be able to potentiate p53-dependent apoptosis in RSV-treated cancer cell lines, when NS398 is present [39].

Resveratrol and Sirtuins

The "silent information regulator" (SIR) genes (sirtuins) consist of a "highly conserved"family of proteins, with at least one sirtuin present in almost every species, ranging from prokaryotic organisms to eukaryotic ones. Seven sirtuin genes -SIRT1 to SIRT7- have been isolated among mammals [40]. There is a substantial body of evidence that sirtuins constitute a very perplexed biological response system, which has a major impact on many other molecular pathways, such as aging, apoptosis and inflammation in complex manners. Resveratrol has been the first compound discovered, able to mimic calorie restriction by stimulating sirtuins [41, 42]. Calorie restriction (CR) is a process that alters the concentrations of many genes implicated in different biological steps, including cell growth, metabolism, oxidative stress and DNA damage repair [43]. The molecular results induced by CR extend over two pivotal pathways with lifespan modulation: "insulin/insulin-like growth factor (IGF) signaling and target of rapamycin (TOR) signaling" [44]. It is suggested to induce gene expression patterns in multiple tissues. Indeed, treatment with resveratrol has reduced tumorigenesis in SIRT1+/−;p53+/− mice, and this protective effect has been attributed to SIRT1 activation [45]. Resveratrol and ERK1/2 Identification of a receptor site for resveratrol in cancer cell lines, by the implication of "extracellular signal-regulated kinases 1 and 2" (ERK1/2) of the "resveratrol signal downstream into p53-dependent apoptosis" reinforces the potential of this substance as an anti-cancer weapon [46]. This receptor could serve in studies of future resveratrol analogues. Regarding tumor cells, integrins are suggested to play a much more perplexed role than simply the transduction of signals originating from integrin-matrix protein interplays. It is noteworthy that dysregulation of the β-3 integrins has been implicated in the pathogenesis of cancer. Cancer cell growth together angiogenesis, like those associated with "vascular endothelial growth factor" pathway are increased in "β-3 -null mice" [40]. On the contrary, integrin β-3 over-expression may decrease tumor growth of human glioma cell in rodents [47]. The above-mentioned paradigms are indicative of the notion that increases in integrin β-3 expression in tumor cell lines could act as a therapeutic strategy in combating the process of carcinogenesis. ERK1/2 are MAPK iso-enzymes, which serve as "inducible components" of the "normal cellular signal transduction" process. ERK1/2 activation pathway may be switched on in the setting of growth factor stimulated cells or by inflammation. MAPK-kinase activates ERK1/2 straight-forward [48]. RSV activates MAPK at low concentrations, but higher concentrations of resveratrol may impede MAPK in tumor cell lines [49]. It has been demonstrated that RSV promotes ERK1/2 activation in prostate, breast, head and neck, glial, and ovarian cancer cell lines [48, 49]. This activation of ERK1/2 by RSV may be hindered by a MAPK kinase (MEK) inhibitor, PD98059 [49].

Resveratrol and AMPK/mTOR Pathway

AMPK is associated with the "phosphatidyl-inositol-3 kinase/AKT/mTOR signaling pathway", a cascade, which seems of vital importance for cell growth when mitogenic stimuli are present [50, 51]. AMPK activation impedes "phosphorylation and activation of the mTORC1 complex", which is partly mediated by "the upstream kinase AKT," whose activation results in diminishing the "AMP:ATP ratio" [52]. Resveratrol has also been shown to modulate AMPK in BCR-ABL transformed cells and to exhibit antileukemic effects [53 - 55]. Treatment of either "imatinib mesylate-sensitive or imatinib mesylate-resistant

CML cells" with resveratrol has resulted in apoptosis [56].

Resveratrol, NF-κB and STAT3

The "transcription factor nuclear factor (NF)-κB" regulates many genes involved in growth regulation and inflammation [57, 58]. In "vitro and *in vivo*" studies have documented that constitutive activation of NF-κB results in inhibition of "chemotherapy-induced apoptosis" in a number of cancer cells [59, 60]. "Signal transducer and activator of transcription 3" (STAT3) is also a uniformly existing –like NF-κB- member of the STAT family of transcription factors, which is activated by "tyrosine phosphorylation" by means of "upstream" receptors like "epidermal growth factor" (EGF), "platelet-derived growth factor and cytokines," for example interleukin-6 (IL-6) [61]. Recent studies have demonstrated STAT3 to grant cancer resistance to chemotherapeutic regimens [62 - 65]. STAT3 is one of the major compounds implicated in carcinogenesis [66, 67]. Cell proliferation, angiogenesis, apoptosis, and immune system evasion account for the oncogenic properties of STAT3 [68, 69]. Constitutively active STAT3 has been involved in promoting resistance to apoptosis, probably through the induction of Bcl- xL and cyclin D1 [70 - 74]. Its major role in carcinogenesis is by means of the induction of genes which inhibit apoptosis and interact with cell proliferation and angiogenesis. "Constitutive activation of STAT3" is implicated in a variety of cancers, including breast, brain, colon, gastric, esophageal, ovarian, nasopharyngeal, pancreatic, prostate, head and neck cancer, multiple myeloma, lymphomas and leukemia [75 - 77]. Nevertheless, it is not completely understood why STAT3 is "constitutively active" in cancer cell lines.

Recent data have shown that resveratrol may overcome chemo-resistance in cancer cells by interacting with apoptosis, down-regulating drug transporters and down-regulating proteins implicated in cancer cell proliferation. Besides, resveratrol is documented to beat chemo-resistance by impeding NF- κB and STAT3 pathway [78 - 80]. Resveratrol has been suggested to enhance the apoptotic and "anti-proliferative" properties of bortezomib and thalidomide in multiple myeloma cell lines. Such effect has been related to the inactivation of NF-κB and STAT3 pathways. Resveratrol administration has also been related to the accumulation of "sub-G (1) population", augmentation in Bax, and activation of caspase-3. This has been related to the down-regulation of many proliferative and anti-apoptotic gene products, including "cyclin D1, cIAP-2, XIAP, survivin, Bcl-2, Bcl-xL, Bfl-1/A1, and TRAF2" [81]. Data have demonstrated that RSV inhibited NF-κB activation by impeding IκBα phosphorylation and IKK activation. These data have been further confirmed by the inhibition of NF-κB and STAT-3 in patients with multiple myeloma [82].

Resveratrol and Autophagy

Autophagy is an evolutionarily conserved intracellular process, characterized by lysosomal degradation of proteins, which is essential for survival of eukaryotic cells under metabolic stress. It has also been suggested to be a type of "programmed cell death" [83 - 86]. Pterostilbene, an analog of resveratrol, is documented to exert both autophagy and apoptosis in bladder cancer cells as well as in breast cancer cells [87, 88]. It has also been found to cause accumulation of autophagic vacuoles, as well as promote cell death, *via* a mechanism involving lysosomal membrane permeabilization in human melanoma, colon, lung and breast cancer cell lines [89, 90]. Pterostilbene has been documented to produce autophagy in human leukemia cells, too [90].

Resveratrol, VEGF and PI3K/Akt

A new RSV analogue, HS-1793, has recently been demonstrated to inhibit "vascular endothelial growth factor" (VEGF) in human prostate cancer cell lines. HS-1793 has been suggested to impede phosphorylation of PI3K and Akt in human prostate cancer cell lines [91]. Resveratrol itself has been suggested to inhibit the PI3K and Akt pathway in acute lymphoblastic leukemia cells [92]. Also, it has mainly produced degradation of HIF-1a protein by means of the proteasome. Furthermore, HS-1793 has shown more powerful results than rRSV on the cytotoxic effects on PC-3 cells [93]. Resveratrol is found to act with anti-angiogenic effects, through the inhibition of VEGF [94]. Also, this resveratrol analogue, HS-1793, has been shown to induce cell cycle arrest and programmed cell death, in human breast cancer cell lines. In particular, it has been demonstrated to cause G2/M cell arrest in human breast cancer cells [94].

Resveratrol and ATF3 Transcription Factor

Recently, ATF3 has been identified as a new "target" of RSV in colorectal cancer cells [95]. ATF3, a member of the "ATF/CREB family of transcription factors", is characterized as an adaptive response gene [96]. Latest data are indicative of ATF3 functioning as a cancer suppressor gene in colorectal tumorigenesis [97]. First, ATF3 function is substantially decreased in tumor tissues, comparing to normal tissues [98]. Second, ATF3over-expression is reported to produce inhibition of proliferation, promotion of apoptosis, inhibition of invasion and decrease of tumor formation *in vivo* [99 - 102]. Finally, ATF3 is demonstrated to increase apoptosis by substances known to possess anti-cancerous properties [103 - 105]. Thus, it is suggested that ATF3 has an anti- carcinogenic role in colorectal cancer [106].

Resveratrol, p53, COX-2, ERK1/2, STAT3 and PI3K/Akt

The biochemical pathways regarding development of the "COX-2-MAPK (ERK1/2) complex" and activation of p53 in "resveratrol-treated" cell lines have been elucidated [35 - 37]. Specifically, data have documented that post-translational modification of COX-2 as tyrosine phosphorylation, modulates COX-2 task in cerebral cells [26]. Studies also indicate that resveratrol-induced COX-2 associates with activated ERK1/2 in the nucleus of cancer cells [35 - 37]. Activated nuclear ERK1/2 has been demonstrated to make structures with transcriptionally active proteins like "signal transducing and activator of transcription" (STAT) proteins, and the p53 [107]. Activated ERK1/2 in this composition acts by means of serine phosphorylation and thus activation of the related proteins. After being activated by resveratrol and after being translocated to the nucleus, ERK1/2 can play a pivotal role in the "resveratrol-induced p53-dependent" apoptotic cell death. Inhibition of ERK1/2 either pharmacologically by PD98050 or by inhibition of COX- 2 by NS398 has been shown to block the nuclear composition of COX-2 with phosphorylated p53 and with ERK1/2 [35 - 37]. Induced COX-2 has been demonstrated either ERK1/2- or p38-dependent [35 - 37, 108]. Besides, apart from the p38 and ERK1/2 pathways, the PI3K/Akt pathway has been suggested to be implicated in the activity of COX-2. The PI3K/Akt cascade is switched on by COX-2 or a product of COX-2 known as PGE2 [109]. Some transcriptional factors and co-activators such as Akt, NF-κB, and STAT proteins which are located in the cytoplasm, translocate to the nucleus after post-translational modification [110]. COX-2 may as well incur post-translational modification. However, uncertainty exists in regards to whether ERK1/2 has a role in the post-translational phosphorylation of nuclear COX-2. COX-2 may make complexes with "caveolin-1" (Cav-1) and with small ubiquitin-like modifier (SUMO) family, which modulate the function of several transcription factors by changing their interplays with specific partners as well as their stability [111]. Nuclear COX-2 may bind to the "promoter region" of at least one gene, a finding pointing to the COX-2 being transcriptionally active or acting as a co-factor for proteins, and that a consideration of this protein simply as a critical enzyme in the production of prostaglandins may be too short-headed. Undoubtedly, more information is warranted about the potential transcriptional role of this protein. It would be useful to determine whether constitutively expressed COX-2 has transcriptional roles, too [111].

Resveratrol and Reactive Oxygen Species (ROS)

RSV suppresses ROS in mitochondria by acting as a capable "scavenger of superoxide anion, hydrogen peroxide, and hydroxyl radical, inhibiting lipid peroxidation", and helping to refill cellular glutathione levels [112 - 115].

Resveratrol may also "exert its antioxidant activity in an indirect manner, by modulating the expression of mitochondrial proteins or by increasing the expression of ROS scavenging enzymes". Resveratrol is capable of activating the antioxidant pathway triggered by Nrf2 in keratinocytes and in cultured coronary arterial endothelial cells [116, 117].

In endothelial cells, RSV reduces mitochondrial ROS generation by enhancing SIRT3 levels within the mitochondria [118]. Regarding scavenging enzymes, RSV is able to upregulate glutathione peroxidase, catalase and MnSOD in endothelial cells, *via* a "SIRT1-dependent manner" [116, 117]. RSV treatment has been proven to prevent the severity of doxorubicin-induced cardiotoxicity by decreasing oxidative stress, as demonstrated by increased levels of superoxide dismutase (SOD) and decreased levels of malondialdehyde, a finding suggestive of its free radical scavenging ability [118]. It was reported that the increase in reactive oxygen species (ROS) production stimulated by doxorubicin was localized to the mitochondria in cardiomyocytes and resveratrol pre-treatment prevented doxorubicin-induced mitochondrial damage [118].

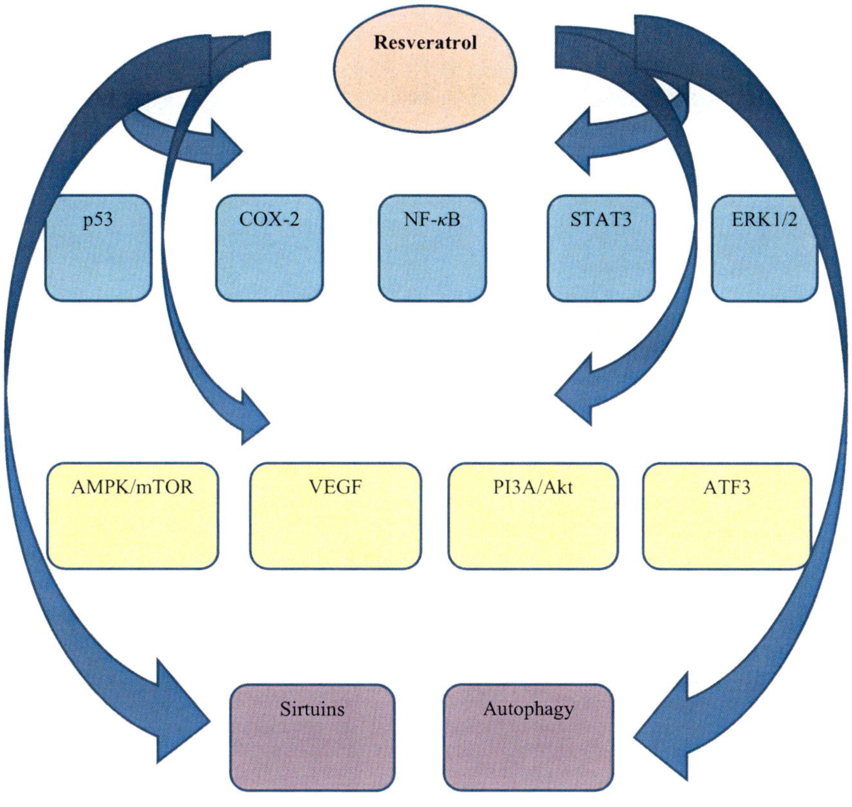

Fig. (1). Resveratrol interacts with many biological pathways to exert its anti-cancer potential.

Curcumin

Curcumin, a yellow component which belongs to the superfamily of polyphenols, is the most potent substance of turmeric, an Indian spice derived from the Curcuma longa plant [119, 120]. Turmeric comprises mainly of three components, curcumin, demethoxycurcumin and bisdemethoxycurcumin, among which curcumin prevails in terms of abundancy and potency [121 - 124]. Curcumin comprises approximately 2%-5% of the turmeric [125]. Many studies have documented that curcumin exerts anti- oxidant, anti-inflammatory and anti-cancer effects [125 - 132]. The transcription factor nuclear factor (NF)-κB, is a uniformly transcription factor which regulates many genes implicated in growth, inflammation, tumorigenesis, and apoptosis [133 - 136]. In "vitro and *in vivo*" data have documented that "constitutive activation of NF-κB" results in inhibition of chemotherapy-induced programmed cell death in many cancer cells [137 - 139]. STAT3 is one of the major substances implicated in carcinogenesis [140, 141]. Recent studies have demonstrated that STAT3 may account for cancer resistance to chemotherapeutic regimens [142 - 145]. Curcumin impedes the STAT3 and NF-κB signaling pathways, which- as it has already been mentioned- have key-roles in the development of cancer [144, 145] (Fig. **2**). "Constitutive activation of the STAT3 and NF-κB signaling pathways" has been demonstrated in prostate cancer cells [144, 145].

Curcumin and Other Transcription Factors

Curcumin has also been suggested to induce apoptosis and cause down- regulation of EGFR, Akt, cMET cyclin D1, in CL-5 xenograft tumors [146]. In addition, it has been documented to inhibit lung cell invasion and metastasis through up-regulation of HLJ1 expression in cancer cells [147]. Apart from its action on STAT3 and NF-κB pathways, curcumin has been documented to impede cell proliferation and stimulate apoptosis *via* modulation of other transcription factors, such as AP-1, Erg-1, p53, β-catenin, Notch-1, Hif-1, and PPAR-α [147]. Curcumin and Sp-1 Sp-1, a transcription factor highly expressed in breast, gastric and thyroid tumor cells as compared to normal cells, has been demonstrated to interact with co-activators and co-repressors and thereby activate multiple biological functions, including cell cycle and carcinogenesis [147]. It has also been implicated in nuclear factors, protein-protein interaction, and sequence-specific DNA binding [147]. Sp-1 has been related to housekeeping gene expression, such as vascular epithelial growth factors (VEGF), "urokinase plasminogen activator" (uPA), "urokinase plasminogen activator receptor" (uPAR) and "epithelial growth factor receptor" (EGFR), which are suggested to be implicated in cell differentiation, tumor angiogenesis and metastasis [148, 149]. Hence, inhibition of Sp-1 and its "housekeeping gene expressions" might be important in preventing cancer formation and invasion [149, 150]. Studies have shown that curcumin could inhibit activation of Sp-1 and its downstream genes, including "ADEM10, calmodulin, EPHB2, HDAC4, and SEPP1 in a concentration- dependent manner in colorectal cancer cells" [150 - 152]. Also, other studies have suggested that curcumin could inhibit Sp-1 activity in bladder cancer and could impede DNA binding activity of Sp-1 in non-small cell lung carcinoma (NSCLC) cells [151, 152]. In addition, curcumin has significantly reduced colony formation in colon tumor cells. Data have documented that down-regulation of Sp-1 prevents the colony formation in a patient derived fibrocarcinoma cell line [152].

Curcumin and Adhesion Molecules

Curcumin has been demonstrated to inhibit focal adhesion kinase (FAK) phosphorylation and enhance the expressions of several extracellular matrix components, which play a pivotal role in invasion and metastasis [153]. Curcumin has been shown to enhance cell adhesion ability, through induction of extra-cellular matrix substances like collagen, fibronectin and laminin [153]. Taken together, these data have suggested that curcumin suppresses FAK activity by means of inhibition of its phosphorylation sites and also induces extra-cellular matrix components to enhance cell adhesion ability, thus, preventing detachment of cancer cells and cell migration. Inhibition of FAK expression leads to increased cell adhesion, which may be the potential mechanism of the anti-invasive effect of curcumin [153]. Curcumin has been shown to reduce CD24 expression in colorectal cancer cell lines. These studies suggest that curcumin could act against metastasis, through down-regulation of Sp-1, FAK, and CD24 and by promoting E-cadherin expression in colorectal cancer cells [153 - 155]. Also, Zhou *et al.* have evaluated eleven curcumin-associated substances, comprising of a "benzyl piperidone moiety" in various cancer cell lines and have found that some of them have been related to inhibition of "phospho-Akt and phospho-extracellular signal-regulated kinase" [156].

Curcumin, Endoplasmic Reticulum Stress and Autophagy

In addition to its involvement in cellular homeostasis, autophagy has been serving as a type of apoptosis as well as a cell -protective role in cases of nutrient deprivation [157 - 159]. Recent studies have suggested that unfolded protein response signaling may also affect interactions within the cancer microenvironment [160]. The "unfolded protein response, autophagy and ER stress- induced apoptosis" have been documented to regulate cancer cell future [160]. Furthermore, different anti-cancer treatments may activate this signaling in cancer cells, by either increasing cancer cell death or by acting as a chemotherapy resistance mechanism [160 - 162]. ER stress and autophagy may play a key-role in apoptosis, which is induced by the curcumin analogue B19 in an epithelial ovarian tumor cell line and in hepatocellular carcinoma cells and that autophagy impediment may enhance curcumin analogue-induced apoptosis by inducing major ER stress [162 - 164]. In other words, this curcumin analogue may induce ER-stress, autophagy and apoptosis in ovarian cancer cell lines *in vitro* [164, 165]. Other researchers have demonstrated that autophagy could act as "programmed cell death" and could efficiently suppress growth of malignant glioma cells after curcumin treatment [166].

Curcumin and ROS

Curcumin has been demonstrated to minimize ATP biosynthesis, to increase F0F1-ATPase in a dose-dependent manner and to impede respiration at concentrations >50 μM in rodents' liver [167]. Nevertheless, it is noteworthy that curcumin inhibits the F0F1-ATPase in rat brain mitochondria, indicating that different effects of this compound can be observed in the same organelle from different tissues [167]. A plausible explanation has been elucidated in *Escherichia coli*, where curcumin directly inhibits F1 ATPase activity by intervening with the "beta subunit catalytic site conformational transitions" [167, 168].

Curcumin Bioavailability

Curcumin lacks toxicity and has lately gained much interest as a potential anti-cancer agent. Its main disadvantage is its low oral bioavailability due to its extensive first-pass metabolism and its poor aqueous solubility [169, 170]. It is widely known that the enhanced permeability and retention effect of nanomaterials may increase the concentration of chemotherapeutics in cancer cells. "Liposomes, carbon nanotubes, dendrimers, and micelles" have been utilized as carriers for SN38, doxorubicin, paclitaxel, and cisplatin to increase drug concentrations in cancer cells and to decrease adverse effects [171 - 173]. Another con of utilizing nanomaterials as drug carriers is the increased solubility of chemotherapeutic agents. Self-assembling peptide nanofibers have attracted much attention due to their "good biocompatibility, easy modification, and design flexibility" through a "bottom-up" approach [174, 175]. They have been utilized in various cell cultures as well as "drug delivery systems" to increase solubility of hydrophobic agents, to increase accumulation in cancer cells, and to decrease adverse effects [176 - 178]. Hydrophobic antitumor drugs such as paclitaxel, camptothecin and ellipticine "encapsulated into peptide nanofibers" have better anti-cancer properties [179 - 181]. "Two dimensional structure of peptide nanofibers" is superior to the three-dimensional structure of nanoparticles for drug carriers [182 - 185]. Indeed, Law *et al.* demonstrated that "peptide-based nanofibers" have "improved biocompatibility, better tumor targeting within a shorter time and a more rapid elimination compared with spherical nanomaterials (poly[lactic-co-glycolic acid], gold, polystyrene, cadmium and selenium quantum dots), and carbon rods" [186]. Recently, some synthetic analogues of curcumin have been prepared and evaluated for potential pharmacological effects [187 - 194]. Some analogues have demonstrated to possess promising properties in various cancer models [195 - 197]. Recently, data have documented that curcumin-related compounds with a "benzyl piperidone" have improved absorption and biological properties [198 - 203]. Other studies have also demonstrated the potential anti-cancer properties of curcumin analogues [198 - 203]. The incorporation of curcumin into nano-formulations for enhanced water-solubility has indeed revolutionized the bioavailability of curcumin [198 - 203]. Nano-formulations have accomplished improved delivery and better concentrations of curcumin into the cell *in vitro*, while their prolonged release formulas along with their higher degree of compatibility seem to be very promising for their effects *in vivo* [198 - 203].

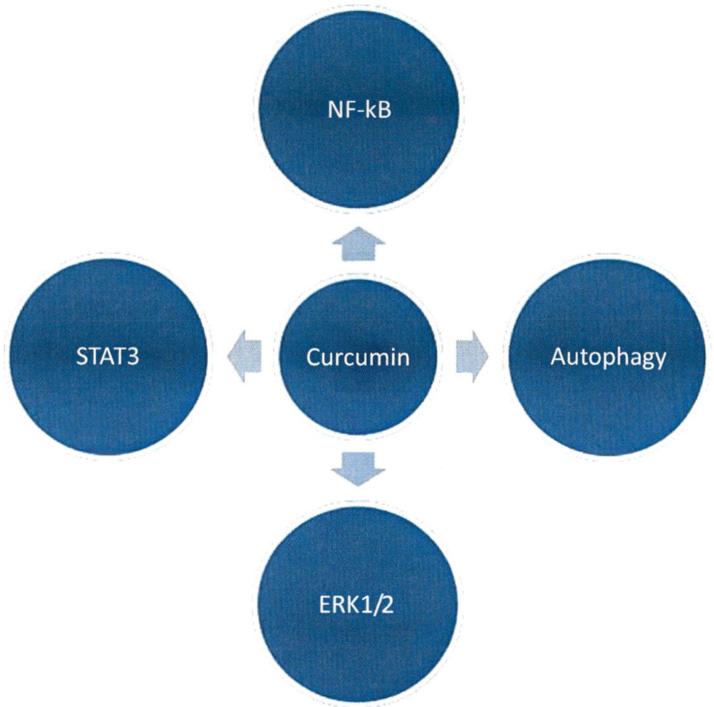

Fig. (2). Anti-cancer mechanisms of action of curcumin involve the NF- κB, STAT3, ERK1/2 and autophagy pathways.

Honey

Honey comprises of sugars, water, vitamins and minerals [204]. Sugars in honey consist predominantly of monosaccharides and oligosaccharides [205]. The most abundant sugar in honey is fructose [204 - 206]. Some of the vitamins in honey are ascorbic acid, pantothenic acid, niacin and riboflavin; while minerals present are calcium, copper, iron, magnesium, manganese, phosphorus, potassium and zinc [207]. Honey contains at least 181 constituents [208, 209]. Other substances found in honey are amino acids, antibiotic-rich inhibin, proteins and polyphenols, flavonoids, organic acids, carotenoid-derived compounds and nitric oxide (NO) metabolites [210 - 212]. Several phenolic constituents of honey like "caffeic acid, caffeic acid phenyl esters, chrysin, galangin, quercetin, kaempferol, acacetin, pinocembrin, pinobanksin, and apigenin" are extensively studied due to their application in cancer treatment [213].

Evidence indicates that some varieties of honey contain kynurenic acid, which partly accounts for its antimicrobial properties [214]. The presence of enzymes like "glucose oxidase, diastase, invertase, phosphatase, catalase and peroxidase" has also been demonstrated in honey [215]. Anti-carcinogenic and anti-proliferative properties of honey have been recently reported and are mainly attributed to its phenolic compounds [216 - 218]. Chrysin has been suggested to possess anti-cancerous activities as well as to inhibit growth of melanoma cells *in vitro* [219, 220]. Acacia honey-induced cytotoxicity in melanoma cell lines has been attributed to the presence of chrysin [221]. Recently, it has been shown that honey comprises of substances that impede the proliferation of a human prostate cell line (PC-3) through promotion of programmed cell death [222]. Data exist confirming that the anti-proliferative effects of honey are attributed to chrysin, too [223]. The sensitization effect of chrysin on tumor necrosis factor-alpha (TNF-α)-induced apoptosis is mainly achieves by means of increased activation of caspase 8, the initial caspase in "the death receptor signaling pathway", which typically enhances programmed cell death [224] (Fig. **3**). Honey has shown anti-tumorous features in the experimental bladder cancer [225]. In one study, administration of honey investigated on "N- ethylmaleimide liver injury" in rodents has demonstrated that decrease of glutathione concentration plays a crucial role in "N-ethylmaleimide induced liver injury" and that the hepato-protective features of honey could be accomplished *via* "sulfhydryl-sensitive processes" [226]. In hepatocellular carcinoma, honey may show promising results on accounts of the inhibition of the "proliferation, protease activity and gelatinase activity of HepG2 cells" [227].

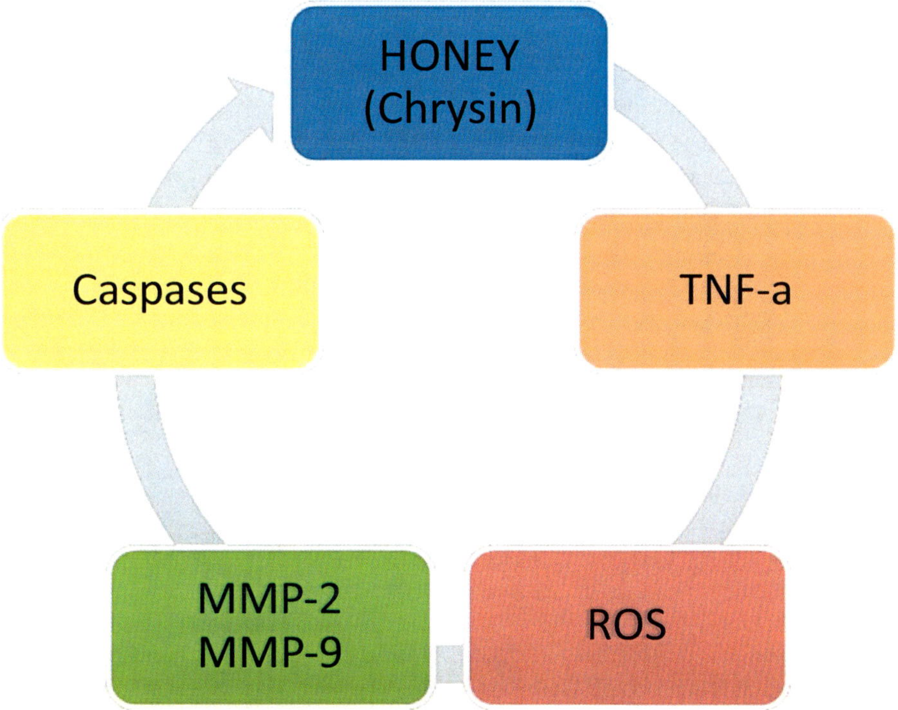

Fig. (3). Honey, mainly through its ingredient chrysin, interacts with caspases, TNF-a, MMP-2, MMP-9 and ROS to exert its anti-cancer effects.

It has been suggested that the proliferative effect of honey on cancer cell lines has been due to features, such as the impediment of DNA synthesis without signs of cytotoxicity and the down- regulation of "MMP-2 and MMP-9" [228]. Nevertheless, most studies regarding the anti-cancerous role of honey have been conducted *in vitro* and not *in vivo*. The study on cell lines could not guarantee the same result in the human body, where many confounding physiological factors are involved. Also, honey has been proved to induce CYP3A4 enzyme activity [229, 230]. Hence, honey may have drug interactions in patients using various drugs. Moreover, there is strong evidence supporting the fact that anti-proliferative and apoptotic properties of honey vary depending on the botanical origin and the phenolic constituents of honey [231]. Many studies have advocated that apoptosis may be induced by oxidative stress, which in turn counteracts with "reactive oxygen species" (ROS). Changes in the "redox status of the cell" to a more oxidizing milieu occur before the final phase of caspase activation in various models of apoptotic cell death [230, 231]. Recent results have shown honey to be able to induce programmed cell death by means of "reactive oxygen species" and "mitochondria-dependent mechanisms" in colon tumor cells [231].

In addition, as the composition of honey varies in regards to its floral origin and milieu factors, it can be reasonably expected that honey properties from different floral sources are different. These findings are in agreement with Jaganathan and Mandal who have suggested that honeys with higher phenolic composition were more capable of inducing apoptosis in colon cancer cells [231]. Also, honey with higher phenolic composition has been documented to impede "Ehrlich ascites carcinoma", when compared to various samples [232]. However, Jaganathan and Mandal have reported that honeys from India induced apoptosis through ROS and mitochondria-dependent mechanism in colon cancer cells [233]. Recently, Fauzi *et al.* have revealed mitochondrial pathway to be implicated in tualang honey-induced programmed cell death of breast and cervical tumor cell lines [234]. Honey-induced apoptosis in human colon cancer cells was related to the activation of caspase-3 and DNA laddering [235, 236]. That is, honey may have the ability to regulate methylation of oncogenes and thus contribute to anti-cancerous actions [236].

Pomegranate and its Potential Therapeutic Role in Various Types of Cancer

Among other non-toxic nutritional agents, it has been suggested that pomegranate may further battle cancer through chemoprevention. In addition, its use in combination with a specific scheme of chemotherapeutic agents has been shown to be beneficial for cancer patients [237 - 239]. Punica Granatum, the botanical name for Pomegranate, is a fruit-borne deciduous shrub, which is currently being studied from researchers worldwidely as a potential weapon against cancer [239]. It has been cultivated throughout the Mediterranean region, Southeast Asia, and Northern India to Iran since antiquity. Its nourishment has also been broadened in parts of the United States of America, especially, in the drier regions of California and Arizona [239, 240]. Its exclusive composition is made up of three parts: the seeds, the juice and the peels, which constitute the inner network of membranes. Its structure along with its chemical content can vary according to the geographical place, the season or the gathering time [239, 240]. Literally, researchers are studying the potential therapeutic effects of pomegranates' bioactive constituents in different types of malignant diseases, particularly in prostate, breast and colon cancer [239, 240]. Many favorable outcomes seem to be the result of the presence of anthocyanins, ellagic acid ellagitanins, punicic acid, flavonoids, anthocyanidins, estrogenic flavonols and flavones, which come from a different part of the pomegranate plant [239, 240] (Table **1**).

Ongoing research studies aim to understand the metabolism and absorption of the aforementioned constituents, since not all of them have the same bioavailability, and thus the same medical benefit [239, 240]. In particular, the bioactive substances can either enter the bloodstream and act directly as antioxidants, or they may be metabolized by the gut microflora into smaller chemically molecules that have potent anti-oncogenic properties. Basically, many commercially available pomegranate extracts contain large concentrations of ellagic acid, since it is believed that it has antioxidant as well as anti-carcinogenic properties [241] (Fig. **4**). However, Ephraim Lansky, a notable researcher, who unfolded the potential medicinal properties of this mystic plant, states that the beneficial effects of the pomegranate are elicited by the synergistic action of all of its ingredients, which seems to be superior than that of its single substances [239, 241, 242]. Numerous studies have demonstrated the importance of pomegranate's use in diabetes, obesity, cardiovascular diseases, male infertility, Alzheimer disease, *etc*. In this section we will focus on the pomegranate's power to reduce cancer burden, and specifically, the lung, colon, breast and prostate tumor cells [239, 240].

Table 1. Pomegranate's basic constituents.

Pomegranate Component	Basic Constituents
Juice	Ellagic acid
Seed oil	Ellagic acid, Punicic acid
Pericarp	Flavonons, anthocyanidins
Leaves	Flavone glycosides, including luteolin
Roots and bark	Ellagitannins, including punicalin and punicalagin

Pomegranate in Inhibiting Breast Cancer

Breast Cancer is the second cause of death from cancer in women, in the USA. Among the four different histological types of cancer, the one that is estrogen dependent is the luminal type and responds well to estrogen-receptor blockers, such as tamoxifen. However, in some cases resistance to tamoxifen has been developed [243]. Pioneer studies have demonstrated that pomegranate fruits extracts may act synergistically with tamoxifen in inhibiting cancer growth in resistant estrogen-dependent breast tumor cells [244 - 246]. Interestingly, Banerjee *et al*., showed that pomegranate fruits extracts inhibit cell growth and viability in sensitive and resistant in tamoxifen MCF cells [244]. Specifically, the proliferative and anti-apoptotic effects of estrogen on MCF cells are minimized by the fruit's extracts and the combination of pomegranate with a "selective estrogen receptor modulator" (SERM), such as tamoxifen, results in growth inhibition and enhancement of cell accumulation in the sub-Go (apoptosis) and G1 phases of cell cycle, with a decrease in S phase [244, 245]. This may be due to the mimicking action of pomegranate fruits extracts to tamoxifen's action, which is down-regulation of proliferation and Bcl-2 expression on estrogen receptor-positive breast cancer cells [245]. The exact mechanism of action however needs to be further elucidated. Another study aimed to examine the anti-oncogenic properties of pomegranate's seed oil (PSO) on human tumor cells, including breast tumor cells [246]. It has been showed that PSO leads to an inhibition of cell viability for two breast cancer cell lines with an enhancement of the cell cycle Go /G1 phase compared to untreated cells [246]. Another characteristic is the reduced levels of "vascular endothelial growth factor" (VEGF), a growth factor responsible for angiogenesis and metastasis, as well as of nine pro-inflammatory cytokines (IL-2, IL-6, IL-12, IL-17, CXCL10, MIP-1a, MIP-1b, MCP-1, and TNF-a), in a "dose-dependent" manner. The suppression of the above-mentioned cytokines can inactivate NF-kB and STAT molecular pathways [244 - 246].

Ellagitannins, a predominant substance of pomegranate juice, are hydrolyzed to ellagic acid (EA), which is then in turn metabolized by human colonic microflora to "3,8-dihydroxy-6H-dibenzo[b,d]pyran-6-one" derivatives [urolithin A (UA) and B (UB)] [246]. These ET-derived compounds not only directly block estrogen receptors on MCF-cells but they also exhibit anti-aromatase activity, by promptly blocking the aromatase, the enzyme that converts androgens to estrogens, as well [246]. Urothilin B has the greatest anti- proliferative activity. Remarkably, it has been observed that UB lessens estrogen or testosterone induced tumor cell proliferation, in an aromatase dependent as well as independent manner [246]. Hence, the goal in treating hormone- dependent breast cancer is to reduce the estrogen levels by different means (*e.g.* not receiving oral contraceptive pills or hormone-replacement therapy, administering estrogen receptor blockers or aromatase enzyme blockers, such as tamoxifen and anastrozole, respectively) [246]. The pomegranate fruit, through ellagitannis, may slow down the growth of breast cancer possessing anti- estrogenic and anti-testosterone properties/effects [246].

Pomegranate and its Role in Colon Cancer

Another leading cause of tumor-related death is colorectal cancer (CRC) [247, 248]. CRC may be treated promptly if the diagnosis is established in the early stages. However, as the disease progresses and spreads, the treatment with chemotherapeutic drugs becomes not only ineffective but can also be toxic, since it halts the anti-oxidants and produces more oxidative stress [248]. Nowadays, it is known that diet rich in fiber decreases the risk of developing colorectal cancer [248]. Consequently, scientists have been directed to investigate the chemo-preventive effects of many other different dietary agents, which contain polyphenols, major non-nutrient components, and are promising agents in reducing the prevalence of colon cancer [248]. Pomegranate juice, due to its peculiar biochemical composition is one of them, considering its pro- apoptotic, anti-proliferative, anti-inflammatory as well as anti-oxidant properties [248]. Recently, the effectiveness of pomegranate juice on colon cancer cells lines has been reviewed [248]. It has been demonstrated that ellagitanins and urolithins, ellagic acid derivatives, which have been found in human colon tissues from CRC patients after consumption of pomegranate, may pause the "HT-29 cell proliferation" through apoptotic mechanisms [248, 249]. Specifically, Larrosa *et al.*, have focused on the anti-oncogenic properties of punicalagin and ellagic acid, components that trigger programmed cell death by causing mitochondrial cytochrome c leakage in the cytosol and by inducing pro-caspase 3 and caspase 9, members of "the caspase family of proteases" [248]. Furthermore, a "downregulation of the anti-apoptotic Bcl-XL protein was achieved with 30 μmol/L ellagic acid and 100 μmol/L punicalagin", respectively [248]. Further studies have depicted the inverse interrelation between augmented COX-2 expression and increased cell proliferation, while pre-treatment of colon cancer cell lines with a COX-2 inhibitor suppresses cell proliferation [248, 249]. As a deduction, researchers hypothesized that pomegranate juice might influence COX-2 expression in an inversely manner. Indeed, pretreatment with the pomegranate juice, containing powerful ingredients such as punicalagin, flavonols and anthocyanins, led to an overall decreased expression of COX-2 [249]. This is partially accomplished through down-modulation of "phosphatidylinositol 3-kinase (PI3K)/protein kinase B (AKT)/nuclear factor- kappa B (NF-κB)" signaling pathways [248, 249]. Another recognized signaling pathway that can regulate the expression of cyclooxygenase-2 is the "mitogen-activated protein kinase" (MAPK) pathways, (SAPK, p38 and ERK1/2) [249].

Another study conducted in rats did reveal that feeding the mammals with pomegranate juice diminished the aberrant cryptic foci(ACF) by 91% in male F-344 rodents [249]. It was also noted that the activity of "hepatic glutathione S transferase" (GST), an enzyme that scavenges the free radicals produced from oxidative stress, was three times higher, thus supporting the anti-oxidant aspects of pomegranate [249]. Moreover, pathologists from the Kanazawa Medical University in Japan ran a study where it was proven that the conjugated linolenic acid (CLN), found in the pomegranate's seed oil (PGO), reverses azoxymethane (AOM) induced colon tumorigenesis in rats [249]. Administration of PGO was linked with an enhanced content of "conjugated linoleic acid" (CLA) in the lipid fraction of "colonic mucosa" as well as with an increased expression of "peroxisome proliferator-activated receptor (PPAR) γ protein" in the intact mucosa [249]. It has also been documented that PPARγ activation by pharmacological agents may impede cancer growth. More studies need to be conducted in order to clarify pomegranate's action on a molecular basis and figure out a chemical format that can be utilized for chemoprevention against colon cancer [249].

Pomegranate and its Role in HCC

Malignant hepatoma, another name for hepatocellular carcinoma (HCC), is a devastating disease with dismal prognosis [250]. The available treatment options, like surgical resection or liver transplantation have their own cons [250]. For this reason, studies have been undertaken to demonstrate the chemo- preventive role of pomegranate on HCC, the most common form of liver tumorigenesis, which is driven primarily by oxidative stress and inflammation, mainly due to hepatitis B and C viral infections, alcohol abuse, metabolic diseases and exposure to dietary carcinogens, *e.g.* aflatoxins and nitrosamines [250]. Actually, it is believed that pomegranate emulsion modulates two inter-connected molecular signaling pathways; canonical NF-kB and Wnt/b-catenin. In an experimental study, rodents were administered PE (1 or 10 g/kg) 4 weeks before and 18 weeks following diethylnitrosamine (DENA) initiation. Increased levels of "nitric oxide synthase, 3-nitrotyrosine, heat shock protein 70 and 90, cyclooxygenase-2 and NF-kB" were noted. Liver carcinogenesis, through suppression of all aforementioned inflammatory markers, following DENA exposure in rats was substantially reversed by treatment with pomegranate emulsion [250]. There is substantial evidence that the inhibitory effects on liver inflammation and carcinogenesis are probably mediated through modulation of NF-kB signaling pathway and up-relegation of "pro-apoptotic protein Bax" and down- modulation of "anti-apoptotic protein Bcl-2" [251]. In a more analytical study, researchers observed that DENA induced nodular hyperplasia in rats' liver, a premalignant condition, which was remarkably inhibited be PE, and at the same time the hepatocellular architecture was reformed [252]. This finding is achieved by suppressing the induction of GGT-positive foci, up-regulating antioxidant genes, inducing Nrf2 expression and subsequently exerting anti-oxidant effects. GGT, a membrane bound enzyme is an indicator of structural hepatocellular alteration and tumor initiation, since it promotes glutathione metabolism to toxic radicals. PE eradicates GGT-positive foci and mitigates the abnormal focal areas [253]. Hence, "the mRNA levels of Nqo1, Gsta2, Gsta5, Gstm1, Gstm7, Ugt1a1 and Ugt2b17 were elevated in parallel with suppression of lipid and protein oxidation by co administration of PE with DENA". An implication of the transcription factor Nfr2 has been figured out, since it induces many anti-oxidant and phase 2 detoxifying enzymes, such as NQO1, GST-alpha2, GST-alpha5 and GST-m1 [253]. Hence,

Nrf2 could play a pivotal role in chemoprevention of HCC [254]. Besides, there is a blockage and inactivation of Wnt/b-catenin signaling cascade [250]. Precisely, the super-fruit's polyphenols induce up-regulation of GSK- 3b, which entails b-catenin protein degradation and thus decreased hepatic expression of cyclin D1, "a b -catenin-dependent target gene" [250]. Therefore, it would be of much value if larger studies were conducted to further enlighten the feasible chemo-prophylactic and or chemotherapeutic characteristics of Punica Granatum [249 - 251].

Pomegranate and its Role in Bladder Cancer

A team of researchers in China investigated the anti-tumor effects of Punica Granatum's polyphenols on EJ Bladder cancer cells [255]. EJ cell treatment with the aforementioned polyphenols elicited poor cell viability through "caspase-dependent pro-apoptosis appearance" and high levels of p53 protein [255]. The raising / raised levels of p53 protein engendered miR-34a expression, whose direct targets were c-Myc, a "proto-oncogene" with a pivotal role in cell growth, differentiation and apoptosis and CD44, a transmembrane glycoprotein, that regulates cell growth, survival as well as differentiation, by playing a critical role in signal-transduction processes [255, 256]. Conclusively, polyphenols, the valuable micronutrients of pomegranate, possess properties that may freeze bladder cancer cell EJ proliferation through "p53/miR-34a axis" [255]. "Urinary Bladder Urothelial Carcinoma" (UBUC) is the most prevalent histological type of bladder cancer, which accounts for more than 90% of bladder cancers in developed countries [256]. As a rule, curettage is the appropriate treatment [256]. However, in order to prevent recurrence and metastases, extracts of pomegranate have been investigated for their anti- neoplastic effects [256]. Experiments in Taiwan, where the prevalence of UBUC is extremely high, showed that pomegranate fruit extracts delay urothelial bladder cancer cell proliferation and induce cell apoptosis, through multiple molecular mechanisms [256] Pointedly, treatment of UBUC T24 and J82 with pomegranate fruits extracts halts cell proliferation. This is mainly attained through S-phase arrest by inactivation of cyclin A and DDK1 [257]. Thus, having cells propagated in G2 phase, there is a profound delayed entry into mitosis [257]. Apoptosis of UBUC is another possible mechanism by which PFEs may exert their anti-cancerous activity [258]. Characteristically, pro-apoptotic caspases 3-, 8-, 9- and 12- were activated by pomegranate fruits extracts treatment, indicating that mitochondrial pathway, death receptor pathway as well as ER stress, might evoke apoptosis. So, in conclusion, inhibition of urinary bladder urothelial carcinoma cell by "the ethanol extract of pomegranate fruit" can be achieved *via* cell cycle arrest and apoptotic cell death [258].

Pomegranate and its Role in Prostate Cancer

Prostate cancer is the second cause of tumor-related death in men worldwide [259]. According to epidemiological data one in six men will be diagnosed with this urological malignancy and 1 in 35 will die of the disease [259]. The appropriate treatment is prostatectomy, which is sometimes accompanied by radiation therapy. Besides this, local recurrence with distal metastasis has been seen, a reason why numerous studies have been focused on chemo-preventive properties of dietary products, such as pomegranate [259]. The Center for Human Nutrition in the University of California studied the effects of pomegranate's bioactive polyphenols on androgen-independent prostate cancer. DU-145 and PC3 androgen-independent Pca cells were treated with ellagic acid (EA) and urothilin A (UA), the active metabolites that are generated in the intestine through the colonic microflora 30 minutes to 5 hours post consumption of pomegranate juice [259]. These hydrolyzable ellagitannins both inhibit proliferation of androgen-independent prostate cancer cell lines, through cell cycle control and apoptosis [260]. Specifically, UA stops the transition from G2 to M phase, through inhibition of cyckin B1/cdc2 complex, which is critical for the initiation of mitosis [260]. Furthermore, EA seems to decrease the "expression of cyclin B1 and cyclin D1" and in that way ''holds" cancer cells in S phase. EA and UA act synergistically in inhibiting growth of cancerous cells by regulating cell cycle and inducing apoptotic cell death [259, 260].

Moreover, luteolin (L), ellagic acid (EA) and punicic acid (P), all components of pomegranate juice, have a great role in inhibiting metastasis. This is accomplished by stimulating and inhibiting adhesion (E-cadherin, claudin 1) and migration molecules (fascin, hyaluronan-mediated motility receptor), respectively [260]. Stromal cell-derived factor 1a (SDF1a), is also suppressed by these compounds, which explains the inhibitory effect of the aforementioned components on metastasis of the prostate cancer cells to the bone [260]. The experiments were conducted in prostate cancer cells that were or were not sensitive to androgen deprivation and suppression of tumor growth was demonstrated [262]. Upregulation of tumor suppression miRNAs (such as Pten), and down-regulation of many oncogenic miRNAs including miR-29b and miR-181b were also noted [260]. EA, L and P are potent anti-metastatic molecules that may explain the anti-tumor effects of pomegranate [260]. An additional molecular pathway that is implicated in androgen-independent prostate cancer growth is the constitutive activation of NF-kB, which is inhibited by pomegranate extract *in vitro*. Low levels of Bcl-2 and cyclin D1 inhibit S phase progression [260]. As a result, there is reduced proliferation and angiogenesis and increased apoptosis [260]. There is clear evidence that pomegranate extract downregulates cdk2, cdk4, cdk6 and cyclin D1, D2 and E and also upregulates "cdk inhibitors, such as p21 and p27"

[260]. Thus, there is G1-phase arrest and apoptosis and blockage of PC3 cells to S transition [260]. *in vivo* studies have proved that pomegranate extract not only causes regression of the tumor but also increases the overall survival of the treated rodents, through "inhibition of PI3K/Akt/mTOR signaling cascade" [260]. A clinical trial (Phase II) examined the clinical outcomes of pomegranate juice in patients with prostate cancer, who have received primary treatment and showed significant PSA doubling time prolongation [261]. Furthermore, *in vitro* studies with the patients' serum confirmed restriction of cancer cell growth and apoptosis [261]. However, further studies need to be done in order to figure out the exact mechanism by which the PSA doubling time is accomplished [261].

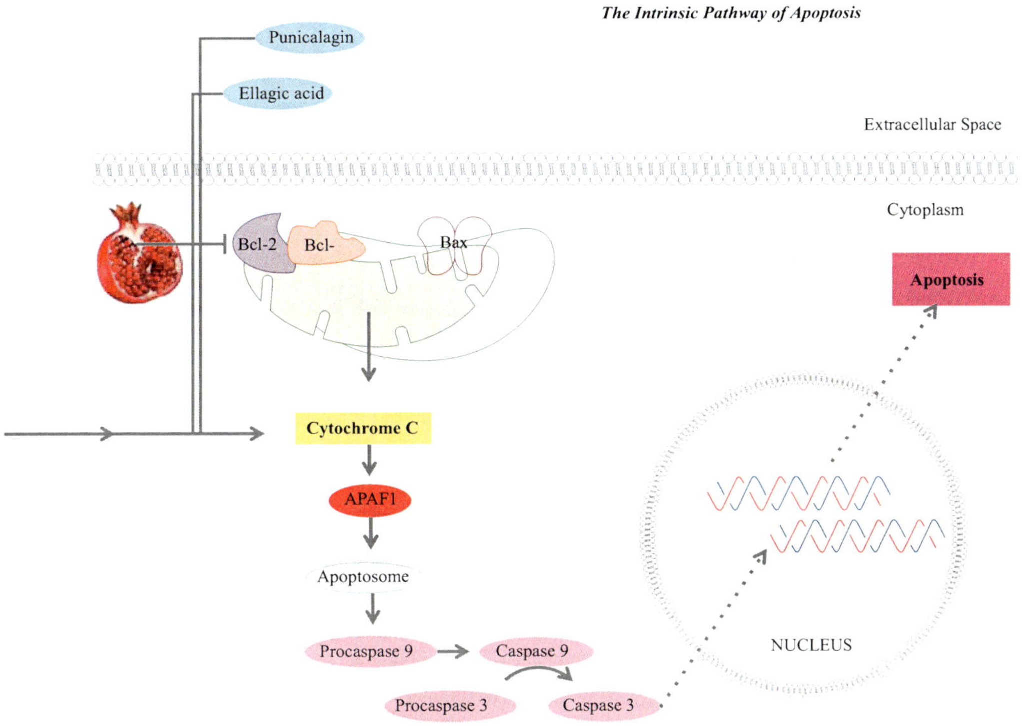

Fig. (4). Pomegranate's basic constituents and their role in apoptosis.

Pomegranate and its Role in Skin and Lung Cancer as Well as in Leukemia Cell Lines

Skin Cancer Besides the sun safety measures, other chemo-preventive agents need to be investigated in order to lower the annual incident of skin cancer, the most prevalent type of cancer in the United States [241].

Polyphenols' biological activities, found in plants, such as pomegranate are promising agents. Researchers, using normal human keratinocytes monolayer cell cultures, examined the effect of Punica Granatum on altering the endogenous antioxidant response to ultraviolet (UV) radiation and the resultant reactive oxygen species (ROS) [241 - 243]. It was noted that pomegranate extract not only induced cell cycle arrest in G1 phase, but it also inhibited the UV-A mediated increase in the phosphorylation of ERK1/2 MAP Kinase, STAT-3, and the AKT/mTOR/p70S6Kinase pro-survival pathway [242]. Additionally, NF-kB nuclear translocation and UV-B induced phosphorylation and activation of MAP Kinases was inhibited, while there was a remarkable decrease in lipid peroxidation and down-regulation of Matrix Metalloproteinases (MMPs) -1, -2, -7, and -9 [243]. The same research team treated skin fibroblasts following UV exposure with PFE and observed increased cell viability through decreased NF-kB activity [253]. DNA repair and inactivation of pro-apoptotic caspase -3 are other molecular mechanisms that may be protective from skin cancer [243]. Additionally, decreased levels of COX-2 and "ornithine decarboxylase (ODC) activity', a "rate- limiting enzyme in the biosynthesis of polyamines", which has a crucial role in cancer development, were observed in mouse models exposed to UV-B radiation, along with a reduction of edema, hyperplasia and leukocytic infiltration in the murine skin [241 - 243].

Lung Cancer

Pomegranate may reduce not only lung inflammation by inhibiting myeloperoxidase activity in neutrophils, as it has been suggested by *in vitro* studies, but may halt lung cancer growth p21 and p27, cyclin-dependent kinase inhibitors, which are highly expressed in human lung carcinoma A549 xenografts after treatment with PE, cause arrest of the cell cycle [241]. As a consequence, popular signaling pathways, such as MAPKinase phosphorylation, PI3K/AKT and NF-kB activity are downregulated, with subsequent low levels of proliferation markers, *e.g.* Ki67 and PCNA [241]. Moreover, experiments in mouse lung tumors models, not only validated the above results but they also showed an inhibitory effect of pomegranate on mTOR signaling cascade and a subsequent decrease in the expression of markers of proliferation and angiogenesis (c-Met, PCNA, Ki-67, iNOS, CD31 and VEGF) [241]. Leukemia Pomegranate's bioactive constituents may have anti-tumor properties on leukemic cell lines. PSP001, a polysaccharide, has been shown to foster apoptosis as well as cell cycle arrest in lymphoid and myeloid leukemia cells. This has been documented through *in vitro* studies with nanoparticles of partially purified pomegranate ellagitannins (PPE), which were proven to elicit early stage apoptosis and necrosis. PPE exert not only the afore-stated mechanisms, but they can also endorse differentiation in human leukemia cells [241].

CONCLUSIONS

Resveratrol, curcumin, honey and pomegranate possess anti-cancer properties, which have been demonstrated both *in vitro* and *in vivo* in many experimental models. Mechanisms such as p53, NF-κB STAT3, ERK1/2, AMPK/mTOR, VEGF, sirtuins, and the process of autophagy have been documented to be implicated in their anti-tumor potential. However, many more studies are needed in order to determine the exact dosage of each of the above-mentioned compounds for the prevention and treatment of various types of cancers. In this view, bioavailability of those substances has to be improved, whilst the maximum dosage that does not cause any serious adverse effects in humans has to be established. In particular, future studies should focus on ameliorating formulas of the above-mentioned compounds, by improving their absorption and kinetics as well as by minimizing adverse side effects, which are estimated to be often in doses that are really helpful for humans in the clinical setting. Regarding the anti-cancer potential of resveratrol, curcumin, honey and pomegranate, there is still a long way to go.

CONFLICT OF INTEREST

The authors (editor) declares no conflict of interest, financial or otherwise.

ACKNOWLEDGMENT

Declared none.

REFERENCES

[1] Baur JA, Sinclair DA. Therapeutic potential of resveratrol: the *in vivo* evidence. Nat Rev Drug Discov 2006; 5(6): 493-506.
[http://dx.doi.org/10.1038/nrd2060] [PMID: 16732220]

[2] Burns J, Yokota T, Ashihara H, Lean ME, Crozier A. Plant foods and herbal sources of resveratrol. J Agric Food Chem 2002; 50(11): 3337-40.
[http://dx.doi.org/10.1021/jf0112973] [PMID: 12010007]

[3] Langcake P, Pryce RJ. The production of resveratrol by Vitis vinifera and other members of the Vitaceae as a response to infection or injury. Physiol Plant Pathol 1976; 9: 77-86.
[http://dx.doi.org/10.1016/0048-4059(76)90077-1]

[4] Athar M, Back JH, Kopelovich L, Bickers DR, Kim AL. Multiple molecular targets of resveratrol: Anti-carcinogenic mechanisms. Arch Biochem Biophys 2009; 486(2): 95-102.
[http://dx.doi.org/10.1016/j.abb.2009.01.018] [PMID: 19514131]

[5] Jang M, Cai L, Udeani GO, *et al.* Cancer chemopreventive activity of resveratrol, a natural product derived from grapes. Science 1997; 275(5297): 218-20.
[http://dx.doi.org/10.1126/science.275.5297.218] [PMID: 8985016]

[6] Lin HY, Shih A, Davis FB, *et al.* Resveratrol induced serine phosphorylation of p53 causes apoptosis in a mutant p53 prostate cancer cell line. J Urol 2002; 168(2): 748-55.
[http://dx.doi.org/10.1016/S0022-5347(05)64739-8] [PMID: 12131363]

[7] Shih A, Zhang SL, Cao JH, *et al.* Inhibitory effect of EGF on resveratrol-induced apoptosis in prostate cancer cells is mediated by protein kinase C-alpha. Mol Cancer Ther 2004; 11: 1355-64. [PMID: 15542774]

[8] Zhang S, Cao HJ, Davis FB, Tang HY, Davis PJ, Lin HY. Oestrogen inhibits resveratrol-induced post-translational modification of p53 and apoptosis in breast cancer cells. Br J Cancer 2004; 91(1): 178-85. [http://dx.doi.org/10.1038/sj.bjc.6601902] [PMID: 15188005]

[9] Tang HY, Shih A, Cao JH, *et al.* Inducible COX-2 facilitates p53-dependent apoptosis in human breast cancer cells. Mol Cancer Ther 2006; 5: 2034-40. [http://dx.doi.org/10.1158/1535-7163.MCT-06-0216] [PMID: 16928824]

[10] Lin HY, Tang HY, Keating T, *et al.* Resveratrol is pro-apoptotic and thyroid hormone is anti-apoptotic in glioma cells: both actions are integrin and ERK mediated. Carcinogenesis 2008; 29(1): 62-9. [http://dx.doi.org/10.1093/carcin/bgm239] [PMID: 17984113]

[11] Lin HY, Sun M, Tang HY, *et al.* Resveratrol causes COX-2- and p53-dependent apoptosis in head and neck squamous cell cancer cells. J Cell Biochem 2008; 104(6): 2131-42. [http://dx.doi.org/10.1002/jcb.21772] [PMID: 18446786]

[12] Lin C, Crawford DR, Lin S, *et al.* Inducible COX-2-dependent apoptosis in human ovarian cancer cells. Carcinogenesis 2011; 32(1): 19-26. [http://dx.doi.org/10.1093/carcin/bgq212] [PMID: 21187340]

[13] Mahyar-Roemer M, Roemer K. p21 Waf1/Cip1 can protect human colon carcinoma cells against p53-dependent and p53-independent apoptosis induced by natural chemopreventive and therapeutic agents. Oncogene 2001; 20(26): 3387-98. [http://dx.doi.org/10.1038/sj.onc.1204440] [PMID: 11423989]

[14] Lavin MF, Gueven N. The complexity of p53 stabilization and activation. Cell Death Differ 2006; 13(6): 941-50. [http://dx.doi.org/10.1038/sj.cdd.4401925] [PMID: 16601750]

[15] Agarwal ML, Taylor WR, Chernov MV, Chernova OB, Stark GR. The p53 network. J Biol Chem 1998; 273(1): 1-4. [http://dx.doi.org/10.1074/jbc.273.1.1] [PMID: 9417035]

[16] Colman MS, Afshari CA, Barrett JC. Regulation of p53 stability and activity in response to genotoxic stress. Mutat Res 2000; 462(2-3): 179-88. [http://dx.doi.org/10.1016/S1383-5742(00)00035-1] [PMID: 10767629]

[17] Barlev NA, Liu L, Chehab NH, *et al.* Acetylation of p53 activates transcription through recruitment of coactivators/histone acetyltransferases. Mol Cell 2001; 8(6): 1243-54. [http://dx.doi.org/10.1016/S1097-2765(01)00414-2] [PMID: 11779500]

[18] Dornan D, Shimizu H, Perkins ND, Hupp TR. DNA-dependent acetylation of p53 by the transcription coactivator p300. J Biol Chem 2003; 278(15): 13431-41. [http://dx.doi.org/10.1074/jbc.M211460200] [PMID: 12499368]

[19] Appella E, Anderson CW. Post-translational modifications and activation of p53 by genotoxic stresses. Eur J Biochem 2001; 268(10): 2764-72. [http://dx.doi.org/10.1046/j.1432-1327.2001.02225.x] [PMID: 11358490]

[20] Nakata R, Takahashi S, Inoue H. Recent advances in the study on resveratrol. Biol Pharm Bull 2012; 35(3): 273-9. [http://dx.doi.org/10.1248/bpb.35.273] [PMID: 22382311]

[21] Murakami M, Nakashima K, Kamei D, *et al.* Cellular prostaglandin E2 production by membrane-bound prostaglandin E synthase-2 *via* both cyclooxygenases-1 and -2. J Biol Chem 2003; 278(39): 37937-47. [http://dx.doi.org/10.1074/jbc.M305108200] [PMID: 12835322]

[22] Parfenova H, Parfenov VN, Shlopov BV, *et al.* Dynamics of nuclear localization sites for COX-2 in vascular endothelial cells. Am J Physiol Cell Physiol 2001; 281(1): C166-78.
[PMID: 11401839]

[23] Ueno N, Murakami M, Tanioka T, *et al.* Coupling between cyclooxygenase, terminal prostanoid synthase, and phospholipase A2. J Biol Chem 2001; 276(37): 34918-27.
[http://dx.doi.org/10.1074/jbc.M100429200] [PMID: 11418589]

[24] Zahner G, Wolf G, Ayoub M, *et al.* Cyclooxygenase-2 overexpression inhibits platelet-derived growth factor-induced mesangial cell proliferation through induction of the tumor suppressor gene p53 and the cyclin-dependent kinase inhibitors p21waf-1/cip-1 and p27kip-1. J Biol Chem 2002; 277(12): 9763-71.
[http://dx.doi.org/10.1074/jbc.M106307200] [PMID: 11756433]

[25] Yuan B, Ohyama K, Bessho T, Toyoda H. Contribution of inducible nitric oxide synthase and cyclooxygenase-2 to apoptosis induction in smooth chorion trophoblast cells of human fetal membrane tissues. Biochem Biophys Res Commun 2006; 341(3): 822-7.
[http://dx.doi.org/10.1016/j.bbrc.2006.01.042] [PMID: 16442500]

[26] Hinz B, Ramer R, Eichele K, Weinzierl U, Brune K. Up-regulation of cox-2 expression is involved in R(+)-methanandamide-induced apoptosis of human neuroglioma cells. Mol Pharmacol 2004; 66: 1643-51.
[http://dx.doi.org/10.1124/mol.104.002618] [PMID: 15361550]

[27] Han JA, Kim JI, Ongusaha PP, *et al.* P53-mediated induction of Cox-2 counteracts p53- or genotoxic stress-induced apoptosis. EMBO J 2002; 21(21): 5635-44.
[http://dx.doi.org/10.1093/emboj/cdf591] [PMID: 12411481]

[28] Moalic-Juge S, Liagre B, Duval R, *et al.* The anti-apoptotic property of NS-398 at high dose can be mediated in part through NF-kappaB activation, hsp70 induction and a decrease in caspase-3 activity in human osteosarcoma cells. Int J Oncol 2002; 20(6): 1255-62.
[PMID: 12012007]

[29] Swamy MV, Herzog CR, Rao CV. Inhibition of COX-2 in colon cancer cell lines by celecoxib increases the nuclear localization of active p53. Cancer Res 2003; 63(17): 5239-42.
[PMID: 14500353]

[30] Lin HY, Tang HY, Davis FB, Davis PJ. Resveratrol and apoptosis. Ann N Y Acad Sci 2011; 1215: 79-88.
[http://dx.doi.org/10.1111/j.1749-6632.2010.05846.x] [PMID: 21261644]

[31] Williams JL, Nath N, Chen J, *et al.* Growth inhibition of human colon cancer cells by nitric oxide Resveratrol-induced apoptosis (NO)-donating aspirin is associated with cyclooxygenase-2 induction and beta-catenin/T-cell factor signaling, nuclear factor-kappaB, and NO synthase 2 inhibition: implications for chemoprevention. Cancer Res 2002; 63: 7613-8.
[PMID: 14633677]

[32] King JG Jr, Khalili K. Inhibition of human brain tumor cell growth by the anti-inflammatory drug, flurbiprofen. Oncogene 2001; 20(47): 6864-70.
[http://dx.doi.org/10.1038/sj.onc.1204907] [PMID: 11687965]

[33] Jang M, Pezzuto JM. Effects of resveratrol on 12-O-tetradecanoylphorbol-13-acetate-induced oxidative events and gene expression in mouse skin. Cancer Lett 1998; 134(1): 81-9.
[http://dx.doi.org/10.1016/S0304-3835(98)00250-X] [PMID: 10381133]

[34] Banerjee S, Bueso-Ramos C, Aggarwal BB. Suppression of 7,12-dimethylbenz(a)anthracene-induced mammary carcinogenesis in rats by resveratrol: role of nuclear factor-kappaB, cyclooxygenase 2, and matrix metalloprotease 9. Cancer Res 2002; 62(17): 4945-54.
[PMID: 12208745]

[35] Li ZG, Hong T, Shimada Y, *et al.* Suppression of N-nitrosomethylbenzylamine (NMBA)-induced

esophageal tumorigenesis in F344 rats by resveratrol. Carcinogenesis 2002; 23(9): 1531-6. [http://dx.doi.org/10.1093/carcin/23.9.1531] [PMID: 12189197]

[36] Lin HY, Sun M, Tang HY, *et al.* Resveratrol causes COX-2- and p53-dependent apoptosis in head and neck squamous cell cancer cells. J Cell Biochem 2008; 104(6): 2131-42. [http://dx.doi.org/10.1002/jcb.21772] [PMID: 18446786]

[37] Kelly GS. A review of the sirtuin system, its clinical implications, and the potential role of dietary activators like resveratrol: part 2. Altern Med Rev 2010; 15(4): 313-28. [PMID: 21194247]

[38] Huang J, Plass C, Gerhauser C. Cancer chemoprevention by targeting the epigenome. Curr Drug Targets 2011; 12(13): 1925-56. [http://dx.doi.org/10.2174/138945011798184155] [PMID: 21158707]

[39] Howells LM, Berry DP, Elliott PJ, *et al.* Phase I randomized, double-blind pilot study of micronized resveratrol (SRT501) in patients with hepatic metastases--safety, pharmacokinetics, and pharmacodynamics. Cancer Prev Res (Phila) 2011; 4(9): 1419-25. [http://dx.doi.org/10.1158/1940-6207.CAPR-11-0148] [PMID: 21680702]

[40] Vang O, Ahmad N, Baile CA, *et al.* What is new for an old molecule? Systematic review and recommendations on the use of resveratrol. PLoS One 2011; 6(6): e19881. [http://dx.doi.org/10.1371/journal.pone.0019881] [PMID: 21698226]

[41] Zhao G, Guo S, Somel M, Khaitovich P. Evolution of human longevity uncoupled from caloric restriction mechanisms. PLoS One 2014; 9(1): e84117. [http://dx.doi.org/10.1371/journal.pone.0084117] [PMID: 24400080]

[42] Villalba JM, Alcaín FJ. Sirtuin activators and inhibitors. Biofactors 2012; 38(5): 349-59. [http://dx.doi.org/10.1002/biof.1032] [PMID: 22730114]

[43] Kanamori M, Vanden Berg SR, Bergers G, Berger MS, Pieper RO. Integrin beta3 overexpression suppresses tumor growth in a human model of gliomagenesis: implications for the role of beta3 overexpression in glioblastoma multiforme. Cancer Res 2004; 64(8): 2751-8. [http://dx.doi.org/10.1158/0008-5472.CAN-03-3354] [PMID: 15087390]

[44] Bergh JJ, Lin HY, Lansing L, *et al.* L-Thyroxine induces mitogen-activated protein kinase activation *via* binding to integrin alpha(V)beta3. Endocrinology 2005; 146: 2864-71. [http://dx.doi.org/10.1210/en.2005-0102] [PMID: 15802494]

[45] Lassus P, Roux P, Zugasti O, Philips A, Fort P, Hibner U. Extinction of rac1 and Cdc42Hs signalling defines a novel p53-dependent apoptotic pathway. Oncogene 2000; 19(20): 2377-85. [http://dx.doi.org/10.1038/sj.onc.1203553] [PMID: 10828879]

[46] Dong Z. Molecular mechanism of the chemopreventive effect of resveratrol. Mutat Res 2003; 523-524: 145-50. [http://dx.doi.org/10.1016/S0027-5107(02)00330-5] [PMID: 12628512]

[47] Hardie DG. AMPK and Raptor: matching cell growth to energy supply. Mol Cell 2008; 30(3): 263-5. [http://dx.doi.org/10.1016/j.molcel.2008.04.012] [PMID: 18471972]

[48] Hahn-Windgassen A, Nogueira V, Chen CC, Skeen JE, Sonenberg N, Hay N. Akt activates the mammalian target of rapamycin by regulating cellular ATP level and AMPK activity. J Biol Chem 2005; 280(37): 32081-9. [http://dx.doi.org/10.1074/jbc.M502876200] [PMID: 16027121]

[49] Wu Y, Liu F. Targeting mTOR: evaluating the therapeutic potential of resveratrol for cancer treatment. Anticancer Agents Med Chem 2013; 13(7): 1032-8. [http://dx.doi.org/10.2174/18715206113139990113] [PMID: 23272906]

[50] Sen R, Baltimore D. Multiple nuclear factors interact with the immunoglobulin enhancer sequences. Cell 1986; 46(5): 705-16. [http://dx.doi.org/10.1016/0092-8674(86)90346-6] [PMID: 3091258]

[51] Puissant A, Grosso S, Jacquel A, *et al.* Imatinib mesylate-resistant human chronic myelogenous leukemia cell lines exhibit high sensitivity to the phytoalexin resveratrol. FASEB J 2008; 22(6): 1894-904.
[http://dx.doi.org/10.1096/fj.07-101394] [PMID: 18245170]

[52] Puissant A, Robert G, Fenouille N, *et al.* Resveratrol promotes autophagic cell death in chronic myelogenous leukemia cells *via* JNK-mediated p62/SQSTM1 expression and AMPK activation. Cancer Res 2010; 70(3): 1042-52.
[http://dx.doi.org/10.1158/0008-5472.CAN-09-3537] [PMID: 20103647]

[53] Vakana E, Platanias LC. AMPK in BCR-ABL expressing leukemias. Regulatory effects and therapeutic implications. Oncotarget 2011; 2(12): 1322-8.
[http://dx.doi.org/10.18632/oncotarget.413] [PMID: 22249159]

[54] Bharti AC, Aggarwal BB. Chemopreventive agents induce suppression of nuclear factor-kappaB leading to chemosensitization. Ann N Y Acad Sci 2002; 973: 392-5.
[http://dx.doi.org/10.1111/j.1749-6632.2002.tb04671.x] [PMID: 12485899]

[55] Bharti AC, Aggarwal BB. Nuclear factor-kappa B and cancer: its role in prevention and therapy. Biochem Pharmacol 2002; 64(5-6): 883-8.
[http://dx.doi.org/10.1016/S0006-2952(02)01154-1] [PMID: 12213582]

[56] Wang CY, Mayo MW, Baldwin AS Jr. TNF- and cancer therapy-induced apoptosis: potentiation by inhibition of NF-kappaB. Science 1996; 274(5288): 784-7.
[http://dx.doi.org/10.1126/science.274.5288.784] [PMID: 8864119]

[57] Ghiringhelli F, Rebe C, Hichami A, Delmas D. Immunomodulation and anti-inflammatory roles of polyphenols as anticancer agents. Anticancer Agents Med Chem 2012; 12(8): 852-73.
[http://dx.doi.org/10.2174/187152012802650048] [PMID: 22292769]

[58] Spatafora C, Tringali C. Natural-derived polyphenols as potential anticancer agents. Anticancer Agents Med Chem 2012; 12(8): 902-18.
[http://dx.doi.org/10.2174/187152012802649996] [PMID: 22292766]

[59] Fulda S. Regulation of cell death and survival by resveratrol: implications for cancer therapy. Anticancer Agents Med Chem 2012; 12(8): 874-9.
[http://dx.doi.org/10.2174/187152012802650129] [PMID: 22292764]

[60] Levy DE, Darnell JE Jr. Stats: transcriptional control and biological impact. Nat Rev Mol Cell Biol 2002; 3(9): 651-62.
[http://dx.doi.org/10.1038/nrm909] [PMID: 12209125]

[61] Bharti AC, Shishodia S, Reuben JM, *et al.* Nuclear factor-kappaB and STAT3 are constitutively active in CD138+ cells derived from multiple myeloma patients, and suppression of these transcription factors leads to apoptosis. Blood 2004; 103(8): 3175-84.
[http://dx.doi.org/10.1182/blood-2003-06-2151] [PMID: 15070700]

[62] Real PJ, Sierra A, De Juan A, Segovia JC, Lopez-Vega JM, Fernandez-Luna JL. Resistance to chemotherapy *via* Stat3-dependent overexpression of Bcl-2 in metastatic breast cancer cells. Oncogene 2002; 21(50): 7611-8.
[http://dx.doi.org/10.1038/sj.onc.1206004] [PMID: 12400004]

[63] Bhardwaj A, Sethi G, Vadhan-Raj S, *et al.* Resveratrol inhibits proliferation, induces apoptosis, and overcomes chemoresistance through down-regulation of STAT3 and nuclear factor-kappaB-regulated antiapoptotic and cell survival gene products in human multiple myeloma cells. Blood 2007; 109(6): 2293-302.
[http://dx.doi.org/10.1182/blood-2006-02-003988] [PMID: 17164350]

[64] Ahn KS, Sethi G, Sung B, Goel A, Ralhan R, Aggarwal BB. Guggulsterone, a farnesoid X receptor antagonist, inhibits constitutive and inducible STAT3 activation through induction of a protein tyrosine phosphatase SHP-1. Cancer Res 2008; 68(11): 4406-15.

[http://dx.doi.org/10.1158/0008-5472.CAN-07-6696] [PMID: 18519703]

[65] Gamero AM, Young HA, Wiltrout RH. Inactivation of Stat3 in tumor cells: releasing a brake on immune responses against cancer? Cancer Cell 2004; 5(2): 111-2.
[http://dx.doi.org/10.1016/S1535-6108(04)00028-5] [PMID: 14998485]

[66] Bowman T, Garcia R, Turkson J, Jove R. STATs in oncogenesis. Oncogene 2000; 19(21): 2474-88.
[http://dx.doi.org/10.1038/sj.onc.1203527] [PMID: 10851046]

[67] Kortylewski M, Kujawski M, Wang T, *et al.* Inhibiting Stat3 signaling in the hematopoietic system elicits multicomponent antitumor immunity. Nat Med 2005; 11(12): 1314-21.
[http://dx.doi.org/10.1038/nm1325] [PMID: 16288283]

[68] Catlett-Falcone R, Landowski TH, Oshiro MM, *et al.* Constitutive activation of Stat3 signaling confers resistance to apoptosis in human U266 myeloma cells. Immunity 1999; 10(1): 105-15.
[http://dx.doi.org/10.1016/S1074-7613(00)80011-4] [PMID: 10023775]

[69] Zushi S, Shinomura Y, Kiyohara T, *et al.* STAT3 mediates the survival signal in oncogenic ras-transfected intestinal epithelial cells. Int J Cancer 1998; 78(3): 326-30.
[http://dx.doi.org/10.1002/(SICI)1097-0215(19981029)78:3<326::AID-IJC12>3.0.CO;2-4] [PMID: 9766567]

[70] Huang M, Page C, Reynolds RK, Lin J. Constitutive activation of stat 3 oncogene product in human ovarian carcinoma cells. Gynecol Oncol 2000; 79(1): 67-73.
[http://dx.doi.org/10.1006/gyno.2000.5931] [PMID: 11006034]

[71] Röder S, Steimle C, Meinhardt G, Pahl HL. STAT3 is constitutively active in some patients with Polycythemia rubra vera. Exp Hematol 2001; 29(6): 694-702.
[http://dx.doi.org/10.1016/S0301-472X(01)00637-3] [PMID: 11378264]

[72] Mora LB, Buettner R, Seigne J, *et al.* Constitutive activation of Stat3 in human prostate tumors and cell lines: direct inhibition of Stat3 signaling induces apoptosis of prostate cancer cells. Cancer Res 2002; 62(22): 6659-66.
[PMID: 12438264]

[73] Greten FR, Weber CK, Greten TF, *et al.* Stat3 and NF-kappaB activation prevents apoptosis in pancreatic carcinogenesis. Gastroenterology 2002; 123(6): 2052-63.
[http://dx.doi.org/10.1053/gast.2002.37075] [PMID: 12454861]

[74] Kirito K, Nagashima T, Ozawa K, Komatsu N. Constitutive activation of Stat1 and Stat3 in primary erythroleukemia cells. Int J Hematol 2002; 75(1): 51-4.
[http://dx.doi.org/10.1007/BF02981979] [PMID: 11843291]

[75] Schaefer LK, Ren Z, Fuller GN, Schaefer TS. Constitutive activation of Stat3alpha in brain tumors: localization to tumor endothelial cells and activation by the endothelial tyrosine kinase receptor (VEGFR-2). Oncogene 2002; 21(13): 2058-65.
[http://dx.doi.org/10.1038/sj.onc.1205263] [PMID: 11960378]

[76] Hsiao JR, Jin YT, Tsai ST, Shiau AL, Wu CL, Su WC. Constitutive activation of STAT3 and STAT5 is present in the majority of nasopharyngeal carcinoma and correlates with better prognosis. Br J Cancer 2003; 89(2): 344-9.
[http://dx.doi.org/10.1038/sj.bjc.6601003] [PMID: 12865928]

[77] Song L, Turkson J, Karras JG, Jove R, Haura EB. Activation of Stat3 by receptor tyrosine kinases and cytokines regulates survival in human non-small cell carcinoma cells. Oncogene 2003; 22(27): 4150-65.
[http://dx.doi.org/10.1038/sj.onc.1206479] [PMID: 12833138]

[78] To KF, Chan MW, Leung WK, *et al.* Constitutional activation of IL-6-mediated JAK/STAT pathway through hypermethylation of SOCS-1 in human gastric cancer cell line. Br J Cancer 2004; 91(7): 1335-41.
[http://dx.doi.org/10.1038/sj.bjc.6602133] [PMID: 15354212]

[79] Lin Q, Lai R, Chirieac LR, *et al.* Constitutive activation of JAK3/STAT3 in colon carcinoma tumors and cell lines: inhibition of JAK3/STAT3 signaling induces apoptosis and cell cycle arrest of colon carcinoma cells. Am J Pathol 2005; 167(4): 969-80.
[http://dx.doi.org/10.1016/S0002-9440(10)61187-X] [PMID: 16192633]

[80] Gupta SC, Kannappan R, Reuter S, Kim JH, Aggarwal BB. Chemosensitization of tumors by resveratrol. Ann N Y Acad Sci 2011; 1215: 150-60.
[http://dx.doi.org/10.1111/j.1749-6632.2010.05852.x] [PMID: 21261654]

[81] Kroemer G, Levine B. Autophagic cell death: the story of a misnomer. Nat Rev Mol Cell Biol 2008; 9(12): 1004-10.
[http://dx.doi.org/10.1038/nrm2529] [PMID: 18971948]

[82] Eskelinen EL. Maturation of autophagic vacuoles in Mammalian cells. Autophagy 2005; 1(1): 1-10.
[http://dx.doi.org/10.4161/auto.1.1.1270] [PMID: 16874026]

[83] Chen RJ, Ho CT, Wang YJ. Pterostilbene induces autophagy and apoptosis in sensitive and chemoresistant human bladder cancer cells. Mol Nutr Food Res 2010; 54(12): 1819-32.
[http://dx.doi.org/10.1002/mnfr.201000067] [PMID: 20603834]

[84] Wang Y, Ding L, Wang X, *et al.* Pterostilbene simultaneously induces apoptosis, cell cycle arrest and cyto-protective autophagy in breast cancer cells. Am J Transl Res 2012; 4(1): 44-51.
[PMID: 22347521]

[85] Mena S, Rodríguez ML, Ponsoda X, Estrela JM, Jäättela M, Ortega AL. Pterostilbene-induced tumor cytotoxicity: a lysosomal membrane permeabilization-dependent mechanism. PLoS One 2012; 7(9): e44524.
[http://dx.doi.org/10.1371/journal.pone.0044524] [PMID: 22957077]

[86] Wang Q, Mora-Jensen H, Weniger MA, *et al.* ERAD inhibitors integrate ER stress with an epigenetic mechanism to activate BH3-only protein NOXA in cancer cells. Proc Natl Acad Sci USA 2009; 106(7): 2200-5.
[http://dx.doi.org/10.1073/pnas.0807611106] [PMID: 19164757]

[87] Kapetanovic IM, Muzzio M, Huang Z, Thompson TN, McCormick DL. Pharmacokinetics, oral bioavailability, and metabolic profile of resveratrol and its dimethylether analog, pterostilbene, in rats. Cancer Chemother Pharmacol 2011; 68(3): 593-601.
[http://dx.doi.org/10.1007/s00280-010-1525-4] [PMID: 21116625]

[88] Siedlecka-Kroplewska K, Jozwik A, Boguslawski W, *et al.* Pterostilbene induces accumulation of autophagic vacuoles followed by cell death in HL60 human leukemia cells. J Physiol Pharmacol 2013; 64(5): 545-56.
[PMID: 24304568]

[89] Kim DH, Hossain MA, Kim MY, *et al.* A novel resveratrol analogue, HS-1793, inhibits hypoxia-induced HIF-1α and VEGF expression, and migration in human prostate cancer cells. Int J Oncol 2013; 43(6): 1915-24.
[http://dx.doi.org/10.3892/ijo.2013.2116] [PMID: 24100602]

[90] Li Y, Wicha MS, Schwartz SJ, Sun D. Implications of cancer stem cell theory for cancer chemoprevention by natural dietary compounds. J Nutr Biochem 2011; 22(9): 799-806.
[http://dx.doi.org/10.1016/j.jnutbio.2010.11.001] [PMID: 21295962]

[91] Liu MM, Huang Y, Wang J. Developing phytoestrogens for breast cancer prevention. Anticancer Agents Med Chem 2012; 12(10): 1306-13.
[http://dx.doi.org/10.2174/187152012803833062] [PMID: 22583409]

[92] Kim JA, Kim DH, Hossain MA, *et al.* HS-1793, a resveratrol analogue, induces cell cycle arrest and apoptotic cell death in human breast cancer cells. Int J Oncol 2014; 44(2): 473-80.
[http://dx.doi.org/10.3892/ijo.2013.2207] [PMID: 24316714]

[93] Whitlock NC, Bahn JH, Lee SH, Eling TE, Baek SJ. Resveratrol-induced apoptosis is mediated by

early growth response-1, Krüppel-like factor 4, and activating transcription factor 3. Cancer Prev Res (Phila) 2011; 4(1): 116-27.
[http://dx.doi.org/10.1158/1940-6207.CAPR-10-0218] [PMID: 21205742]

[94] Lu D, Chen J, Hai T. The regulation of ATF3 gene expression by mitogen-activated protein kinases. Biochem J 2007; 401(2): 559-67.
[http://dx.doi.org/10.1042/BJ20061081] [PMID: 17014422]

[95] Yan C, Boyd DD. ATF3 regulates the stability of p53: a link to cancer. Cell Cycle 2006; 5(9): 926-9.
[http://dx.doi.org/10.4161/cc.5.9.2714] [PMID: 16628010]

[96] Fan F, Jin S, Amundson SA, *et al.* ATF3 induction following DNA damage is regulated by distinct signaling pathways and over-expression of ATF3 protein suppresses cells growth. Oncogene 2002; 21(49): 7488-96.
[http://dx.doi.org/10.1038/sj.onc.1205896] [PMID: 12386811]

[97] Huang X, Li X, Guo B. KLF6 induces apoptosis in prostate cancer cells through up-regulation of ATF3. J Biol Chem 2008; 283(44): 29795-801.
[http://dx.doi.org/10.1074/jbc.M802515200] [PMID: 18755691]

[98] Lu D, Wolfgang CD, Hai T. Activating transcription factor 3, a stress-inducible gene, suppresses Ras-stimulated tumorigenesis. J Biol Chem 2006; 281(15): 10473-81.
[http://dx.doi.org/10.1074/jbc.M509278200] [PMID: 16469745]

[99] Yamaguchi K, Lee SH, Kim JS, Wimalasena J, Kitajima S, Baek SJ. Activating transcription factor 3 and early growth response 1 are the novel targets of LY294002 in a phosphatidylinositol 3-kinas--independent pathway. Cancer Res 2006; 66(4): 2376-84.
[http://dx.doi.org/10.1158/0008-5472.CAN-05-1987] [PMID: 16489044]

[100] Turchi L, Aberdam E, Mazure N, *et al.* Hif-2alpha mediates UV-induced apoptosis through a novel ATF3-dependent death pathway. Cell Death Differ 2008; 15(9): 1472-80.
[http://dx.doi.org/10.1038/cdd.2008.74] [PMID: 18511933]

[101] Bottone FG Jr, Moon Y, Kim JS, Alston-Mills B, Ishibashi M, Eling TE. The anti-invasive activity of cyclooxygenase inhibitors is regulated by the transcription factor ATF3 (activating transcription factor 3). Mol Cancer Ther 2005; 4(5): 693-703.
[http://dx.doi.org/10.1158/1535-7163.MCT-04-0337] [PMID: 15897233]

[102] Whitlock NC, Baek SJ. The anticancer effects of resveratrol: modulation of transcription factors. Nutr Cancer 2012; 64(4): 493-502.
[http://dx.doi.org/10.1080/01635581.2012.667862] [PMID: 22482424]

[103] Schlessinger K, Levy DE. Malignant transformation but not normal cell growth depends on signal transducer and activator of transcription 3. Cancer Res 2005; 65(13): 5828-34.
[http://dx.doi.org/10.1158/0008-5472.CAN-05-0317] [PMID: 15994959]

[104] Masuda M, Suzui M, Yasumatu R, *et al.* Constitutive activation of signal transducers and activators of transcription 3 correlates with cyclin D1 overexpression and may provide a novel prognostic marker in head and neck squamous cell carcinoma. Cancer Res 2002; 62(12): 3351-5.
[PMID: 12067972]

[105] Shih A, Lin HY, Davis FB, Davis PJ. Thyroid hormone promotes serine phosphorylation of p53 by mitogen-activated protein kinase. Biochemistry 2001; 40(9): 2870-8.
[http://dx.doi.org/10.1021/bi001978b] [PMID: 11258898]

[106] Cheng HF, Harris RC. Cyclooxygenase-2 expression in cultured cortical thick ascending limb of Henle increases in response to decreased extracellular ionic content by both transcriptional and post-transcriptional mechanisms. Role of p38-mediated pathways. J Biol Chem 2002; 277(47): 45638-43.
[http://dx.doi.org/10.1074/jbc.M206040200] [PMID: 12237297]

[107] Markowitz SD. Aspirin and colon cancer--targeting prevention? N Engl J Med 2007; 356(21): 2195-8.
[http://dx.doi.org/10.1056/NEJMe078044] [PMID: 17522404]

[108] Bai D, Ueno L, Vogt PK. Akt-mediated regulation of NFkappaB and the essentialness of NFkappaB for the oncogenicity of PI3K and Akt. Int J Cancer 2009; 125(12): 2863-70. [http://dx.doi.org/10.1002/ijc.24748] [PMID: 19609947]

[109] Perrone G, Zagami M, Altomare V, Battista C, Morini S, Rabitti C. COX-2 localization within plasma membrane caveolae-like structures in human lobular intraepithelial neoplasia of the breast. Virchows Arch 2007; 451(6): 1039-45. [http://dx.doi.org/10.1007/s00428-007-0506-4] [PMID: 17851687]

[110] Roychowdhury LS, Engelmann M, Wolf G, Horn TF. "Oxyresveratrol and resveratrol are potent antioxidants and free radical scavengers: effect on nitrosative and oxidative stress derived from microglial cells," Nitric Oxide. Biol Chem 2003; 9(2): 64-76.

[111] Guha P, Dey A, Dhyani MV, *et al.* Calpain and caspase orchestrated death signal to accomplish apoptosis induced by resveratrol and its novel analog hydroxstilbene-1 in cancer cells. J Pharmacol Experiment Therapeutics 2010; 334(2): 381-94. [http://dx.doi.org/10.1124/jpet.110.167668] [PMID: 20484155]

[112] Leonard SS, Xia C, Jiang BH, *et al.* Resveratrol scavenges reactive oxygen species and effects radical-induced cellular responses. Biochem Biophys Res Commun 2003; 309(4): 1017-26. [http://dx.doi.org/10.1016/j.bbrc.2003.08.105] [PMID: 13679076]

[113] Hung LM, Su MJ, Chu WK, Chiao CW, Chan WF, Chen JK. The protective effect of resveratrols on ischaemia-reperfusion injuries of rat hearts is correlated with antioxidant efficacy. Br J Pharmacol 2002; 135(7): 1627-33. [http://dx.doi.org/10.1038/sj.bjp.0704637] [PMID: 11934802]

[114] Soeur J, Eilstein J, Léreaux G, Jones C, Marrot L. Skin resistance to oxidative stress induced by resveratrol: from Nrf2 activation to GSH biosynthesis. Free Radic Biol Med 2015; 78: 213-23. [http://dx.doi.org/10.1016/j.freeradbiomed.2014.10.510] [PMID: 25451641]

[115] Ungvari Z, Bagi Z, Feher A, *et al.* Resveratrol confers endothelial protection *via* activation of the antioxidant transcription factor Nrf2. Am J Physiol Heart Circ Physiol 2010; 299(1): H18-24. [http://dx.doi.org/10.1152/ajpheart.00260.2010] [PMID: 20418481]

[116] Ungvari Z, Labinskyy N, Mukhopadhyay P, *et al.* Resveratrol attenuates mitochondrial oxidative stress in coronary arterial endothelial cells. Am J Physiol Heart Circ Physiol 2009; 297(5): H1876-81. [http://dx.doi.org/10.1152/ajpheart.00375.2009] [PMID: 19749157]

[117] Tatlidede E, Sehirli O, Velioğlu-Oğünc A, *et al.* Resveratrol treatment protects against doxorubicin-induced cardiotoxicity by alleviating oxidative damage. Free Radic Res 2009; 43(3): 195-205. [http://dx.doi.org/10.1080/10715760802673008] [PMID: 19169920]

[118] Arafa MH, Mohammad NS, Atteia HH, Abd-Elaziz HR. Protective effect of resveratrol against doxorubicin-induced cardiac toxicity and fibrosis in male experimental rats. J Physiol Biochem 2014; 70(3): 701-11. [http://dx.doi.org/10.1007/s13105-014-0339-y] [PMID: 24939721]

[119] Goel A, Kunnumakkara AB, Aggarwal BB. Curcumin as "Curecumin": from kitchen to clinic. Biochem Pharmacol 2008; 75(4): 787-809. [http://dx.doi.org/10.1016/j.bcp.2007.08.016] [PMID: 17900536]

[120] Kim YS, Ahn Y, Hong MH, *et al.* Curcumin attenuates inflammatory responses of TNF-α-stimulated human endothelial cells. J Cardiovasc Pharmacol 2007; 50(1): 41-9. [http://dx.doi.org/10.1097/FJC.0b013e31805559b9] [PMID: 17666914]

[121] Da-Lozzo EJ, Moledo RC, Faraco CD, Ortolani-Machado CF, Bresolin TM, Silveira JL. Curcumin/xanthan-galactomannan hydrogels: rheological analysis and biocompatibility. Carbohydr Polym 2013; 93(1): 279-84. [http://dx.doi.org/10.1016/j.carbpol.2012.02.036] [PMID: 23465931]

[122] Hsu CH, Cheng AL. Clinical studies with curcumin. Adv Exp Med Biol 2007; 595: 471-80.

[http://dx.doi.org/10.1007/978-0-387-46401-5_21] [PMID: 17569225]

[123] Park W, Amin AR, Chen ZG, Shin DM. New perspectives of curcumin in cancer prevention. Cancer Prev Res (Phila) 2013; 6(5): 387-400.
[http://dx.doi.org/10.1158/1940-6207.CAPR-12-0410] [PMID: 23466484]

[124] Strimpakos AS, Sharma RA. Curcumin: preventive and therapeutic properties in laboratory studies and clinical trials. Antioxid Redox Signal 2008; 10(3): 511-45.
[http://dx.doi.org/10.1089/ars.2007.1769] [PMID: 18370854]

[125] Kapakos G, Youreva V, Srivastava AK. Cardiovascular protection by curcumin: molecular aspects. Indian J Biochem Biophys 2012; 49(5): 306-15.
[PMID: 23259317]

[126] Agrawal DK, Mishra PK. Curcumin and its analogues: potential anticancer agents. Med Res Rev 2010; 30(5): 818-60.
[PMID: 20027668]

[127] Du ZY, Jiang YF, Tang ZK, *et al.* Antioxidation and tyrosinase inhibition of polyphenolic curcumin analogs. Biosci Biotechnol Biochem 2011; 75(12): 2351-8.
[http://dx.doi.org/10.1271/bbb.110547] [PMID: 22146732]

[128] Masuda T, Maekawa T, Hidaka K, Bando H, Takeda Y, Yamaguchi H. Chemical studies on antioxidant mechanism of curcumin: analysis of oxidative coupling products from curcumin and linoleate. J Agric Food Chem 2001; 49(5): 2539-47.
[http://dx.doi.org/10.1021/jf001442x] [PMID: 11368633]

[129] Menon VP, Sudheer AR. Antioxidant and anti-inflammatory properties of curcumin. Adv Exp Med Biol 2007; 595: 105-25.
[http://dx.doi.org/10.1007/978-0-387-46401-5_3] [PMID: 17569207]

[130] Chainani-Wu N. Safety and anti-inflammatory activity of curcumin: a component of tumeric (Curcuma longa). J Altern Complement Med 2003; 9(1): 161-8.
[http://dx.doi.org/10.1089/107555303321223035] [PMID: 12676044]

[131] Jurenka JS. Anti-inflammatory properties of curcumin, a major constituent of Curcuma longa: a review of preclinical and clinical research. Altern Med Rev 2009; 14(2): 141-53.
[PMID: 19594223]

[132] Kuttan R, Bhanumathy P, Nirmala K, George MC. Potential anticancer activity of turmeric (Curcuma longa). Cancer Lett 1985; 29(2): 197-202.
[http://dx.doi.org/10.1016/0304-3835(85)90159-4] [PMID: 4075289]

[133] Huang MT, Smart RC, Wong CQ, Conney AH. Inhibitory effect of curcumin, chlorogenic acid, caffeic acid, and ferulic acid on tumor promotion in mouse skin by 12-O-tetradecanoylphorbol-13-acetate. Cancer Res 1988; 48(21): 5941-6.
[PMID: 3139287]

[134] Huang MT, Wang ZY, Georgiadis CA, Laskin JD, Conney AH. Inhibitory effects of curcumin on tumor initiation by benzo[a]pyrene and 7,12-dimethylbenz[a]anthracene. Carcinogenesis 1992; 13(11): 2183-6.
[http://dx.doi.org/10.1093/carcin/13.11.2183] [PMID: 1423891]

[135] Huang MT, Lou YR, Ma W, Newmark HL, Reuhl KR, Conney AH. Inhibitory effects of dietary curcumin on forestomach, duodenal, and colon carcinogenesis in mice. Cancer Res 1994; 54(22): 5841-7.
[PMID: 7954412]

[136] Rao CV, Rivenson A, Simi B, Reddy BS. Chemoprevention of colon carcinogenesis by dietary curcumin, a naturally occurring plant phenolic compound. Cancer Res 1995; 55(2): 259-66.
[PMID: 7812955]

[137] Gupta SC, Sundaram C, Reuter S, Aggarwal BB. Inhibiting NF-κB activation by small molecules as a

therapeutic strategy. Biochim Biophys Acta 2010; 1799(10-12): 775-87.
[http://dx.doi.org/10.1016/j.bbagrm.2010.05.004] [PMID: 20493977]

[138] Bharti AC, Aggarwal BB. Nuclear factor-kappa B and cancer: its role in prevention and therapy. Biochem Pharmacol 2002; 64(5-6): 883-8.
[http://dx.doi.org/10.1016/S0006-2952(02)01154-1] [PMID: 12213582]

[139] Kasdagly M, Radhakrishnan S, Reddivari L, Veeramachaneni DN, Vanamala J. Colon carcinogenesis: influence of Western diet-induced obesity and targeting stem cells using dietary bioactive compounds. Nutrition 2014; 30(11-12): 1242-56.
[http://dx.doi.org/10.1016/j.nut.2014.02.016] [PMID: 25280404]

[140] Anthwal A, Thakur BK, Rawat MS, Rawat DS, Tyagi AK, Aggarwal BB. Synthesis, characterization and *in vitro* anticancer activity of C-5 curcumin analogues with potential to inhibit TNF-a-induced NF-kB activation. Biomed Res Int 2014; 2014: 524161.

[141] Chiablaem K, Lirdprapamongkol K, Keeratichamroen S, Surarit R, Svasti J. Curcumin suppresses vasculogenic mimicry capacity of hepatocellular carcinoma cells through STAT3 and PI3K/AKT inhibition. Anticancer Res 2014; 34(4): 1857-64.
[PMID: 24692720]

[142] Abdulghani J, Gu L, Dagvadorj A, *et al.* Stat3 promotes metastatic progression of prostate cancer. Am J Pathol 2008; 172(6): 1717-28.
[http://dx.doi.org/10.2353/ajpath.2008.071054] [PMID: 18483213]

[143] Chen HW, Lee JY, Huang JY, *et al.* Curcumin inhibits lung cancer cell invasion and metastasis through the tumor suppressor HLJ1. Cancer Res 2008; 68(18): 7428-38.
[http://dx.doi.org/10.1158/0008-5472.CAN-07-6734] [PMID: 18794131]

[144] Lee JY, Lee YM, Chang GC, *et al.* Curcumin induces EGFR degradation in lung adenocarcinoma and modulates p38 activation in intestine: the versatile adjuvant for gefitinib therapy. PLoS ONE 2011; 6, 8: e23756.

[145] Shishodia S, Singh T, Chaturvedi MM. Modulation of transcription factors by curcumin. Adv Exp Med Biol 2007; 595: 127-48.
[http://dx.doi.org/10.1007/978-0-387-46401-5_4] [PMID: 17569208]

[146] Davie JR, He S, Li L, *et al.* Nuclear organization and chromatin dynamics--Sp1, Sp3 and histone deacetylases. Adv Enzyme Regul 2008; 48(1): 189-208.
[http://dx.doi.org/10.1016/j.advenzreg.2007.11.016] [PMID: 18187045]

[147] Shi Q, Le X, Abbruzzese JL, *et al.* Constitutive Sp1 activity is essential for differential constitutive expression of vascular endothelial growth factor in human pancreatic adenocarcinoma. Cancer Res 2001; 61(10): 4143-54.
[PMID: 11358838]

[148] Zannetti A, Del Vecchio S, Carriero MV, *et al.* Coordinate up-regulation of Sp1 DNA-binding activity and urokinase receptor expression in breast carcinoma. Cancer Res 2000; 60(6): 1546-51.
[PMID: 10749121]

[149] Kitadai Y, Yasui W, Yokozaki H, *et al.* The level of a transcription factor Sp1 is correlated with the expression of EGF receptor in human gastric carcinomas. Biochem Biophys Res Commun 1992; 189(3): 1342-8.
[http://dx.doi.org/10.1016/0006-291X(92)90221-6] [PMID: 1282800]

[150] Leupold JH, Yang HS, Colburn NH, Asangani I, Post S, Allgayer H. Tumor suppressor Pdcd4 inhibits invasion/intravasation and regulates urokinase receptor (u-PAR) gene expression *via* Sp-transcription factors. Oncogene 2007; 26(31): 4550-62.
[http://dx.doi.org/10.1038/sj.onc.1210234] [PMID: 17297470]

[151] Cho SG, Yi Z, Pang X, *et al.* Kisspeptin-10, a KISS1-derived decapeptide, inhibits tumor angiogenesis by suppressing Sp1-mediated VEGF expression and FAK/Rho GTPase activation. Cancer Res 2009;

69(17): 7062-70.
[http://dx.doi.org/10.1158/0008-5472.CAN-09-0476] [PMID: 19671799]

[152] Chadalapaka G, Jutooru I, Chintharlapalli S, *et al.* Curcumin decreases specificity protein expression in bladder cancer cells. Cancer Res 2008; 68(13): 5345-54.
[http://dx.doi.org/10.1158/0008-5472.CAN-07-6805] [PMID: 18593936]

[153] Aigner S, Ramos CL, Hafezi-Moghadam A, *et al.* CD24 mediates rolling of breast carcinoma cells on P-selectin. FASEB J 1998; 12(12): 1241-51.
[PMID: 9737727]

[154] Lou Z, O'Reilly S, Liang H, Maher VM, Sleight SD, McCormick JJ. Down-regulation of overexpressed sp1 protein in human fibrosarcoma cell lines inhibits tumor formation. Cancer Res 2005; 65(3): 1007-17.
[PMID: 15705902]

[155] Chen CC, Sureshbabul M, Chen HW, *et al.* Curcumin Suppresses Metastasis *via* Sp-1, FAK Inhibition, and E-Cadherin Upregulation in Colorectal Cancer. Evidence-Based Complem Altern Med 2013; e541685.

[156] Zhou DY, Zhang K, Conney AH, *et al.* Synthesis and evaluation of curcumin-related compounds containing benzyl piperidone for their effects on human cancer cells. Chem Pharm Bull (Tokyo) 2013; 61(11): 1149-55.
[http://dx.doi.org/10.1248/cpb.c13-00507] [PMID: 23985704]

[157] Zhang K, Kaufman RJ. From endoplasmic-reticulum stress to the inflammatory response. Nature 2008; 454(7203): 455-62.
[http://dx.doi.org/10.1038/nature07203] [PMID: 18650916]

[158] Marciniak SJ, Ron D. Endoplasmic reticulum stress signaling in disease. Physiol Rev 2006; 86(4): 1133-49.
[http://dx.doi.org/10.1152/physrev.00015.2006] [PMID: 17015486]

[159] Carew JS, Nawrocki ST, Cleveland JL. Modulating autophagy for therapeutic benefit. Autophagy 2007; 3(5): 464-7.
[http://dx.doi.org/10.4161/auto.4311] [PMID: 17495516]

[160] Dalby KN, Tekedereli I, Lopez-Berestein G, Ozpolat B. Targeting the prodeath and prosurvival functions of autophagy as novel therapeutic strategies in cancer. Autophagy 2010; 6(3): 322-9.
[http://dx.doi.org/10.4161/auto.6.3.11625] [PMID: 20224296]

[161] Lin SR, Fu YS, Tsai MJ, Cheng H, Weng CF. Natural Compounds from Herbs that can Potentially Execute as Autophagy Inducers for Cancer Therapy. Int J Mol Sci 2017; 18(7): e1412.
[http://dx.doi.org/10.3390/ijms18071412] [PMID: 28671583]

[162] de Oliveira MR, Jardim FR, Setzer WN, Nabavi SM, Nabavi SF. Curcumin, mitochondrial biogenesis, and mitophagy: Exploring recent data and indicating future needs. Biotechnol Adv 2016; 34(5): 813-26.
[http://dx.doi.org/10.1016/j.biotechadv.2016.04.004] [PMID: 27143655]

[163] Qu W, Xiao J, Zhang H, *et al.* B19, a novel monocarbonyl analogue of curcumin, induces human ovarian cancer cell apoptosis *via* activation of endoplasmic reticulum stress and the autophagy signaling pathway. Int J Biol Sci 2013; 9(8): 766-77.
[http://dx.doi.org/10.7150/ijbs.5711] [PMID: 23983610]

[164] Grabacka MM, Gawin M, Pierzchalska M. Phytochemical modulators of mitochondria: the search for chemopreventive agents and supportive therapeutics. Pharmaceuticals (Basel) 2014; 7(9): 913-42.
[http://dx.doi.org/10.3390/ph7090913] [PMID: 25192192]

[165] Zhou T, Ye L, Bai Y, *et al.* Autophagy and apoptosis in hepatocellular carcinoma induced by EF25-(GSH)2: a novel curcumin analog. PLoS One 2014; 9(9): e107876.
[http://dx.doi.org/10.1371/journal.pone.0107876] [PMID: 25268357]

[166] Aoki H, Takada Y, Kondo S, Sawaya R, Aggarwal BB, Kondo Y. Evidence that curcumin suppresses the growth of malignant gliomas *in vitro* and *in vivo* through induction of autophagy: role of Akt and extracellular signal-regulated kinase signaling pathways. Mol Pharmacol 2007; 72(1): 29-39. [http://dx.doi.org/10.1124/mol.106.033167] [PMID: 17395690]

[167] Dhillon N, Aggarwal BB, Newman RA, *et al.* Phase II trial of curcumin in patients with advanced pancreatic cancer. Clin Cancer Res 2008; 14(14): 4491-9. [http://dx.doi.org/10.1158/1078-0432.CCR-08-0024] [PMID: 18628464]

[168] Zlotogorski A, Dayan A, Dayan D, Chaushu G, Salo T, Vered M. Nutraceuticals as new treatment approaches for oral cancer--I: Curcumin. Oral Oncol 2013; 49(3): 187-91. [http://dx.doi.org/10.1016/j.oraloncology.2012.09.015] [PMID: 23116961]

[169] Adams BK, Ferstl EM, Davis MC, *et al.* Synthesis and biological evaluation of novel curcumin analogs as anti-cancer and anti-angiogenesis agents. Bioorg Med Chem 2004; 12(14): 3871-83. [http://dx.doi.org/10.1016/j.bmc.2004.05.006] [PMID: 15210154]

[170] Sekiya M, Hisasaka R, Iwamoto-Kihara A, Futai M, Nakanishi-Matsui M. A unique mechanism of curcumin inhibition on F1 ATPase. Biochem Biophys Res Commun 2014; 452(4): 940-4. [http://dx.doi.org/10.1016/j.bbrc.2014.09.027] [PMID: 25230139]

[171] Zheng J, Ramirez VD. Inhibition of mitochondrial proton F0F1-ATPase/ATP synthase by polyphenolic phytochemicals. Br J Pharmacol 2000; 130(5): 1115-23. [http://dx.doi.org/10.1038/sj.bjp.0703397] [PMID: 10882397]

[172] Lee KH, Ab Aziz FH, Syahida A, *et al.* Synthesis and biological evaluation of curcumin-like diarylpentanoid analogues for anti-inflammatory, antioxidant and anti-tyrosinase activities. Eur J Med Chem 2009; 44(8): 3195-200. [http://dx.doi.org/10.1016/j.ejmech.2009.03.020] [PMID: 19359068]

[173] Zhao C, Yang J, Wang Y, *et al.* Synthesis of mono-carbonyl analogues of curcumin and their effects on inhibition of cytokine release in LPS-stimulated RAW 264.7 macrophages. Bioorg Med Chem 2010; 18(7): 2388-93. [http://dx.doi.org/10.1016/j.bmc.2010.03.001] [PMID: 20338767]

[174] Bala V, Rao S, Boyd BJ, Prestidge CA. Prodrug and nanomedicine approaches for the delivery of the camptothecin analogue SN38. J Control Release 2013; 172(1): 48-61. [http://dx.doi.org/10.1016/j.jconrel.2013.07.022] [PMID: 23928356]

[175] Das M, Singh RP, Datir SR, Jain S. Intranuclear drug delivery and effective *in vivo* cancer therapy *via* estradiol-PEG-appended multiwalled carbon nanotubes. Mol Pharm 2013; 10(9): 3404-16. Epub ahead of print [http://dx.doi.org/10.1021/mp4002409] [PMID: 23905512]

[176] Teow HM, Zhou Z, Najlah M, Yusof SR, Abbott NJ, D'Emanuele A. Delivery of paclitaxel across cellular barriers using a dendrimer-based nanocarrier. Int J Pharm 2013; 441(1-2): 701-11. [http://dx.doi.org/10.1016/j.ijpharm.2012.10.024] [PMID: 23089576]

[177] Qi X, Li N, Gu H, *et al.* Amphiphilic oligomer-based micelles as cisplatin nanocarriers for cancer therapy. Nanoscale 2013; 5(19): 8925-9. [http://dx.doi.org/10.1039/c3nr03262k] [PMID: 23933624]

[178] Arzuman L, Beale P, Chan C, Yu JQ, Huq F. Synergism from combinations of tris(benzimidazole) monochloroplatinum(II) chloride with capsaicin, quercetin, curcumin and cisplatin in human ovarian cancer cell lines. Anticancer Res 2014; 34(10): 5453-64. [PMID: 25275041]

[179] Huq F, Yu JQ, Beale P, *et al.* Combinations of platinums and selected phytochemicals as a means of overcoming resistance in ovarian cancer. Anticancer Res 2014; 34(1): 541-5. [PMID: 24403514]

[180] Nune M, Kumaraswamy P, Krishnan UM, Sethuraman S. Self-assembling peptide nanofibrous

scaffolds for tissue engineering: novel approaches and strategies for effective functional regeneration. Curr Protein Pept Sci 2013; 14(1): 70-84.
[http://dx.doi.org/10.2174/1389203711314010010] [PMID: 23544748]

[181] Hu Y, Wang H, Wang J, *et al.* Supramolecular hydrogels inspired by collagen for tissue engineering. Org Biomol Chem 2010; 8(14): 3267-71.
[http://dx.doi.org/10.1039/c002609c] [PMID: 20502821]

[182] Wang H, Wei J, Yang C, *et al.* The inhibition of tumor growth and metastasis by self-assembled nanofibers of taxol. Biomaterials 2012; 33(24): 5848-53.
[http://dx.doi.org/10.1016/j.biomaterials.2012.04.047] [PMID: 22607913]

[183] Zhang P, Cheetham AG, Lin YA, Cui H. Self-assembled Tat nanofibers as effective drug carrier and transporter. ACS Nano 2013; 7(7): 5965-77.
[http://dx.doi.org/10.1021/nn401667z] [PMID: 23758167]

[184] Soukasene S, Toft DJ, Moyer TJ, *et al.* Antitumor activity of peptide amphiphile nanofiber-encapsulated camptothecin. ACS Nano 2011; 5(11): 9113-21.
[http://dx.doi.org/10.1021/nn203343z] [PMID: 22044255]

[185] Cho H, Indig GL, Weichert J, Shin HC, Kwon GS. *in vivo* cancer imaging by poly(ethylene glycol)---poly(ε-caprolactone) micelles containing a near-infrared probe. Nanomedicine (Lond) 2012; 8(2): 228-36.
[http://dx.doi.org/10.1016/j.nano.2011.06.009] [PMID: 21704593]

[186] Wagh A, Singh J, Qian S, Law B. A short circulating peptide nanofiber as a carrier for tumoral delivery. Nanomedicine (Lond) 2013; 9(4): 449-57.
[http://dx.doi.org/10.1016/j.nano.2012.10.009] [PMID: 23178287]

[187] Yadav B, Taurin S, Rosengren RJ, *et al.* Synthesis and cytotoxic potential of heterocyclic cyclohexanone analogues of curcumin. Bioorg Med Chem 2010; 18(18): 6701-7.
[http://dx.doi.org/10.1016/j.bmc.2010.07.063] [PMID: 20728364]

[188] Sun A, Shoji M, Lu YJ, Liotta DC, Snyder JP. Synthesis of EF24-tripeptide chloromethyl ketone: a novel curcumin-related anticancer drug delivery system. J Med Chem 2006; 49(11): 3153-8.
[http://dx.doi.org/10.1021/jm051141k] [PMID: 16722634]

[189] Somers-Edgar TJ, Taurin S, Larsen L, Chandramouli A, Nelson MA, Rosengren RJ. Mechanisms for the activity of heterocyclic cyclohexanone curcumin derivatives in estrogen receptor negative human breast cancer cell lines. Invest New Drugs 2011; 29(1): 87-97.
[http://dx.doi.org/10.1007/s10637-009-9339-0] [PMID: 19816657]

[190] Robinson TP, Ehlers T, Hubbard RB IV, *et al.* Design, synthesis, and biological evaluation of angiogenesis inhibitors: aromatic enone and dienone analogues of curcumin. Bioorg Med Chem Lett 2003; 13(1): 115-7.
[http://dx.doi.org/10.1016/S0960-894X(02)00832-6] [PMID: 12467629]

[191] Ohori H, Yamakoshi H, Tomizawa M, *et al.* Synthesis and biological analysis of new curcumin analogues bearing an enhanced potential for the medicinal treatment of cancer. Mol Cancer Ther 2006; 5(10): 2563-71.
[http://dx.doi.org/10.1158/1535-7163.MCT-06-0174] [PMID: 17041101]

[192] Subramaniam D, May R, Sureban SM, *et al.* Diphenyl difluoroketone: a curcumin derivative with potent *in vivo* anticancer activity. Cancer Res 2008; 68(6): 1962-9.
[http://dx.doi.org/10.1158/0008-5472.CAN-07-6011] [PMID: 18339878]

[193] Liang G, Shao L, Wang Y, *et al.* Exploration and synthesis of curcumin analogues with improved structural stability both *in vitro* and *in vivo* as cytotoxic agents. Bioorg Med Chem 2009; 17(6): 2623-31.
[http://dx.doi.org/10.1016/j.bmc.2008.10.044] [PMID: 19243951]

[194] Karthikeyan NS, Sathiyanarayanan KI, Aravindan PG, Giridharan P. Synthesis, crystal structure and

anti-cancer properties of cyclic monocarbonyl analogs of curcumin. Med Chem Res 2011; 20: 81-7. [http://dx.doi.org/10.1007/s00044-009-9284-7]

[195] Dayton A, Selvendiran K, Kuppusamy ML, *et al.* Cellular uptake, retention and bioabsorption of HO-3867, a fluorinated curcumin analog with potential antitumor properties. Cancer Biol Ther 2010; 10(10): 1027-32. [http://dx.doi.org/10.4161/cbt.10.10.13250] [PMID: 20798598]

[196] Lee HE, Choi ES, Jung JY, *et al.* Inhibition of specificity protein 1 by dibenzylideneacetone, a curcumin analogue, induces apoptosis in mucoepidermoid carcinomas and tumor xenografts through Bim and truncated Bid 2013.

[197] Novaković M, Pešić M, Trifunović S, *et al.* Diarylheptanoids from the bark of black alder inhibit the growth of sensitive and multi-drug resistant non-small cell lung carcinoma cells. Phytochemistry 2014; 97: 46-54. [http://dx.doi.org/10.1016/j.phytochem.2013.11.001] [PMID: 24290194]

[198] Zheng A, Li H, Wang X, *et al.* Anticancer Effect of a Curcumin Derivative B63: ROS Production and Mitochondrial Dysfunction. Curr Cancer Drug Targets 2013. Nov 25. [Epub ahead of print]

[199] Ucisik MH, Küpcü S, Schuster B, Sleytr UB. Characterization of CurcuEmulsomes: nanoformulation for enhanced solubility and delivery of curcumin. J Nanobiotechnology 2013; 11: 37. [http://dx.doi.org/10.1186/1477-3155-11-37] [PMID: 24314310]

[200] Sahu A, Kasoju N, Goswami P, Bora U. Encapsulation of curcumin in Pluronic block copolymer micelles for drug delivery applications. J Biomater Appl 2011; 25(6): 619-39. [http://dx.doi.org/10.1177/0885328209357110] [PMID: 20207782]

[201] Basniwal RK, Khosla R, Jain N. Improving the anticancer activity of curcumin using nanocurcumin dispersion in water. Nutr Cancer 2014; 66(6): 1015-22. [http://dx.doi.org/10.1080/01635581.2014.936948] [PMID: 25068616]

[202] Esmaili M, Ghaffari SM, Moosavi-Movahedi Z, *et al.* Beta casein-micelle as a nano vehicle for solubility enhancement of curcumin; food industry application. Lebensm Wiss Technol 2011; 44: 2166-72. [http://dx.doi.org/10.1016/j.lwt.2011.05.023]

[203] Anuchapreeda S, Fukumori Y, Okonogi S, Ichikawa H. Preparation of Lipid Nanoemulsions Incorporating Curcumin for Cancer Therapy. J Nanotechnol 2012; •••: 1-11. [http://dx.doi.org/10.1155/2012/270383]

[204] White JW, Doner LW. Honey composition and properties: Beekeeping in the United States. Agric Handbook 1980; 335: 82-91.

[205] Bogdanov S, Jurendic T, Sieber R, Gallmann P. Honey for nutrition and health: a review. J Am Coll Nutr 2008; 27(6): 677-89. [http://dx.doi.org/10.1080/07315724.2008.10719745] [PMID: 19155427]

[206] Ajibola A, Idowu GO, Amballi AA, Oyefuga OH, Iquot IS. Improvement of some haematological parameters in albino rats with pure natural honey. J Biol Sci Res 2007; 2: 67-9.

[207] Ajibola A, Chamunorwa JP, Erlwanger KH. Nutraceutical values of natural honey and its contribution to human health and wealth. Nutr Metab (Lond) 2012; 9(61): 61. [http://dx.doi.org/10.1186/1743-7075-9-61] [PMID: 22716101]

[208] Gheldof N, Wang XH, Engeseth NJ. Identification and quantification of antioxidant components of honeys from various floral sources. J Agric Food Chem 2002; 50(21): 5870-7. [http://dx.doi.org/10.1021/jf0256135] [PMID: 12358452]

[209] Wang J, Li QX. Chemical composition, characterization, and differentiation of honey botanical and geographical origins. Adv Food Nutr Res 2011; 62: 89-137. [http://dx.doi.org/10.1016/B978-0-12-385989-1.00003-X] [PMID: 21504822]

[210] Arriaga E, Navarro-Calvo AL, Díaz-Carbajal E. Botanical characterisation of Mexican honeys from a subtropical region (Oaxaca) based on pollen analysis. Grana 2011; 50: 40-5.
[http://dx.doi.org/10.1080/00173134.2010.537767]

[211] Beretta G, Gelmini F, Lodi V, Piazzalunga A, Maffei Facino R. Profile of nitric oxide (NO) metabolites (nitrate, nitrite and N-nitroso groups) in honeys of different botanical origins: nitrate accumulation as index of origin, quality and of therapeutic opportunities. J Pharm Biomed Anal 2010; 53(3): 343-9.
[http://dx.doi.org/10.1016/j.jpba.2010.04.010] [PMID: 20451343]

[212] Beretta G, Caneva E, Facino RM. Kynurenic acid in honey from arboreal plants: MS and NMR evidence. Planta Med 2007; 73(15): 1592-5.
[http://dx.doi.org/10.1055/s-2007-993740] [PMID: 17999354]

[213] Crane E. History of honey.Honey: a comprehensive survey. London: William Heinemann 1975; pp. 439-88.

[214] Mandal MD, Mandal S. Honey: its medicinal property and antibacterial activity. Asian Pac J Trop Biomed 2011; 1(2): 154-60.
[http://dx.doi.org/10.1016/S2221-1691(11)60016-6] [PMID: 23569748]

[215] Jaganathan SK, Mandal M. Antiproliferative effects of honey and of its polyphenols: A review. J Biomed Biotechnol 2009; 2009: 830616.

[216] Li X, Huang Q, Ong CN, Yang XF, Shen HM. Chrysin sensitizes tumor necrosis factor-alpha-induced apoptosis in human tumor cells *via* suppression of nuclear factor-kappaB. Cancer Lett 2010; 293(1): 109-16.
[http://dx.doi.org/10.1016/j.canlet.2010.01.002] [PMID: 20133051]

[217] Orsolic N. Bee honey and cancer. J Anal At Spectrom 2009; 1: 93-103.

[218] Orsolic N, Basic I. Cancer chemoprevention by propolis and its polyphenolic compounds in experimental animals.Recent Progress in Medicinal Plants. USA: Studium Press LLC 2007; pp. 55-113.

[219] Mademtzoglou D, Haza AI, Coto LA, Morales P. Protective Effects of Rosemary, Heather and heterofloral Honeys towards Cytotoxicity of Acrylamide in Human Hepatoma Cells. Rev Complut Cienc Vet 2010; 4: 12-32.

[220] Pichichero E, Cicconi R, Mattei M, Muzi MG, Canini A. Acacia honey and chrysin reduce proliferation of melanoma cells through alterations in cell cycle progression. Int J Oncol 2010; 37(4): 973-81.
[PMID: 20811719]

[221] Samarghandian S, Afshari JT, Davoodi S. Chrysin reduces proliferation and induces apoptosis in the human prostate cancer cell line pc-3. Clinics (Sao Paulo) 2011; 66(6): 1073-9.
[http://dx.doi.org/10.1590/S1807-59322011000600026] [PMID: 21808878]

[222] Tsiapara AV, Jaakola M, Chinou I, *et al.* Bioactivity of Greek honey extracts on breast cancer (MCF-7), prostate cancer (PC-3) and endometrial cancer (Ishikawa) cells: Profile analysis of extracts. Food Chem 2009; 116: 702.
[http://dx.doi.org/10.1016/j.foodchem.2009.03.024]

[223] Rossiter K, Cooper AJ, Voegeli D, Lwaleed BA. Honey promotes angiogeneic activity in the rat aortic ring assay. J Wound Care 2010; 19(10): 440-, 442-446.
[http://dx.doi.org/10.12968/jowc.2010.19.10.79091] [PMID: 20948492]

[224] Swellam T, Miyanaga N, Onozawa M, *et al.* Antineoplastic activity of honey in an experimental bladder cancer implantation model: *in vivo* and *in vitro* studies. Int J Urol 2003; 10(4): 213-9.
[http://dx.doi.org/10.1046/j.0919-8172.2003.00602.x] [PMID: 12657101]

[225] Korkmaz A, Kolankaya D. Anzer honey prevents N-ethylmaleimide-induced liver damage in rats. Exp

Toxicol Pathol 2009; 61(4): 333-7.
[http://dx.doi.org/10.1016/j.etp.2008.07.005] [PMID: 18990552]

[226] Abdel Aziz A, Rady HM, Amer MA, Kiwan HS. Effect of Some honey bee extracts on the proliferation, proteolytic and gelatinolytic activities of the hepatocellular carcinoma Hepg2 cell line. Aust J Basic Appl Sci 2009; 3: 2754-69.

[227] Tonks A, Cooper RA, Price AJ, Molan PC, Jones KP. Stimulation of TNF-alpha release in monocytes by honey. Cytokine 2001; 14(4): 240-2.
[http://dx.doi.org/10.1006/cyto.2001.0868] [PMID: 11448125]

[228] Tushar T, Vinod T, Rajan S, Shashindran C, Adithan C. Effect of honey on CYP3A4, CYP2D6 and CYP2C19 enzyme activity in healthy human volunteers. Basic Clin Pharmacol Toxicol 2007; 100(4): 269-72.
[http://dx.doi.org/10.1111/j.1742-7843.2006.00039.x] [PMID: 17371531]

[229] Smith NF, Mani S, Schuetz eg, *et al.* Induction of CYP3A4 by vinblastine: Role of the nuclear receptor NR1I2. Ann Pharmacother 2010; 44(11): 1709-17.
[http://dx.doi.org/10.1345/aph.1P354] [PMID: 20959500]

[230] Ibuki Y, Toyooka T, Goto R. Inhibition of apoptosis by menadione on exposure to UVA. Cell Biol Toxicol 2006; 22(5): 351-60.
[http://dx.doi.org/10.1007/s10565-006-0089-x] [PMID: 16845610]

[231] Jaganathan SK, Mandal M. Involvement of non-protein thiols, mitochondrial dysfunction, reactive oxygen species and p53 in honey-induced apoptosis. Invest New Drugs 2010; 28(5): 624-33.
[http://dx.doi.org/10.1007/s10637-009-9302-0] [PMID: 19705065]

[232] Jaganathan SK, Mandal M. Honey constituents and their apoptotic effect in colon cancer cells. J Apiprod Apimed Sci 2009; 1: 29-36.
[http://dx.doi.org/10.3896/IBRA.4.01.2.02]

[233] Jaganathan SK, Mondhe D, Wani ZA, Pal HC, Mandal M. Effect of Honey and Eugenol on Ehrlich Ascites andvSolid Carcinoma. J Biomed Biotechnol 2010; 2010: 989163.

[234] Fauzi AN, Norazmi MN, Yaacob NS. Tualang honey induces apoptosis and disrupts the mitochondrial membrane potential of human breast and cervical cancer cell lines. Food Chem Toxicol 2011; 49(4): 871-8.
[http://dx.doi.org/10.1016/j.fct.2010.12.010] [PMID: 21167897]

[235] Bardy J, Slevin NJ, Mais KL, Molassiotis A. A systematic review of honey uses and its potential value within oncology care. J Clin Nurs 2008; 17(19): 2604-23.
[http://dx.doi.org/10.1111/j.1365-2702.2008.02304.x] [PMID: 18808626]

[236] Foret S, Kucharski R, Pellegrini M, *et al.* DNA methylation dynamics, metabolic fluxes, gene splicing, and alternative phenotypes in honey bees. Proc Natl Acad Sci USA 2012; 109(13): 4968-73.
[http://dx.doi.org/10.1073/pnas.1202392109] [PMID: 22416128]

[237] Jurenka JS. Therapeutic applications of pomegranate (Punica granatum L.): a review. Altern Med Rev 2008; 13(2): 128-44.
[PMID: 18590349]

[238] Viladomiu M, Hontecillas R, Lu P, Bassaganya-Riera J. Preventive and prophylactic mechanisms of action of pomegranate bioactive constituents 2013.
[http://dx.doi.org/10.1155/2013/789764]

[239] Adhami VM, Khan N, Mukhtar H. Cancer chemoprevention by pomegranate: laboratory and clinical evidence. Nutr Cancer 2009; 61(6): 811-5.
[http://dx.doi.org/10.1080/01635580903285064] [PMID: 20155621]

[240] Lansky EP, Newman RA. Punica granatum (pomegranate) and its potential for prevention and treatment of inflammation and cancer. J Ethnopharmacol 2007; 109(2): 177-206.
[http://dx.doi.org/10.1016/j.jep.2006.09.006] [PMID: 17157465]

[241] Gullett NP, Mazurak VC, Hebbar G, Ziegler TR. Nutritional interventions for cancer-induced cachexia. Curr Probl Cancer 2011; 35(2): 58-90.
[http://dx.doi.org/10.1016/j.currproblcancer.2011.01.001] [PMID: 21420558]

[242] Banerjee S, Kambhampati S, Haque I, Banerjee SK. Pomegranate sensitizes Tamoxifen action in ER-α positive breast cancer cells. J Cell Commun Signal 2011; 5(4): 317-24.
[http://dx.doi.org/10.1007/s12079-011-0138-y] [PMID: 21706446]

[243] Costantini S, Rusolo F, De Vito V, *et al.* Potential anti-inflammatory effects of the hydrophilic fraction of pomegranate (Punica granatum L.) seed oil on breast cancer cell lines. Molecules 2014; 19(6): 8644-60.
[http://dx.doi.org/10.3390/molecules19068644] [PMID: 24962397]

[244] Syed DN, Chamcheu JC, Adhami VM, Mukhtar H. Pomegranate extracts and cancer prevention: molecular and cellular activities. Anticancer Agents Med Chem 2013; 13(8): 1149-61.
[http://dx.doi.org/10.2174/1871520611313080003] [PMID: 23094914]

[245] Jaganathan SK, Vellayappan MV, Narasimhan G, Supriyanto E. Role of pomegranate and citrus fruit juices in colon cancer prevention. World J Gastroenterol 2014; 20(16): 4618-25.
[http://dx.doi.org/10.3748/wjg.v20.i16.4618] [PMID: 24782614]

[246] Larrosa M, Tomás-Barberán FA, Espín JC. The dietary hydrolysable tannin punicalagin releases ellagic acid that induces apoptosis in human colon adenocarcinoma Caco-2 cells by using the mitochondrial pathway. J Nutr Biochem 2006; 17(9): 611-25.
[http://dx.doi.org/10.1016/j.jnutbio.2005.09.004] [PMID: 16426830]

[247] Kohno H, Suzuki R, Yasui Y, Hosokawa M, Miyashita K, Tanaka T. Pomegranate seed oil rich in conjugated linolenic acid suppresses chemically induced colon carcinogenesis in rats. Cancer Sci 2004; 95(6): 481-6.
[http://dx.doi.org/10.1111/j.1349-7006.2004.tb03236.x] [PMID: 15182427]

[248] Bhatia D, Thoppil RJ, Mandal A, Samtani KA, Darvesh AS, Bishayee A. The dietary hydrolysable tannin punicalagin releases ellagic acid that induces apoptosis in human colon adenocarcinoma Caco-2 cells by using the mitochondrial pathway. J Nutr Biochem 2006; 17(9): 611-25.

[249] Bishayee A, Thoppil RJ, Darvesh AS, Ohanyan V, Meszaros JG, Bhatia D. Pomegranate phytoconstituents blunt the inflammatory cascade in a chemically induced rodent model of hepatocellular carcinogenesis. J Nutr Biochem 2013; 24(1): 178-87.
[http://dx.doi.org/10.1016/j.jnutbio.2012.04.009] [PMID: 22841394]

[250] Bishayee A, Bhatia D, Thoppil RJ, Darvesh AS, Nevo E, Lansky EP. Pomegranate-mediated chemoprevention of experimental hepatocarcinogenesis involves Nrf2-regulated antioxidant mechanisms. Carcinogenesis 2011; 32(6): 888-96.
[http://dx.doi.org/10.1093/carcin/bgr045] [PMID: 21389260]

[251] Manach C, Scalbert A, Morand C, Rémésy C, Jiménez L. Polyphenols: food sources and bioavailability. Am J Clin Nutr 2004; 79(5): 727-47.
[PMID: 15113710]

[252] Zhou B, Yi H, Tan J, Wu Y, Liu G, Qiu Z. Anti-proliferative effects of polyphenols from pomegranate rind (Punica granatum L.) on EJ bladder cancer cells *via* regulation of p53/miR-34a axis. Phytother Res 2015; 29(3): 415-22.
[http://dx.doi.org/10.1002/ptr.5267] [PMID: 25572695]

[253] Hoffman B, Liebermann DA. Apoptotic signaling by c-MYC. Oncogene 2008; 27(50): 6462-72.
[http://dx.doi.org/10.1038/onc.2008.312] [PMID: 18955973]

[254] Ponta Helmut, Sherman Larry, Peter A. Herrlich. CD44: From adhesion molecules to signalling regulators. Nature Reviews Molecular Cell Biology 2006; 4: 33-45.

[255] Lee ST, Lu MH, Chien LH, Wu TF, Huang LC, Liao GI. Suppression of urinary bladder urothelial carcinoma cell by the ethanol extract of pomegranate fruit through cell cycle arrest and apoptosis.

BMC Complement Altern Med 2013; 13: 364.
[http://dx.doi.org/10.1186/1472-6882-13-364] [PMID: 24359437]

[256] De Boer L, Oakes V, Beamish H, *et al.* Cyclin A/cdk2 coordinates centrosomal and nuclear mitotic events. Oncogene 2008; 27(31): 4261-8.
[http://dx.doi.org/10.1038/onc.2008.74] [PMID: 18372919]

[257] Vicinanza R, Zhang Y, Henning SM, Heber D. Pomegranate juice metabolites, ellagic acid and urolithin a, synergistically inhibit androgen-independent prostate cancer cell growth *via* distinct effects on cell cycle control and apoptosis. Evid Based Complement Alternat Med 2013; 247504.

[258] Wang L, Ho J, Glackin C, Martins-Green M. Specific pomegranate juice components as potential inhibitors of prostate cancer metastasis. Transl Oncol 2012; 5(5): 344-55.
[http://dx.doi.org/10.1593/tlo.12190] [PMID: 23066443]

[259] Rettig MB, Heber D, An J, *et al.* Pomegranate extract inhibits androgen-independent prostate cancer growth through a nuclear factor-kappaB-dependent mechanism. Mol Cancer Ther 2008; 7(9): 2662-71.
[http://dx.doi.org/10.1158/1535-7163.MCT-08-0136] [PMID: 18790748]

[260] Wang L, Martins-Green M. Pomegranate and its components as alternative treatment for prostate cancer. Int J Mol Sci 2014; 15(9): 14949-66.
[http://dx.doi.org/10.3390/ijms150914949] [PMID: 25158234]

[261] Pantuck AJ, Leppert JT, Zomorodian N, *et al.* Phase II study of pomegranate juice for men with rising prostate-specific antigen following surgery or radiation for prostate cancer. Clin Cancer Res 2006; 12(13): 4018-26.
[http://dx.doi.org/10.1158/1078-0432.CCR-05-2290] [PMID: 16818701]

CHAPTER 4

Clinical Anticancer Nanodrug: Advantages and Challenges

Yi Hou[1,2,#], Mengjia Dou[1,2,#], Debra L. Zynger[3], Wei Rao[1,*] and **Jing Liu[1,4]**

[1] *Beijing Key Lab of CryoBiomedical Engineering and Key Lab of Cryogenics, Technical Institute of Physics and Chemistry, Chinese Academy of Sciences, Beijing, 100190, China*

[2] *Department of Thermal energy and Power engineering, University of Science & Technology Beijing, 100083, China*

[3] *Department of Pathology, Division of Genitourinary Pathology, The Ohio State University, Columbus, OH 43210, USA*

[4] *Department of Biomedical Engineering, Tsinghua University, Beijing, 100084, China*

Abstract: Low targeting ability, drug resistance and toxicity are main hurdles for clinical applications of conventional anticancer drugs. Collectively, nanodrug, through loading traditional chemotherapeutics into nanocarrier, provides an effective platform to overcome these obstacles. Anticancer nanomedicine is an emerging method of therapy that focuses on specific drug delivery to improve treatment efficacy while reducing detrimental side effects to normal tissues. By taking advantage of the physical and chemical properties of nanocarrier, a nanodrug can achieve effective transport and delivery of active species to the targeted site of action within the biological systems for cancer treatment. In particular, nanocarriers can be modified in their structure, composition and other properties to overcome drug resistance, improve membrane permeability, prolong duration of action and even lower the toxicity. Furthermore, a nanodrug, by incorporation of certain functionalities could conjugate with thermal therapy or radiotherapy to achieve dual treatment. Due to these advantages, various types of nanodrugs have been approved for clinical cancer therapy and a number of nanodrugs are currently in clinical trials, demonstrating the great potential of their effectiveness in cancer destruction. In this chapter, the fundamentals of clinical nanodrugs for anticancer therapy are highlighted, and recently developed typical clinical nanocarrier platforms are summarized, with a focus on their advanced pharmacokinetic characteristics versus conventional anticancer chemotherapeutics. However, it is evident that multidisciplinary and international collaboration is needed to stimulate clinical translation of nanotherapeutics. Ultimately, nanodrugs may become a new therapeutic modality to treat cancer in the near future.

* **Corresponding author Wei Rao:** Technical Institute of Physics and Chemistry, Chinese Academy of Sciences, 29 Zhongguancun East Road, Haidian District, Beijing, 100190, China; Tel: +86-10-82543719; Fax: +86-10-82543766; E-mail: weirao@mail.ipc.ac.cn

# Contributed Equally

Keywords: Anticancer nanomedicine, Clinical trials, Drug delivery, Drug resistance, Nanodrug, Nanocarrier, Pharmacokinetic characteristic, Physico-chemical characterization, Targeting, Toxicity.

INTRODUCTION

Nanodrugs refer to the use of nanotechnology to develop a new class of pharmaceutical formulation. With the decrease of particle size, the surface area of the nanocarrier increases, which is helpful for the dissolution rate of active pharmaceutical components and enhances the bioavailability and absorption rate of drugs. Large dose of drug can achieve bioavailability using smaller volumes. In addition, nanotechnology contributes to the development of new diagnostic and therapeutic methods, especially targeted drug delivery systems owing to the significantly promotion of the cancer drugs and cancer gene therapies' therapeutic efficacy compared with non-targeted drug delivery systems. Targeted delivery of antitumor drugs is the core of nano-antineoplastic drugs. This is one of the most active directions, with the goal to reduce the adverse drug reactions and increase the effectiveness of chemotherapy. In the nanodrug delivery system, the following strategies are often used. Firstly, decoration with polyethylene glycol and other modifications help avoid the reticuloendothelial system uptake and achieve long-cycle circulation. Secondly, drug accumulation in the tumor tissue can be promoted by controlling the particle size enhanced permeation and retention (Fig. **1**). Thirdly, ligand-receptor, antigen-antibody-mediated chemical modification of the nanocarrier surface improves the active targeting of tumor cells and promotes intracellular transport of nanodrugs (Fig. **2**). Additionally, the multi-drug resistance of tumor cells can be overcome through modification of the surface of nanocarriers.

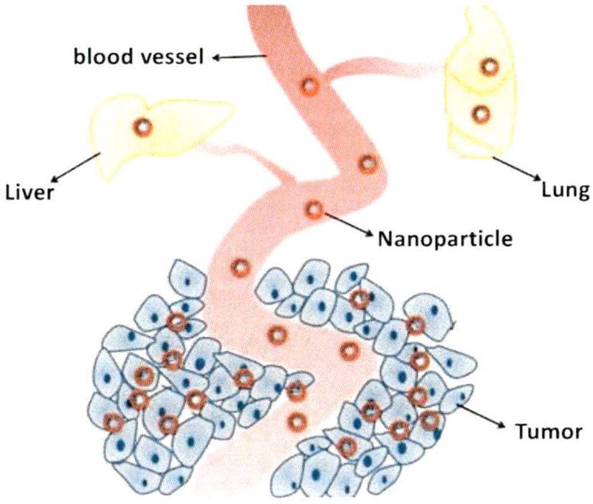

Fig. (1). The specific accumulation of nanodrugs in tumor.

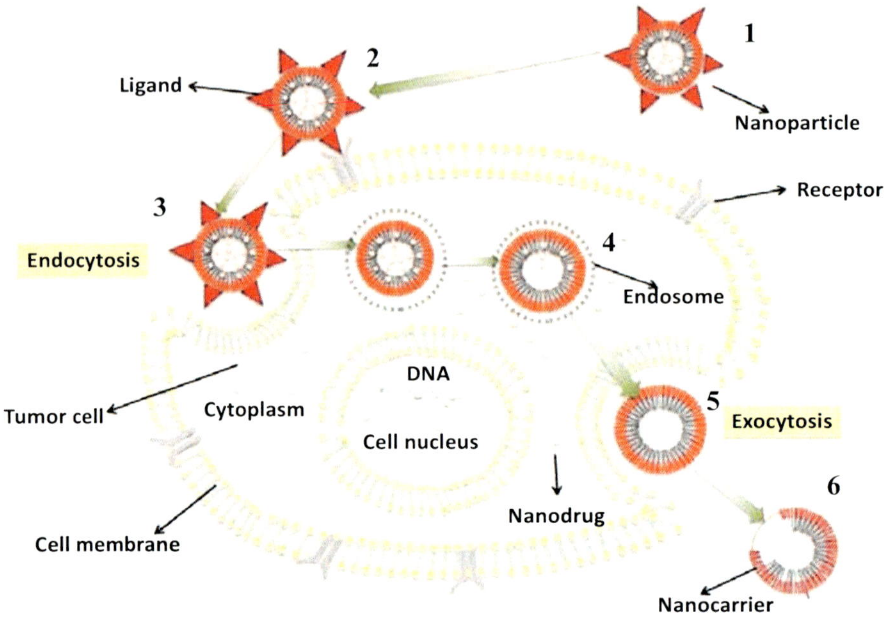

Fig. (2). The targeted delivery of nanodrugs acting on tumor cells **1**. The polymeric nanocarrier containing the desired chemotherapeutic drugs. **2**. The targeted delivery of nanodrugs to the tumor cell. **3**. The nanodrug enters the cell by endocytosis. **4**. The nanodrug release the chemotherapeutic drugs. **5**. The nanocarrier exits the cancer cells by exocytosis. **6**. The nanocarrier degrades.

According to incomplete statistics, the U.S. FOOD & DRUG Administration (FDA) has approved more than 50 kinds of small molecule targeted anticancer drugs to date, and there are more than 70 kinds anticancer nanodrugs in clinical trial, such as the approved drug Doxil, Myocet, Abraxane and ThermoDox, with Aroplatin in clinical research. As it is close to industrialization, the most promising direction of nanotechnology is nanodrugs, especially as the research and development of anticancer nanodrugs have become the frontier and hotspot of current international medical science [1, 2].

PHYSICAL AND CHEMICAL CHARACTERISTICS OF NANODRUGS FOR CANCER THERAPEUTIC APPLICATIONS

Size

The size of the nanoparticle is a physical measure of the scale of the particle. Many researchers tend to define nanomedicine within 1000 nm, and a large number of studies have shown that when the particle size of the nanodrugs is between 10 and 1000 nm, it has a great influence on the physical and chemical properties, pharmacokinetics and pharmacodynamics, showing significant differences with conventional preparations. The size of particles not only influences its biodistribution *in vivo*, but also substantially affects in the mechanism of cellular internalization.

The size of nanoparticles determines the *in vivo* biodistribution and circulating. On the one hand, smaller particles bring on the decreased specificity problems, because although volume reduction helps nanodrugs to arrive at targeted tumor cells more easily, it also leads to nanodrugs diffusion into circumambient ordinary cells. On the other hand, larger ones are more discriminating and extravasate with more difficulty, but in the tumor, their distribution is heterogeneous [3]. According to a previous study, particles of more than 15 μm cumulate primarily in the three parts, including of spleen, liver and bone marrow [4]. Sizes less than 5 nm will be removed through extravasation or renal clearance immediately. Rigid, 100–200 nm is the most promising spherical particles scales in prolonging the drugs' circulation *in vivo*, because it is large enough to avoid absorption by the liver as well as small enough to prevent filtration in the spleen [5 - 7]. Furthermore, the enhanced permeability and retention effect (EPR effect) also depends on the size of particle, that is nanodrugs within this size scale are preferentially cumulating in tissues with leaky vasculature or high vascular permeability especially tumors [8 - 11]. When it comes to therapeutics of maintaining high drug concentrations, aggregation of nanoparticles in tumor tissue is extremely important. The velocity of murine macrophages uptaking and blood clearing the larger nanoparticles is twice quicker than the smaller ones. It is further confirmed by comparing the quantity of tumor necrosis factor-R (TNF-R) accumulating in the tumor within 24h. The results showed that drugs encapsulated in the smaller nanoparticles are twice that in the larger nanoparticles [12]. Size of typical anticancer nanodrugs is summarized in Table **1**.

Table 1. Size of typical clinical anticancer nanodrugs.

Product name	**Particle size**	**Nanocarrier**	**Payload**	**Phase of development**	**Reference**
Myocet	190nm	Non-PEGylated	Doxorubicin	Approved	[18, 19]
Doxil	100nm	PEGylated liposomes	Doxorubicin	Approved	[20, 21]
NanoTherm	15nm	Magnetic(iron-oxide)nanoparticles	N/A	Approved	[22, 23]
Abraxane	100-200nm	Albumin nanoparticle	Paclitaxel	Approved	[24, 25]
Daunoxome	45nm	Non-PEGylated	The citrate salt of daunorubicin	Approved	[26]
SP1049C	~25nm	Pluronic polymer	Doxorubicin	Phase 2	[27, 28]
Genexol-PM	30~60nm	Micelle polymer	Paclitaxel	Phase 2	[29, 30]
Prolindac	6~15nm	Platinum	Diaminocyclohexane	Phase 2	[31, 32]
IT-101	~40nm	Cyclodextrin	20(S)-camptothecin	Phase 1	[33, 34]

(Table 1) contd.....

Product name	Particle size	Nanocarrier	Payload	Phase of development	Reference
PK2	8.4 nm	N-(2-hydroxypropyl)methacrylamide	Doxorubicin	Phase 1	[35]

The mechanism of cellular internalization includes phagocytosis, macropinocytosis, caveolar-mediated endocytosis or clathrin-mediated endocytosis. In order to make nanodrugs target different tumor cells' subcellular structures (*e.g.* DNA) [13], it's important to understand malignant cells' uptake mechanisms and to investigate the ability of more complex nanocarriers to target both malignant cells and their specific subcellular compartments [14 - 17].

Structure

The crystal structure of nanoparticles is influenced by the geometric size. Data reports that when the diameter of metal cobalt is less than 50nm, the crystal structure is face-centered cube (fcc) rather than hexagonal close-packed (hcp) when in solid form. The macroscopic properties of the NCS are not only decided by the properties of each specific particle but also by the interaction between interconnected nanocrystals and coupling between organic molecules and nanocrystals [36]. The schematics showing of nanocrystal superlattices of spherical and faceted particles are demonstrated in Fig. (**3**).

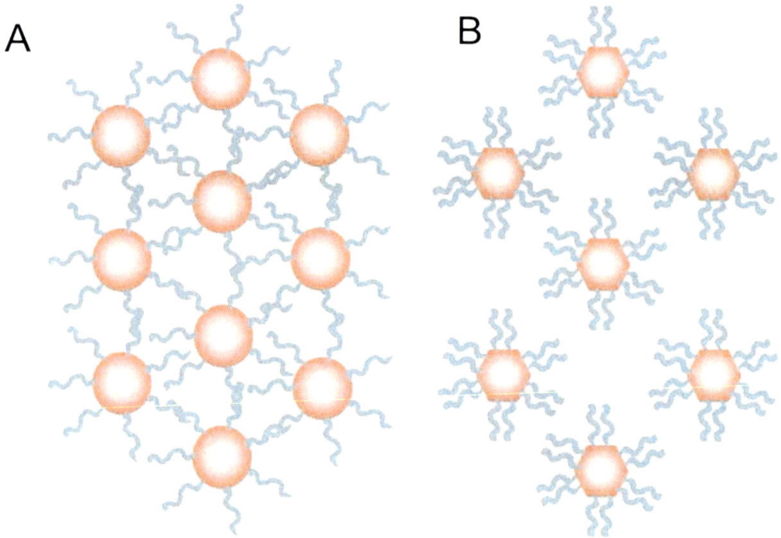

Fig. (3). Schematics showing nanocrystal superlattices of (**A**) spherical and (**B**) faceted particles.

With further insight into nanodrug characteristic, it has been possible to design the structure of nanoparticles from atomic or molecular scale. Typical examples include mixed type, core/shell type, hollow type, multi-layer type, mesoporous type. Mixed type nanoparticles for drugs and carriers are mainly via organic compound. Halas *et al* obtained Au/SiO_2 core/shell type nanoparticles by self-assembly and achieved that the surface of the SiO_2 particle is covered with a continuous Au shell. The mesoporous nanoparticles can be formed on the surface of inorganic nanoparticles by atomization drying technology and it is potential in drug carrier, catalyst and other aspect [37]. Different structures of nanoparticles are illustrated in Fig. (**4**).

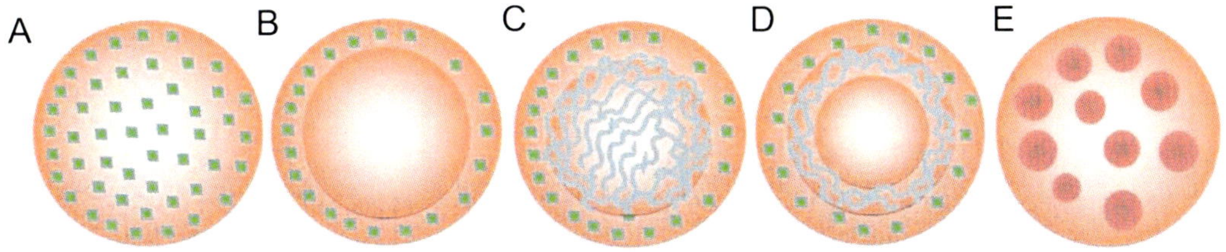

Fig. (4). Typical schematic structure of nanoparticles. (**A**) Mixed type. (**B**) Core/shell type. (**C**) Hollow type. (**D**) Multi-layer type (**E**) Mesoporous type.

Dose

Conventional drug needs high doses to guarantee the bioavailability. Intravenous injection of common poorly soluble drug can result in an embolization of blood vessels due to the accumulation of the insoluble drugs, and often leads to local toxicity due to higher drug concentrations at the site of deposition. Consequently, it is of great significance to improve the solubility of hydrophobic drugs, increase the drug systemic circulation half-life, release drug at sustained rate and lower the frequency of administration [38 - 39].

Different nanodrugs have various applicable dose. For example, dose of dichloro (1, 2-diaminocyclohexane) platinum (II)-loaded and hyaluronic acid polymer-coated nanoparticle (DACHPt/HANP) is administrated to be 2g/kg in clinical dosage [40]. The recommended dosage of Marqibo for the treatment of acute lymphoblastic leukemia is 2.25 mg/m^2 once every seven days. If provided excessive this medication, it may result in problems like vomiting, double vision, or seizures [41].

Surface Chemistry

Surface chemistry as well as surface charge is two foremost parameters that determine the interactions between nanoparticles with physiological environments. The surface charge has great influence on the fate of nanodrugs at different stages, such as physical stability, blood stability, interactions with extracellular matrix, cellular uptake and endocytosis [42, 43]. In addition, the surface charge is the main determinant of colloidal behavior, it will be through the aggregation or agglomeration of nanoparticles to affect the size and morphology of nanoparticles. Leroueil *et al.* found that when the cell membrane was destroyed by cationic nanoparticles [44], they were perforated on the membrane, making the membrane thinner, or corroding the membrane. Ruenraroengsak and Tetley [45] compared the cytotoxicity of neutral, positive and negative surface charge nanoparticles on the alveolar epithelial cell, and found that positive nanoparticles treatment of cell activity was significantly reduced, resulting in more ROS (reactive oxygen species), mitochondrial damage serious; neutral and negative nanoparticles produced some ROS, but almost no cytotoxicity and lead to weak mitochondrial damage. Shao et.al. found positively charged nanoparticles have higher interaction force with cells [46]. It is possible for excess positive charge of a nanodrug to induce the occlusion of blood vessels and capillaries [47]. In addition, positively charged nanodrugs are easy to interact with extracellular matrix charged negatively, which may impede their further penetration [48]. At the cellular level, the positive surface charge is directly proportional to the cellular binding, because living cells have a negatively charged plasma membrane with high electrostatic attractive to the positively charged nanoparticles. Therefore, the uptake quantity of nanoparticles often increases with elevation of the zeta potential values. However, excess positive charge can induce toxicity and lead to immunological reactions [42, 49]. In addition to surface charge, nanoparticles with different surface modification may determine intracellular trafficking pathways by various endocytic mechanisms. For instance, to avoid the lysosomal degradation pathways, the surface of nanoparticles decorated with folic acid, albumin or cholesterol favor caveolin-mediated endocytosis. Alternatively, transferrin and cell penetrating peptide encourage uptake by clathrin-mediated endocytosis and macropinocytosis, individually [50].

Nanoparticles reduce the surface potential energy by surface modification, while introducing new groups also bring new properties to the particles, such as catalytic, surface charge, adsorption and desorption of solution molecules. As these properties are directly related to the composition, structure, and solution environment of surface molecules, these properties are dynamic in biological systems. For example, surface aminated polystyrene particles (60 nm) in the PBS buffer have positive electromotive force, while in the complete medium environment, the electric potential decreased sharply negative, indicating that the adsorption of serum proteins shield the surface of positive charge [51].

Targeting

Targeted drug delivery founded on nanotechnology carries nanodrugs to specific diseased parts. Compared with conventional drug delivery, because of prolonged circulation and localized accumulation in the tumour tissues, targeted drug delivery helps to reduce the frequency of taking a drug, side-effects of the drug and prevents damage to normal cells by the nanodrug. However, this method reduces the ability to regulate drug dosage, increases the cost, and decreases the productivity [52]. Targeting properties of nanodrugs can be divided into passive and active targeting according to the way they are transported *in vivo*.

Passive Targeting

Passive targeting reached clinical trials in the mid-1980s for the first time and marketed in the mid-1990s. It takes full advantage of the characteristic features of tumor biology as mentioned: the enhanced EPR effect. In passive targeting, nanodrug are carried through the normal physiological processes, including leakage from the capillaries to the target organ or the occurrence of cancer, inflammation of the site, or by the liver, spleen and bone marrow and other organs of the reticuloendothelial system absorption [53 - 58]. The drugs' circulation time that can be improved by decorating the nanoparticles with special coating directly determines drugs success. At present, there have been a variety of passive-targeted nanodrugs approved or entering the clinical trial stage, such as Genexol-PM、NK911、NK105、NC600 and so on [59].

Active Targeting

Active targeting moieties that have been used in drug delivery include antibodies, protein and derivatives, peptides, carbohydrate and derivatives, aptamers and small molecules [60 - 62] among which conjugate nanocarriers containing chemotherapeutics with molecules that bind to overexpressed antigens or receptors on the target cells are used frequently to achieve drug delivery.

Though passive targeting approaches occupy an important position in the clinical therapy, they still have some limitations, including the random nature of this approach resulting in the difficulty of controlling this process and drugs inefficient diffusion owing to the parts of tumor cells' unfeasibility. At present, the active targeting molecules mainly include small molecules, polypeptide fragments, proteins, aptamers and polysaccharides [59].

The utilization of passive and active targeting makes drugs loaded on nanoparticles show an extended period circulating in the body before they are attacked to its specific targets through the use of cell-specific ligands, magnetic positioning, or pH responsive materials. Because of this specificity for tumor cells, their side effects are greatly reduced compared with ordinary drugs.

Stimuli-Responsiveness

Stimuli-responsiveness of nanodrugs means the ability of altering their properties in response to the changes of environment during drug delivery in cancer therapy [63], such as temperature, pH, optical or electrical signals [64]. With excellent stimuli-responsiveness, nanodrugs for cancer therapy perform well not only *in vivo* pharmacokinetic profiles and tumor homing ability, but also promote the absorption of nanodrugs in cells and/or control drug release selectively in the targeted region, thus achieving high efficacy and reduced side effects in cancer-specific treatment [65 - 68].

The most representative stimulation is pH. Polymeric nanodrugs mainly take two strategies: the use of polymers (polyacids or polybases) with ionizable groups that undergo conformational and/or solubility changes in response to environmental pH variation and the design of polymeric systems with acid-sensitive bonds whose cleavage enables the release of molecules anchored at the polymer backbone, the modification of the charge of the polymer or the exposure of targeting ligands [69, 70]. And now pH-sensitive liposomes have also been developed(*e.g.* poly-styrene-co-maleic acid [71]) which maintain stability at a physiological pH (pH 7.4), while entering into endosomes (pH 6.0–6.5) of the diseased cells Liposomal contents are released into the cytoplasmic due to structural destabilization [72]. Consequently, the change of pH triggers liposomal drug release spontaneously [73].

Similarly, the recent research of the temperature-sensitive liposome has showed a great superiority in cancer therapy. For instance, when exposed to a sub-lethal temperature of 39.5°C, ThermoDox releases the anti-cancer agent [70, 74] to achieve a higher drug concentration in the target area. And as a temperature-sensitive nanodrug delivery system, it is undergoing evaluation in a phase III clinical trial aiming to the breast cancer [70, 74].

ADVANCED NANOCARRIER PLATFORMS FOR CLINICAL ANTICANCER DRUG

Liposome Drug

As a spherical vesicle that has at least one lipid bilayer and a hydrophilic (polar) cavity encircled by a bilayer of amphiphilic phospholipids, liposome particles (Fig. **5**) can be prepared by disrupting biological membranes (such as by sonication) [75 - 78]. In the liposome particles, hydrophobic chemicals combine with the bilayer while hydrophilic solutes dissolved in the liposome are unable to pass through the bilayer easily. When delivering the nanodrugs to the targeted tumor cells, liposome particles empty the liposome contents by fusion of the lipid bilayer's to other bilayers (*e.g.* the cell membrane) [79].

The most representative liposomal formulation is Doxil. Doxil is created by loaded doxorubicin onto PEGylated liposomes [80].

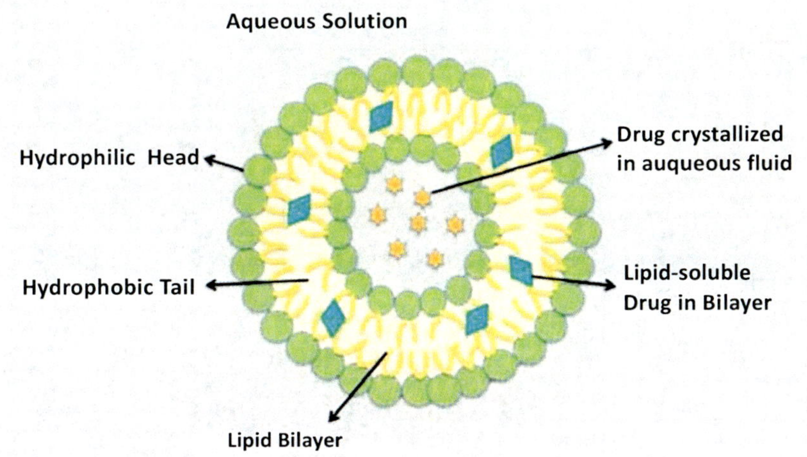

Fig. (5). Schematic image of liposome drugs.

The development of Doxil has overcome limitations, maximized drug accumulation in malignant cells and achieved longer circulation time and higher stability biocompatibility, possibility to delivery multiple drugs [81, 82]. At the same time, its toxicity can't be ignored, including dose-limiting mucosal and mild myelosuppression, cutaneous toxicities, reduced cardiotoxicity compared with free doxorubicin and minimal alopecia. Moreover, in some regions, it has been approved for cancer treatment. It is indicated for the treatment of patients with ovarian cancer whose disease has progressed or recurred after platinum-based chemotherapy and AIDS-related Kaposi's sarcoma in patients after failure of prior systemic chemotherapy or intolerance to such therapy. In addition, Doxil in combination with bortezomib is introduced for the treatment of multiple myeloma patients who have not previously received bortezomib but have received more than one prior therapy [41]. Except Doxil, Myocet, DaunoXome, Depocyt and Margibo, Marqibo and MEPACT are the other liposomal formulations for the anti-cancer nanomedicines [3]. Their applications and advantages are shown in Table **2**.

Table 2. Liposomal formulations for the anticancer nanodrugs.

Product	PEGylated/ Non-PEGylated	Drug	Advantages	Targeted cancer	Ref.
Doxil	PEGylated	Doxorubicin	Longer circulation time Higher drug loading efficiency Higher stability Higher biocompatibility	Recurrent ovarian cancer	[81, 82]
Myocet	Non-PEGylated	Doxorubicin	Decrease circulating half-life Generate a sustained delivery pattern Less toxic Better toleration	Metastatic breast cancer	[76, 83]
DaunoXome	Non-PEGylated	The citrate salt of daunorubicin	Increase aqueous stability, cytotoxicity, avoid the mononuclear phagocyte system) a favorable circulation half-life of 2-4 hours	Advanced HIV-associated Kaposi sarcoma	[83 - 87]
DepoCyt	Non-PEGylated	Cytarabine	Increase the half-life, Prolong tumor exposure More convenient administration	Lymphomatous meningitis	[88, 89]
Marqibo	N/A	Vincristine sulfate	Extend circulation period Specificity for optimized drug delivery system Increase target-tissue dose Dose intensification without toxicity enhancement	Philadelphia	[90]
MEPACT	N/A	Mifamurtide	Enhances activation Improve the drug's persistence in the lungs	Non-metastatic, resectable osteosarcoma	[91]

Polymeric Nanodrugs

At present, polymeric nanoparticles have received increasing attention as drug carriers in the field of drug delivery and controlled release. Polymeric nanodrugs can be modified in their structures, compositions and property to achieve specific applications. Furthermore, responsiveness of the nanodrugs in biological environment under different temperature, pH, enzymatic *etc.* can be modulated through incorporated with certain functionalities [70]. A variety of polymers have been used in this field because the polymeric nanoparticles are also effective in the drug delivery to the specific target sites, such as tissues or organs, thereby increasing the efficacy of reducing side effects. Compared with liposomes, it can enhance the stability of drugs, especially protein drugs, and has good sustained-release, controlled release properties [92]. According to the hydrophilic ability of loading drug, polymeric nanodrugs are classified into three types (Fig. **6**).

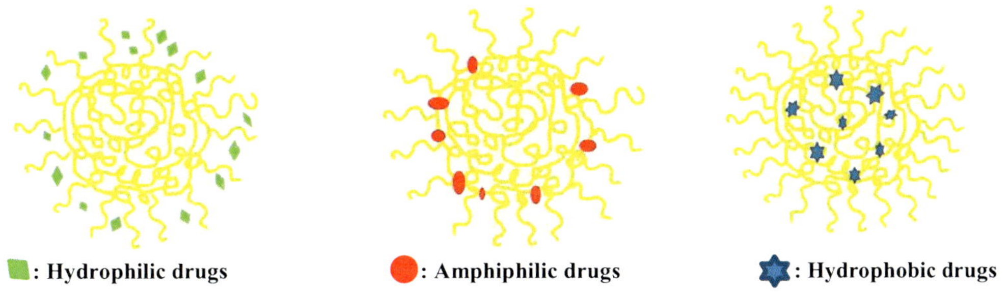

Fig. (6). Polymeric nanoparticles as drug carries.

Oncaspar (**asparaginase**) is a common polymeric nanodrug used as an important component of a multi-agent chemotherapeutic regimen for the first line treatment of patients with Acute Lymphoblastic Leukemia (ALL) with hypersensitivity to native forms of L-asparaginase. The conjugation of an antitumor protein neocarzinostatin (NCS) with styrene-malic acid (SMA), created Zinostatin Stimalamer which was approved to treat primary unresectable hepatocellular carcinoma in Japan in 1994. Due to this conjugation, the nanoparticles increase the drug's efficacy and half-life as well as decrease its toxicity. Also, Neulasta (pegfilgrastim) is an important adjunct therapy to myelosuppressive anticancer drugs which is used to treat patients with non-myeloid malignancies [3].

Albumin-Based Nanodrugs

Owing to its good biocompatibility, low toxicity and irritation, and no immune response, albumin-based nanodrugs are emerging in numerous applications of cancer diagnosis and therapy. Hyperplastic tumors tend to accumulate albumin to provide nutrition and energy for their rapid growth, so albumin can be used to promote the tumor's absorption of low molecular weight compounds and peptides, and applied to prepare nanoparticles which encapsulates diagnostic agents or anti-cancer drugs [93]. Significant quantities of bovine serum albumin and human serum albumin are obtained to prepare nanocarries for drug delivery. The three-dimensional structure of human serum albumin contains multiple binding sites for hydrophobic substances, and albumin can be tightly but reversibly bound to the substance by non-covalent interactions to achieve drugs' transportation in the body and release on the cell surface, which ensures human serum albumin to be a natural carrier of hydrophobic substances [94]. While gene recombinant human serum albumin not only has similar structure and physicochemical properties compared to human serum albumin, but also avoids cross-infection caused by pathogens which are carried by albumin in blood and is produced without limitation, so it provides material security for Albumin-based nanodrugs' research and application [95]. Typical methods for albumin-based nanoparticles' development includes desolvation, emulsification, thermal gelation, nano-spray drying, nanoparticle albumin-bound and self-assembly [96]. Abraxane is an injectable paclitaxel condensate with nanoparticle albumin, approved for clinical therapy in 2005 by the FDA. After binding to albumin as a nanocarrier, paclitaxel overcomes its poor drug solubility and has reduced cardiotoxicity [97]. Nowadays, it is mainly used to treat breast cancer, pancreatic cancer, lung cancer and so on [98, 99]. Aldoxorubicin is a kind of albumin-binding prodrug that binds doxorubicin by small molecules, which enhances efficacy and reduces side effects [96].

Inorganic Nanoparticles for Combination Therapy

Inorganic nanoparticle is another type of nanoparticle already approved for clinical use. Inorganic nanoparticles encompass a vast range of nanoparticle platforms synthesized from metals, metal oxides and metal sulfides. Inorganic nanoparticles not only hold relatively stable over a wide range of pH and temperature compared with liposomes and solid polymeric nanoparticles, but also show an efficient light-to-heat conversion, consequently local cellular can be destructed by lower laser energies [100]. However, they performed poorly in biodegradation and dissolution rates resulting in the difficulty in their degradation and elimination from the body [101].

As an aqueous colloidal dispersion of iron-oxide nanoparticles used to enhance the effect of chemotherapy, NanoTherm has received market approval in Europe for glioblastoma treatment and is currently under clinical investigation for other solid malignancies. Magnetic-thermal therapy directly exposes NanoTherm to tumor cells and applies alternating magnetic field at the same time leading the rapid rotation of nanoparticles which heats the tumor via the friction. Magnetic resonance imaging (MRI) and computerized tomography (CT) are used to keep watch on particle accumulation and direct treatment. Tumor cells are susceptible to chemotherapy with this novel therapeutic strategy [60].

Table **3** presents a summary of a few clinically relevant nanodrugs of different types which are currently under investigation in clinical trials or approved for cancer therapy.

Table 3. Examples of clinical anticancer nanodrugs.

Type of Nanocarrier	**Product name [1]**	**Description of nanoparticle**	**Indication**	**Phase of development**	**Therapeutic efficacy**
Liposomes	Myocet [18-19]	Doxorubicin (mPEG-DSPE, HSPC, and cholesterol)	EGFR2-positive metastatic breast cancer	Approved	Less cardiotoxic, Better tolerated Equally efficacious doxorubicin
	Doxil [20-21]	Doxorubicin(Phosphatidylcholine and cholesterol)	HIV-related Kaposi sarcoma, metastatic breast and ovarian cancer	Approved in 2005 by FDA	Longer circulation time Higher stability Higher biocompatibility
	ThermoDox [108]	Thermosensitive doxorubicin delivery, in combination with radiofrequency ablation or microwave hyperthermia	Hepatocellular carcinoma, breast cancer, tumors of the liver and other types of solid tumors	Phase 3	Rapid release of anticancer drugs Improve tolerance
	NKTR102 [109-110]	PEG micelle with irinotecan	Breast and colorectal cancer,Ovarian cancer	Phase 3,Phase 2	Increased survival rate Enhance irinotecan active metabolite
Polymeric nanodrugs	Oncaspar [111]	Polymeric PEG-L-asparaginase	Acute lymphoblastic leukemia	Approved	Increase the drug's half-life Enhance efficacy, Decrease toxicity.
	Zinostatin stimalamer [112-114]	Copolymer styrene maleicacid-conjugated neocarzinostatin	Unresectable hepatocellular carcinoma	Approved	Potent antitumor activity *in vitro* and *in vivo*
	CRLX-101 [115-116]	Cyclodextrin-PEG micelle with camptothecin	Various solid malignancies	Phase2,Phase 1	Larger efficacyLower toxicity

(Table 3) contd.....

Type of Nanocarrier	Product name [1]	Description of nanoparticle	Indication	Phase of development	Therapeutic efficacy
Albumin-based nanodrugs	Abraxane [24-25]	Albumin-bound form of paclitaxel	Metastatic breast cancer, locally advanced or metastatic non-small cell, lung cancer. Pancreatic and breast	Approved in January 2005 by FDA	45% probability of surviving No severe hypersensitivity reactions
Inorganic nanodrug	NanoTherm [22-23]	Nanodrug conjugates injected in the tumor site and heated selectively by a magnetic field, resulting in temperature-dependent release of drug	Local ablation of glioblastoma multiform	Approved	A stable depot within the tumor tissue, Convert energy into heat efficiently

PHARMACOKINETIC CHARACTERISTICS OF NANODRUGS

A nanodrug system has several pharmacokinetic superiorities compared to a conventional drug, such as specific drug delivery, improved membrane permeability, overcome multidrug resistance and long duration of action [102]. The pharmacokinetic manifestations of nanoparticles *in vivo* mainly depend on their chemical and physical properties such as size, charge and surface properties. Through the construction of nano-size structure, nanoparticles can increase the dissolution rate of drugs, thereby enhancing the absorption of drugs and bioavailability. Allen *et al.* found that nano-liposomes with a diameter of 120 nm were 10% ~ 20% higher accumulated in tumor tissue than that of 170nm [103]. Torchillin [104] *et al.* have also come to a similar conclusion. Therefore, by appropriate specific surface modification to synthesize small particle size nanoparticles, the absorption and distribution of nanoparticles in tumor tissue can be significantly increased, which can enhance the efficacy of the drug. Levchenko *et.al.* studied the distribution of nanoparticles in mice [105]. The results showed that the blood removal rate and liver absorption rate of the nanoparticles with negative charge (40mV) were significantly higher than those of the surface with neutral charge (10mV). Nanoparticles of positive surface charge in the lung and liver accumulation also showed a faster blood clearance time. In addition, the hydrophilicity / hydrophobicity of the nanoparticles on the *in vivo* distribution and metabolic dynamics of the nanoparticles mainly change the dispersion of the nanoparticles in the body and the protein binding rate. Polyethylene glycol (PEG) can be used to attach hydrophilic groups to the surface of nanoparticles to improve the hydrophilicity of nanoparticles and reduce their protein binding rate *in vivo*. Discher *et al.* [106] found that rod type polymer nanoparticles are not easily engulfed into cells, and have longer biological half-life. An illustration is shown in Fig. (**7**) to explain the influence of physicochemical characteristics of PK.

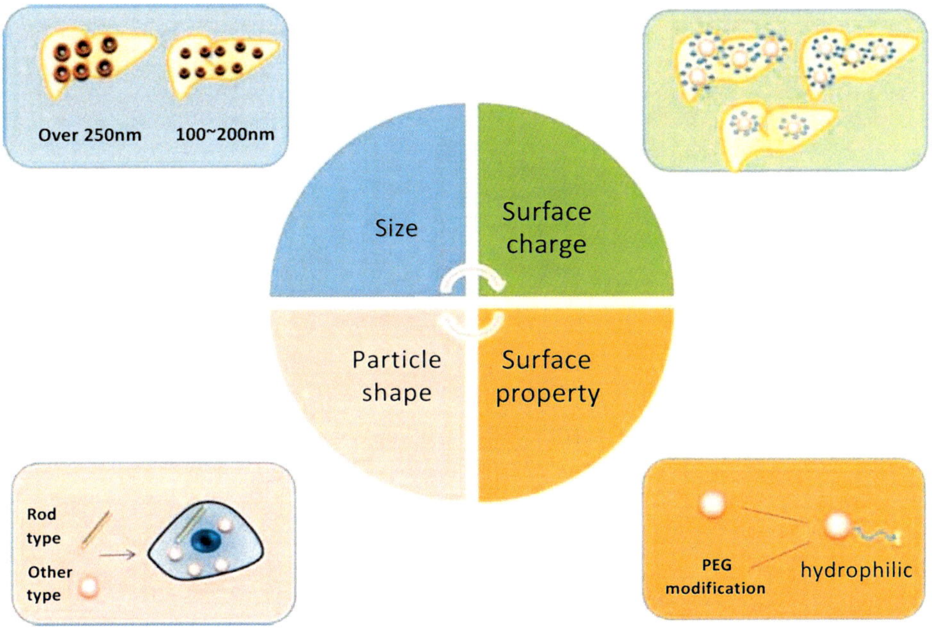

Fig. (7). The illustration of the influence of physicochemical characteristics of PK.

Specific Drug Delivery

One avail of nanoparticles is to serve as a novel drug delivery system, delivering to tissues throughout the body. This is achieved by sheltering the membrane barrier, limiting characteristics of the therapeutic drug molecule, keep possession of drug stability, and the properties of the colloidal drug carrier. It is possible that the camouflage of the drug allows access across the previously impermeable membrane. Once the nanoparticle reaches the targeted tissue, release of the drug may occur by desorption, diffusion through the nanoparticle matrix or polymer wall or nanoparticle erosion, or combination of other mechanisms [107].

Improved Membrane Permeability

The pharmacological effects of drugs are not only related to the unique chemical composition of drugs, but also with the physical state of the drug preparation. For solid drugs, especially insoluble solid drugs, the ultrafine particles are of great significance to improve the bioavailability of drugs. Drugs made of nanoparticles, greatly increase the exposure to the medium surface area and promote the dissolution of drugs. Due to the adhesion of nanocarriers and small particle size, both conducive to local drug retention, and improved drug contact with the intestinal wall, increase contact area, and increased penetration of the interstitial space, greatly improve the absorption rate, extent of the effective ingredient and permeability and solubility, thereby increasing the bioavailability [26].

Overcoming Multidrug Resistance

Multidrug resistance (MDR) presents the main obstacle affecting the clinical success rate of surgery in cancer treatment which may leads to high mortality and recurrence rate of cancers [117]. MDR in cancer describes the capacity of cancer cells to survive or become resistant from treatment of related and unrelated classes of drugs [118]. The presence of cancer drug resistance has become a central issue of cancer research. The mechanisms of MDR include decreasing uptake of drugs, reducing intracellular drug concentration by activation of the efflux transporters, modifying cellular pathways via altering cell cycle checkpoints, enhancing metabolism of drugs, inducing emergency response genes to impair apoptotic pathways and alter DNA repair mechanisms. Over-expression of three typical ATP-binding cassette (ABC) transporters, P-glycoprotein (P-gp), multidrug resistance proteins (MRP) and breast cancer resistance protein (BCRP), have been considered to be the most fundamental causes of MDR in cancer [119].

A main strategy is employing drugs to inhibit ABC transporters, but there still are obstacles in that almost all ABC transporter inhibitors and chemotherapeutic drugs lack selectivity to tumor cells, thus resulting in low therapeutic efficiency and high toxicity. With the ability of mediating the endocytosis as well as circumventing the efflux pump effect of ABC transporters, mesoporous silica nanomaterials (MSN) are being vigorously investigated to overcome this problem. Besides the above strategy for anti-MDR, it is new and promising to employ the efflux circumventing delivery strategy and the multi-modal combination therapy strategy.

Lipid nanoparticles, polymeric nanoparticles, metal nanoparticles, dendrimers have been already been reported to be an effective platform to avoid drug resistance. For example, it is reported that solid lipid nanoparticles (SLNs), polymeric nanoparticles and micelles all can release drug in acidic microenvironment of multidrug resistance cells and can deliver anticancer drugs in order to overcome P-gp mediated multidrug resistance. Lipid bilayers encapsulated hydrophobic and hydrophilic drugs have long circulating time, liposomes have the ability in overcoming drug resistance due to its preferential accumulation in tumor tissues via enhanced permeability and retention effect. In addition, inorganic nanoparticles can overcome multi-drug resistance in preventing over expression of ABC transporters [120]. Besides, FDA approved nanodrugs stimulates more exploration for other nanoscale particles in gene delivery. For example, incorporating nanoparticles to TRAIL (TNF-related apoptosis-inducing ligand) for inducing apoptosis and interfering cancer stem cells self-renewal pathways altered drug resistance [121]. In addition, Taratula developed double targeting NLCs (nanostructured lipid carriers) which are loaded with an anticancer drug such as Doxorubicin or Paclitaxel. NLCs have the ability in inducing cell death and siRNA targeting MRP1 mRNA and BCL2 mRNA to inhibit pump and nonpump cellular resistance in lung cancer cells respectively, thus overcoming drug resistance [122]. Zhang *et al.* investigated a novel nanostructured lipid-dextran sulfate hybrid carriers and they found it could overcome MDR of Mitoxantrone hydrochloride via enhanced cytotoxicity and invaded cells by clathrin-mediated endocytosis with high drug accumulation in BCRP overexpressing MCF-7/MX cells [123].

Long Duration of Action

After nanoparticles enter into the blood, a large number of them will be rapidly taken in by the reticuloendothelial system and blood concentration will decrease rapidly, followed by drug release from the nanoparticles and blood re-entry; this mechanism lengthens the time of drug action. For example, the half-life of paclitaxel-loaded solid lipid nanoparticles modified with Brij78 and Pluronic F68 was 4.88 h and 10.06 h, while respectively the half-life of paclitaxel was 1.36 h. Both nanoparticles could prolong the *in vivo* duration time of paclitaxel [124]. Additionally, in the experiment of Daunorubicin long circulating liposomes, the long chain of PEG formed a sterically hindered and hydrophobic protective layer on the surface of liposomes and could prevent the recognition and uptake of liposomes by the reticuloendothelial system. Therefore, the half-life and AUC (area under concentration-time curve) were 17.6 and 96 times of that of common injection respectively, which significantly prolonged blood circulation time (Table **4**) allowing more daunorubicin to reach the target site [125].

ASSESSING THE SAFETY OF CLINICAL NANODRUG

Degradation and Clearance of Nanocarrier

The residual nanoparticle carrier can be made of a biologically inert material and nonbiological material. Most developed nanoparticle-drug systems are made of biodegradable materials (*i.e.* phospholipids, lipids, dextran, and chitosan), which allow the release of the drug after safe degradation of the nanoparticle carrier. On the opposite, nonbiological carriers, such as inorganic nanoparticles, are relatively stable under a range of temperature and pH which may raise concerns concerning their lack of biodegradation after drug delivery. Consequently, it is necessary for nonbiological carriers to remove safely from the body or store in a stable state within the body (*i.e.* within inactive macrophages). The physico-chemical properties of a nanomaterial and the formation of a "protein corona" have a significant influence on nanomedicine clearance. Opsonization is one of the principal mechanisms underlying nanoparticle degradation and metabolic clearance. This is the process by which an external organism or particle gets coated with opsonin proteins existing in human blood serum. The opsonin receptors in the cell membrane make it more evident to phagocytic cells of the mononuclear phagocytic system; monocytes in blood and macrophages in tissue are included. The circulation half-life is different for different nanodrugs, for example, doxorubicin is 3 h, paclitaxel is 12 h, DACH-platinum is 70 h and arginine deaminase is 7 days in the human body [12].

Table 4. Pharmacokinetic parameters of daunorubicin long circulating liposomes (Lip) and conventional injections (Con) [126].

Parameters	Lip	Con	Con/Lip	Lip/Con
T1/2α	1.94±0.14	0.11±0.06	--	17.6
T1/2β	13±8	7.3±2.8	--	1.8
Vc/L	0.086±0.008	8.6±1.2	100	
AUC/μg/(ml·h)	35211	3.6±1.0	--	96
CL/L/h	0.028±0.0008	2.8±0.7	99.3	

*is half-life of drug distribution, T1/2β is half-life of drug elimination, AUC is the area under concentration-time curve, representing the degree of drug absorption and bioavailability.

Toxicity

At present, the toxicity of nanodrug is still under detailed evaluation. Generally speaking, part of nanomaterials have been proved to be toxic to human cell lines and tissues, leading in increased oxidative stress, cell death and inflammatory cytokine production [127]. In addition to questions that non-degradable or slowly degradable nanoparticles accumulate in body organs as mentioned above, their interference or potential interaction with biological processes inside the body is another concern.

Toxicity of conventional anticancer drugs usually show dose-response relationship, as to nanodrugs, size-response, nanostructure-response, surface chemistry-response are also needed to take into consideration. The physical and chemical properties of the nanoparticles themselves (size, purity, the surface properties including surface charge, hydrophobicity property and surface modification, and the external environmental conditions determining the interaction of the nanomaterials with the cells) all have influence on the cytotoxicity of the nanoparticles. By optimizing the physical and chemical properties of nanodrug, the direct effect or interaction of nanoparticles and organisms can be improved. When the size of the nanoparticles decreases, the surface area of the nanoparticles increases exponentially, so that more atoms or molecules are distributed on the surface of the nanoparticles rather than inside, and the nanodrug become more reactive with the cells in physical or chemical reactions [128 - 130]. Another key factor affecting the cytotoxicity of nanoparticles is the dose and exposure time. Studies have shown that the larger the exposure dose of nanoparticles or the longer the exposure time, the greater the cyotototoxicity produced [131 - 132]. Therefore, it is necessary to reduce the exposure dose and time of the nanoparticles. In the studies of Ruenraroengsak and Tetley [45, 133], it is shown that positively charged nanoparticles are more cytotoxic than negatively charged nanoparticles, which are more likely to cause hemolysis and platelet aggregation, whereas neutral nanoparticles are shown the least toxicity. When toxic nanoparticles are applied in biomedicine, they can be better utilized by modifying their surface charges to reduce their toxicity or enhance their biocompatibility. In addition, the hydrophobicity of nanoparticles on the surface of the nanoparticles also affected the cytotoxicity induced by them [134]. Experiments on the interaction between hydrophobic Au NPs and cells showed that the hydrophobic gold nanoparticles induced more obvious membrane destruction and autophagy.

However, compared to the conventional anticancer drugs, nanoparticles with targeting function can increase the drug concentration and lower the concentration of other parts of the body, thereby greatly reducing the systemic toxicity of drugs. After drug nanoparticlization, the absolute increased absorption can reduce drug delivery dose, achieving the goal of reducing drug side effects. Furthermore, the sustained release effects of nanoparticles reduce the blood drug concentration fluctuation, so as to improve the safety of the drug.

For example, mitoxantrone (MTO) is an anti-cancer drug representative and belongs to nonspecific anti-cancer drugs, mainly used in clinical practice for acute leukemia, breast cancer and malignant lymphoma. Its obvious side effects are myelosuppression and cardiac toxicity, especially leukopenia. In order to reduce the toxic side effects, three nano-carriers, namely albumin nanospheres, chitosan nanospheres, and solid lipid nanospheres were designed. Experimental studies have shown that these three nanospheres of MTO have lymphatic targeting properties; in addition, the effect of breast cancer and lymphatic metastasis are better than the original non-nanodrugs [135, 136]. Doxorubicin can be used to treat malignancies such as breast cancer. However, it has dangerous side effect, especially cardiomyopathy. However, liposomal doxorubicin: Myocet, demonstrates comparatively favorable toxicity including less myelosuppression, vomiting, alopecia and nausea [137]. A retrospective case series reported that DepoCyt was well tolerated in total. But 12.5% patients couldn't be easily identified and pretreated, father developed serious treatment-related neurological complications.

THE CHALLENGES AND FUTURE DIRECTIONS FOR CLINICAL TRANSLATION

Challenges for Clinical Transformation

Despite a large number of investments and the scientific publications regarding "nanotechnology for cancer treatment" in the last 10 years, the translation of anticancer nanodrugs into clinical practice has been slow when compared to that for small-molecule drugs. In fact, the majority of nanodrugs designed for clinical use are just at the stage of *in vivo* evaluation, and even fewer have reached ideal clinical trials [138 - 139]. The most nanodrugs applied in patients are far of giving the expected clinical results.

Thus, this field has many challenges for clinical translation. Furthermore, our understanding of the relationship between nanoparticle structure and behavior in biologically complex environments like the human body is far from complete. This limitation makes it difficult to predict nanocarrier biodistribution, design rational nanomedicines and develop nano-specific modeling tools for broad application.

Future Directions

Development of Predictive Animal Models

It's worth noting that murine models, including cancer cell line choose and murine genomic profile, are not capable of mimicking the real human biological environment. Researchers do not pay more attention on evaluating or select the most proper tumor model or genomic profile cell lines, therefore, the experimental data could not be translated exactly into the clinic.

Larger animals can be considered and recommended for using more frequently in the pre-clinical trials. For example, rabbits, pigs, dogs, or nonhuman primates. Although the budgets limitation, the results from larger animals is more reliable and be beneficial more in development of effective nanodrugs.

Real-Time Physico-Chemical Characterization in Human Body

Another weakness in the nanomedicine translation lies in lack of the real-time physico-chemical characterization. The physical and chemical characteristics of nanomaterials are size, surface charge, ligand density and *et al*, but there is still a lack of knowledge on the most suitable characterization techniques to be employed. In addition, nanomedicine characterization is often performed under conditions that has difficult in reflecting the complexity of the physiological environment adequately. In order to accelerate the transition from lab to clinic therapy, it is vital for investigators to employ methods that are appropriate, standardized, and widely accepted.

The Integration of Different Scientific Disciplines

Numerous anticancer nanodrugs are being investigated currently in clinical trials. Cancer nanodrugs lie at the interface of various scientific disciplines including physics, chemistry, biology, physical-chemistry, clinical medicine, nanoscience, medical oncology, engineering, and pharmacology (Fig. **5**). It is of great significance to integrate all the relevant disciplines to break through the limits of discipline-specific knowledge, and combine the practical needs of medicine with the huge potential of nanotechnology better. Ultimately, it has become clear that the critical steps for clinical translation of nanotherapy require further interdisciplinary and international collaboration to stimulate clinical translation of nanotherapeutics.

OUTLOOK AND CONCLUSIONS

In recent years, with the development of new nanoparticles for the treatment of cancer, there has been an unprecedented evolution in the field of nanodrugs. Compared with conventional drugs, nanodrugs have unique biological properties. Their small size and large surface area-to-volume ratio make it easy to absorb, bind and carry compounds with higher efficiency comparing with free drugs. Furthermore, its specific drug delivery broadens our horizons profoundly and increases the possibilities for improving cancer therapy efficacy greatly. Their tunable size, shape, and surface chemistry also enable them to have higher stability, higher carrier capacity, lower toxicity, better biocompatibility and the ability to incorporate both hydrophilic and hydrophobic materials [101]. However, as a special commodity closely related to people's health, the safety of nanodrugs should not be neglected when presenting their attractive nanobiological effects. The common problems in the study of nanomedicine safety includes the interactions of biological macromolecules and cells in nanocarrier system, the correlation between the pharmacokinetic characteristics of nanomedicine and toxicity, the relationship of particle size, surface charge, hydrophilic lipophilicity, and bioactivity of nanoparticles. At present, the safety evaluation of nanomaterials is focused on the *in vitro* cytotoxicity of nanodrugs. Nanodrugs' toxicity in the target tissue, the body's immune system, the organization's tumorigenicity, and the impact of various barrier functions still require further investigation.

CONFLICT OF INTEREST

The author(s) confirm that this chapter content has no conflict of interest.

ACKNOWLEDGMENTS

Declared none.

REFERENCES

[1] Marchal S, El Hor A, Millard M, Gillon V, Bezdetnaya L. Anticancer drug delivery: an update on clinically applied nanotherapeutics. Drugs 2015; 75(14): 1601-11. [http://dx.doi.org/10.1007/s40265-015-0453-3] [PMID: 26323338]

[2] Wicki A, Witzigmann D, Balasubramanian V, Huwyler J. Nanomedicine in cancer therapy: challenges, opportunities, and clinical applications. J Control Release 2015; 200: 138-57. [http://dx.doi.org/10.1016/j.jconrel.2014.12.030] [PMID: 25545217]

[3] Bregoli L, Movia D, Gavigan-Imedio JD, Lysaght J, Reynolds J, Prina-Mello A. Nanomedicine applied to translational oncology: A future perspective on cancer treatment. Nanomedicine (Lond) 2016; 12(1): 81-103. [http://dx.doi.org/10.1016/j.nano.2015.08.006] [PMID: 26370707]

[4] Illum L, Davis SS, Wilson CG, Thomas NW, Frier M, Hardy JG. Blood clearance and organ deposition of intravenously administered colloidal particles - the effects of particle-size, nature and shape. Int J Pharm 1982; 12: 135-46. [http://dx.doi.org/10.1016/0378-5173(82)90113-2]

[5] Petros RA, DeSimone JM. Strategies in the design of nanoparticles for therapeutic applications. Nat Rev Drug Discov 2010; 9(8): 615-27. [http://dx.doi.org/10.1038/nrd2591] [PMID: 20616808]

[6] Vinogradov SV, Bronich TK, Kabanov AV. Nanosized cationic hydrogels for drug delivery: preparation, properties and interactions with cells. Adv Drug Deliv Rev 2002; 54(1): 135-47. [http://dx.doi.org/10.1016/S0169-409X(01)00245-9] [PMID: 11755709]

[7] Choi HS, Liu W, Misra P, *et al.* Renal clearance of quantum dots. Nat Biotechnol 2007; 25(10): 1165-70. [http://dx.doi.org/10.1038/nbt1340] [PMID: 17891134]

[8] Cabral H, Matsumoto Y, Mizuno K, *et al.* Accumulation of sub-100 nm polymeric micelles in poorly permeable tumours depends on size. Nat Nanotechnol 2011; 6(12): 815-23. [http://dx.doi.org/10.1038/nnano.2011.166] [PMID: 22020122]

[9] Farokhzad OC, Langer R. Impact of nanotechnology on drug delivery. ACS Nano 2009; 3(1): 16-20. [http://dx.doi.org/10.1021/nn900002m] [PMID: 19206243]

[10] Fang J, Nakamura H, Maeda H. The EPR effect: Unique features of tumor blood vessels for drug delivery, factors involved, and limitations and augmentation of the effect. Adv Drug Deliv Rev 2011; 63(3): 136-51. [http://dx.doi.org/10.1016/j.addr.2010.04.009] [PMID: 20441782]

[11] Chauhan VP, Stylianopoulos T, Martin JD, *et al.* Normalization of tumour blood vessels improves the delivery of nanomedicines in a size-dependent manner. Nat Nanotechnol 2012; 7(6): 383-8. [http://dx.doi.org/10.1038/nnano.2012.45] [PMID: 22484912]

[12] Alexis F, Pridgen E, Molnar LK, Farokhzad OC. Factors affecting the clearance and biodistribution of

polymeric nanoparticles. Mol Pharm 2008; 5(4): 505-15.
[http://dx.doi.org/10.1021/mp800051m] [PMID: 18672949]

[13] Rajendran L, Knölker HJ, Simons K. Subcellular targeting strategies for drug design and delivery. Nat Rev Drug Discov 2010; 9(1): 29-42.
[http://dx.doi.org/10.1038/nrd2897] [PMID: 20043027]

[14] Conner SD, Schmid SL. Regulated portals of entry into the cell. Nature 2003; 422(6927): 37-44.
[http://dx.doi.org/10.1038/nature01451] [PMID: 12621426]

[15] Rejman J, Oberle V, Zuhorn IS, Hoekstra D. Size-dependent internalization of particles via the pathways of clathrin- and caveolae-mediated endocytosis. Biochem J 2004; 377(Pt 1): 159-69.
[http://dx.doi.org/10.1042/bj20031253] [PMID: 14505488]

[16] Hillaireau H, Couvreur P. Nanocarriers' entry into the cell: relevance to drug delivery. Cell Mol Life Sci 2009; 66(17): 2873-96.
[http://dx.doi.org/10.1007/s00018-009-0053-z] [PMID: 19499185]

[17] Huang JG, Leshuk T, Gu FX. Emerging nanomaterials for targeting subcellular organelles. Nano Today 2011; 6: 478-92.
[http://dx.doi.org/10.1016/j.nantod.2011.08.002]

[18] Batist G. Improving the therapeutic index when using Myocet™ in the treatment of metastatic breast cancer. Breast 2001; 10: 16-21.
[http://dx.doi.org/10.1016/S0960-9776(01)80003-5]

[19] Lao J, Madani J, Puértolas T, *et al.* Liposomal Doxorubicin in the treatment of breast cancer patients: a review 2013.
[http://dx.doi.org/10.1155/2013/456409]

[20] Smith JA, Mathew L, Burney M, Nyshadham P, Coleman RL. Equivalency challenge: Evaluation of Lipodox® as the generic equivalent for Doxil® in a human ovarian cancer orthotropic mouse model. Gynecol Oncol 2016; 141(2): 357-63.
[http://dx.doi.org/10.1016/j.ygyno.2016.02.033] [PMID: 26946092]

[21] Barenholz Y. Doxil®--the first FDA-approved nano-drug: lessons learned. J Control Release 2012; 160(2): 117-34.
[http://dx.doi.org/10.1016/j.jconrel.2012.03.020] [PMID: 22484195]

[22] Maier-Hauff K, Ulrich F, Nestler D, *et al.* Efficacy and safety of intratumoral thermotherapy using magnetic iron-oxide nanoparticles combined with external beam radiotherapy on patients with recurrent glioblastoma multiforme. J Neurooncol 2011; 103(2): 317-24.
[http://dx.doi.org/10.1007/s11060-010-0389-0] [PMID: 20845061]

[23] Rivera Gil P, Hühn D, del Mercato LL, Sasse D, Parak WJ. Nanopharmacy: Inorganic nanoscale devices as vectors and active compounds. Pharmacol Res 2010; 62(2): 115-25.
[http://dx.doi.org/10.1016/j.phrs.2010.01.009] [PMID: 20097288]

[24] Montero AJ, Adams B, Diaz-Montero CM, Glück S. Nab-paclitaxel in the treatment of metastatic breast cancer: a comprehensive review. Expert Rev Clin Pharmacol 2011; 4(3): 329-34.
[http://dx.doi.org/10.1586/ecp.11.7] [PMID: 22114779]

[25] Green MR, Manikhas GM, Orlov S, *et al.* Abraxane, a novel Cremophor-free, albumin-bound particle form of paclitaxel for the treatment of advanced non-small-cell lung cancer. Ann Oncol 2006; 17(8): 1263-8.
[http://dx.doi.org/10.1093/annonc/mdl104] [PMID: 16740598]

[26] Yang X. Naodrugs. Tsinghua University Press 2007.

[27] Danson S, Ferry D, Alakhov V, *et al.* Phase I dose escalation and pharmacokinetic study of pluronic polymer-bound doxorubicin (SP1049C) in patients with advanced cancer. Br J Cancer 2004; 90(11): 2085-91.
[PMID: 15150584]

[28] Schluep T, Cheng J, Khin KT, Davis ME. Pharmacokinetics and biodistribution of the camptothecin-polymer conjugate IT-101 in rats and tumor-bearing mice. Cancer Chemother Pharmacol 2006; 57(5): 654-62.
[http://dx.doi.org/10.1007/s00280-005-0091-7] [PMID: 16133526]

[29] Kim DW, Kim SY, Kim HK, *et al.* Multicenter phase II trial of Genexol-PM, a novel Cremophor-free, polymeric micelle formulation of paclitaxel, with cisplatin in patients with advanced non-small-cell lung cancer. Ann Oncol 2007; 18(12): 2009-14.
[http://dx.doi.org/10.1093/annonc/mdm374] [PMID: 17785767]

[30] Kim SC, Kim DW, Shim YH, *et al.* *In vivo* evaluation of polymeric micellar paclitaxel formulation: toxicity and efficacy. J Control Release 2001; 72(1-3): 191-202.
[http://dx.doi.org/10.1016/S0168-3659(01)00275-9] [PMID: 11389998]

[31] Campone M, Rademaker-Lakhai JM, Bennouna J, *et al.* Phase I and pharmacokinetic trial of AP5346, a DACH-platinum-polymer conjugate, administered weekly for three out of every 4 weeks to advanced solid tumor patients. Cancer Chemother Pharmacol 2007; 60(4): 523-33.
[http://dx.doi.org/10.1007/s00280-006-0397-0] [PMID: 17308894]

[32] Rice JR, Gerberich JL, Nowotnik DP, Howell SB. Preclinical efficacy and pharmacokinetics of AP5346, a novel diaminocyclohexane-platinum tumor-targeting drug delivery system. Clin Cancer Res 2006; 12(7 Pt 1): 2248-54.
[http://dx.doi.org/10.1158/1078-0432.CCR-05-2169] [PMID: 16609041]

[33] Homsi J, Simon GR, Garrett CR, *et al.* Phase I trial of poly-L-glutamate camptothecin (CT-2106) administered weekly in patients with advanced solid malignancies. Clin Cancer Res 2007; 13(19): 5855-61.
[http://dx.doi.org/10.1158/1078-0432.CCR-06-2821] [PMID: 17908979]

[34] Schluep T, Hwang J, Cheng J, *et al.* Preclinical efficacy of the camptothecin-polymer conjugate IT-101 in multiple cancer models. Clin Cancer Res 2006; 12(5): 1606-14.
[http://dx.doi.org/10.1158/1078-0432.CCR-05-1566] [PMID: 16533788]

[35] Hopewel JW, Duncan R, Wilding D, Chakrabarti K. Preclinical evaluation of the cardiotoxicity of PK2: a novel HPMA copolymer-doxorubicin-galactosamine conjugate antitumour agent. Hum Exp Toxicol 2001; 20(9): 461-70.
[http://dx.doi.org/10.1191/096032701682693017] [PMID: 11776408]

[36] Wang ZL. Structural Analysis of Self-Assembling Nanocrystal Superlattices. International Joint Conference on Ijcai. 13-30.

[37] Caruso F. Nanoengineering of Particle Surfaces. Adv Mater 2001; 13: 11-22.
[http://dx.doi.org/10.1002/1521-4095(200101)13:1<11::AID-ADMA11>3.0.CO;2-N]

[38] Zhang L, Gu FX, Chan JM, Wang AZ, Langer RS, Farokhzad OC. Nanoparticles in medicine: therapeutic applications and developments. Clin Pharmacol Ther 2008; 83(5): 761-9.
[http://dx.doi.org/10.1038/sj.clpt.6100400] [PMID: 17957183]

[39] Jain DK. Nanotechnology: A safe and effective drug delivery system. Asian Journal of Pharmaceutical & Clinical Research 2010; 3: 159-65.

[40] Liu L, Ye Q, Lu M, *et al.* A new approach to reduce toxicities and to improve bioavailabilities of platinum-containing anti-cancer nanodrugs. Sci Rep 2015; 5: 10881.
[http://dx.doi.org/10.1038/srep10881] [PMID: 26039249]

[41] Marqibo Rx. 2017. Available at: http://www.empr.com/marqibo/drug/8210/

[42] He C, Hu Y, Yin L, Tang C, Yin C. Effects of particle size and surface charge on cellular uptake and biodistribution of polymeric nanoparticles. Biomaterials 2010; 31(13): 3657-66.
[http://dx.doi.org/10.1016/j.biomaterials.2010.01.065] [PMID: 20138662]

[43] Verma A, Stellacci F. Effect of surface properties on nanoparticle-cell interactions. Small 2010; 6(1):

12-21.
[http://dx.doi.org/10.1002/smll.200901158] [PMID: 19844908]

[44] Leroueil PR, Berry SA, Duthie K, *et al.* Banaszak Holl. Wide Varieties of Cationic Nanoparticles Induce Defects in Supported Lipid Bilayers. Nano Lett 2014; 8(2): 420-4.
[http://dx.doi.org/10.1021/nl0722929] [PMID: 18217783]

[45] Ruenraroengsak P, Tetley TD. Differential bioreactivity of neutral, cationic and anionic polystyrene nanoparticles with cells from the human alveolar compartment: robust response of alveolar type 1 epithelial cells. Part Fibre Toxicol 2015; 12: 19.
[http://dx.doi.org/10.1186/s12989-015-0091-7] [PMID: 26133975]

[46] Shao XR, Wei XQ, Song X, *et al.* Independent effect of polymeric nanoparticle zeta potential/surface charge, on their cytotoxicity and affinity to cells. Cell Prolif 2015; 48(4): 465-74.
[http://dx.doi.org/10.1111/cpr.12192] [PMID: 26017818]

[47] Bertrand N, Wu J, Xu X, Kamaly N, Farokhzad OC. Cancer nanotechnology: the impact of passive and active targeting in the era of modern cancer biology. Adv Drug Deliv Rev 2014; 66: 2-25.
[http://dx.doi.org/10.1016/j.addr.2013.11.009] [PMID: 24270007]

[48] Stylianopoulos T, Poh MZ, Insin N, *et al.* Diffusion of particles in the extracellular matrix: the effect of repulsive electrostatic interactions. Biophys J 2010; 99(5): 1342-9.
[http://dx.doi.org/10.1016/j.bpj.2010.06.016] [PMID: 20816045]

[49] Fröhlich E. The role of surface charge in cellular uptake and cytotoxicity of medical nanoparticles. Int J Nanomedicine 2012; 7: 5577-91.
[http://dx.doi.org/10.2147/IJN.S36111] [PMID: 23144561]

[50] Felber AE, Dufresne MH, Leroux JC. pH-sensitive vesicles, polymeric micelles, and nanospheres prepared with polycarboxylates. Adv Drug Deliv Rev 2012; 64(11): 979-92.
[http://dx.doi.org/10.1016/j.addr.2011.09.006] [PMID: 21996056]

[51] Xia T, Kovochich M, Liong M, Zink JI, Nel AE. Cationic polystyrene nanosphere toxicity depends on cell-specific endocytic and mitochondrial injury pathways. ACS Nano 2008; 2(1): 85-96.
[http://dx.doi.org/10.1021/nn700256c] [PMID: 19206551]

[52] Saltzman WM, Torchilin VP. Drug delivery systems. AccessScience 2014.

[53] Torchilin VP. Drug targeting. Eur J Pharm Sci 2000; 11 (Suppl. 2): S81-91.
[http://dx.doi.org/10.1016/S0928-0987(00)00166-4] [PMID: 11033430]

[54] Oliveira MC, Fattal E, Ropert C, Malvy C, Couvreur P. Delivery of antisense oligonucleotides by means of pH-sensitive liposomes. J Control Release 1997; 48: 179-84.
[http://dx.doi.org/10.1016/S0168-3659(97)00042-4]

[55] Löbenberg R, Araujo L, von Briesen H, Rodgers E, Kreuter J. Body distribution of azidothymidine bound to hexyl-cyanoacrylate nanoparticles after i.v. injection to rats. J Control Release 1998; 50(1-3): 21-30.
[http://dx.doi.org/10.1016/S0168-3659(97)00105-3] [PMID: 9685869]

[56] Goodwin S, Peterson C, Hoh C, Bittner C. Targeting and retention of magnetic targeted carriers (MTCs) enhancing intra-arterial chemotherapy. J Magn Magn Mater 1999; 194: 132-9.
[http://dx.doi.org/10.1016/S0304-8853(98)00584-8]

[57] Alexiou C, Arnold W, Hulin P, *et al.* Magnetic mitoxantrone nanoparticle detection by histology, X-ray and MRI after magnetic tumor targeting. J Magn Magn Mater 2001; 225: 187-93.
[http://dx.doi.org/10.1016/S0304-8853(00)01256-7]

[58] Lu B. New dosage form and new technology of traditional chinese medicine. Chemical Industry Press 2008; pp. 297-8.

[59] Zhong Y, Meng F, Deng C, Zhong Z. Ligand-directed active tumor-targeting polymeric nanoparticles for cancer chemotherapy. Biomacromolecules 2014; 15(6): 1955-69.

[http://dx.doi.org/10.1021/bm5003009] [PMID: 24798476]

[60] Peer D, Karp JM, Hong S, Farokhzad OC, Margalit R, Langer R. Nanocarriers as an emerging platform for cancer therapy. Nat Nanotechnol 2007; 2(12): 751-60.
[http://dx.doi.org/10.1038/nnano.2007.387] [PMID: 18654426]

[61] Yu MK, Park J, Jon S. Targeting strategies for multifunctional nanoparticles in cancer imaging and therapy. Theranostics 2012; 2(1): 3-44.
[http://dx.doi.org/10.7150/thno.3463] [PMID: 22272217]

[62] Steichen SD, Caldorera-Moore M, Peppas NA. A review of current nanoparticle and targeting moieties for the delivery of cancer therapeutics. Eur J Pharm Sci 2013; 48(3): 416-27.
[http://dx.doi.org/10.1016/j.ejps.2012.12.006] [PMID: 23262059]

[63] Guragain S, Bastakoti BP, Malgras V, Nakashima K, Yamauchi Y. Multi-Stimuli-Responsive Polymeric Materials. Chemistry 2015; 21(38): 13164-74.
[http://dx.doi.org/10.1002/chem.201501101] [PMID: 26219746]

[64] Stuart MA, Huck WT, Genzer J, *et al.* Emerging applications of stimuli-responsive polymer materials. Nat Mater 2010; 9(2): 101-13.
[http://dx.doi.org/10.1038/nmat2614] [PMID: 20094081]

[65] Yuliang Z. Sensing system for mimicking cancer cell-drug interaction. Science Bulletin 2015; 60: 1218-9.
[http://dx.doi.org/10.1007/s11434-015-0835-7]

[66] Das D, Ghosh P, Ghosh A, *et al.* Stimulus-responsive, Biodegradable, Biocompatible, Covalently Crosslinked Hydrogel Based on Dextrin and Poly (N-isopropyl acrylamide) for In-vitro/In-vivo Controlled Drug Release. ACS Appl Mater Inter 2015; p. 7.

[67] Zhang W, Gao C. Recent advances in cell imaging and cytotoxicity of intracellular stimuli-responsive nanomaterials. Science Bulletin 2015; 60: 1973-9.
[http://dx.doi.org/10.1007/s11434-015-0952-3]

[68] Yang K, Feng L, Liu Z. Stimuli responsive drug delivery systems based on nano-graphene for cancer therapy. Adv Drug Deliv Rev 2016; 105(Pt B): 228-41.
[http://dx.doi.org/10.1016/j.addr.2016.05.015] [PMID: 27233212]

[69] Ganta S, Devalapally H, Shahiwala A, Amiji M. A review of stimuli-responsive nanocarriers for drug and gene delivery. J Control Release 2008; 126(3): 187-204.
[http://dx.doi.org/10.1016/j.jconrel.2007.12.017] [PMID: 18261822]

[70] Mura S, Nicolas J, Couvreur P. Stimuli-responsive nanocarriers for drug delivery. Nat Mater 2013; 12(11): 991-1003.
[http://dx.doi.org/10.1038/nmat3776] [PMID: 24150417]

[71] Banerjee S, Sen K, Pal TK, Guha SK. Poly(styrene-co-maleic acid)-based pH-sensitive liposomes mediate cytosolic delivery of drugs for enhanced cancer chemotherapy. Int J Pharm 2012; 436(1-2): 786-97.
[http://dx.doi.org/10.1016/j.ijpharm.2012.07.059] [PMID: 22884831]

[72] Karanth H, Murthy RS. pH-sensitive liposomes--principle and application in cancer therapy. J Pharm Pharmacol 2007; 59(4): 469-83.
[http://dx.doi.org/10.1211/jpp.59.4.0001] [PMID: 17430630]

[73] Gao ZG, Lee DH, Kim DI, Bae YH. Doxorubicin loaded pH-sensitive micelle targeting acidic extracellular pH of human ovarian A2780 tumor in mice. J Drug Target 2005; 13(7): 391-7.
[http://dx.doi.org/10.1080/10611860500376741] [PMID: 16308207]

[74] Babu A, Templeton AK, Munshi A, Ramesh R. Nanodrug delivery systems: a promising technology for detection, diagnosis, and treatment of cancer. AAPS PharmSciTech 2014; 15(3): 709-21.
[http://dx.doi.org/10.1208/s12249-014-0089-8] [PMID: 24550101]

[75] Davis ME, Chen ZG, Shin DM. Nanoparticle therapeutics: an emerging treatment modality for cancer. Nat Rev Drug Discov 2008; 7(9): 771-82.
[http://dx.doi.org/10.1038/nrd2614] [PMID: 18758474]

[76] Dawidczyk CM, Kim C, Park JH, *et al.* State-of-the-art in design rules for drug delivery platforms: lessons learned from FDA-approved nanomedicines. J Control Release 2014; 187: 133-44.
[http://dx.doi.org/10.1016/j.jconrel.2014.05.036] [PMID: 24874289]

[77] Mansour HM, Park CW, Bawa R. Book Chapter: Design and Development of Approved Nanopharmaceutical Products 1-33.

[78] Torchilin VP. Multifunctional nanocarriers. Adv Drug Deliv Rev 2006; 58(14): 1532-55.
[http://dx.doi.org/10.1016/j.addr.2006.09.009] [PMID: 17092599]

[79] Cevc G, Richardsen H. Lipid vesicles and membrane fusion. Adv Drug Deliv Rev 1999; 38(3): 207-32.
[http://dx.doi.org/10.1016/S0169-409X(99)00030-7] [PMID: 10837758]

[80] Gabizon A, Shmeeda H, Barenholz Y. Pharmacokinetics of pegylated liposomal Doxorubicin: review of animal and human studies. Clin Pharmacokinet 2003; 42(5): 419-36.
[http://dx.doi.org/10.2165/00003088-200342050-00002] [PMID: 12739982]

[81] Sanna V, Pala N, Sechi M. Targeted therapy using nanotechnology: focus on cancer. Int J Nanomedicine 2014; 9: 467-83.
[PMID: 24531078]

[82] Tan S, Li X, Guo Y, Zhang Z. Lipid-enveloped hybrid nanoparticles for drug delivery. Nanoscale 2013; 5(3): 860-72.
[http://dx.doi.org/10.1039/c2nr32880a] [PMID: 23292080]

[83] Waterhouse DN, Tardi PG, Mayer LD, Bally MB. A comparison of liposomal formulations of doxorubicin with drug administered in free form: changing toxicity profiles. Drug Saf 2001; 24(12): 903-20.
[http://dx.doi.org/10.2165/00002018-200124120-00004] [PMID: 11735647]

[84] Forssen EA. The design and development of DaunoXome® for solid tumor targeting *in vivo.* Adv Drug Deliv Rev 1997; 24: 133-50.
[http://dx.doi.org/10.1016/S0169-409X(96)00453-X]

[85] Allen TM, Martin FJ. Advantages of liposomal delivery systems for anthracyclines. Semin Oncol 2004; 31(6) (Suppl. 13): 5-15.
[http://dx.doi.org/10.1053/j.seminoncol.2004.08.001] [PMID: 15717735]

[86] Fassas A, Anagnostopoulos A. The use of liposomal daunorubicin (DaunoXome) in acute myeloid leukemia. Leuk Lymphoma 2005; 46(6): 795-802.
[http://dx.doi.org/10.1080/10428190500052438] [PMID: 16019523]

[87] Wang R, Billone PS, Mullett WM. Nanomedicine in action: an overview of cancer nanomedicine on the market and in clinical trials. J Nanomater 2013; 2013: 101-7.

[88] Phuphanich S, Maria B, Braeckman R, Chamberlain M. A pharmacokinetic study of intra-CSF administered encapsulated cytarabine (DepoCyt) for the treatment of neoplastic meningitis in patients with leukemia, lymphoma, or solid tumors as part of a phase III study. J Neurooncol 2007; 81(2): 201-8.
[http://dx.doi.org/10.1007/s11060-006-9218-x] [PMID: 16941075]

[89] Angst MS, Drover DR. Pharmacology of drugs formulated with DepoFoam: a sustained release drug delivery system for parenteral administration using multivesicular liposome technology. Clin Pharmacokinet 2006; 45(12): 1153-76.
[http://dx.doi.org/10.2165/00003088-200645120-00002] [PMID: 17112293]

[90] Silverman JA, Deitcher SR. Marqibo® (vincristine sulfate liposome injection) improves the

pharmacokinetics and pharmacodynamics of vincristine. Cancer Chemother Pharmacol 2013; 71(3): 555-64.
[http://dx.doi.org/10.1007/s00280-012-2042-4] [PMID: 23212117]

[91] Fogler WE, Fidler IJ. Comparative interaction of free and liposome-encapsulated nor-muramyl dipeptide or muramyl tripeptide phosphatidylethanolamine (3H-labelled) with human blood monocytes. Int J Immunopharmacol 1987; 9(2): 141-50.
[http://dx.doi.org/10.1016/0192-0561(87)90088-9] [PMID: 3583507]

[92] Langer R. Biomaterials in drug delivery and tissue engineering: one laboratory's experience. Acc Chem Res 2000; 33(2): 94-101.
[http://dx.doi.org/10.1021/ar9800993] [PMID: 10673317]

[93] Curnis F, Sacchi A, Longhi R, Colombo B, Gasparri A, Corti A. Abstract 5617: A new alphaV/beta3 integrin selective carrier for nanodrug delivery to tumors based on isoDGR-tagged albumin. Cancer Res 2013; 73: 5617-17.
[http://dx.doi.org/10.1158/1538-7445.AM2013-5617]

[94] Fasano M, Curry S, Terreno E, *et al.* The extraordinary ligand binding properties of human serum albumin. 2005.

[95] Peters T Jr. Serum albumin. Adv Protein Chem 1985; 37: 161-245.
[http://dx.doi.org/10.1016/S0065-3233(08)60065-0] [PMID: 3904348]

[96] Loureiro A, Azoia NG, Gomes AC, Cavaco-Paulo A. Albumin-based nanodevices as drug carriers. Curr Pharm Des 2016; 22(10): 1371-90.
[http://dx.doi.org/10.2174/1381612822666160125114900] [PMID: 26806342]

[97] Gradishar WJ, Tjulandin S, Davidson N, *et al.* Phase III trial of nanoparticle albumin-bound paclitaxel compared with polyethylated castor oil-based paclitaxel in women with breast cancer. J Clin Oncol 2005; 23(31): 7794-803.
[http://dx.doi.org/10.1200/JCO.2005.04.937] [PMID: 16172456]

[98] Business. FDA approves Celgene's Abraxane for lung cancer. Thomson Reuters.

[99] Pacilli NF. A Approves a Drug for Late-Stage Pancreatic Cancer 2013. https://prezi.com/3tbilw4sth_o/fda-approves-a-drug-for-late-stage-pancreatic-cancer/

[100] Huang X, Jain PK, El-Sayed IH, El-Sayed MA. Plasmonic photothermal therapy (PPTT) using gold nanoparticles. Lasers Med Sci 2008; 23(3): 217-28.
[http://dx.doi.org/10.1007/s10103-007-0470-x] [PMID: 17674122]

[101] Thakor AS, Gambhir SS. Nanooncology: the future of cancer diagnosis and therapy. CA Cancer J Clin 2013; 63(6): 395-418.
[http://dx.doi.org/10.3322/caac.21199] [PMID: 24114523]

[102] Onoue S, Yamada S, Chan HK. Nanodrugs: pharmacokinetics and safety. Int J Nanomedicine 2014; 9: 1025-37.
[http://dx.doi.org/10.2147/IJN.S38378] [PMID: 24591825]

[103] Moreira JN, Gaspar R, Allen TM. Targeting Stealth liposomes in a murine model of human small cell lung cancer. Biochim Biophys Acta 2001; 1515(2): 167-76.
[http://dx.doi.org/10.1016/S0005-2736(01)00411-4] [PMID: 11718672]

[104] Torchilin VP, Lukyanov AN, Gao Z, Papahadjopoulos-Sternberg B. Immunomicelles: targeted pharmaceutical carriers for poorly soluble drugs. Proc Natl Acad Sci USA 2003; 100(10): 6039-44.
[http://dx.doi.org/10.1073/pnas.0931428100] [PMID: 12716967]

[105] Levchenko TS, Rammohan R, Lukyanov AN, Whiteman KR, Torchilin VP. Liposome clearance in mice: the effect of a separate and combined presence of surface charge and polymer coating. Int J Pharm 2002; 240(1-2): 95-102.
[http://dx.doi.org/10.1016/S0378-5173(02)00129-1] [PMID: 12062505]

[106] Boyd RL. Esthetic orthodontic treatment using the invisalign appliance for moderate to complex malocclusions. J Dent Educ 2008; 72(8): 948-67.
[PMID: 18676803]

[107] Lockman PR, Mumper RJ, Khan MA, Allen DD. Nanoparticle technology for drug delivery across the blood-brain barrier. Drug Dev Ind Pharm 2002; 28(1): 1-13.
[http://dx.doi.org/10.1081/DDC-120001481] [PMID: 11858519]

[108] May JP, Li SD. Hyperthermia-induced drug targeting. Expert Opin Drug Deliv 2013; 10(4): 511-27.
[http://dx.doi.org/10.1517/17425247.2013.758631] [PMID: 23289519]

[109] Azim HA Jr, Awada A. Clinical development of new formulations of cytotoxics in solid tumors. Curr Opin Oncol 2012; 24(3): 325-31.
[http://dx.doi.org/10.1097/CCO.0b013e328351fb29] [PMID: 22343388]

[110] Adkins CE, Nounou MI, Hye T, *et al.* NKTR-102 Efficacy versus irinotecan in a mouse model of brain metastases of breast cancer. BMC Cancer 2015; 15: 685.
[http://dx.doi.org/10.1186/s12885-015-1672-4] [PMID: 26463521]

[111] Dinndorf PA, Gootenberg J, Cohen MH, Keegan P, Pazdur R. FDA drug approval summary: pegaspargase (oncaspar) for the first-line treatment of children with acute lymphoblastic leukemia (ALL). Oncologist 2007; 12(8): 991-8.
[http://dx.doi.org/10.1634/theoncologist.12-8-991] [PMID: 17766659]

[112] Okusaka T, Okada S, Ueno H, *et al.* Transcatheter arterial embolization with zinostatin stimalamer for hepatocellular carcinoma. Oncology 2002; 62(3): 228-33.
[http://dx.doi.org/10.1159/000059570] [PMID: 12065870]

[113] Greish K, Fang J, Inutsuka T, Nagamitsu A, Maeda H. Macromolecular therapeutics: advantages and prospects with special emphasis on solid tumour targeting. Clin Pharmacokinet 2003; 42: 1089-105.
[http://dx.doi.org/10.2165/00003088-200342130-00002]

[114] Masuda E, Maeda H. Antitumor resistance induced by zinostatin stimalamer (ZSS), a polymer-conjugated neocarzinostatin (NCS) derivative. I. Meth A tumor eradication and tumor-neutralizing activity in mice pretreated with ZSS or NCS. Cancer Immunol Immunother 1995; 40(5): 329-38.
[http://dx.doi.org/10.1007/BF01519634] [PMID: 7600566]

[115] Alphandéry E, Grand-Dewyse P, Lefèvre R, Mandawala C, Durand-Dubief M. Cancer therapy using nanoformulated substances: scientific, regulatory and financial aspects. Expert Rev Anticancer Ther 2015; 15(10): 1233-55.
[http://dx.doi.org/10.1586/14737140.2015.1086647] [PMID: 26402250]

[116] Young C, Schluep T, Hwang J, Eliasof S. CRLX101 (formerly IT-101)-A Novel Nanopharmaceutical of Camptothecin in Clinical Development. Curr Bioact Compd 2011; 7(1): 8-14.
[http://dx.doi.org/10.2174/157340711795163866] [PMID: 22081768]

[117] He Q, Shi J. MSN anti-cancer nanomedicines: chemotherapy enhancement, overcoming of drug resistance, and metastasis inhibition. Adv Mater 2014; 26(3): 391-411.
[http://dx.doi.org/10.1002/adma.201303123] [PMID: 24142549]

[118] Conde J, de la Fuente JM, Baptista PV. Nanomaterials for reversion of multidrug resistance in cancer: a new hope for an old idea? Front Pharmacol 2013; 4: 134.
[http://dx.doi.org/10.3389/fphar.2013.00134] [PMID: 24298255]

[119] Han N, Zhao Q, Wan L, *et al.* Hybrid lipid-capped mesoporous silica for stimuli-responsive drug release and overcoming multidrug resistance. ACS Appl Mater Interfaces 2015; 7(5): 3342-51.
[http://dx.doi.org/10.1021/am5082793] [PMID: 25584634]

[120] Kapse-Mistry S, Govender T, Srivastava R, Yergeri M, Yergeri M. Nanodrug delivery in reversing multidrug resistance in cancer cells. Front Pharmacol 2014; 5: 159.
[PMID: 25071577]

[121] Naoum GE, Tawadros F, Farooqi AA, *et al.* Role of nanotechnology and gene delivery systems in TRAIL-based therapies. Ecancermedicalscience 2016; 10: 660.
[http://dx.doi.org/10.3332/ecancer.2016.660] [PMID: 27594905]

[122] Taratula O, Kuzmov A, Shah M, Garbuzenko OB, Minko T. Nanostructured lipid carriers as multifunctional nanomedicine platform for pulmonary co-delivery of anticancer drugs and siRNA. J Control Release 2013; 171(3): 349-57.
[http://dx.doi.org/10.1016/j.jconrel.2013.04.018] [PMID: 23648833]

[123] Zhang P, Ling G, Pan X, *et al.* Novel nanostructured lipid-dextran sulfate hybrid carriers overcome tumor multidrug resistance of mitoxantrone hydrochloride 2011.

[124] Chen DB, Wan Liang LU, Yang TZ, Jing LI, Zhang Q. Preparation and characterization of long-circulating solid lipid nanoparticles containing paclitaxel. J Beijing Med Univ 2002.

[125] Zhang H, Qi XR, Zhang Q. [Study on pharmaceutical characterization and pharmacokinetics of daunorubicin long-circulating liposomes in rat]. Yao Xue Xue Bao 2002; 37(4): 299-303.
[PMID: 12579828]

[126] Xu H. Nano-medicine. Tsinghua University Press 2004.

[127] Oberdörster G, Maynard A, Donaldson K, *et al.* Principles for characterizing the potential human health effects from exposure to nanomaterials: elements of a screening strategy. Part Fibre Toxicol 2005; 2: 8.
[http://dx.doi.org/10.1186/1743-8977-2-8] [PMID: 16209704]

[128] Akhavan O, Ghaderi E, Akhavan A. Size-dependent genotoxicity of graphene nanoplatelets in human stem cells. Biomaterials 2012; 33(32): 8017-25.
[http://dx.doi.org/10.1016/j.biomaterials.2012.07.040] [PMID: 22863381]

[129] Zhang H, Peng C, Yang J, *et al.* Uniform ultrasmall graphene oxide nanosheets with low cytotoxicity and high cellular uptake. ACS Appl Mater Interfaces 2013; 5(5): 1761-7.
[http://dx.doi.org/10.1021/am303005j] [PMID: 23402618]

[130] Gliga AR, Skoglund S, Wallinder IO, Fadeel B, Karlsson HL. Size-dependent cytotoxicity of silver nanoparticles in human lung cells: the role of cellular uptake, agglomeration and Ag release. Part Fibre Toxicol 2014; 11: 11.
[http://dx.doi.org/10.1186/1743-8977-11-11] [PMID: 24529161]

[131] Zhang Y, Ali SF, Dervishi E, *et al.* Cytotoxicity effects of graphene and single-wall carbon nanotubes in neural phaeochromocytoma-derived PC12 cells. ACS Nano 2010; 4(6): 3181-6.
[http://dx.doi.org/10.1021/nn1007176] [PMID: 20481456]

[132] Fajardo C, Saccà ML, Costa G, Nande M, Martin M. Impact of Ag and Al_2O_3 nanoparticles on soil organisms: *In vitro* and soil experiments. Science of the Total Environment 2014; 473–474: 254-61.

[133] Lewinski N, Colvin V, Drezek R. Cytotoxicity of nanoparticles. Small 2008; 4(1): 26-49.
[http://dx.doi.org/10.1002/smll.200700595] [PMID: 18165959]

[134] Manshian BB, Moyano DF, Corthout N, *et al.* High-content imaging and gene expression analysis to study cell-nanomaterial interactions: the effect of surface hydrophobicity. Biomaterials 2014; 35(37): 9941-50.
[http://dx.doi.org/10.1016/j.biomaterials.2014.08.031] [PMID: 25218858]

[135] Lu B, Xiong SB, Yang H, Yin XD, Zhao RB. Mitoxantrone-loaded BSA nanospheres and chitosan nanospheres for local injection against breast cancer and its lymph node metastases. I: Formulation and *in vitro* characterization. Int J Pharm 2006; 307(2): 168-74.
[http://dx.doi.org/10.1016/j.ijpharm.2005.09.027] [PMID: 16310729]

[136] Lu B, Xiong SB, Yang H, Yin XD, Chao RB. Solid lipid nanoparticles of mitoxantrone for local injection against breast cancer and its lymph node metastases. Eur J Pharm Sci 2006; 28(1-2): 86-95.
[http://dx.doi.org/10.1016/j.ejps.2006.01.001] [PMID: 16472996]

[137] Chamberlain MC. Neurotoxicity of intra-CSF liposomal cytarabine (DepoCyt) administered for the treatment of leptomeningeal metastases: a retrospective case series. J Neurooncol 2012; 109(1): 143-8. [http://dx.doi.org/10.1007/s11060-012-0880-x] [PMID: 22539243]

[138] Venditto VJ, Szoka FC Jr. Cancer nanomedicines: so many papers and so few drugs! Adv Drug Deliv Rev 2013; 65(1): 80-8. [http://dx.doi.org/10.1016/j.addr.2012.09.038] [PMID: 23036224]

[139] Valencia PM, Farokhzad OC, Karnik R, Langer R. Microfluidic technologies for accelerating the clinical translation of nanoparticles. Nat Nanotechnol 2012; 7(10): 623-9. [http://dx.doi.org/10.1038/nnano.2012.168] [PMID: 23042546]

CHAPTER 5

Adiponectin-Based Therapeutics for Cancer Treatment

Dahui Li, Handi Cao, Jie Chen, Ming Liu and **Yu Wang**[*]

State Key Laboratory of Pharmaceutical Biotechnology and Department of Pharmacology and Pharmacy, The University of Hong Kong, Hong Kong, China

Abstract: Adiponectin is an adipokine predominantly produced from adipocytes and exerts potent growth inhibitory activity in a wide range of cancer cells. Decreased expression and/or function of adiponectin are associated with increased risks and aggressive development of cancers with poor prognoses. To restore the expression level of endogenous adiponectin and to activate its functional pathways represent promising strategies for the prevention and treatment of cancer, especially in obese patients. However, the development of recombinant adiponectin as a therapeutic agent has been hampered by the complexity of its structures and the highly diversified functionality of this magic molecule. Here, the application of different adiponectin-based prophylactics and therapeutics in cancer treatment will be thoroughly reviewed and scrutinized.

Keywords: Adiponectin, Adipokine, Cancer, Obesity, Glycosylation, Peptidomimetic, Tumor microenvironment, Angiogenesis, Therapeutics.

INTRODUCTION

Adiponectin (other names including adipocyte complement-related protein of 30 kDa (Acrp30) [1], AdipoQ [2], adipose most abundant gene transcript-1 (apM1) [3], and gelatin-binding protein of 28 kDa (GBP28) [4]) is a soluble matrix protein synthesized predominantly in the adipocytes of mammals [5 - 7]. It belongs to the expanding C1q/TNF-related protein (CTRP) family that contains 15 additional paralogs, designated as CTRP1–15 [8 - 10]. Most adiponectin paralogs are ubiquitously expressed in and secreted from multiple tissues [11]. Homologues of mammalian adiponectin have been found in plants [12 - 14], and bacteria [15 - 19]. The gene of human adiponectin (*ADIPOQ*) is located on chromosome 3q27 and encodes a 244-amino acid polypeptide [20, 21]. Polymorphisms of *ADIPOQ* affect circulating adiponectin concentration and/or

[*] **Corresponding author Yu Wang:** Department of Pharmacology and Pharmacy, Faculty of Medicine Building, 21 Sassoon Road, Pokfulam, Hong Kong, China; Tel: (852) 39176864; Fax: (852) 28170859; E-mail: yuwanghk@hku.hk

function [22, 23], and modulates the metabolic phenotypes of obesity [24 - 26], the susceptibility to type 2 diabetes [27 - 29], the risks of coronary and cerebral artery diseases [30, 31], the severity of non-alcoholic fatty liver disease [32, 33], as well as the aggressive development of various cancers [34 - 38].

Unlike many other adipokines released from adipose tissue, adiponectin is abundantly present in the circulation and acts as a hormonal factor to regulate energy metabolism, immunity, and cellular homeostasis [39, 40]. In animals, replenishment of adiponectin decreases glucose production and restores insulin sensitivity [41 - 44], reduces visceral adiposity [45, 46], protects against hepatic steatosis and inflammatory liver diseases [47 - 51], attenuates the development of atherosclerotic vascular disease [52 - 56], and inhibits cancer development [57 - 62]. The diversified biological functions and therapeutic potentials of adiponectin have been thoroughly summarized by many excellent reviews [63 - 72]. The present chapter mainly focuses on the approaches, challenges and controversies as well as the future aspects in the development of adiponectin-based therapeutics for cancer diseases.

STUCTURAL POLYMORPHISM OF OLIGOMERIC ADIPONECTIN

The primary structure of adiponectin contains an NH_2-terminal signal peptide and a species-specific variable region, followed by a collagen-like domain with 22 Gly-X-Pro or Gly-X-Y repeats and a COOH-terminal globular domain (Fig. **1**); the latter exhibits similar three-dimensional folding topology with the pro-inflammatory cytokine, tumor necrosis factor alpha (TNF-α) [73, 74]. In adipocytes, adiponectin is synthesized as a monomeric subunit that undergoes oligomeric assembly to form trimers, hexamers, and high molecular weight (HMW) multimers (Fig. **1**) [75]. Trimerization of adiponectin is triggered by the hydrophobic interactions between the globular domains [74, 76, 77]. Two trimers are cross-linked to form hexamers *via* the disulfide bridges between the cysteine 39 residue of the variable region [78 - 80]. A number of chaperones, including endoplasmic reticulum protein 44 (ERp44), ER oxidoreductase 1-like protein alpha (Ero1-Lα) and disulfide-bond-A oxidoreductase-like protein (DsbA-L), interact with the variable region of adiponectin and facilitate the reduction of oxidized trimers and hexamers, thus assuring an efficient assembly and release of HMW oligomers [81 - 86]. Multiple trimers assemble to form the higher order structures of HMW adiponectin, resembling a "bouquet" of collagenous stalk and "blossoms" of the globular heads [87]. Hydroxylation and glycosylation of several conserved lysine residues within the collagen-like domain are indispensable for the formation of HMW adiponectin complexes (Fig. **1**) [6, 78 - 80, 88 - 90].

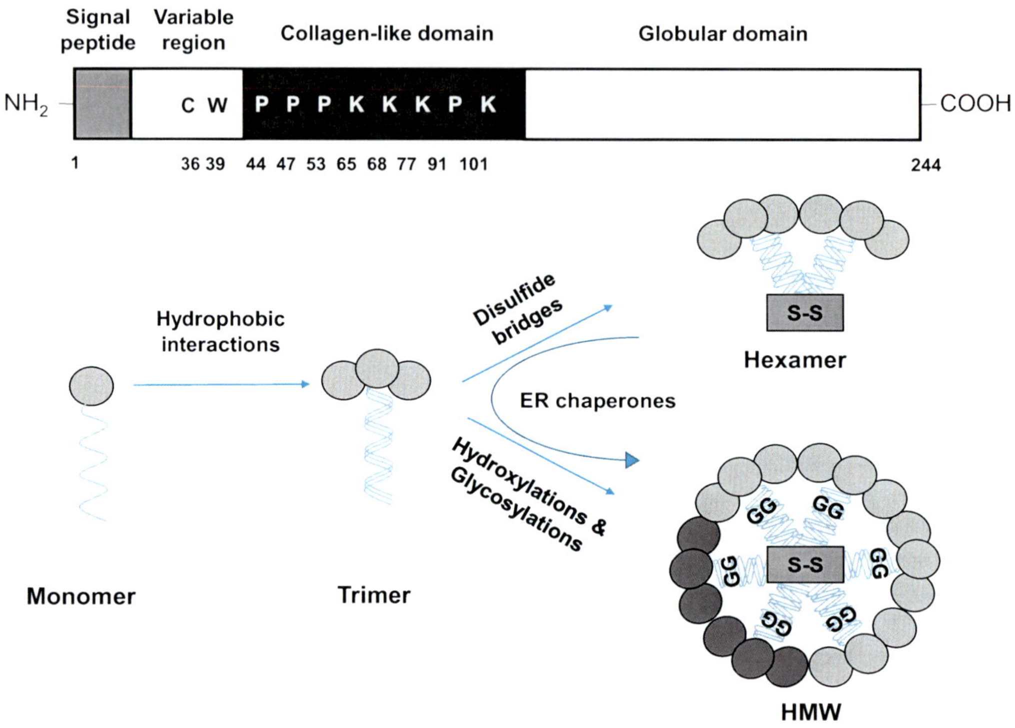

Fig. (1). Schematic illustrations of the primary structure (upper panel) and the oligomeric complex assembly (lower panel) of human adiponectin.

From the NH_2- to COOH-terminus, the primary sequence of adiponectin consists of a signal peptide, a variable region, a collagen-like domain and a globular domain. Adiponectin is extensively modified during the process of oligomeric assembly: The cysteine (C) 36 residue forms intra- and inter-trimer disulfide bridges, depicted as "S-S" [76]; The tryptophan (W) 39 residue modulates the interactions with ERp44 and the oxidative folding of trimers and hexamers [85]; A number of proline (P) and lysine (K) residues located within the collagen-like domain are post-translationally modified by hydroxylation [88 - 90]; Glycosylation of the hydroxylated lysines plays a crucial role in the formation of adiponectin HMW oligomers [6, 89]; A number of endoplasmic reticulum (ER) chaperones are responsible for the reduction of fully oxidized trimers and hexamers for HMW assembly [86].

Once released into the blood, the trimers, hexamers and HMW adiponectin oligomers do not spontaneously interconvert and exhibit non-overlapping biological activities by targeting distinct receptors and signaling pathways [73, 76, 78, 80, 91 - 97]. The circulating level of HMW adiponectin is considered the most relevant biomarker for disease-associated adipocyte dysfunctions [63, 98 - 101]. High levels of HMW adiponectin are associated with favorable metabolic profile [92, 102, 103]. The circulating concentration of adiponectin ranges from ~10 to 30 μg/ml, approximately 1000-fold higher than other hormonal factors and accounting for ~0.1% of total protein in human plasma [5, 102, 104, 105]. Under obesity and associated pathological conditions, the plasma levels of adiponectin are significantly decreased – a phenomenon referred to as hypoadiponectinaemia, which is caused by the reduced gene expression and/or protein assembly/secretion of this molecule in adipocytes [2, 104, 106 - 111]. Thus, the protective effects of adiponectin on the pathogenesis of insulin resistance, atherosclerosis, inflammation and cancers are significantly compromised by weight gain [112]. In animals, replenishment of adiponectin leads to remarkable remedial activities against various obesity-related medical complications [44, 113]. However, the structural polymorphism and the high abundancy of this molecule in circulation have posted major challenges to produce large amounts of homogenous and bioactive adiponectin for therapeutic applications in human.

RECEPTORS AND SIGNALING PATHWAYS OF ADIPONECTIN

Adiponectin receptors 1 (AdipoR1) and 2 (AdipoR2) are seven transmembrane domain receptors originally identified by molecular cloning strategies using the globular domain of adiponectin as a bait probe [114, 115]. Consequently, both receptors show relatively higher affinities to the COOH-terminal globular fragment than the full-length molecule of adiponectin [63, 94]. The two receptors share 67% identity in protein sequences and show an atypical topology different from the conventional G-protein-coupled receptor, with an internal NH_2-terminus and external COOH-terminus [63, 94]. Adiponectin stimulates the phosphorylation of AMP-activated protein kinase (AMPK) through AdipoR1, whereas its effects on peroxisome proliferator-activated receptor alpha (PPARα) are mediated by AdipoR2 [114, 116].

AdipoR1 and AdipoR2 are widely expressed in many organs/tissues but differ in the pattern of distribution and the relative abundance [114, 117]. For example, AdipoR1 is relatively more abundant in skeletal muscle whereas AdipoR2 is predominantly expressed in liver [117]. Moreover, the expression levels of AdipoR1 and AdipoR2 are affected by status of adiposity and diabetes [118], plasma insulin and adiponectin levels [118, 119], food intake (increase with fasting and decrease with feeding) [120, 121], physical activity [120, 122, 123], and age [124]. Human and murine AdipoR share over 95% sequence homology [114, 116]. Genetic polymorphisms of human *ADIPOR1* and *ADIPOR2* do not display a significant correlation with the development of obesity-related metabolic abnormalities [125 - 133].

Disruption of AdipoR1 and/or AdipoR2 partially abolishes the metabolic actions of adiponectin, leading to the development of insulin resistance especially in skeletal muscle and adipose tissues [118, 134]. However, the phenotypes of adiponectin-deficient mice have not been thoroughly replicated by the AdipoR1/R2 dual-deficient mice [116]. Moreover, a number of studies have suggested a pro-inflammatory role of AdipoR1/R2 [135, 136]. AdipoR1 and AdipoR2, as well as other members of the progestin and AdipoQ receptor superfamily, possess potent ceramidase activities to produce ceramide and phosphorylated sphingoid, which are causatively involved in the development of insulin resistance, atherosclerosis and heart failure [137 - 140]. In this regard, a key question to be answered is whether or not by binding to AdipoR, adiponectin acts to inhibit their activities and if so, which oligomeric form(s) of adiponectin function as the antagonists.

T-cadherin (also known as CDH13, cadherin 13, and H-cadherin) is a "truncated" cadherin anchored on the plasma membrane of cells *via* a glycosylphosphatidylinositol linkage [141, 142]. It acts as an adiponectin receptor and binds to the hexamers and HMW multimers [143]. Due to the lacking of the transmembrane and intracellular domains, T-cadherin is considered to have no direct effects on adiponectin-mediated signal transduction in cells, but regulates the circulating levels and tissue distribution of this adipokine [144, 145]. Genetic studies have shown that single nucleotide polymorphisms in *CDH13*, the gene encoding human T-cadherin, contribute to low levels of adiponectin [146 - 151].

In T-cadherin-deficient mice, adiponectin accumulates in the circulation at higher than normal levels [145, 147 - 149, 152]. Meanwhile, a significant reduction in the tissue content of adiponectin leads to an increased susceptibility of these mice to injuries caused by inflammatory stimuli [145, 153]. Notably, T-cadherin is abundantly expressed in injured vascular endothelial and smooth muscle cells, such as those of the atherosclerotic plaques [141, 154, 155]. Without this molecule, heart and vascular tissues become insensitive to adiponectin, despite a constant expression of AdipoR1/R2 in T-cadherin-deficient mice [141]. Consistently, the vascular dysfunction and ischemia-induced heart problems have been observed in T-cadherin-deficient mice, similar to those of the adiponectin-deficient mice [154, 156, 157]. In human, genetic polymorphisms of *CDH13* are significantly associated with various cardiometabolic and vascular phenotypes [144, 146, 151, 154, 158, 159].

In summary, the evidence supports a synergistic role of T-cadeherin in adiponectin-mediated beneficial functions. However, without elucidating the detailed structural mechanisms underlying the interactions between adiponectin and T-cadeherin, further development of therapeutics based on this "third promising receptor" remain elusive.

ADIPONECTIN AND ITS RECEPTORS IN OBESITY-RELATED CANCER DISEASES

Excess adiposity is an independent risk factor for the development of cancer and closely associated with late-stage disease and poor prognosis [160 - 163]. Adipose tissue is the largest endocrine organ in human body [164]. During obesity, excessive expansion of the fat depots leads to chronic adipose tissue inflammation and the augmented productions of inflammatory adipokines, which act directly on tumor cells to promote their survival and proliferation [91, 165 - 169]. In addition, dysregulated metabolic homeostasis, characterized by hyperinsulinemia, hyperglycemia and dyslipidemia, indirectly enhances tumor growth and development in obese patients [170 - 172]. Increasing insulin and insulin-like growth factors (IGF) contribute to carcinogenesis [173, 174]. Moreover, adipose tissue located in proximity to or within the tumor microenvironment contributes to the local production of carcinogenic matrix proteins and the continuous supply of cancer stem cells [175, 176].

In obese subjects, hypoadiponectinaemia (<4 μg/ml) is associated with an increased risk of developing cancers including but not limited to those of the breast [177, 178], endometrium [179 - 182], prostate [183], colon [184 - 186], stomach [187 - 189], pancreas [190], liver [191, 192], kidney [193 - 195], leukemia [196, 197], lymphoma and myeloma [198 - 201]. Moreover, tumors from patients with low plasma adiponectin levels are larger, exhibiting higher histologic grade, more aggressive invasion and metastasis, as well as poorer prognosis [182, 183, 189, 202 - 211]. High plasma adiponectin levels, especially those of the HMW oligomers, are associated with a decreased cancer risk [180, 203, 212 - 216]. Genetic polymorphisms of the *ADN* gene affect its circulating levels, the tumor grade, clinical stage and aggressiveness in cancer patients [36, 126, 217 - 221]. Taken in conjunction, hypoadiponectinaemia not only represents a useful biomarker for early detection but also plays a causative role in the development of obesity-related cancer diseases.

The adiponectin receptors AdipoR1 and AdipoR2 are expressed in a plethora of malignant tissues including breast, endometrium, prostate, esophagus, stomach, colon, liver, pancreas, and lung [222 - 230]. Contrary to adiponectin, AdipoR1 and AdipoR2 exhibit an inconsistent pattern of change in both the expression and distribution across different types of tumors and relative to the normal/benign tissues. For instance, the expression of AdipoR1 and AdipoR2 is downregulated in cancerous in comparison to healthy prostate tissues [220], but increased in tumor tissues of both non-small and small cell lung cancer [222 - 230]. In gastric cancer, the expression levels of AdipoR1 and AdipoR2 are lower in tumor than normal tissue and acts as a marker of better prognosis [231]. However, higher expression of AdipoR2 is associated with moderately differentiated when compared to well-differentiated gastric tumors [232]. Pancreatic cancer patients show positive or strong expression of AdipoR1/R2 in the tumor tissues [228]. Significantly increased or decreased expressions of AdipoR1 and AdipoR2 are found in colorectal cancer tissues *versus* normal colon epithelium [233 - 237]. A higher expression of AdipoR1 is found in tumor tissue and cell lines of breast cancer [225, 238, 239], and associated with a more invasive phenotype [239, 240]. In addition to the highly variable expression patterns of AdipoR1/R2 in tumor tissues, studies of their genetic polymorphisms in different populations of cancer patients preclude the application of these receptors as either biomarkers or suitable targets for therapeutic development in cancer diseases [126, 127, 231, 241 - 246].

ANTI-TUMORIGENIC ACTIVITY OF ADIPONECTIN

Unlike most of the inflammatory adipokines that are causally linked to obesity-related diseases, adiponectin possesses potent insulin-sensitizing, anti-inflammatory, anti-angiogenic and anti-tumorigenic activities [6, 47, 62, 63, 247 - 249]. Adiponectin prevents cancer development by improving the energy metabolism systemically and suppressing the actions of pro-inflammatory or pro-tumorigenic modulators in the tumor microenvironment locally [250 - 253]. It selectively binds to various carcinogenic growth factors to prevent their interactions with the respective receptors [247, 254]. Most importantly, adiponectin released from adipocytes acts as an endocrine and a paracrine factor to directly inhibit the survival, growth and invasion of cells in the tumor microenvironment, *via* both receptor-dependent and independent mechanisms [59, 191, 216, 248].

Inhibition of Tumor Cell Growth by Adiponectin

Adiponectin elicits potent cytostatic actions to suppress the proliferation of various types of cells, including endothelial and smooth muscle cells, myelomonocytic cells, hepatic stellate cells, satellite cell, myoblast and various different types of cancer or stem cells [57, 59, 247, 254 - 261]. Adiponectin inhibits the *in vitro* growth of human breast cancer MDA-MB-231, T47D MCF7 and SK-BR-3 cells, as well as the non-cancerous MCF10A and human mammary epithelial cells [61, 238, 262 - 274]. Animal studies demonstrate that supplement therapy with mammalian adiponectin suppresses the tumor development in mice implanted with human breast cancer cells [61, 238, 275].

Adiponectin, in particular the HMW form at a sub-physiological concentration, inhibits leptin-, IGF-1- or dihydrotestosterone-stimulated proliferation of prostate cancer cells [276, 277]. It enhances the inhibitory effects of doxorubicin, a cytotoxic chemotherapy agent, on prostate cancer cell growth [276, 277]. Treatment with adiponectin reduces the viability and inhibits growth factor-induced proliferation and invasion of endometrial cancer cells [224, 276, 278, 279], hepatic carcinoma cells [280, 281], colorectal cancer cells [282], gastric cancer cells [189], esophageal cancer cells [283], pancreatic cancer cells [284], and lung cancer cells [227]. Prolonged exposure to adiponectin induces cell cycle arrest and apoptosis in multiple types of myeloma cells [285, 286]. It enhances the sensitivity of human chronic myelogenous leukemia cells to imatinib treatment [207, 287 - 290].

Depending on the experimental model, cytostatic/apoptotic effects of adiponectin are mediated by the increased activation of AMPK [230, 291, 292], the reduced signal transduction through extracellular-signal-regulated kinases 1 and 2, p38 or c-Jun N-terminal kinases [117, 270, 293], the inhibition of Wnt signaling or Akt and glycogen synthase kinase 3β/β-catenin pathways [61, 294, 295], and/or the enhanced expression of Bax, p53 and p21, which are important regulators of growth arrest and apoptosis [230, 274, 296, 297].

Inhibition of Tumor Angiogenesis by Adiponectin

In vascular endothelial cells, the cross-talks between adiponectin-mediated activation of AMPK (a cytostatic factor to inhibit both growth and death [298 - 301]) and inhibition of Akt/protein kinase B (a signal to promote survival and growth [302]) determine the outcome of vascularization or angiogenesis [303 - 306]. By activating endothelial nitric oxide synthase *via* AMPK, adiponectin elicits potent anti-apoptotic and anti-oxidant activities during ischemia-reperfusion injury, thus facilitating revascularization [6, 252, 255, 305, 307, 308]. In the setting of tumor development, a condition that differs significantly from tissue ischemia [309 - 311], adiponectin inhibits the activation, proliferation and migration of vascular endothelial cells, and prevents new blood vessel formation [57, 153, 252, 255, 305, 307, 308, 312 - 315]. By contrast, the globular domain of adiponectin increases endothelial activation, proliferation, migration and angiogenesis, largely through its binding to AdipoR1 or AdipoR2 [306, 307, 316 - 322]. In human vascular endothelial cells, globular adiponectin stimulates the formation of capillary-like structures and acts as a chemoattractant [306]. It also restores the function of endothelial progenitor cells under high glucose conditions [318]. While the evidence support certain pharmacological properties of the globular domain, the existence and presence of this adiponectin fragment *in vivo* has not been confirmed under both physiological and pathological conditions [1, 3, 4, 321, 323, 324].

T-cadherin plays a role in the crosstalks between adiponectin and vasculogenic factors as well as AdipoR in the tumor microenvironment [311, 325 - 327], and enhances endothelial barrier function [328]. It may act as a co-receptor by competing with AdipoR1 and AdipoR2 receptors for adiponectin binding or interfering with adiponectin signal transduction [329]. Nevertheless, the role of T-cadherin in tumor angiogenesis remain to be elucidated.

ADIPONECTIN-DERIVED ANTI-CANCER THERAPEUTICS

Despite that adiponectin elicits potent activities against obesity-related pathologies [43, 44, 330], it is extremely challenging to convert the full-length molecule into a peptidomimetic drug for the therapeutic applications in human [331, 332]. Therefore, efforts have been directed towards the identification and testing of the active moieties on adiponectin. Based on earlier studies [52, 333 - 335], Otvos *et al.* designed a series of overlapping peptides across the entire globular domain of adiponectin and tested their agonistic effects on AdipoRs [336]. A lead peptidomimetic, ADP355 (H-DAsn-Ile-Pro-Nva-Leu-T-r-DSer-Phe-Ala-DSer-NH2) with four non-natural amino acid replacements, was generated to elicit cytostatic and anti-oncogenic activities primarily through AdipoR1 [337]. Subsequently, a number of derivatives of ADP355 were designed for testing the anti-proliferative activities in MCF-7 breast cancer cells and K562 chronic myeloid leukemia cells [331, 338]. A bell-shaped dose-response curve was observed for all the ADP adiponectin peptidomimetics, *i.e.* the growth inhibition reversed at high dose [338]. The variable activities of the ADP peptidomimetics are probably due to cell type-specific mechanisms and involve additional signaling biopolymers apart from AdipoRs [331]. Although multimerization of the peptide ligands enhances the cellular activities, none of the ADP peptidomimetics exerts favorable effects on the inhibition of endothelial cell mitogenesis [339].

While the pharmacological properties of the ADP355 derivatives are worthwhile further exploration, the rationale for their applications in cancer treatment needs to be thoroughly re-reviewed. First of all, although there is evidence that activation of AdipoR receptors limits the proliferation of cancer cell lines *in vitro* [222, 230, 238], studies have not been able to demonstrate any robust inhibitory effects of the globular domain on the proliferation of breast, colorectal, prostate and leukemia cancer cells [238, 263, 264, 269, 270, 274, 277, 340 - 342]. On the other hand, the globular form of adiponectin has been reported to be associated with the development of colorectal and breast cancer [324, 343]. Second, most of studies have utilized bacterially-produced recombinant protein representing the globular domain, which is not detected under pathophysiological conditions; thus its roles in cancer development are insufficiently connected to the *in vivo* functions of the properly glycosylated adiponectin [323, 333]. Third, the structural similarity between the globular domain and TNF-α may not necessarily be translated into the anti-tumor but rather the pro-tumor and inflammatory activities, partly through the stimulation of AdipoR1/R2 [139, 173, 324, 329, 344 - 346]. The latter is associated with an increased ceramidase activity and production of the anti-apoptotic metabolites, sphingolipid ceramide [139, 347].

A recent study attempted to identify non-peptidic AdipoR agonists from a natural product library, using ADP355 as a reference compound [348]. The most active agonists are matairesinol, arctiin, (-)-arctigenin and gramine for AdipoR1, parthenolide, taxifoliol, deoxyschizandrin and syringing for AdipoR2. Most of the hit compounds possess potent anti-oxidative and anti-proliferative properties [112, 349]. However, as the involvement and role of AdipoR in cancer cell proliferation and tumor growth remain to be established [139, 350], results derived from the receptor-based screening experiment need to be carefully scrutinized to differentiate the agonistic or antagonistic effects [348]. AdipoR1 and AdipoR2 form homo- and hetero-multimers, which influence the ligand binding and thus drug compound screening [63, 91, 351]. Multimerization by binding with different ligands may activate distinct signaling pathways, *via* modulating the alternative interactions with the intracellular adaptor protein, APPL1 (adaptor protein, phosphotyrosine interacting with PH domain and leucine zipper 1) [352 - 354]. The latter mediates the signaling of AdipoR to promote the proliferation and migration of cancer cells [329, 355].

Okada-Iwabu *et al.* screened a compound library and identified several small-molecule AdipoR agonists [356]. One of them, AdipoRon, binds to AdipoR1 and AdipoR2 at low micromolar concentrations and activates AMPK as well as the transcriptional coactivator, peroxisome proliferator–activated receptor gamma (PPARγ) coactivator 1–alpha [115, 357]. Like adiponectin, AdipoRon improves energy metabolism and insulin sensitivity, in turn extending life span in obese animals [357]. Meanwhile, AdipoR- or adiponectin-independent effects of AdipoRon have been identified and reported [357 - 363]. Nevertheless, the effects of the low molecular weight agonists of AdipoR on tumor development remain to be explored.

ALTERNATIVE STRATEGIES FOR ADIPONECTIN REPLACEMENT THERAPIES

Despite the existing controversies on AdipoR and their agonists, there are compelling evidence supporting a key role of hypoadiponectinaemia in the pathogenesis of malignant diseases associated with obesity [216, 279, 349, 364]. Increasing circulating adiponectin levels counteracts metabolic dysfunction and slows cancer progression in experimental models [286, 365]. Given that it is difficult to convert the full-length adiponectin protein into a viable drug, restoring the balanced production and mimicking the cancer-protective effects of adiponectin have attracted significant interests for potential clinical applications. For example, metformin, a biguanide derivative and first-line oral medication for type 2 diabetes mellitus, stimulates AMPK to inhibit a number of anabolic/mitogenic pathways activated by growth factors and nutrients, thus eliciting similar anti-tumorigenic effects as adiponectin [366 - 379].

Circulating adiponectin levels are modulated by lifestyle, dietary components and pharmacological agents [380 - 386]. Body weight loss, particularly in the form of visceral fat reduction, is effective in boosting the plasma adiponectin levels [387, 388]. Calorie restriction induces AMPK signaling in tumor tissues and exerts anticancer effects, concurrent with an augmented adiponectin levels [389]. Specific dietary components for cancer prevention, such as omega-3 and vitamin D, increase adiponectin expression and secretion [390 - 392]. Astragaloside II and isoastragaloside I, the active ingredients of anticancer medicinal herb Radix Astragali, increase adiponectin production from adipocytes [393]. The molecular mechanisms underlying the modulation of adiponectin levels by body weight reduction or dietary components remain unclear. Nevertheless, it is suggested that obesity-induced endoplasmic reticulum stress leads to a decreased production of adiponectin, especially the HMW form [394, 395]. Thus, targeting the machinery responsible for adiponectin assembly and secretion represents an attractive strategy for elevating the circulating levels of this molecule [396].

Adiponectin is secreted from adipocytes into the bloodstream as three oligomeric isoforms. Both clinical and animal studies suggest that the HMW is the predominant isoform mediating the beneficial effects of adiponectin [79, 397, 398]. L-4F, an apolipoprotein A-I mimetic peptide, increases serum concentrations of HMW adiponectin in obese mice [399], and reduces tumor burden through induction of myeloma cell apoptosis [400]. Administration of the PPARγ agonist thiazolidinediones (TZDs, a class of drugs used to improve lipid and glucose metabolism in type 2 diabetes [401]) in both diabetic patients and rodents results in a selective elevation in serum levels of the HMW oligomer *via* stimulating the biosynthesis and secretion pathways of adiponectin [109, 391,

402, 403]. *In vitro* and *in vivo* treatment with TZDs inhibits the growth, migration and invasion of cancer cells [403 - 408]. Leuprolide, a gonadotropin-releasing hormone agonist, increases plasma adiponectin levels in nondiabetic men with prostate cancer when combined with androgen blockade, bicalutamide [409]. While the above evidence suggest that pharmacological intervention is feasible to modulate circulating adiponectin levels, a specific target to effectively sustain the expression and/or production of adiponectin remain to be identified.

Adiponectin shares homology with collagen VIII and X, complement factor C1q and TNF-α [73, 261]. Adiponectin may act as soluble defense collagen by negatively regulating the functions of inflammatory cells [261]. Apart from the globular domain, there are other pharmacological moieties of adiponectin remain to be explored and examined. For example, a number of groups have adopted chemical synthetic approaches to produce non-globular adiponectin fragments. Among them, the adiponectin variable domain [410], and the post-translationally modified collagen-like domain [411], might prove beneficial for the treatment of obesity-related malignant diseases. A full uncover of the structure/function relationships of different domains of adiponectin will facilitate the conversion of this magic molecule into a viable drug.

CONCLUDING REMARKS

Obesity is a global epidemic problem with widespread health consequences, including an increase in the incidence and death of many lifestyle-related malignant diseases. The need for cancer prevention and therapies is of upmost importance. Adiponectin represents a critical link between obesity and many types of cancer development. Epidemiological, genetic, and animal studies have demonstrated the protective effects of adiponectin against a cluster of obesity-related metabolic, cardiovascular and malignant complications. Various strategies to increase the circulating level and function of adiponectin have been successfully applied in animals and proved to be effective in enhancing the insulin sensitivity and energy homeostasis, alleviating the obesity-associated inflammatory injuries, as well as preventing the formation and development of various cancers. However, the direct translation of animal studies to human turns out to be a challenge. Unlike other metabolic hormones (such as insulin), adiponectin is a highly abundant plasma protein in humans. Long-term replacement of bioactive adiponectin is not cost-effective and feasible at this stage. Different oligomeric forms of adiponectin and the quality of the preparation significantly affect experimental outcome and have impeded the clinical use of adiponectin. For the same reason, our current understanding of the mode of adiponectin function at the molecular level has been compromised. Considering the highly dynamic and complexed structural-signaling mechanisms, approaches to up-regulate the endogenous adiponectin levels are more appealing for cancer prevention and treatment. Alternatively, the identification of the true functional moieties of adiponectin will greatly enhance the process of future drug discovery and development, especially in therapeutic applications for obesity-related medical complications.

CONFLICT OF INTEREST

The authors declare no conflict of interest, financial or otherwise.

ACKNOWLEDGEMENTS

The work was supported in part by a seed grant from the University of Hong Kong, Research Grant Council grants (HKU779712M and 17153016) from University Grants Committee, HKSAR, and the National Basic Research Program of China (973 Program 2015CB553603).

REFERENCES

[1] Scherer PE, Williams S, Fogliano M, Baldini G, Lodish HF. A novel serum protein similar to C1q, produced exclusively in adipocytes. J Biol Chem 1995; 270(45): 26746-9. [http://dx.doi.org/10.1074/jbc.270.45.26746] [PMID: 7592907]

[2] Hu E, Liang P, Spiegelman BM. AdipoQ is a novel adipose-specific gene dysregulated in obesity. J Biol Chem 1996; 271(18): 10697-703. [http://dx.doi.org/10.1074/jbc.271.18.10697] [PMID: 8631877]

[3] Maeda K, Okubo K, Shimomura I, Funahashi T, Matsuzawa Y, Matsubara K. cDNA cloning and expression of a novel adipose specific collagen-like factor, apM1 (AdiPose Most abundant Gene transcript 1). Biochem Biophys Res Commun 1996; 221(2): 286-9. [http://dx.doi.org/10.1006/bbrc.1996.0587] [PMID: 8619847]

[4] Nakano Y, Tobe T, Choi-Miura NH, Mazda T, Tomita M. Isolation and characterization of GBP28, a novel gelatin-binding protein purified from human plasma. J Biochem 1996; 120(4): 803-12. [http://dx.doi.org/10.1093/oxfordjournals.jbchem.a021483] [PMID: 8947845]

[5] Arita Y, Kihara S, Ouchi N, *et al.* Paradoxical decrease of an adipose-specific protein, adiponectin, in obesity. Biochem Biophys Res Commun 1999; 257(1): 79-83. [http://dx.doi.org/10.1006/bbrc.1999.0255] [PMID: 10092513]

[6] Wang Y, Lam KS, Yau MH, Xu A. Post-translational modifications of adiponectin: mechanisms and functional implications. Biochem J 2008; 409(3): 623-33. [http://dx.doi.org/10.1042/BJ20071492] [PMID: 18177270]

[7] Shen L, Evans IM, Souza D, Dreifaldt M, Dashwood MR, Vidya MA. Adiponectin: An Endothelium-Derived Vasoprotective Factor? Curr Vasc Pharmacol 2016; 14(2): 168-74. [http://dx.doi.org/10.2174/1570161114666151202210128] [PMID: 26638793]

[8] Wong GW, Wang J, Hug C, Tsao TS, Lodish HF. A family of Acrp30/adiponectin structural and functional paralogs. Proc Natl Acad Sci USA 2004; 101(28): 10302-7. [http://dx.doi.org/10.1073/pnas.0403760101] [PMID: 15231994]

[9] Wong GW, Krawczyk SA, Kitidis-Mitrokostas C, Revett T, Gimeno R, Lodish HF. Molecular, biochemical and functional characterizations of C1q/TNF family members: adipose-tissue-selective expression patterns, regulation by PPAR-gamma agonist, cysteine-mediated oligomerizations, combinatorial associations and metabolic functions. Biochem J 2008; 416(2): 161-77.

[http://dx.doi.org/10.1042/BJ20081240] [PMID: 18783346]

[10] Seldin MM, Tan SY, Wong GW. Metabolic function of the CTRP family of hormones. Rev Endocr Metab Disord 2014; 15(2): 111-23.
[http://dx.doi.org/10.1007/s11154-013-9255-7] [PMID: 23963681]

[11] Schäffler A, Buechler C. CTRP family: linking immunity to metabolism. Trends Endocrinol Metab 2012; 23(4): 194-204.
[http://dx.doi.org/10.1016/j.tem.2011.12.003] [PMID: 22261190]

[12] Narasimhan ML, Coca MA, Jin J, *et al.* Osmotin is a homolog of mammalian adiponectin and controls apoptosis in yeast through a homolog of mammalian adiponectin receptor. Mol Cell 2005; 17(2): 171-80.
[http://dx.doi.org/10.1016/j.molcel.2004.11.050] [PMID: 15664187]

[13] Anil Kumar S, Hima Kumari P, Shravan Kumar G, Mohanalatha C, Kavi Kishor PB. Osmotin: a plant sentinel and a possible agonist of mammalian adiponectin. Front Plant Sci 2015; 6: 163.
[http://dx.doi.org/10.3389/fpls.2015.00163] [PMID: 25852715]

[14] Miele M, Costantini S, Colonna G. Structural and functional similarities between osmotin from Nicotiana tabacum seeds and human adiponectin. PLoS One 2011; 6(2): e16690.
[http://dx.doi.org/10.1371/journal.pone.0016690] [PMID: 21311758]

[15] Sylvestre P, Couture-Tosi E, Mock M. A collagen-like surface glycoprotein is a structural component of the Bacillus anthracis exosporium. Mol Microbiol 2002; 45(1): 169-78.
[http://dx.doi.org/10.1046/j.1365-2958.2000.03000.x] [PMID: 12100557]

[16] Kubler-Kielb J, Vinogradov E, Hu H, Leppla SH, Robbins JB, Schneerson R. Saccharides cross-reactive with Bacillus anthracis spore glycoprotein as an anthrax vaccine component. Proc Natl Acad Sci USA 2008; 105(25): 8709-12.
[http://dx.doi.org/10.1073/pnas.0803897105] [PMID: 18562275]

[17] Réty S, Salamitou S, Garcia-Verdugo I, *et al.* The crystal structure of the Bacillus anthracis spore surface protein BclA shows remarkable similarity to mammalian proteins. J Biol Chem 2005; 280(52): 43073-8.
[http://dx.doi.org/10.1074/jbc.M510087200] [PMID: 16249180]

[18] Kailas L, Terry C, Abbott N, *et al.* Surface architecture of endospores of the Bacillus cereus/anthracis/thuringiensis family at the subnanometer scale. Proc Natl Acad Sci USA 2011; 108(38): 16014-9.
[http://dx.doi.org/10.1073/pnas.1109419108] [PMID: 21896762]

[19] Vilas-Bôas GT, Peruca AP, Arantes OM. Biology and taxonomy of Bacillus cereus, Bacillus anthracis, and Bacillus thuringiensis. Can J Microbiol 2007; 53(6): 673-87.
[http://dx.doi.org/10.1139/W07-029] [PMID: 17668027]

[20] Saito K, Tobe T, Minoshima S, *et al.* Organization of the gene for gelatin-binding protein (GBP28). Gene 1999; 229(1-2): 67-73.
[http://dx.doi.org/10.1016/S0378-1119(99)00041-4] [PMID: 10095105]

[21] Takahashi M, Arita Y, Yamagata K, *et al.* Genomic structure and mutations in adipose-specific gene, adiponectin. International journal of obesity and related metabolic disorders : journal of the International Association for the Study of Obesity 2000; 24: 861-8.
[http://dx.doi.org/10.1038/sj.ijo.0801244]

[22] Breitfeld J, Stumvoll M, Kovacs P. Genetics of adiponectin. Biochimie 2012; 94(10): 2157-63.
[http://dx.doi.org/10.1016/j.biochi.2012.03.004] [PMID: 22449616]

[23] Menzaghi C, Trischitta V, Doria A. Genetic influences of adiponectin on insulin resistance, type 2 diabetes, and cardiovascular disease. Diabetes 2007; 56(5): 1198-209.
[http://dx.doi.org/10.2337/db06-0506] [PMID: 17303804]

[24] Riestra P, Gebreab SY, Xu R, *et al.* Gender-specific associations between ADIPOQ gene

polymorphisms and adiponectin levels and obesity in the Jackson Heart Study cohort. BMC Med Genet 2015; 16: 65.
[http://dx.doi.org/10.1186/s12881-015-0214-x] [PMID: 26290432]

[25] Wu J, Liu Z, Meng K, Zhang L. Association of adiponectin gene (ADIPOQ) rs2241766 polymorphism with obesity in adults: a meta-analysis. PLoS One 2014; 9(4): e95270.
[http://dx.doi.org/10.1371/journal.pone.0095270] [PMID: 24740426]

[26] Yu Z, Han S, Cao X, Zhu C, Wang X, Guo X. Genetic polymorphisms in adipokine genes and the risk of obesity: a systematic review and meta-analysis. Obesity (Silver Spring) 2012; 20(2): 396-406.
[http://dx.doi.org/10.1038/oby.2011.148] [PMID: 21660081]

[27] Fan Y, Wang K, Xu S, *et al.* Association between ADIPOQ +45T>G polymorphism and type 2 diabetes: a systematic review and meta-analysis. Int J Mol Sci 2014; 16(1): 704-23.
[http://dx.doi.org/10.3390/ijms16010704] [PMID: 25561226]

[28] Han LY, Wu QH, Jiao ML, *et al.* Associations between single-nucleotide polymorphisms (+45T>G, +276G>T, -11377C>G, -11391G>A) of adiponectin gene and type 2 diabetes mellitus: a systematic review and meta-analysis. Diabetologia 2011; 54(9): 2303-14.
[http://dx.doi.org/10.1007/s00125-011-2202-9] [PMID: 21638131]

[29] Li Y, Yang Y, Shi L, Li X, Zhang Y, Yao Y. The association studies of ADIPOQ with type 2 diabetes mellitus in Chinese populations. Diabetes Metab Res Rev 2012; 28(7): 551-9.
[http://dx.doi.org/10.1002/dmrr.2309] [PMID: 22539443]

[30] Shen D, Xing S, Chen C. Adiponectin gene polymorphisms contributes to ischemic stroke risk: A meta-analysis. J Renin Angiotensin Aldosterone Syst 2015; 16(1): 178-84.
[http://dx.doi.org/10.1177/1470320314552311] [PMID: 25501307]

[31] Zhou D, Jin Y, Yao F, Duan Z, Wang Q, Liu J. Association between the adiponectin +45T>G genotype and risk of cardiovascular disease: a meta-analysis. Heart Lung Circ 2014; 23(2): 159-65.
[http://dx.doi.org/10.1016/j.hlc.2013.07.010] [PMID: 23972466]

[32] Gupta AC, Misra R, Sakhuja P, Singh Y, Basir SF, Sarin SK. Association of adiponectin gene functional polymorphisms (-11377C/G and +45T/G) with nonalcoholic fatty liver disease. Gene 2012; 496(1): 63-7.
[http://dx.doi.org/10.1016/j.gene.2011.12.023] [PMID: 22269154]

[33] Li XL, Sui JQ, Lu LL, *et al.* Gene polymorphisms associated with non-alcoholic fatty liver disease and coronary artery disease: a concise review. Lipids Health Dis 2016; 15: 53.
[http://dx.doi.org/10.1186/s12944-016-0221-8] [PMID: 26965314]

[34] Al-Harithy RN, Al-Zahrani MH. The adiponectin gene, ADIPOQ, and genetic susceptibility to colon cancer. Oncol Lett 2012; 3(1): 176-80.
[PMID: 22740876]

[35] Bieńkiewicz J, Smolarz B, Malinowski A. Association Between Single Nucleotide Polymorphism +276G > T (rs1501299) in ADIPOQ and Endometrial Cancer. Pathol Oncol Res 2016; 22(1): 135-8.
[http://dx.doi.org/10.1007/s12253-015-9985-9] [PMID: 26386690]

[36] Xu Y, He B, Pan Y, *et al.* The roles of ADIPOQ genetic variations in cancer risk: evidence from published studies. Mol Biol Rep 2013; 40(2): 1135-44.
[http://dx.doi.org/10.1007/s11033-012-2154-2] [PMID: 23065236]

[37] Ye J, Jiang L, Wu C, Liu A, Mao S, Ge L. Three ADIPOR1 Polymorphisms and Cancer Risk: A Meta-Analysis of Case-Control Studies. PLoS One 2015; 10(6): e0127253.
[http://dx.doi.org/10.1371/journal.pone.0127253] [PMID: 26047008]

[38] Vasseur F, Meyre D, Froguel P. Adiponectin, type 2 diabetes and the metabolic syndrome: lessons from human genetic studies. Expert Rev Mol Med 2006; 8(27): 1-12.
[http://dx.doi.org/10.1017/S1462399406000147] [PMID: 17112391]

[39] Stern JH, Rutkowski JM, Scherer PE. Adiponectin, Leptin, and Fatty Acids in the Maintenance of

Metabolic Homeostasis through Adipose Tissue Crosstalk. Cell Metab 2016; 23(5): 770-84. [http://dx.doi.org/10.1016/j.cmet.2016.04.011] [PMID: 27166942]

[40] Guerre-Millo M. Adipose tissue and adipokines: for better or worse. Diabetes Metab 2004; 30(1): 13-9. [http://dx.doi.org/10.1016/S1262-3636(07)70084-8] [PMID: 15029093]

[41] Combs TP, Pajvani UB, Berg AH, *et al.* A transgenic mouse with a deletion in the collagenous domain of adiponectin displays elevated circulating adiponectin and improved insulin sensitivity. Endocrinology 2004; 145(1): 367-83. [http://dx.doi.org/10.1210/en.2003-1068] [PMID: 14576179]

[42] Combs TP, Berg AH, Obici S, Scherer PE, Rossetti L. Endogenous glucose production is inhibited by the adipose-derived protein Acrp30. J Clin Invest 2001; 108(12): 1875-81. [http://dx.doi.org/10.1172/JCI14120] [PMID: 11748271]

[43] Berg AH, Combs TP, Du X, Brownlee M, Scherer PE. The adipocyte-secreted protein Acrp30 enhances hepatic insulin action. Nat Med 2001; 7(8): 947-53. [http://dx.doi.org/10.1038/90992] [PMID: 11479628]

[44] Yamauchi T, Kamon J, Waki H, *et al.* The fat-derived hormone adiponectin reverses insulin resistance associated with both lipoatrophy and obesity. Nat Med 2001; 7(8): 941-6. [http://dx.doi.org/10.1038/90984] [PMID: 11479627]

[45] Masaki T, Chiba S, Yasuda T, *et al.* Peripheral, but not central, administration of adiponectin reduces visceral adiposity and upregulates the expression of uncoupling protein in agouti yellow (Ay/a) obese mice. Diabetes 2003; 52(9): 2266-73. [http://dx.doi.org/10.2337/diabetes.52.9.2266] [PMID: 12941765]

[46] Yokota T, Meka CS, Medina KL, *et al.* Paracrine regulation of fat cell formation in bone marrow cultures *via* adiponectin and prostaglandins. J Clin Invest 2002; 109(10): 1303-10. [http://dx.doi.org/10.1172/JCI0214506] [PMID: 12021245]

[47] Wang Y, Zhou M, Lam KS, Xu A. Protective roles of adiponectin in obesity-related fatty liver diseases: mechanisms and therapeutic implications. Arq Bras Endocrinol Metabol 2009; 53(2): 201-12. [http://dx.doi.org/10.1590/S0004-27302009000200012] [PMID: 19466213]

[48] Moschen AR, Wieser V, Tilg H. Adiponectin: key player in the adipose tissue-liver crosstalk. Curr Med Chem 2012; 19(32): 5467-73. [http://dx.doi.org/10.2174/092986712803833254] [PMID: 22876924]

[49] Ouchi N, Kihara S, Arita Y, *et al.* Adipocyte-derived plasma protein, adiponectin, suppresses lipid accumulation and class A scavenger receptor expression in human monocyte-derived macrophages. Circulation 2001; 103(8): 1057-63. [http://dx.doi.org/10.1161/01.CIR.103.8.1057] [PMID: 11222466]

[50] Silva TE, Colombo G, Schiavon LL. Adiponectin: A multitasking player in the field of liver diseases. Diabetes Metab 2014; 40(2): 95-107. [http://dx.doi.org/10.1016/j.diabet.2013.11.004] [PMID: 24486145]

[51] Xu A, Wang Y, Keshaw H, Xu LY, Lam KS, Cooper GJ. The fat-derived hormone adiponectin alleviates alcoholic and nonalcoholic fatty liver diseases in mice. J Clin Invest 2003; 112(1): 91-100. [http://dx.doi.org/10.1172/JCI200317797] [PMID: 12840063]

[52] Yamauchi T, Kamon J, Waki H, *et al.* Globular adiponectin protected ob/ob mice from diabetes and ApoE-deficient mice from atherosclerosis. J Biol Chem 2003; 278(4): 2461-8. [http://dx.doi.org/10.1074/jbc.M209033200] [PMID: 12431986]

[53] Wang X, Pu H, Ma C, *et al.* Adiponectin abates atherosclerosis by reducing oxidative stress. Med Sci Monit 2014; 20: 1792-800. [http://dx.doi.org/10.12659/MSM.892299] [PMID: 25275545]

[54] Funahashi T, Nakamura T, Shimomura I, *et al.* Role of adipocytokines on the pathogenesis of atherosclerosis in visceral obesity. Intern Med 1999; 38(2): 202-6.
[http://dx.doi.org/10.2169/internalmedicine.38.202] [PMID: 10225688]

[55] Kumada M, Kihara S, Sumitsuji S, *et al.* Association of hypoadiponectinemia with coronary artery disease in men. Arterioscler Thromb Vasc Biol 2003; 23(1): 85-9.
[http://dx.doi.org/10.1161/01.ATV.0000048856.22331.50] [PMID: 12524229]

[56] Ouchi N, Kihara S, Funahashi T, Matsuzawa Y, Walsh K. Obesity, adiponectin and vascular inflammatory disease. Curr Opin Lipidol 2003; 14(6): 561-6.
[http://dx.doi.org/10.1097/00041433-200312000-00003] [PMID: 14624132]

[57] Bråkenhielm E, Veitonmäki N, Cao R, *et al.* Adiponectin-induced antiangiogenesis and antitumor activity involve caspase-mediated endothelial cell apoptosis. Proc Natl Acad Sci USA 2004; 101(8): 2476-81.
[http://dx.doi.org/10.1073/pnas.0308671100] [PMID: 14983034]

[58] Ishikawa M, Kitayama J, Yamauchi T, *et al.* Adiponectin inhibits the growth and peritoneal metastasis of gastric cancer through its specific membrane receptors AdipoR1 and AdipoR2. Cancer Sci 2007; 98(7): 1120-7.
[http://dx.doi.org/10.1111/j.1349-7006.2007.00486.x] [PMID: 17459059]

[59] Dalamaga M, Diakopoulos KN, Mantzoros CS. The role of adiponectin in cancer: a review of current evidence. Endocr Rev 2012; 33(4): 547-94.
[http://dx.doi.org/10.1210/er.2011-1015] [PMID: 22547160]

[60] Boura P, Loukides S, Grapsa D, Achimastos A, Syrigos K. The diverse roles of adiponectin in non-small-cell lung cancer: current data and future perspectives. Future Oncol 2015; 11(15): 2193-203.
[http://dx.doi.org/10.2217/fon.15.96] [PMID: 26235182]

[61] Wang Y, Lam JB, Lam KS, *et al.* Adiponectin modulates the glycogen synthase kinase-3beta/bet--catenin signaling pathway and attenuates mammary tumorigenesis of MDA-MB-231 cells in nude mice. Cancer Res 2006; 66(23): 11462-70.
[http://dx.doi.org/10.1158/0008-5472.CAN-06-1969] [PMID: 17145894]

[62] Hebbard L, Ranscht B. Multifaceted roles of adiponectin in cancer. Best Pract Res Clin Endocrinol Metab 2014; 28(1): 59-69.
[http://dx.doi.org/10.1016/j.beem.2013.11.005] [PMID: 24417946]

[63] Kadowaki T, Yamauchi T, Kubota N, Hara K, Ueki K, Tobe K. Adiponectin and adiponectin receptors in insulin resistance, diabetes, and the metabolic syndrome. J Clin Invest 2006; 116(7): 1784-92.
[http://dx.doi.org/10.1172/JCI29126] [PMID: 16823476]

[64] Siasos G, Tousoulis D, Kollia C, *et al.* Adiponectin and cardiovascular disease: mechanisms and new therapeutic approaches. Curr Med Chem 2012; 19(8): 1193-209.
[http://dx.doi.org/10.2174/092986712799320583] [PMID: 22257055]

[65] Ruan H, Dong LQ. Adiponectin signaling and function in insulin target tissues. J Mol Cell Biol 2016; 8(2): 101-9.
[http://dx.doi.org/10.1093/jmcb/mjw014] [PMID: 26993044]

[66] Yamauchi T, Hara K, Kubota N, *et al.* Dual roles of adiponectin/Acrp30 *in vivo* as an anti-diabetic and anti-atherogenic adipokine. Curr Drug Targets Immune Endocr Metabol Disord 2003; 3(4): 243-54.
[http://dx.doi.org/10.2174/1568008033340090] [PMID: 14683455]

[67] Vansaun MN. Molecular pathways: adiponectin and leptin signaling in cancer. Clin Cancer Res 2013; 19(8): 1926-32.
[http://dx.doi.org/10.1158/1078-0432.CCR-12-0930] [PMID: 23355630]

[68] Kawano J, Arora R. The role of adiponectin in obesity, diabetes, and cardiovascular disease. J Cardiometab Syndr 2009; 4(1): 44-9.
[http://dx.doi.org/10.1111/j.1559-4572.2008.00030.x] [PMID: 19245516]

[69] Díez JJ, Iglesias P. The role of the novel adipocyte-derived hormone adiponectin in human disease. Eur J Endocrinol 2003; 148(3): 293-300.
[http://dx.doi.org/10.1530/eje.0.1480293] [PMID: 12611609]

[70] Montecucco F, Mach F. Update on therapeutic strategies to increase adiponectin function and secretion in metabolic syndrome. Diabetes Obes Metab 2009; 11(5): 445-54.
[http://dx.doi.org/10.1111/j.1463-1326.2008.00986.x] [PMID: 19220391]

[71] Fisman EZ, Tenenbaum A. Adiponectin: a manifold therapeutic target for metabolic syndrome, diabetes, and coronary disease? Cardiovasc Diabetol 2014; 13: 103.
[http://dx.doi.org/10.1186/1475-2840-13-103] [PMID: 24957699]

[72] Parker-Duffen JL, Walsh K. Cardiometabolic effects of adiponectin. Best Pract Res Clin Endocrinol Metab 2014; 28(1): 81-91.
[http://dx.doi.org/10.1016/j.beem.2013.09.001] [PMID: 24417948]

[73] Simpson F, Whitehead JP. Adiponectin--it's all about the modifications. Int J Biochem Cell Biol 2010; 42(6): 785-8.
[http://dx.doi.org/10.1016/j.biocel.2009.12.021] [PMID: 20044026]

[74] Shapiro L, Scherer PE. The crystal structure of a complement-1q family protein suggests an evolutionary link to tumor necrosis factor. Curr Biol 1998; 8(6): 335-8.
[http://dx.doi.org/10.1016/S0960-9822(98)70133-2] [PMID: 9512423]

[75] Chandran M, Phillips SA, Ciaraldi T, Henry RR. Adiponectin: more than just another fat cell hormone? Diabetes Care 2003; 26(8): 2442-50.
[http://dx.doi.org/10.2337/diacare.26.8.2442] [PMID: 12882876]

[76] Pajvani UB, Du X, Combs TP, *et al.* Structure-function studies of the adipocyte-secreted hormone Acrp30/adiponectin. Implications fpr metabolic regulation and bioactivity. J Biol Chem 2003; 278(11): 9073-85.
[http://dx.doi.org/10.1074/jbc.M207198200] [PMID: 12496257]

[77] Waki H, Yamauchi T, Kamon J, *et al.* Impaired multimerization of human adiponectin mutants associated with diabetes. Molecular structure and multimer formation of adiponectin. J Biol Chem 2003; 278(41): 40352-63.
[http://dx.doi.org/10.1074/jbc.M300365200] [PMID: 12878598]

[78] Tsao TS, Tomas E, Murrey HE, *et al.* Role of disulfide bonds in Acrp30/adiponectin structure and signaling specificity. Different oligomers activate different signal transduction pathways. J Biol Chem 2003; 278(50): 50810-7.
[http://dx.doi.org/10.1074/jbc.M309469200] [PMID: 14522956]

[79] Fisher ffM, Trujillo ME, Hanif W, *et al.* Serum high molecular weight complex of adiponectin correlates better with glucose tolerance than total serum adiponectin in Indo-Asian males. Diabetologia2005; 48:1084-7.

[80] Wang Y, Lam KS, Chan L, *et al.* Post-translational modifications of the four conserved lysine residues within the collagenous domain of adiponectin are required for the formation of its high molecular weight oligomeric complex. J Biol Chem 2006; 281(24): 16391-400.
[http://dx.doi.org/10.1074/jbc.M513907200] [PMID: 16621799]

[81] Qiang L, Wang H, Farmer SR. Adiponectin secretion is regulated by SIRT1 and the endoplasmic reticulum oxidoreductase Ero1-L α. Mol Cell Biol 2007; 27(13): 4698-707.
[http://dx.doi.org/10.1128/MCB.02279-06] [PMID: 17452443]

[82] Wang ZV, Schraw TD, Kim JY, *et al.* Secretion of the adipocyte-specific secretory protein adiponectin critically depends on thiol-mediated protein retention. Mol Cell Biol 2007; 27(10): 3716-31.
[http://dx.doi.org/10.1128/MCB.00931-06] [PMID: 17353260]

[83] Zhou L, Liu M, Zhang J, Chen H, Dong LQ, Liu F. DsbA-L alleviates endoplasmic reticulum stress-induced adiponectin downregulation. Diabetes 2010; 59(11): 2809-16.

[http://dx.doi.org/10.2337/db10-0412] [PMID: 20699416]

[84] Liu M, Xiang R, Wilk SA, *et al.* Fat-specific DsbA-L overexpression promotes adiponectin multimerization and protects mice from diet-induced obesity and insulin resistance. Diabetes 2012; 61(11): 2776-86.
[http://dx.doi.org/10.2337/db12-0169] [PMID: 22807031]

[85] Radjainia M, Huang B, Bai B, *et al.* A highly conserved tryptophan in the N-terminal variable domain regulates disulfide bond formation and oligomeric assembly of adiponectin. FEBS J 2012; 279(14): 2495-507.
[http://dx.doi.org/10.1111/j.1742-4658.2012.08630.x] [PMID: 22583869]

[86] Hampe L, Radjainia M, Xu C, *et al.* Regulation and Quality Control of Adiponectin Assembly by Endoplasmic Reticulum Chaperone ERp44. J Biol Chem 2015; 290(29): 18111-23.
[http://dx.doi.org/10.1074/jbc.M115.663088] [PMID: 26060250]

[87] Radjainia M, Wang Y, Mitra AK. Structural polymorphism of oligomeric adiponectin visualized by electron microscopy. J Mol Biol 2008; 381(2): 419-30.
[http://dx.doi.org/10.1016/j.jmb.2008.06.015] [PMID: 18614177]

[88] Peake PW, Hughes JT, Shen Y, Charlesworth JA. Glycosylation of human adiponectin affects its conformation and stability. J Mol Endocrinol 2007; 39(1): 45-52.
[http://dx.doi.org/10.1677/JME-07-0030] [PMID: 17601884]

[89] Wang Y, Xu A, Knight C, Xu LY, Cooper GJ. Hydroxylation and glycosylation of the four conserved lysine residues in the collagenous domain of adiponectin. Potential role in the modulation of its insulin-sensitizing activity. J Biol Chem 2002; 277(22): 19521-9.
[http://dx.doi.org/10.1074/jbc.M200601200] [PMID: 11912203]

[90] Richards AA, Stephens T, Charlton HK, *et al.* Adiponectin multimerization is dependent on conserved lysines in the collagenous domain: evidence for regulation of multimerization by alterations in posttranslational modifications. Mol Endocrinol 2006; 20(7): 1673-87.
[http://dx.doi.org/10.1210/me.2005-0390] [PMID: 16497731]

[91] Tilg H, Moschen AR. Adipocytokines: mediators linking adipose tissue, inflammation and immunity. Nat Rev Immunol 2006; 6(10): 772-83.
[http://dx.doi.org/10.1038/nri1937] [PMID: 16998510]

[92] Schraw T, Wang ZV, Halberg N, Hawkins M, Scherer PE. Plasma adiponectin complexes have distinct biochemical characteristics. Endocrinology 2008; 149(5): 2270-82.
[http://dx.doi.org/10.1210/en.2007-1561] [PMID: 18202126]

[93] Peake PW, Kriketos AD, Campbell LV, Shen Y, Charlesworth JA. The metabolism of isoforms of human adiponectin: studies in human subjects and in experimental animals. Eur J Endocrinol 2005; 153(3): 409-17.
[http://dx.doi.org/10.1530/eje.1.01978] [PMID: 16131604]

[94] Brochu-Gaudreau K, Rehfeldt C, Blouin R, Bordignon V, Murphy BD, Palin MF. Adiponectin action from head to toe. Endocrine 2010; 37(1): 11-32.
[http://dx.doi.org/10.1007/s12020-009-9278-8] [PMID: 20963555]

[95] Yamauchi T, Kamon J, Minokoshi Y, *et al.* Adiponectin stimulates glucose utilization and fatty-acid oxidation by activating AMP-activated protein kinase. Nat Med 2002; 8(11): 1288-95.
[http://dx.doi.org/10.1038/nm788] [PMID: 12368907]

[96] Banga A, Bodles AM, Rasouli N, Ranganathan G, Kern PA, Owens RJ. Calcium is involved in formation of high molecular weight adiponectin. Metab Syndr Relat Disord 2008; 6(2): 103-11.
[http://dx.doi.org/10.1089/met.2007.0033] [PMID: 18510435]

[97] Palanivel R, Fang X, Park M, *et al.* Globular and full-length forms of adiponectin mediate specific changes in glucose and fatty acid uptake and metabolism in cardiomyocytes. Cardiovasc Res 2007; 75(1): 148-57.

[http://dx.doi.org/10.1016/j.cardiores.2007.04.011] [PMID: 17499232]

[98] Koh EH, Park JY, Park HS, *et al.* Essential role of mitochondrial function in adiponectin synthesis in adipocytes. Diabetes 2007; 56(12): 2973-81.
[http://dx.doi.org/10.2337/db07-0510] [PMID: 17827403]

[99] Han KL, Choi JS, Lee JY, *et al.* Therapeutic potential of peroxisome proliferators--activated receptor-alpha/gamma dual agonist with alleviation of endoplasmic reticulum stress for the treatment of diabetes. Diabetes 2008; 57(3): 737-45.
[http://dx.doi.org/10.2337/db07-0972] [PMID: 18065517]

[100] Lara-Castro C, Luo N, Wallace P, Klein RL, Garvey WT. Adiponectin multimeric complexes and the metabolic syndrome trait cluster. Diabetes 2006; 55(1): 249-59.
[http://dx.doi.org/10.2337/diabetes.55.01.06.db05-1105] [PMID: 16380500]

[101] Scherer PE. Adipose tissue: from lipid storage compartment to endocrine organ. Diabetes 2006; 55(6): 1537-45.
[http://dx.doi.org/10.2337/db06-0263] [PMID: 16731815]

[102] Lara-Castro C, Fu Y, Chung BH, Garvey WT. Adiponectin and the metabolic syndrome: mechanisms mediating risk for metabolic and cardiovascular disease. Curr Opin Lipidol 2007; 18(3): 263-70.
[http://dx.doi.org/10.1097/MOL.0b013e32814a645f] [PMID: 17495599]

[103] Pajvani UB, Hawkins M, Combs TP, *et al.* Complex distribution, not absolute amount of adiponectin, correlates with thiazolidinedione-mediated improvement in insulin sensitivity. J Biol Chem 2004; 279(13): 12152-62.
[http://dx.doi.org/10.1074/jbc.M311113200] [PMID: 14699128]

[104] Ziemke F, Mantzoros CS. Adiponectin in insulin resistance: lessons from translational research. Am J Clin Nutr 2010; 91(1): 258S-61S.
[http://dx.doi.org/10.3945/ajcn.2009.28449C] [PMID: 19906806]

[105] Ryan AS, Berman DM, Nicklas BJ, *et al.* Plasma adiponectin and leptin levels, body composition, and glucose utilization in adult women with wide ranges of age and obesity. Diabetes Care 2003; 26(8): 2383-8.
[http://dx.doi.org/10.2337/diacare.26.8.2383] [PMID: 12882866]

[106] Fisher FM, McTernan PG, Valsamakis G, *et al.* Differences in adiponectin protein expression: effect of fat depots and type 2 diabetic status. Horm Metab Res 2002; 34(11-12): 650-4.
[http://dx.doi.org/10.1055/s-2002-38246] [PMID: 12660876]

[107] Lihn AS, Bruun JM, He G, Pedersen SB, Jensen PF, Richelsen B. Lower expression of adiponectin mRNA in visceral adipose tissue in lean and obese subjects. Mol Cell Endocrinol 2004; 219(1-2): 9-15.
[http://dx.doi.org/10.1016/j.mce.2004.03.002] [PMID: 15149722]

[108] Degawa-Yamauchi M, Moss KA, Bovenkerk JE, *et al.* Regulation of adiponectin expression in human adipocytes: effects of adiposity, glucocorticoids, and tumor necrosis factor alpha. Obes Res 2005; 13(4): 662-9.
[http://dx.doi.org/10.1038/oby.2005.74] [PMID: 15897474]

[109] Bodles AM, Banga A, Rasouli N, Ono F, Kern PA, Owens RJ. Pioglitazone increases secretion of high-molecular-weight adiponectin from adipocytes. Am J Physiol Endocrinol Metab 2006; 291(5): E1100-5.
[http://dx.doi.org/10.1152/ajpendo.00187.2006] [PMID: 16803857]

[110] Xu A, Chan KW, Hoo RL, *et al.* Testosterone selectively reduces the high molecular weight form of adiponectin by inhibiting its secretion from adipocytes. J Biol Chem 2005; 280(18): 18073-80.
[http://dx.doi.org/10.1074/jbc.M414231200] [PMID: 15760892]

[111] Galic S, Oakhill JS, Steinberg GR. Adipose tissue as an endocrine organ. Mol Cell Endocrinol 2010; 316(2): 129-39.

[http://dx.doi.org/10.1016/j.mce.2009.08.018] [PMID: 19723556]

[112] Nigro E, Scudiero O, Monaco ML, *et al.* New insight into adiponectin role in obesity and obesity-related diseases. BioMed Res Int 2014; 2014.
[http://dx.doi.org/10.1155/2014/658913]

[113] Zhao L, Fu Z, Wu J, *et al.* Globular adiponectin ameliorates metabolic insulin resistance via AMPK-mediated restoration of microvascular insulin responses. J Physiol 2015; 593(17): 4067-79.
[http://dx.doi.org/10.1113/JP270371] [PMID: 26108677]

[114] Yamauchi T, Kamon J, Ito Y, *et al.* Cloning of adiponectin receptors that mediate antidiabetic metabolic effects. Nature 2003; 423(6941): 762-9.
[http://dx.doi.org/10.1038/nature01705] [PMID: 12802337]

[115] Tanabe H, Fujii Y, Okada-Iwabu M, *et al.* Crystal structures of the human adiponectin receptors. Nature 2015; 520(7547): 312-6.
[http://dx.doi.org/10.1038/nature14301] [PMID: 25855295]

[116] Yamauchi T, Nio Y, Maki T, *et al.* Targeted disruption of AdipoR1 and AdipoR2 causes abrogation of adiponectin binding and metabolic actions. Nat Med 2007; 13(3): 332-9.
[http://dx.doi.org/10.1038/nm1557] [PMID: 17268472]

[117] Kadowaki T, Yamauchi T. Adiponectin and adiponectin receptors. Endocr Rev 2005; 26(3): 439-51.
[http://dx.doi.org/10.1210/er.2005-0005] [PMID: 15897298]

[118] Tsuchida A, Yamauchi T, Ito Y, *et al.* Insulin/Foxo1 pathway regulates expression levels of adiponectin receptors and adiponectin sensitivity. J Biol Chem 2004; 279(29): 30817-22.
[http://dx.doi.org/10.1074/jbc.M402367200] [PMID: 15123605]

[119] Civitarese AE, Jenkinson CP, Richardson D, *et al.* Adiponectin receptors gene expression and insulin sensitivity in non-diabetic Mexican Americans with or without a family history of Type 2 diabetes. Diabetologia 2004; 47(5): 816-20.
[http://dx.doi.org/10.1007/s00125-004-1359-x] [PMID: 15105989]

[120] Bullen JW Jr, Bluher S, Kelesidis T, Mantzoros CS. Regulation of adiponectin and its receptors in response to development of diet-induced obesity in mice. Am J Physiol Endocrinol Metab 2007; 292(4): E1079-86.
[http://dx.doi.org/10.1152/ajpendo.00245.2006] [PMID: 17164441]

[121] Liu BH, Wang PH, Wang YC, Cheng WM, Mersmann HJ, Ding ST. Fasting regulates the expression of adiponectin receptors in young growing pigs. J Anim Sci 2008; 86(12): 3377-84.
[http://dx.doi.org/10.2527/jas.2008-0971] [PMID: 18676723]

[122] Blüher M, Williams CJ, Klöting N, *et al.* Gene expression of adiponectin receptors in human visceral and subcutaneous adipose tissue is related to insulin resistance and metabolic parameters and is altered in response to physical training. Diabetes Care 2007; 30(12): 3110-5.
[http://dx.doi.org/10.2337/dc07-1257] [PMID: 17878241]

[123] Blüher M, Bullen JW Jr, Lee JH, *et al.* Circulating adiponectin and expression of adiponectin receptors in human skeletal muscle: associations with metabolic parameters and insulin resistance and regulation by physical training. J Clin Endocrinol Metab 2006; 91(6): 2310-6.
[http://dx.doi.org/10.1210/jc.2005-2556] [PMID: 16551730]

[124] Storgaard H, Poulsen P, Ling C, Groop L, Vaag AA. Relationships of plasma adiponectin level and adiponectin receptors 1 and 2 gene expression to insulin sensitivity and glucose and fat metabolism in monozygotic and dizygotic twins. J Clin Endocrinol Metab 2007; 92(7): 2835-9.
[http://dx.doi.org/10.1210/jc.2006-1812] [PMID: 17426101]

[125] Bermúdez VJ, Rojas E, Toledo A, *et al.* Single nucleotide polymorphisms in adiponectin, AdipoR1, and AdipoR2 genes: insulin resistance and type 2 diabetes mellitus candidate genes. Am J Ther 2013; 20(4): 414-21.
[http://dx.doi.org/10.1097/MJT.0b013e318235f206] [PMID: 23656997]

[126] Hu MB, Xu H, Hu JM, *et al.* Genetic polymorphisms in leptin, adiponectin and their receptors affect risk and aggressiveness of prostate cancer: evidence from a meta-analysis and pooled-review. Oncotarget 2016; 7(49): 81049-61.
[PMID: 27768592]

[127] Ou Y, Chen P, Zhou Z, *et al.* Associations between variants on ADIPOQ and ADIPOR1 with colorectal cancer risk: a Chinese case-control study and updated meta-analysis. BMC Med Genet 2014; 15: 137.
[http://dx.doi.org/10.1186/s12881-014-0137-y] [PMID: 25516230]

[128] Soccio T, Zhang YY, Bacci S, *et al.* Common haplotypes at the adiponectin receptor 1 (ADIPOR1) locus are associated with increased risk of coronary artery disease in type 2 diabetes. Diabetes 2006; 55(10): 2763-70.
[http://dx.doi.org/10.2337/db06-0613] [PMID: 17003341]

[129] Stefan N, Machicao F, Staiger H, *et al.* Polymorphisms in the gene encoding adiponectin receptor 1 are associated with insulin resistance and high liver fat. Diabetologia 2005; 48(11): 2282-91.
[http://dx.doi.org/10.1007/s00125-005-1948-3] [PMID: 16205883]

[130] Wang H, Zhang H, Jia Y, *et al.* Adiponectin receptor 1 gene (ADIPOR1) as a candidate for type 2 diabetes and insulin resistance. Diabetes 2004; 53(8): 2132-6.
[http://dx.doi.org/10.2337/diabetes.53.8.2132] [PMID: 15277397]

[131] Kantartzis K, Fritsche A, Machicao F, Häring HU, Stefan N. The -8503 G/A polymorphism of the adiponectin receptor 1 gene is associated with insulin sensitivity dependent on adiposity. Diabetes Care 2006; 29(2): 464.
[http://dx.doi.org/10.2337/diacare.29.02.06.dc05-2020] [PMID: 16443913]

[132] Hara K, Horikoshi M, Kitazato H, *et al.* Absence of an association between the polymorphisms in the genes encoding adiponectin receptors and type 2 diabetes. Diabetologia 2005; 48(7): 1307-14.
[http://dx.doi.org/10.1007/s00125-005-1806-3] [PMID: 15918014]

[133] Damcott CM, Ott SH, Pollin TI, *et al.* Genetic variation in adiponectin receptor 1 and adiponectin receptor 2 is associated with type 2 diabetes in the Old Order Amish. Diabetes 2005; 54(7): 2245-50.
[http://dx.doi.org/10.2337/diabetes.54.7.2245] [PMID: 15983228]

[134] Iwabu M, Yamauchi T, Okada-Iwabu M, *et al.* Adiponectin and AdipoR1 regulate PGC-1alpha and mitochondria by Ca(2+) and AMPK/SIRT1. Nature 2010; 464(7293): 1313-9.
[http://dx.doi.org/10.1038/nature08991] [PMID: 20357764]

[135] Chen HM, Yang CM, Chang JF, Wu CS, Sia KC, Lin WN. AdipoR-increased intracellular ROS promotes cPLA2 and COX-2 expressions *via* activation of PKC and p300 in adiponectin-stimulated human alveolar type II cells. Am J Physiol Lung Cell Mol Physiol 2016; 311(2): L255-69.
[PMID: 27288489]

[136] Nigro E, Matteis M, Roviezzo F, *et al.* Role of adiponectin in sphingosine-1-phosphate induced airway hyperresponsiveness and inflammation. Pharmacol Res 2016; 103: 114-22.
[http://dx.doi.org/10.1016/j.phrs.2015.10.004] [PMID: 26462929]

[137] Kupchak BR, Garitaonandia I, Villa NY, Smith JL, Lyons TJ. Antagonism of human adiponectin receptors and their membrane progesterone receptor paralogs by TNFalpha and a ceramidase inhibitor. Biochemistry 2009; 48(24): 5504-6.
[http://dx.doi.org/10.1021/bi9006258] [PMID: 19453184]

[138] Holland WL, Scherer PE. PAQRs: a counteracting force to ceramides? Mol Pharmacol 2009; 75(4): 740-3.
[http://dx.doi.org/10.1124/mol.109.054817] [PMID: 19158359]

[139] Holland WL, Miller RA, Wang ZV, *et al.* Receptor-mediated activation of ceramidase activity initiates the pleiotropic actions of adiponectin. Nat Med 2011; 17(1): 55-63.
[http://dx.doi.org/10.1038/nm.2277] [PMID: 21186369]

[140] Villa NY, Kupchak BR, Garitaonandia I, *et al.* Sphingolipids function as downstream effectors of a fungal PAQR. Mol Pharmacol 2009; 75(4): 866-75. [http://dx.doi.org/10.1124/mol.108.049809] [PMID: 19066337]

[141] Takeuchi T, Adachi Y, Ohtsuki Y, Furihata M. Adiponectin receptors, with special focus on the role of the third receptor, T-cadherin, in vascular disease. Med Mol Morphol 2007; 40(3): 115-20. [http://dx.doi.org/10.1007/s00795-007-0364-9] [PMID: 17874043]

[142] Doyle DD, Goings GE, Upshaw-Earley J, Page E, Ranscht B, Palfrey HC. T-cadherin is a major glycophosphoinositol-anchored protein associated with noncaveolar detergent-insoluble domains of the cardiac sarcolemma. J Biol Chem 1998; 273(12): 6937-43. [http://dx.doi.org/10.1074/jbc.273.12.6937] [PMID: 9506999]

[143] Hug C, Wang J, Ahmad NS, Bogan JS, Tsao TS, Lodish HF. T-cadherin is a receptor for hexameric and high-molecular-weight forms of Acrp30/adiponectin. Proc Natl Acad Sci USA 2004; 101(28): 10308-13. [http://dx.doi.org/10.1073/pnas.0403382101] [PMID: 15210937]

[144] Schoenenberger AW, Pfaff D, Dasen B, *et al.* Gender-Specific Associations between Circulating T-Cadherin and High Molecular Weight-Adiponectin in Patients with Stable Coronary Artery Disease. PLoS One 2015; 10(6): e0131140. [http://dx.doi.org/10.1371/journal.pone.0131140] [PMID: 26083608]

[145] Matsuda K, Fujishima Y, Maeda N, *et al.* Positive feedback regulation between adiponectin and T-cadherin impacts adiponectin levels in tissue and plasma of male mice. Endocrinology 2015; 156(3): 934-46. [http://dx.doi.org/10.1210/en.2014-1618] [PMID: 25514086]

[146] Ling H, Waterworth DM, Stirnadel HA, *et al.* Genome-wide linkage and association analyses to identify genes influencing adiponectin levels: the GEMS Study. Obesity (Silver Spring) 2009; 17(4): 737-44. [http://dx.doi.org/10.1038/oby.2008.625] [PMID: 19165155]

[147] Morisaki H, Yamanaka I, Iwai N, *et al.* CDH13 gene coding T-cadherin influences variations in plasma adiponectin levels in the Japanese population. Hum Mutat 2012; 33(2): 402-10. [http://dx.doi.org/10.1002/humu.21652] [PMID: 22065538]

[148] Uetani E, Tabara Y, Kawamoto R, *et al.* CDH13 genotype-dependent association of high-molecular weight adiponectin with all-cause mortality: the J-SHIPP study. Diabetes Care 2014; 37(2): 396-401. [http://dx.doi.org/10.2337/dc13-1658] [PMID: 24041676]

[149] Gao H, Kim YM, Chen P, *et al.* Genetic variation in CDH13 is associated with lower plasma adiponectin levels but greater adiponectin sensitivity in East Asian populations. Diabetes 2013; 62(12): 4277-83. [http://dx.doi.org/10.2337/db13-0129] [PMID: 24009259]

[150] Fava C, Danese E, Montagnana M, *et al.* A variant upstream of the CDH13 adiponectin receptor gene and metabolic syndrome in Swedes. Am J Cardiol 2011; 108(10): 1432-7. [http://dx.doi.org/10.1016/j.amjcard.2011.06.068] [PMID: 21872196]

[151] Chung CM, Lin TH, Chen JW, *et al.* A genome-wide association study reveals a quantitative trait locus of adiponectin on CDH13 that predicts cardiometabolic outcomes. Diabetes 2011; 60(9): 2417-23. [http://dx.doi.org/10.2337/db10-1321] [PMID: 21771975]

[152] Teng MS, Hsu LA, Wu S, Sun YC, Juan SH, Ko YL. Association of CDH13 genotypes/haplotypes with circulating adiponectin levels, metabolic syndrome, and related metabolic phenotypes: the role of the suppression effect. PLoS One 2015; 10(4): e0122664. [http://dx.doi.org/10.1371/journal.pone.0122664] [PMID: 25875811]

[153] Konter JM, Parker JL, Baez E, *et al.* Adiponectin attenuates lipopolysaccharide-induced acute lung

injury through suppression of endothelial cell activation. J Immunol 2012; 188(2): 854-63. [http://dx.doi.org/10.4049/jimmunol.1100426] [PMID: 22156343]

[154] Kostopoulos CG, Spiroglou SG, Varakis JN, Apostolakis E, Papadaki HH. Adiponectin/T-cadherin and apelin/APJ expression in human arteries and periadventitial fat: implication of local adipokine signaling in atherosclerosis? Cardiovasc Pathol 2014; 23(3): 131-8. [http://dx.doi.org/10.1016/j.carpath.2014.02.003] [PMID: 24675084]

[155] Pfaff D, Schoenenberger AW, Dasen B, Erne P, Resink TJ, Philippova M. Plasma T-cadherin negatively associates with coronary lesion severity and acute coronary syndrome. Eur Heart J Acute Cardiovasc Care 2015; 4(5): 410-8. [http://dx.doi.org/10.1177/2048872614557229] [PMID: 25344491]

[156] Denzel MS, Scimia MC, Zumstein PM, Walsh K, Ruiz-Lozano P, Ranscht B. T-cadherin is critical for adiponectin-mediated cardioprotection in mice. J Clin Invest 2010; 120(12): 4342-52. [http://dx.doi.org/10.1172/JCI43464] [PMID: 21041950]

[157] Yamauchi T, Iwabu M, Okada-Iwabu M, Kadowaki T. Adiponectin receptors: a review of their structure, function and how they work. Best Pract Res Clin Endocrinol Metab 2014; 28(1): 15-23. [http://dx.doi.org/10.1016/j.beem.2013.09.003] [PMID: 24417942]

[158] Putku M, Kals M, Inno R, *et al.* CDH13 promoter SNPs with pleiotropic effect on cardiometabolic parameters represent methylation QTLs. Hum Genet 2015; 134(3): 291-303. [http://dx.doi.org/10.1007/s00439-014-1521-6] [PMID: 25543204]

[159] Org E, Eyheramendy S, Juhanson P, *et al.* Genome-wide scan identifies CDH13 as a novel susceptibility locus contributing to blood pressure determination in two European populations. Hum Mol Genet 2009; 18(12): 2288-96. [http://dx.doi.org/10.1093/hmg/ddp135] [PMID: 19304780]

[160] James FR, Wootton S, Jackson A, Wiseman M, Copson ER, Cutress RI. Obesity in breast cancer--what is the risk factor? Eur J Cancer 2015; 51(6): 705-20. [http://dx.doi.org/10.1016/j.ejca.2015.01.057] [PMID: 25747851]

[161] Renehan AG, Zwahlen M, Egger M. Adiposity and cancer risk: new mechanistic insights from epidemiology. Nat Rev Cancer 2015; 15(8): 484-98. [http://dx.doi.org/10.1038/nrc3967] [PMID: 26205341]

[162] Capurso C, Capurso A. From excess adiposity to insulin resistance: the role of free fatty acids. Vascul Pharmacol 2012; 57(2-4): 91-7. [http://dx.doi.org/10.1016/j.vph.2012.05.003] [PMID: 22609131]

[163] Rose DP, Vona-Davis L. Influence of obesity on breast cancer receptor status and prognosis. Expert Rev Anticancer Ther 2009; 9(8): 1091-101. [http://dx.doi.org/10.1586/era.09.71] [PMID: 19671029]

[164] Kershaw EE, Flier JS. Adipose tissue as an endocrine organ. J Clin Endocrinol Metab 2004; 89(6): 2548-56. [http://dx.doi.org/10.1210/jc.2004-0395] [PMID: 15181022]

[165] Cabia B, Andrade S, Carreira MC, Casanueva FF, Crujeiras AB. A role for novel adipose tissue-secreted factors in obesity-related carcinogenesis. Obes Rev 2016; 17(4): 361-76. [http://dx.doi.org/10.1111/obr.12377] [PMID: 26914773]

[166] Feldman DE, Chen C, Punj V, Tsukamoto H, Machida K. Pluripotency factor-mediated expression of the leptin receptor (OB-R) links obesity to oncogenesis through tumor-initiating stem cells. Proc Natl Acad Sci USA 2012; 109(3): 829-34. [http://dx.doi.org/10.1073/pnas.1114438109] [PMID: 22207628]

[167] Mauer J, Denson JL, Brüning JC. Versatile functions for IL-6 in metabolism and cancer. Trends Immunol 2015; 36(2): 92-101. [http://dx.doi.org/10.1016/j.it.2014.12.008] [PMID: 25616716]

[168] Weichhaus M, Broom I, Bermano G. The molecular contribution of TNF-α in the link between obesity and breast cancer. Oncol Rep 2011; 25(2): 477-83.
[PMID: 21165572]

[169] Rose DP, Komninou D, Stephenson GD. Obesity, adipocytokines, and insulin resistance in breast cancer. Obes Rev 2004; 5(3): 153-65.
[http://dx.doi.org/10.1111/j.1467-789X.2004.00142.x] [PMID: 15245384]

[170] Deng T, Lyon CJ, Bergin S, Caligiuri MA, Hsueh WA. Obesity, Inflammation, and Cancer. Annu Rev Pathol 2016; 11: 421-49.
[http://dx.doi.org/10.1146/annurev-pathol-012615-044359] [PMID: 27193454]

[171] Gallagher EJ, Fierz Y, Vijayakumar A, Haddad N, Yakar S, LeRoith D. Inhibiting PI3K reduces mammary tumor growth and induces hyperglycemia in a mouse model of insulin resistance and hyperinsulinemia. Oncogene 2012; 31(27): 3213-22.
[http://dx.doi.org/10.1038/onc.2011.495] [PMID: 22037215]

[172] Jung UJ, Choi M-S. Obesity and its metabolic complications: the role of adipokines and the relationship between obesity, inflammation, insulin resistance, dyslipidemia and nonalcoholic fatty liver disease. Int J Mol Sci 2014; 15(4): 6184-223.
[http://dx.doi.org/10.3390/ijms15046184] [PMID: 24733068]

[173] Kelesidis I, Kelesidis T, Mantzoros CS. Adiponectin and cancer: a systematic review. Br J Cancer 2006; 94(9): 1221-5.
[http://dx.doi.org/10.1038/sj.bjc.6603051] [PMID: 16570048]

[174] Calle EE, Kaaks R. Overweight, obesity and cancer: epidemiological evidence and proposed mechanisms. Nat Rev Cancer 2004; 4(8): 579-91.
[http://dx.doi.org/10.1038/nrc1408] [PMID: 15286738]

[175] Zheng Q, Banaszak L, Fracci S, *et al.* Leptin receptor maintains cancer stem-like properties in triple negative breast cancer cells. Endocr Relat Cancer 2013; 20(6): 797-808.
[http://dx.doi.org/10.1530/ERC-13-0329] [PMID: 24025407]

[176] Diedrich J, Gusky HC, Podgorski I. Adipose tissue dysfunction and its effects on tumor metabolism. Horm Mol Biol Clin Investig 2015; 21(1): 17-41.
[http://dx.doi.org/10.1515/hmbci-2014-0045] [PMID: 25781550]

[177] Ye J, Jia J, Dong S, *et al.* Circulating adiponectin levels and the risk of breast cancer: a meta-analysis. Eur J Cancer Prev 2014; 23(3): 158-65.
[http://dx.doi.org/10.1097/CEJ.0b013e328364f293] [PMID: 23929213]

[178] Miyoshi Y, Funahashi T, Kihara S, *et al.* Association of serum adiponectin levels with breast cancer risk. Clin Cancer Res 2003; 9(15): 5699-704.
[PMID: 14654554]

[179] Dal Maso L, Augustin LS, Karalis A, *et al.* Circulating adiponectin and endometrial cancer risk. J Clin Endocrinol Metab 2004; 89(3): 1160-3.
[http://dx.doi.org/10.1210/jc.2003-031716] [PMID: 15001602]

[180] Zheng Q, Wu H, Cao J. Circulating adiponectin and risk of endometrial cancer. PLoS One 2015; 10(6): e0129824.
[http://dx.doi.org/10.1371/journal.pone.0129824] [PMID: 26030130]

[181] Gong TT, Wu QJ, Wang YL, Ma XX. Circulating adiponectin, leptin and adiponectin-leptin ratio and endometrial cancer risk: Evidence from a meta-analysis of epidemiologic studies. Int J Cancer 2015; 137(8): 1967-78.
[http://dx.doi.org/10.1002/ijc.29561] [PMID: 25899043]

[182] Ohbuchi Y, Suzuki Y, Hatakeyama I, *et al.* A lower serum level of middle-molecular-weight adiponectin is a risk factor for endometrial cancer. Int J Clin Oncol 2014; 19(4): 667-73.
[http://dx.doi.org/10.1007/s10147-013-0603-0] [PMID: 23963826]

[183] Goktas S, Yilmaz MI, Caglar K, Sonmez A, Kilic S, Bedir S. Prostate cancer and adiponectin. Urology 2005; 65(6): 1168-72.
[http://dx.doi.org/10.1016/j.urology.2004.12.053] [PMID: 15922427]

[184] Wei EK, Giovannucci E, Fuchs CS, Willett WC, Mantzoros CS. Low plasma adiponectin levels and risk of colorectal cancer in men: a prospective study. J Natl Cancer Inst 2005; 97(22): 1688-94.
[http://dx.doi.org/10.1093/jnci/dji376] [PMID: 16288122]

[185] Kumor A, Daniel P, Pietruczuk M, Małecka-Panas E. Serum leptin, adiponectin, and resistin concentration in colorectal adenoma and carcinoma (CC) patients. Int J Colorectal Dis 2009; 24(3): 275-81.
[http://dx.doi.org/10.1007/s00384-008-0605-y] [PMID: 18979105]

[186] Inamura K, Song M, Jung S, *et al.* Prediagnosis Plasma Adiponectin in Relation to Colorectal Cancer Risk According to KRAS Mutation Status. J Natl Cancer Inst 2015; 108(4): 108.
[PMID: 26598515]

[187] Diakowska D, Markocka-Maczka K, Szelachowski P, Grabowski K. Serum levels of resistin, adiponectin, and apelin in gastroesophageal cancer patients. Dis Markers 2014; 2014: 619649.
[http://dx.doi.org/10.1155/2014/619649]

[188] Seker M, Bilici A, Sonmez B, *et al.* The association of serum adiponectin levels with histopathological variables in gastric cancer patients. Med Oncol 2010; 27(4): 1319-23.
[http://dx.doi.org/10.1007/s12032-009-9382-x] [PMID: 20013320]

[189] Ishikawa M, Kitayama J, Kazama S, Hiramatsu T, Hatano K, Nagawa H. Plasma adiponectin and gastric cancer. Clin Cancer Res 2005; 11(2 Pt 1): 466-72.
[PMID: 15701829]

[190] Bao Y, Giovannucci EL, Kraft P, *et al.* A prospective study of plasma adiponectin and pancreatic cancer risk in five US cohorts. J Natl Cancer Inst 2013; 105(2): 95-103.
[http://dx.doi.org/10.1093/jnci/djs474] [PMID: 23243202]

[191] Perrier S, Jardé T. Adiponectin, an anti-carcinogenic hormone? A systematic review on breast, colorectal, liver and prostate cancer. Curr Med Chem 2012; 19(32): 5501-12.
[http://dx.doi.org/10.2174/092986712803833137] [PMID: 22876928]

[192] Hamdy K, Al Swaff R, Hussein HA, Gamal M. Assessment of serum adiponectin in Egyptian patients with HCV-related cirrhosis and hepatocellular carcinoma. J Endocrinol Invest 2015; 38(11): 1225-31.
[http://dx.doi.org/10.1007/s40618-015-0379-3] [PMID: 26359143]

[193] Spyridopoulos TN, Petridou ET, Skalkidou A, Dessypris N, Chrousos GP, Mantzoros CS. Low adiponectin levels are associated with renal cell carcinoma: a case-control study. Int J Cancer 2007; 120(7): 1573-8.
[http://dx.doi.org/10.1002/ijc.22526] [PMID: 17205522]

[194] Liao LM, Schwartz K, Pollak M, *et al.* Serum leptin and adiponectin levels and risk of renal cell carcinoma. Obesity (Silver Spring) 2013; 21(7): 1478-85.
[http://dx.doi.org/10.1002/oby.20138] [PMID: 23666639]

[195] Wang H, Wu J, Gu W, *et al.* Serum Adiponectin Level May be an Independent Predictor of Clear Cell Renal Cell Carcinoma. J Cancer 2016; 7(10): 1340-6.
[http://dx.doi.org/10.7150/jca.14716] [PMID: 27390609]

[196] Petridou E, Mantzoros CS, Dessypris N, Dikalioti SK, Trichopoulos D. Adiponectin in relation to childhood myeloblastic leukaemia. Br J Cancer 2006; 94(1): 156-60.
[http://dx.doi.org/10.1038/sj.bjc.6602896] [PMID: 16404369]

[197] Ma JJ, Shang J, Wang H, Sui JR, Liu K, Du JX. Serum adiponectin levels are inversely correlated with leukemia: A meta-analysis. J Cancer Res Ther 2016; 12(2): 897-902.
[http://dx.doi.org/10.4103/0973-1482.186695] [PMID: 27461671]

[198] Petridou ET, Sergentanis TN, Dessypris N, *et al.* Serum adiponectin as a predictor of childhood non-Hodgkin's lymphoma: a nationwide case-control study. J Clin Oncol 2009; 27(30): 5049-55. [http://dx.doi.org/10.1200/JCO.2008.19.7525] [PMID: 19738128]

[199] Dalamaga M, Karmaniolas K, Panagiotou A, *et al.* Low circulating adiponectin and resistin, but not leptin, levels are associated with multiple myeloma risk: a case-control study. Cancer Causes Control 2009; 20(2): 193-9. [http://dx.doi.org/10.1007/s10552-008-9233-7] [PMID: 18814045]

[200] Dalamaga M, Karmaniolas K, Nikolaidou A, *et al.* Adiponectin and resistin are associated with risk for myelodysplastic syndrome, independently from the insulin-like growth factor-I (IGF-I) system. Eur J Cancer 2008; 44(12): 1744-53. [http://dx.doi.org/10.1016/j.ejca.2008.04.015] [PMID: 18515085]

[201] Han S, Jeong AL, Lee S, *et al.* Adiponectin deficiency suppresses lymphoma growth in mice by modulating NK cells, CD8 T cells, and myeloid-derived suppressor cells. J Immunol 2013; 190(9): 4877-86. [http://dx.doi.org/10.4049/jimmunol.1202487] [PMID: 23530146]

[202] Horiguchi A, Ito K, Sumitomo M, Kimura F, Asano T, Hayakawa M. Decreased serum adiponectin levels in patients with metastatic renal cell carcinoma. Jpn J Clin Oncol 2008; 38(2): 106-11. [http://dx.doi.org/10.1093/jjco/hym158] [PMID: 18245516]

[203] Chen MW, Ye S, Zhao LL, *et al.* Association of plasma total and high-molecular-weight adiponectin with risk of colorectal cancer: an observational study in Chinese male. Med Oncol 2012; 29(5): 3129-35. [http://dx.doi.org/10.1007/s12032-012-0280-2] [PMID: 22752603]

[204] Otokozawa S, Tanaka R, Akasaka H, *et al.* Associations of Serum Isoflavone, Adiponectin and Insulin Levels with Risk for Epithelial Ovarian Cancer: Results of a Case-control Study. Asian Pac J Cancer Prev 2015; 16(12): 4987-91. [http://dx.doi.org/10.7314/APJCP.2015.16.12.4987] [PMID: 26163627]

[205] Izadi V, Farabad E, Azadbakht L. Serum adiponectin level and different kinds of cancer: a review of recent evidence. ISRN Oncol 2012; 2012: 982769. [http://dx.doi.org/10.5402/2012/982769]

[206] Avcu F, Ural AU, Yilmaz MI, Bingol N, Nevruz O, Caglar K. Association of plasma adiponectin concentrations with chronic lymphocytic leukemia and myeloproliferative diseases. Int J Hematol 2006; 83(3): 254-8. [http://dx.doi.org/10.1532/IJH97.NA0411] [PMID: 16720558]

[207] Fitter S, Vandyke K, Schultz CG, White D, Hughes TP, Zannettino AC. Plasma adiponectin levels are markedly elevated in imatinib-treated chronic myeloid leukemia (CML) patients: a mechanism for improved insulin sensitivity in type 2 diabetic CML patients? J Clin Endocrinol Metab 2010; 95(8): 3763-7. [http://dx.doi.org/10.1210/jc.2010-0086] [PMID: 20466781]

[208] Gao Q, Zheng J. Adiponectin-induced antitumor activity on prostatic cancers through inhibiting proliferation. Cell Biochem Biophys 2014; 70(1): 461-5. [http://dx.doi.org/10.1007/s12013-014-9941-4] [PMID: 24793551]

[209] Pinthus JH, Kleinmann N, Tisdale B, *et al.* Lower plasma adiponectin levels are associated with larger tumor size and metastasis in clear-cell carcinoma of the kidney. Eur Urol 2008; 54(4): 866-73. [http://dx.doi.org/10.1016/j.eururo.2008.02.044] [PMID: 18343565]

[210] Gunter MJ, Wang T, Cushman M, *et al.* Circulating Adipokines and Inflammatory Markers and Postmenopausal Breast Cancer Risk. J Natl Cancer Inst 2015; 107(9): 107. [http://dx.doi.org/10.1093/jnci/djv169] [PMID: 26185195]

[211] Tworoger SS, Eliassen AH, Kelesidis T, *et al.* Plasma adiponectin concentrations and risk of incident

breast cancer. J Clin Endocrinol Metab 2007; 92(4): 1510-6.
[http://dx.doi.org/10.1210/jc.2006-1975] [PMID: 17213279]

[212] Aleksandrova K, Boeing H, Jenab M, *et al.* Total and high-molecular weight adiponectin and risk of colorectal cancer: the European Prospective Investigation into Cancer and Nutrition Study. Carcinogenesis 2012; 33(6): 1211-8.
[http://dx.doi.org/10.1093/carcin/bgs133] [PMID: 22431719]

[213] Cust AE, Kaaks R, Friedenreich C, *et al.* Plasma adiponectin levels and endometrial cancer risk in pre- and postmenopausal women. J Clin Endocrinol Metab 2007; 92(1): 255-63.
[http://dx.doi.org/10.1210/jc.2006-1371] [PMID: 17062769]

[214] Guo MM, Duan XN, Cui SD, *et al.* Circulating High-Molecular-Weight (HMW) Adiponectin Level Is Related with Breast Cancer Risk Better than Total Adiponectin: A Case-Control Study. PLoS One 2015; 10(6): e0129246.
[http://dx.doi.org/10.1371/journal.pone.0129246] [PMID: 26070203]

[215] Kotani K, Wakai K, Shibata A, *et al.* Serum adiponectin multimer complexes and liver cancer risk in a large cohort study in Japan. Asian Pac J Cancer Prev 2009; 10 (Suppl.): 87-90.
[PMID: 20553088]

[216] Wei T, Ye P, Peng X, Wu LL, Yu GY. Circulating adiponectin levels in various malignancies: an updated meta-analysis of 107 studies. Oncotarget 2016; 7(30): 48671-91.
[http://dx.doi.org/10.18632/oncotarget.8932] [PMID: 27119501]

[217] Ye L, Wang G, Tang Y, Bai J. A population-specific correlation between ADIPOQ rs2241766 and rs 1501299 and colorectal cancer risk: a meta-analysis for debate. Int J Clin Oncol 2016.
[PMID: 27704292]

[218] Kaklamani VG, Hoffmann TJ, Thornton TA, *et al.* Adiponectin pathway polymorphisms and risk of breast cancer in African Americans and Hispanics in the Women's Health Initiative. Breast Cancer Res Treat 2013; 139(2): 461-8.
[http://dx.doi.org/10.1007/s10549-013-2546-6] [PMID: 23624817]

[219] Cai X, Gan Y, Fan Y, *et al.* The adiponectin gene single-nucleotide polymorphism rs1501299 is associated with hepatocellular carcinoma risk. Clin Transl Oncol 2014; 16(2): 166-72.
[http://dx.doi.org/10.1007/s12094-013-1056-7] [PMID: 23740135]

[220] Dhillon PK, Penney KL, Schumacher F, *et al.* Common polymorphisms in the adiponectin and its receptor genes, adiponectin levels and the risk of prostate cancer. Cancer Epidemiol Biomarkers Prev 2011; 20(12): 2618-27.
[http://dx.doi.org/10.1158/1055-9965.EPI-11-0434] [PMID: 21960694]

[221] Yang JP, Li X, Wang F, Gao M, Li SL, Chen KS. Association analysis of genetic variants of adiponectin gene and risk of pancreatic cancer. Int J Clin Exp Med 2015; 8(5): 8094-100.
[PMID: 26221375]

[222] Barb D, Williams CJ, Neuwirth AK, Mantzoros CS. Adiponectin in relation to malignancies: a review of existing basic research and clinical evidence. Am J Clin Nutr 2007; 86(3): s858-66.
[PMID: 18265479]

[223] Mistry T, Digby JE, Chen J, Desai KM, Randeva HS. The regulation of adiponectin receptors in human prostate cancer cell lines. Biochem Biophys Res Commun 2006; 348(3): 832-8.
[http://dx.doi.org/10.1016/j.bbrc.2006.07.139] [PMID: 16899222]

[224] Wu X, Yan Q, Zhang Z, Du G, Wan X. Acrp30 inhibits leptin-induced metastasis by downregulating the JAK/STAT3 pathway *via* AMPK activation in aggressive SPEC-2 endometrial cancer cells. Oncol Rep 2012; 27(5): 1488-96.
[PMID: 22327423]

[225] Takahata C, Miyoshi Y, Irahara N, Taguchi T, Tamaki Y, Noguchi S. Demonstration of adiponectin receptors 1 and 2 mRNA expression in human breast cancer cells. Cancer Lett 2007; 250(2): 229-36.

[http://dx.doi.org/10.1016/j.canlet.2006.10.006] [PMID: 17123704]

[226] Takemura Y, Osuga Y, Yamauchi T, *et al.* Expression of adiponectin receptors and its possible implication in the human endometrium. Endocrinology 2006; 147(7): 3203-10. [http://dx.doi.org/10.1210/en.2005-1510] [PMID: 16601138]

[227] Petridou ET, Mitsiades N, Gialamas S, *et al.* Circulating adiponectin levels and expression of adiponectin receptors in relation to lung cancer: two case-control studies. Oncology 2007; 73(3-4): 261-9. [http://dx.doi.org/10.1159/000127424] [PMID: 18424891]

[228] Dalamaga M, Migdalis I, Fargnoli JL, *et al.* Pancreatic cancer expresses adiponectin receptors and is associated with hypoleptinemia and hyperadiponectinemia: a case-control study. Cancer Causes Control 2009; 20(5): 625-33. [http://dx.doi.org/10.1007/s10552-008-9273-z] [PMID: 19051043]

[229] Byeon JS, Jeong JY, Kim MJ, *et al.* Adiponectin and adiponectin receptor in relation to colorectal cancer progression. Int J Cancer 2010; 127(12): 2758-67. [http://dx.doi.org/10.1002/ijc.25301] [PMID: 21351255]

[230] Kim AY, Lee YS, Kim KH, *et al.* Adiponectin represses colon cancer cell proliferation *via* AdipoR1- and -R2-mediated AMPK activation. Mol Endocrinol 2010; 24(7): 1441-52. [http://dx.doi.org/10.1210/me.2009-0498] [PMID: 20444885]

[231] Wu X, Chen P, Ou Y, *et al.* Association of variants on ADIPOQ and AdipoR1 and the prognosis of gastric cancer patients after gastrectomy treatment. Mol Biol Rep 2015; 42(2): 355-61. [http://dx.doi.org/10.1007/s11033-014-3775-4] [PMID: 25270251]

[232] Ayyildiz T, Dolar E, Ugras N, Dizdar OS, Adim SB, Yerci O. Lack of any prognostic relationship between adiponectin receptor (Adipo R1/R2) expression for early/advanced stage gastric cancer. Asian Pac J Cancer Prev 2014; 15(11): 4711-6. [http://dx.doi.org/10.7314/APJCP.2014.15.11.4711] [PMID: 24969908]

[233] Ayyildiz T, Dolar E, Ugras N, *et al.* Adipo-R1 and adipo-R2 expression in colorectal adenomas and carcinomas. Asian Pac J Cancer Prev 2015; 16(1): 367-72. [http://dx.doi.org/10.7314/APJCP.2015.16.1.367] [PMID: 25640382]

[234] Tae CH, Kim SE, Jung SA, *et al.* Involvement of adiponectin in early stage of colorectal carcinogenesis. BMC Cancer 2014; 14: 811. [http://dx.doi.org/10.1186/1471-2407-14-811] [PMID: 25370174]

[235] Williams CJ, Mitsiades N, Sozopoulos E, *et al.* Adiponectin receptor expression is elevated in colorectal carcinomas but not in gastrointestinal stromal tumors. Endocr Relat Cancer 2008; 15(1): 289-99. [http://dx.doi.org/10.1677/ERC-07-0197] [PMID: 18310295]

[236] Yunusova NV, Kondakova IV, Kolomiets LA, *et al.* [Serum adipokines and their receptors in endometrial and colon cancer patients: relationship with tumor invasion and metastasis]. Vopr Onkol 2015; 61(4): 619-23. [Serum adipokines and their receptors in endometrial and colon cancer patients: relationship with tumor invasion and metastasis]. [PMID: 26571833]

[237] Hiyoshi M, Tsuno NH, Otani K, *et al.* Adiponectin receptor 2 is negatively associated with lymph node metastasis of colorectal cancer. Oncol Lett 2012; 3(4): 756-60. [PMID: 22740988]

[238] Körner A, Pazaitou-Panayiotou K, Kelesidis T, *et al.* Total and high-molecular-weight adiponectin in breast cancer: *In vitro* and *in vivo* studies. J Clin Endocrinol Metab 2007; 92(3): 1041-8. [http://dx.doi.org/10.1210/jc.2006-1858] [PMID: 17192291]

[239] Pfeiler G, Hudelist G, Wülfing P, *et al.* Impact of AdipoR1 expression on breast cancer development. Gynecol Oncol 2010; 117(1): 134-8.

[http://dx.doi.org/10.1016/j.ygyno.2009.12.018] [PMID: 20071013]

[240] Jeong YJ, Bong JG, Park SH, Choi JH, Oh HK. Expression of leptin, leptin receptor, adiponectin, and adiponectin receptor in ductal carcinoma in situ and invasive breast cancer. J Breast Cancer 2011; 14(2): 96-103.
[http://dx.doi.org/10.4048/jbc.2011.14.2.96] [PMID: 21847403]

[241] Kaklamani VG, Wisinski KB, Sadim M, *et al.* Variants of the adiponectin (ADIPOQ) and adiponectin receptor 1 (ADIPOR1) genes and colorectal cancer risk. JAMA 2008; 300(13): 1523-31.
[http://dx.doi.org/10.1001/jama.300.13.1523] [PMID: 18827209]

[242] Mahmoudi T, Karimi K, Karimi N, *et al.* Association of adiponectin receptor 1 gene - 106 C > T variant with susceptibility to colorectal cancer. Meta Gene 2016; 9: 210-4.
[http://dx.doi.org/10.1016/j.mgene.2016.07.008] [PMID: 27617220]

[243] Kaklamani V, Yi N, Zhang K, *et al.* Polymorphisms of ADIPOQ and ADIPOR1 and prostate cancer risk. Metabolism 2011; 60(9): 1234-43.
[http://dx.doi.org/10.1016/j.metabol.2011.01.005] [PMID: 21397927]

[244] Yuan C, Yip SP, Wu VW, Kwong DL, Cheuk IW, Ying M. Association between genetic polymorphisms and carotid atherosclerosis in patients treated with radiotherapy for nasopharyngeal carcinoma. Radiat Oncol 2015; 10: 39.
[http://dx.doi.org/10.1186/s13014-015-0341-8] [PMID: 25880731]

[245] Kaklamani VG, Sadim M, Hsi A, *et al.* Variants of the adiponectin and adiponectin receptor 1 genes and breast cancer risk. Cancer Res 2008; 68(9): 3178-84.
[http://dx.doi.org/10.1158/0008-5472.CAN-08-0533] [PMID: 18451143]

[246] Beebe-Dimmer JL, Zuhlke KA, Ray AM, Lange EM, Cooney KA. Genetic variation in adiponectin (ADIPOQ) and the type 1 receptor (ADIPOR1), obesity and prostate cancer in African Americans. Prostate Cancer Prostatic Dis 2010; 13(4): 362-8.
[http://dx.doi.org/10.1038/pcan.2010.27] [PMID: 20697428]

[247] Wang Y, Lam KS, Xu JY, *et al.* Adiponectin inhibits cell proliferation by interacting with several growth factors in an oligomerization-dependent manner. J Biol Chem 2005; 280(18): 18341-7.
[http://dx.doi.org/10.1074/jbc.M501149200] [PMID: 15734737]

[248] Wang Y, Lam KS, Xu A. Adiponectin as a negative regulator in obesity-related mammary carcinogenesis. Cell Res 2007; 17(4): 280-2.
[http://dx.doi.org/10.1038/cr.2007.14] [PMID: 17404599]

[249] Sundaram S, Johnson AR, Makowski L. Obesity, metabolism and the microenvironment: Links to cancer. J Carcinog 2013; 12: 19.
[http://dx.doi.org/10.4103/1477-3163.119606] [PMID: 24227994]

[250] Dalamaga M, Christodoulatos GS. Adiponectin as a biomarker linking obesity and adiposopathy to hematologic malignancies. Horm Mol Biol Clin Investig 2015; 23(1): 5-20.
[http://dx.doi.org/10.1515/hmbci-2015-0016] [PMID: 26057219]

[251] Prieto-Hontoria PL, Perez-Matute P, Fernandez-Galilea M, Bustos M, Martinez JA, Moreno-Aliaga MJ. Role of obesity-associated dysfunctional adipose tissue in cancer: a molecular nutrition approach. Biochim Biophys Acta 2011; 1807: 664-78.

[252] Ouedraogo R, Gong Y, Berzins B, *et al.* Adiponectin deficiency increases leukocyte-endothelium interactions *via* upregulation of endothelial cell adhesion molecules *in vivo*. J Clin Invest 2007; 117(6): 1718-26.
[http://dx.doi.org/10.1172/JCI29623] [PMID: 17549259]

[253] Ouchi N, Kihara S, Arita Y, *et al.* Novel modulator for endothelial adhesion molecules: adipocyte-derived plasma protein adiponectin. Circulation 1999; 100(25): 2473-6.
[http://dx.doi.org/10.1161/01.CIR.100.25.2473] [PMID: 10604883]

[254] Arita Y, Kihara S, Ouchi N, *et al.* Adipocyte-derived plasma protein adiponectin acts as a platelet-

derived growth factor-BB-binding protein and regulates growth factor-induced common postreceptor signal in vascular smooth muscle cell. Circulation 2002; 105(24): 2893-8. [http://dx.doi.org/10.1161/01.CIR.0000018622.84402.FF] [PMID: 12070119]

[255] Mahadev K, Wu X, Donnelly S, Ouedraogo R, Eckhart AD, Goldstein BJ. Adiponectin inhibits vascular endothelial growth factor-induced migration of human coronary artery endothelial cells. Cardiovasc Res 2008; 78(2): 376-84. [http://dx.doi.org/10.1093/cvr/cvn034] [PMID: 18267956]

[256] Ding X, Saxena NK, Lin S, Xu A, Srinivasan S, Anania FA. The roles of leptin and adiponectin: a novel paradigm in adipocytokine regulation of liver fibrosis and stellate cell biology. Am J Pathol 2005; 166(6): 1655-69. [http://dx.doi.org/10.1016/S0002-9440(10)62476-5] [PMID: 15920151]

[257] Zhang W, Shu C, Li Q, Li M, Li X. Adiponectin affects vascular smooth muscle cell proliferation and apoptosis through modulation of the mitofusin-2-mediated Ras-Raf-Erk1/2 signaling pathway. Mol Med Rep 2015; 12(3): 4703-7. [http://dx.doi.org/10.3892/mmr.2015.3899] [PMID: 26059448]

[258] Bohlouli S, Rabzia A, Sadeghi E, Chobsaz F, Khazaei M. *in vitro* Anti-Proliferative Effect of Adiponectin on Human Endometriotic Stromal Cells through AdipoR1 and AdipoR2 Gene Receptor Expression. Iran Biomed J 2016; 20(1): 12-7. [PMID: 26459399]

[259] DeClercq V, McMurray DN, Chapkin RS. Obesity promotes colonic stem cell expansion during cancer initiation. Cancer Lett 2015; 369(2): 336-43. [http://dx.doi.org/10.1016/j.canlet.2015.10.001] [PMID: 26455770]

[260] Zhang L, Wen K, Han X, Liu R, Qu Q. Adiponectin mediates antiproliferative and apoptotic responses in endometrial carcinoma by the AdipoRs/AMPK pathway. Gynecol Oncol 2015; 137(2): 311-20. [http://dx.doi.org/10.1016/j.ygyno.2015.02.012] [PMID: 25703675]

[261] Yokota T, Oritani K, Takahashi I, *et al.* Adiponectin, a new member of the family of soluble defense collagens, negatively regulates the growth of myelomonocytic progenitors and the functions of macrophages. Blood 2000; 96(5): 1723-32. [PMID: 10961870]

[262] Li G, Cong L, Gasser J, Zhao J, Chen K, Li F. Mechanisms underlying the anti-proliferative actions of adiponectin in human breast cancer cells, MCF7-dependency on the cAMP/protein kinase-A pathway. Nutr Cancer 2011; 63(1): 80-8. [PMID: 21108124]

[263] Arditi JD, Venihaki M, Karalis KP, Chrousos GP. Antiproliferative effect of adiponectin on MCF7 breast cancer cells: a potential hormonal link between obesity and cancer. Horm Metab Res 2007; 39(1): 9-13. [http://dx.doi.org/10.1055/s-2007-956518] [PMID: 17226107]

[264] Dieudonne MN, Bussiere M, Dos Santos E, Leneveu MC, Giudicelli Y, Pecquery R. Adiponectin mediates antiproliferative and apoptotic responses in human MCF7 breast cancer cells. Biochem Biophys Res Commun 2006; 345(1): 271-9. [http://dx.doi.org/10.1016/j.bbrc.2006.04.076] [PMID: 16678125]

[265] Pfeiler GH, Buechler C, Neumeier M, *et al.* Adiponectin effects on human breast cancer cells are dependent on 17-beta estradiol. Oncol Rep 2008; 19(3): 787-93. [PMID: 18288417]

[266] Nigro E, Scudiero O, Sarnataro D, *et al.* Adiponectin affects lung epithelial A549 cell viability counteracting TNFα and IL-1ß toxicity through AdipoR1. Int J Biochem Cell Biol 2013; 45(6): 1145-53. [http://dx.doi.org/10.1016/j.biocel.2013.03.003] [PMID: 23500159]

[267] Ruan CC, Li Y, Ma Y, Zhu DL, Gao PJ. YIA 03-02 Adiponectin-mediated epithelial autophagy

attenuates hypertensive renal fibrosis. J Hypertens 2016; 34 Suppl 1 - ISH 2016 Abstract Book: e204.

[268] Treeck O, Lattrich C, Juhasz-Boess I, Buchholz S, Pfeiler G, Ortmann O. Adiponectin differentially affects gene expression in human mammary epithelial and breast cancer cells. Br J Cancer 2008; 99(8): 1246-50.
[http://dx.doi.org/10.1038/sj.bjc.6604692] [PMID: 18827813]

[269] Kang JH, Lee YY, Yu BY, *et al.* Adiponectin induces growth arrest and apoptosis of MDA-MB-231 breast cancer cell. Arch Pharm Res 2005; 28(11): 1263-9.
[http://dx.doi.org/10.1007/BF02978210] [PMID: 16350853]

[270] Grossmann ME, Nkhata KJ, Mizuno NK, Ray A, Cleary MP. Effects of adiponectin on breast cancer cell growth and signaling. Br J Cancer 2008; 98(2): 370-9.
[http://dx.doi.org/10.1038/sj.bjc.6604166] [PMID: 18182989]

[271] Grossmann ME, Ray A, Dogan S, Mizuno NK, Cleary MP. Balance of adiponectin and leptin modulates breast cancer cell growth. Cell Res 2008; 18(11): 1154-6.
[http://dx.doi.org/10.1038/cr.2008.293] [PMID: 18957939]

[272] Nakayama S, Miyoshi Y, Ishihara H, Noguchi S. Growth-inhibitory effect of adiponectin *via* adiponectin receptor 1 on human breast cancer cells through inhibition of S-phase entry without inducing apoptosis. Breast Cancer Res Treat 2008; 112(3): 405-10.
[http://dx.doi.org/10.1007/s10549-007-9874-3] [PMID: 18163210]

[273] Jardé T, Caldefie-Chézet F, Damez M, *et al.* Adiponectin and leptin expression in primary ductal breast cancer and in adjacent healthy epithelial and myoepithelial tissue. Histopathology 2008; 53(4): 484-7.
[http://dx.doi.org/10.1111/j.1365-2559.2008.03121.x] [PMID: 18983614]

[274] Dos Santos E, Benaitreau D, Dieudonne MN, *et al.* Adiponectin mediates an antiproliferative response in human MDA-MB 231 breast cancer cells. Oncol Rep 2008; 20(4): 971-7.
[PMID: 18813842]

[275] Landskroner-Eiger S, Qian B, Muise ES, *et al.* Proangiogenic contribution of adiponectin toward mammary tumor growth *in vivo.* Clin Cancer Res 2009; 15(10): 3265-76.
[http://dx.doi.org/10.1158/1078-0432.CCR-08-2649] [PMID: 19447867]

[276] Renehan AG, Frystyk J, Flyvbjerg A. Obesity and cancer risk: the role of the insulin-IGF axis. Trends Endocrinol Metab 2006; 17(8): 328-36.
[http://dx.doi.org/10.1016/j.tem.2006.08.006] [PMID: 16956771]

[277] Bub JD, Miyazaki T, Iwamoto Y. Adiponectin as a growth inhibitor in prostate cancer cells. Biochem Biophys Res Commun 2006; 340(4): 1158-66.
[http://dx.doi.org/10.1016/j.bbrc.2005.12.103] [PMID: 16403434]

[278] Bohlouli S, Khazaei M, Teshfam M, Hassanpour H. Adiponectin effect on the viability of human endometrial stromal cells and mRNA expression of adiponectin receptors. Int J Fertil Steril 2013; 7(1): 43-8.
[PMID: 24520463]

[279] Nagaraju GP, Rajitha B, Aliya S, *et al.* The role of adiponectin in obesity-associated female-specific carcinogenesis. Cytokine Growth Factor Rev 2016; 31: 37-48.
[http://dx.doi.org/10.1016/j.cytogfr.2016.03.014] [PMID: 27079372]

[280] Chou IP, Lin YY, Ding ST, Chen CY. Adiponectin receptor 1 enhances fatty acid metabolism and cell survival in palmitate-treated HepG2 cells through the PI3 K/AKT pathway. Eur J Nutr 2014; 53(3): 907-17.
[http://dx.doi.org/10.1007/s00394-013-0594-7] [PMID: 24129500]

[281] Sharma D, Wang J, Fu PP, *et al.* Adiponectin antagonizes the oncogenic actions of leptin in hepatocellular carcinogenesis. Hepatology 2010; 52(5): 1713-22.
[http://dx.doi.org/10.1002/hep.23892] [PMID: 20941777]

[282] Xu XT, Xu Q, Tong JL, *et al.* Meta-analysis: circulating adiponectin levels and risk of colorectal cancer and adenoma. J Dig Dis 2011; 12(4): 234-44.
[http://dx.doi.org/10.1111/j.1751-2980.2011.00504.x] [PMID: 21791018]

[283] Alexandre L, Long E, Beales IL. Pathophysiological mechanisms linking obesity and esophageal adenocarcinoma. World J Gastrointest Pathophysiol 2014; 5(4): 534-49.
[http://dx.doi.org/10.4291/wjgp.v5.i4.534] [PMID: 25400997]

[284] Gukovsky I, Li N, Todoric J, Gukovskaya A, Karin M. Inflammation, autophagy, and obesity: common features in the pathogenesis of pancreatitis and pancreatic cancer. Gastroenterology 2013; 144: 1199-209. e4.
[http://dx.doi.org/10.1053/j.gastro.2013.02.007]

[285] Medina EA, Oberheu K, Polusani SR, Ortega V, Velagaleti GV, Oyajobi BO. PKA/AMPK signaling in relation to adiponectin's antiproliferative effect on multiple myeloma cells. Leukemia 2014; 28(10): 2080-9.
[http://dx.doi.org/10.1038/leu.2014.112] [PMID: 24646889]

[286] Hofmann JN, Birmann BM, Teras LR, *et al.* Low Levels of Circulating Adiponectin Are Associated with Multiple Myeloma Risk in Overweight and Obese Individuals. Cancer Res 2016; 76(7): 1935-41.
[http://dx.doi.org/10.1158/0008-5472.CAN-15-2406] [PMID: 26921332]

[287] Wu S, Zheng C, Chen S, *et al.* Adiponectin signals through Adiponectin Receptor 1 to reverse imatinib resistance in K562 human chronic myeloid leukemia cells. Biochem Biophys Res Commun 2015; 456(1): 367-72.
[http://dx.doi.org/10.1016/j.bbrc.2014.11.089] [PMID: 25475722]

[288] Tan G, Shi L, Li Q, Wang M. Adiponectin enhances Imatinib anti-tumour activity in human chronic myeloid leukaemia cells with serum levels associated with Imatinib efficacy in early chronic phase patients. Cell Prolif 2015; 48(4): 486-96.
[http://dx.doi.org/10.1111/cpr.12194] [PMID: 26147296]

[289] Ozturk K, Avcu F, Ural AU. Aberrant expressions of leptin and adiponectin receptor isoforms in chronic myeloid leukemia patients. Cytokine 2012; 57(1): 61-7.
[http://dx.doi.org/10.1016/j.cyto.2011.10.004] [PMID: 22082804]

[290] Fitter S, Vandyke K, Gronthos S, Zannettino AC. Suppression of PDGF-induced PI3 kinase activity by imatinib promotes adipogenesis and adiponectin secretion. J Mol Endocrinol 2012; 48(3): 229-40.
[http://dx.doi.org/10.1530/JME-12-0003] [PMID: 22474082]

[291] Taliaferro-Smith L, Nagalingam A, Zhong D, Zhou W, Saxena NK, Sharma D. LKB1 is required for adiponectin-mediated modulation of AMPK-S6K axis and inhibition of migration and invasion of breast cancer cells. Oncogene 2009; 28(29): 2621-33.
[http://dx.doi.org/10.1038/onc.2009.129] [PMID: 19483724]

[292] Wijesekara N, Krishnamurthy M, Bhattacharjee A, Suhail A, Sweeney G, Wheeler MB. Adiponectin-induced ERK and Akt phosphorylation protects against pancreatic beta cell apoptosis and increases insulin gene expression and secretion. J Biol Chem 2010; 285(44): 33623-31.
[http://dx.doi.org/10.1074/jbc.M109.085084] [PMID: 20709750]

[293] Miyazaki T, Bub JD, Uzuki M, Iwamoto Y. Adiponectin activates c-Jun NH2-terminal kinase and inhibits signal transducer and activator of transcription 3. Biochem Biophys Res Commun 2005; 333(1): 79-87.
[http://dx.doi.org/10.1016/j.bbrc.2005.05.076] [PMID: 15936715]

[294] Liu J, Lam JB, Chow KH, *et al.* Adiponectin stimulates Wnt inhibitory factor-1 expression through epigenetic regulations involving the transcription factor specificity protein 1. Carcinogenesis 2008; 29(11): 2195-202.
[http://dx.doi.org/10.1093/carcin/bgn194] [PMID: 18701434]

[295] Barb D, Neuwirth A, Mantzoros CS, Balk SP. Adiponectin signals in prostate cancer cells through Akt

to activate the mammalian target of rapamycin pathway. Endocr Relat Cancer 2007; 14(4): 995-1005. [http://dx.doi.org/10.1677/ERC-06-0091] [PMID: 18045951]

[296] Luo Z, Saha AK, Xiang X, Ruderman NB. AMPK, the metabolic syndrome and cancer. Trends Pharmacol Sci 2005; 26(2): 69-76. [http://dx.doi.org/10.1016/j.tips.2004.12.011] [PMID: 15681023]

[297] Will K, Kuzinski J, Kalbe C, Palin MF, Rehfeldt C. Effects of leptin and adiponectin on the growth of porcine myoblasts are associated with changes in p44/42 MAPK signaling. Domest Anim Endocrinol 2013; 45(4): 196-205. [http://dx.doi.org/10.1016/j.domaniend.2013.09.002] [PMID: 24209504]

[298] Shackelford DB, Shaw RJ. The LKB1-AMPK pathway: metabolism and growth control in tumour suppression. Nat Rev Cancer 2009; 9(8): 563-75. [http://dx.doi.org/10.1038/nrc2676] [PMID: 19629071]

[299] El-Masry OS, Brown BL, Dobson PR. Effects of activation of AMPK on human breast cancer cell lines with different genetic backgrounds. Oncol Lett 2012; 3(1): 224-8. [PMID: 22740885]

[300] Swinnen JV, Beckers A, Brusselmans K, *et al.* Mimicry of a cellular low energy status blocks tumor cell anabolism and suppresses the malignant phenotype. Cancer Res 2005; 65(6): 2441-8. [http://dx.doi.org/10.1158/0008-5472.CAN-04-3025] [PMID: 15781660]

[301] Youn SH, Lee JS, Lee MS, *et al.* Anticancer properties of pomolic acid-induced AMP-activated protein kinase activation in MCF7 human breast cancer cells. Biol Pharm Bull 2012; 35(1): 105-10. [http://dx.doi.org/10.1248/bpb.35.105] [PMID: 22223345]

[302] Brognard J, Clark AS, Ni Y, Dennis PA. Akt/protein kinase B is constitutively active in non-small cell lung cancer cells and promotes cellular survival and resistance to chemotherapy and radiation. Cancer Res 2001; 61(10): 3986-97. [PMID: 11358816]

[303] Hattori Y, Nakano Y, Hattori S, Tomizawa A, Inukai K, Kasai K. High molecular weight adiponectin activates AMPK and suppresses cytokine-induced NF-kappaB activation in vascular endothelial cells. FEBS Lett 2008; 582(12): 1719-24. [http://dx.doi.org/10.1016/j.febslet.2008.04.037] [PMID: 18455514]

[304] Ouedraogo R, Wu X, Xu SQ, *et al.* Adiponectin suppression of high-glucose-induced reactive oxygen species in vascular endothelial cells: evidence for involvement of a cAMP signaling pathway. Diabetes 2006; 55(6): 1840-6. [http://dx.doi.org/10.2337/db05-1174] [PMID: 16731851]

[305] Ouchi N, Kihara S, Arita Y, *et al.* Adiponectin, an adipocyte-derived plasma protein, inhibits endothelial NF-kappaB signaling through a cAMP-dependent pathway. Circulation 2000; 102(11): 1296-301. [http://dx.doi.org/10.1161/01.CIR.102.11.1296] [PMID: 10982546]

[306] Ouchi N, Kobayashi H, Kihara S, *et al.* Adiponectin stimulates angiogenesis by promoting cross-talk between AMP-activated protein kinase and Akt signaling in endothelial cells. J Biol Chem 2004; 279(2): 1304-9. [http://dx.doi.org/10.1074/jbc.M310389200] [PMID: 14557259]

[307] Adya R, Tan BK, Randeva HS. Differential effects of leptin and adiponectin in endothelial angiogenesis. J Diabetes Res 2015; 2015: 648239.

[308] Motoshima H, Wu X, Mahadev K, Goldstein BJ. Adiponectin suppresses proliferation and superoxide generation and enhances eNOS activity in endothelial cells treated with oxidized LDL. Biochem Biophys Res Commun 2004; 315(2): 264-71. [http://dx.doi.org/10.1016/j.bbrc.2004.01.049] [PMID: 14766203]

[309] Shen L, Miao J, Yuan F, *et al.* Overexpression of adiponectin promotes focal angiogenesis in the

mouse brain following middle cerebral artery occlusion. Gene Ther 2013; 20(1): 93-101. [http://dx.doi.org/10.1038/gt.2012.7] [PMID: 22357512]

[310] Shibata R, Ouchi N, Kihara S, Sato K, Funahashi T, Walsh K. Adiponectin stimulates angiogenesis in response to tissue ischemia through stimulation of amp-activated protein kinase signaling. J Biol Chem 2004; 279(27): 28670-4. [http://dx.doi.org/10.1074/jbc.M402558200] [PMID: 15123726]

[311] Parker-Duffen JL, Nakamura K, Silver M, *et al.* T-cadherin is essential for adiponectin-mediated revascularization. J Biol Chem 2013; 288(34): 24886-97. [http://dx.doi.org/10.1074/jbc.M113.454835] [PMID: 23824191]

[312] Man K, Ng KT, Xu A, *et al.* Suppression of liver tumor growth and metastasis by adiponectin in nude mice through inhibition of tumor angiogenesis and downregulation of Rho kinase/IFN-inducible protein 10/matrix metalloproteinase 9 signaling. Clin Cancer Res 2010; 16(3): 967-77. [http://dx.doi.org/10.1158/1078-0432.CCR-09-1487] [PMID: 20103676]

[313] Xu SQ, Mahadev K, Wu X, *et al.* Adiponectin protects against angiotensin II or tumor necrosis factor alpha-induced endothelial cell monolayer hyperpermeability: role of cAMP/PKA signaling. Arterioscler Thromb Vasc Biol 2008; 28(5): 899-905. [http://dx.doi.org/10.1161/ATVBAHA.108.163634] [PMID: 18292388]

[314] Xu Y, Zhang C, Wang N, *et al.* Adiponectin inhibits lymphotoxin-β receptor-mediated NF-κB signaling in human umbilical vein endothelial cells. Biochem Biophys Res Commun 2011; 404(4): 1060-4. [http://dx.doi.org/10.1016/j.bbrc.2010.12.110] [PMID: 21195057]

[315] Bora PS, Kaliappan S, Lyzogubov VV, *et al.* Expression of adiponectin in choroidal tissue and inhibition of laser induced choroidal neovascularization by adiponectin. FEBS Lett 2007; 581(10): 1977-82. [http://dx.doi.org/10.1016/j.febslet.2007.04.024] [PMID: 17466298]

[316] Adya R, Tan BK, Chen J, Randeva HS. Protective actions of globular and full-length adiponectin on human endothelial cells: novel insights into adiponectin-induced angiogenesis. J Vasc Res 2012; 49(6): 534-43. [http://dx.doi.org/10.1159/000338279] [PMID: 22964477]

[317] Hattori Y, Hattori S, Akimoto K, *et al.* Globular adiponectin activates nuclear factor-kappaB and activating protein-1 and enhances angiotensin II-induced proliferation in cardiac fibroblasts. Diabetes 2007; 56(3): 804-8. [http://dx.doi.org/10.2337/db06-1405] [PMID: 17327451]

[318] Huang PH, Chen JS, Tsai HY, *et al.* Globular adiponectin improves high glucose-suppressed endothelial progenitor cell function through endothelial nitric oxide synthase dependent mechanisms. J Mol Cell Cardiol 2011; 51(1): 109-19. [http://dx.doi.org/10.1016/j.yjmcc.2011.03.008] [PMID: 21439968]

[319] Bobbert P, Antoniak S, Schultheiss HP, Rauch U. Globular adiponectin but not full-length adiponectin induces increased procoagulability in human endothelial cells. J Mol Cell Cardiol 2008; 44(2): 388-94. [http://dx.doi.org/10.1016/j.yjmcc.2007.10.018] [PMID: 18054040]

[320] Addabbo F, Nacci C, De Benedictis L, *et al.* Globular adiponectin counteracts VCAM-1-mediated monocyte adhesion via AdipoR1/NF-κB/COX-2 signaling in human aortic endothelial cells. Am J Physiol Endocrinol Metab 2011; 301(6): E1143-54. [http://dx.doi.org/10.1152/ajpendo.00208.2011] [PMID: 21900123]

[321] Tomizawa A, Hattori Y, Kasai K, Nakano Y. Adiponectin induces NF-kappaB activation that leads to suppression of cytokine-induced NF-kappaB activation in vascular endothelial cells: globular adiponectin vs. high molecular weight adiponectin. Diab Vasc Dis Res 2008; 5(2): 123-7. [http://dx.doi.org/10.3132/dvdr.2008.020] [PMID: 18537100]

[322] Zhang H, Park Y, Zhang C. Coronary and aortic endothelial function affected by feedback between

adiponectin and tumor necrosis factor α in type 2 diabetic mice. Arterioscler Thromb Vasc Biol 2010; 30(11): 2156-63.
[http://dx.doi.org/10.1161/ATVBAHA.110.214700] [PMID: 20814014]

[323] Waki H, Yamauchi T, Kamon J, *et al.* Generation of globular fragment of adiponectin by leukocyte elastase secreted by monocytic cell line THP-1. Endocrinology 2005; 146(2): 790-6.
[http://dx.doi.org/10.1210/en.2004-1096] [PMID: 15528304]

[324] Vetvik KK, Sonerud T, Lindeberg M, *et al.* Globular adiponectin and its downstream target genes are up-regulated locally in human colorectal tumors: ex vivo and *in vitro* studies. Metabolism 2014; 63(5): 672-81.
[http://dx.doi.org/10.1016/j.metabol.2014.02.001] [PMID: 24636346]

[325] Hebbard LW, Garlatti M, Young LJ, Cardiff RD, Oshima RG, Ranscht B. T-cadherin supports angiogenesis and adiponectin association with the vasculature in a mouse mammary tumor model. Cancer Res 2008; 68(5): 1407-16.
[http://dx.doi.org/10.1158/0008-5472.CAN-07-2953] [PMID: 18316604]

[326] Takahashi T, Saegusa S, Sumino H, *et al.* Adiponectin, T-cadherin and tumour necrosis factor-alpha in damaged cardiomyocytes from autopsy specimens. J Int Med Res 2005; 33(2): 236-44.
[http://dx.doi.org/10.1177/147323000503300212] [PMID: 15790136]

[327] Adachi Y, Takeuchi T, Sonobe H, Ohtsuki Y. An adiponectin receptor, T-cadherin, was selectively expressed in intratumoral capillary endothelial cells in hepatocellular carcinoma: possible cross talk between T-cadherin and FGF-2 pathways. Virchows Arch 2006; 448(3): 311-8.
[http://dx.doi.org/10.1007/s00428-005-0098-9] [PMID: 16273386]

[328] Andreeva AV, Han J, Kutuzov MA, Profirovic J, Tkachuk VA, Voyno-Yasenetskaya TA. T-cadherin modulates endothelial barrier function. J Cell Physiol 2010; 223(1): 94-102.
[PMID: 20039275]

[329] Lee MH, Klein RL, El-Shewy HM, Luttrell DK, Luttrell LM. The adiponectin receptors AdipoR1 and AdipoR2 activate ERK1/2 through a Src/Ras-dependent pathway and stimulate cell growth. Biochemistry 2008; 47(44): 11682-92.
[http://dx.doi.org/10.1021/bi801451f] [PMID: 18842004]

[330] Spranger J, Kroke A, Möhlig M, *et al.* Adiponectin and protection against type 2 diabetes mellitus. Lancet 2003; 361(9353): 226-8.
[http://dx.doi.org/10.1016/S0140-6736(03)12255-6] [PMID: 12547549]

[331] Otvos L Jr, Kovalszky I, Olah J, *et al.* Optimization of adiponectin-derived peptides for inhibition of cancer cell growth and signaling. Biopolymers 2015; 104(3): 156-66.
[http://dx.doi.org/10.1002/bip.22627] [PMID: 25683126]

[332] Ding S, Pinkas DM, Barron AE. Synthesis and assembly of functional high molecular weight adiponectin multimers in an engineered strain of Escherichia coli. Biomacromolecules 2012; 13(4): 1035-42.
[http://dx.doi.org/10.1021/bm2017367] [PMID: 22376164]

[333] Fruebis J, Tsao TS, Javorschi S, *et al.* Proteolytic cleavage product of 30-kDa adipocyte complement-related protein increases fatty acid oxidation in muscle and causes weight loss in mice. Proc Natl Acad Sci USA 2001; 98(4): 2005-10.
[http://dx.doi.org/10.1073/pnas.98.4.2005] [PMID: 11172066]

[334] Tomas E, Tsao TS, Saha AK, *et al.* Enhanced muscle fat oxidation and glucose transport by ACRP30 globular domain: acetyl-CoA carboxylase inhibition and AMP-activated protein kinase activation. Proc Natl Acad Sci USA 2002; 99(25): 16309-13.
[http://dx.doi.org/10.1073/pnas.222657499] [PMID: 12456889]

[335] Shimada K, Miyazaki T, Daida H. Adiponectin and atherosclerotic disease. Clin Chim Acta 2004; 344(1-2): 1-12.
[http://dx.doi.org/10.1016/j.cccn.2004.02.020] [PMID: 15149866]

[336] Otvos L Jr. Identification of adipokine receptor agonists and turning them to antagonists. Methods Mol Biol 2013; 1081: 195-209.
[http://dx.doi.org/10.1007/978-1-62703-652-8_12] [PMID: 24014441]

[337] Otvos L Jr, Haspinger E, La Russa F, *et al.* Design and development of a peptide-based adiponectin receptor agonist for cancer treatment. BMC Biotechnol 2011; 11: 90.
[http://dx.doi.org/10.1186/1472-6750-11-90] [PMID: 21974986]

[338] Otvos L Jr, Knappe D, Hoffmann R, *et al.* Development of second generation peptides modulating cellular adiponectin receptor responses. Front Chem 2014; 2: 93.
[http://dx.doi.org/10.3389/fchem.2014.00093] [PMID: 25368867]

[339] Otvos L. Commentary: drugs from bugs. Curr Protein Pept Sci 2015; 16(6): 476-7.
[http://dx.doi.org/10.2174/1389203716061507021350451 [PMID: 26138813]

[340] Fenton JI, Birmingham JM. Adipokine regulation of colon cancer: adiponectin attenuates interleukin-6-induced colon carcinoma cell proliferation *via* STAT-3. Mol Carcinog 2010; 49(7): 700-9.
[PMID: 20564347]

[341] Fujisawa T, Endo H, Tomimoto A, *et al.* Adiponectin suppresses colorectal carcinogenesis under the high-fat diet condition. Gut 2008; 57(11): 1531-8.
[http://dx.doi.org/10.1136/gut.2008.159293] [PMID: 18676419]

[342] Habeeb BS, Kitayama J, Nagawa H. Adiponectin supports cell survival in glucose deprivation through enhancement of autophagic response in colorectal cancer cells. Cancer Sci 2011; 102(5): 999-1006.
[http://dx.doi.org/10.1111/j.1349-7006.2011.01902.x] [PMID: 21299716]

[343] Falk Libby E, Liu J, Li YI, Lewis MJ, Demark-Wahnefried W, Hurst DR. Globular adiponectin enhances invasion in human breast cancer cells. Oncol Lett 2016; 11(1): 633-41.
[PMID: 26870258]

[344] Chiu YC, Shieh DC, Tong KM, *et al.* Involvement of AdipoR receptor in adiponectin-induced motility and alpha2beta1 integrin upregulation in human chondrosarcoma cells. Carcinogenesis 2009; 30(10): 1651-9.
[http://dx.doi.org/10.1093/carcin/bgp156] [PMID: 19549705]

[345] Tang CH, Lu ME. Adiponectin increases motility of human prostate cancer cells *via* adipoR, p38, AMPK, and NF-kappaB pathways. Prostate 2009; 69(16): 1781-9.
[http://dx.doi.org/10.1002/pros.21029] [PMID: 19676095]

[346] Chen MJ, Yeh YT, Lee KT, Tsai CJ, Lee HH, Wang SN. The promoting effect of adiponectin in hepatocellular carcinoma. J Surg Oncol 2012; 106(2): 181-7.
[http://dx.doi.org/10.1002/jso.23059] [PMID: 22287480]

[347] Holland WL, Adams AC, Brozinick JT, *et al.* An FGF21-adiponectin-ceramide axis controls energy expenditure and insulin action in mice. Cell Metab 2013; 17(5): 790-7.
[http://dx.doi.org/10.1016/j.cmet.2013.03.019] [PMID: 23663742]

[348] Sun Y, Zang Z, Zhong L, *et al.* Identification of adiponectin receptor agonist utilizing a fluorescence polarization based high throughput assay. PLoS One 2013; 8(5): e63354.
[http://dx.doi.org/10.1371/journal.pone.0063354] [PMID: 23691032]

[349] Katira A, Tan PH. Evolving role of adiponectin in cancer-controversies and update. Cancer Biol Med 2016; 13(1): 101-19.
[http://dx.doi.org/10.20892/j.issn.2095-3941.2015.0092] [PMID: 27144066]

[350] Takabe K, Paugh SW, Milstien S, Spiegel S. "Inside-out" signaling of sphingosine-1-phosphate: therapeutic targets. Pharmacol Rev 2008; 60(2): 181-95.
[http://dx.doi.org/10.1124/pr.107.07113] [PMID: 18552276]

[351] Almabouada F, Diaz-Ruiz A, Rabanal-Ruiz Y, Peinado JR, Vazquez-Martinez R, Malagon MM. Adiponectin receptors form homomers and heteromers exhibiting distinct ligand binding and

intracellular signaling properties. J Biol Chem 2013; 288(5): 3112-25.
[http://dx.doi.org/10.1074/jbc.M112.404624] [PMID: 23255609]

[352] Mao X, Kikani CK, Riojas RA, *et al.* APPL1 binds to adiponectin receptors and mediates adiponectin signalling and function. Nat Cell Biol 2006; 8(5): 516-23.
[http://dx.doi.org/10.1038/ncb1404] [PMID: 16622416]

[353] Cheng KK, Lam KS, Wang Y, *et al.* Adiponectin-induced endothelial nitric oxide synthase activation and nitric oxide production are mediated by APPL1 in endothelial cells. Diabetes 2007; 56(5): 1387-94.
[http://dx.doi.org/10.2337/db06-1580] [PMID: 17287464]

[354] Park M, Youn B, Zheng XL, Wu D, Xu A, Sweeney G. Globular adiponectin, acting *via* AdipoR1/APPL1, protects H9c2 cells from hypoxia/reoxygenation-induced apoptosis. PLoS One 2011; 6(4): e19143.
[http://dx.doi.org/10.1371/journal.pone.0019143] [PMID: 21552570]

[355] Ding Y, Cao Y, Wang B, *et al.* APPL1-Mediating Leptin Signaling Contributes to Proliferation and Migration of Cancer Cells. PLoS One 2016; 11(11): e0166172.
[http://dx.doi.org/10.1371/journal.pone.0166172] [PMID: 27820851]

[356] Okada-Iwabu M, Iwabu M, Ueki K, Yamauchi T, Kadowaki T. Perspective of Small-Molecule AdipoR Agonist for Type 2 Diabetes and Short Life in Obesity. Diabetes Metab J 2015; 39(5): 363-72.
[http://dx.doi.org/10.4093/dmj.2015.39.5.363] [PMID: 26566493]

[357] Okada-Iwabu M, Yamauchi T, Iwabu M, *et al.* A small-molecule AdipoR agonist for type 2 diabetes and short life in obesity. Nature 2013; 503(7477): 493-9.
[http://dx.doi.org/10.1038/nature12656] [PMID: 24172895]

[358] Hong K, Lee S, Li R, *et al.* Adiponectin Receptor Agonist, AdipoRon, Causes Vasorelaxation Predominantly *Via* a Direct Smooth Muscle Action. Microcirculation 2016; 23(3): 207-20.
[http://dx.doi.org/10.1111/micc.12266] [PMID: 26728950]

[359] Zhang D, Wang X, Wang B, *et al.* Adiponectin regulates contextual fear extinction and intrinsic excitability of dentate gyrus granule neurons through AdipoR2 receptors. Mol Psychiatry 2016.
[PMID: 27137743]

[360] Zhang Y, Zhao J, Li R, *et al.* AdipoRon, the first orally active adiponectin receptor activator, attenuates postischemic myocardial apoptosis through both AMPK-mediated and AMPK-independent signalings. Am J Physiol Endocrinol Metab 2015; 309(3): E275-82.
[http://dx.doi.org/10.1152/ajpendo.00577.2014] [PMID: 26037251]

[361] Malih S, Saidijam M, Mansouri K, *et al.* Promigratory and proangiogenic effects of AdipoRon on bone marrow-derived mesenchymal stem cells: an *in vitro* study. Biotechnol Lett 2016.
[PMID: 27627895]

[362] Lee S, Kwak HB. Role of adiponectin in metabolic and cardiovascular disease. J Exerc Rehabil 2014; 10(2): 54-9.
[http://dx.doi.org/10.12965/jer.140100] [PMID: 24877038]

[363] Wang Y, Wan Y, Ye G, *et al.* Hepatoprotective effects of AdipoRon against d-galactosamine-induced liver injury in mice. Eur J Pharm Sci 2016; 93: 123-31.
[http://dx.doi.org/10.1016/j.ejps.2016.08.017] [PMID: 27516150]

[364] Booth A, Magnuson A, Fouts J, Foster M. Adipose tissue, obesity and adipokines: role in cancer promotion. Horm Mol Biol Clin Investig 2015; 21(1): 57-74.
[http://dx.doi.org/10.1515/hmbci-2014-0037] [PMID: 25781552]

[365] Vona-Davis L, Rose DP. Adipokines as endocrine, paracrine, and autocrine factors in breast cancer risk and progression. Endocr Relat Cancer 2007; 14(2): 189-206.
[http://dx.doi.org/10.1677/ERC-06-0068] [PMID: 17639037]

[366] Hardie DG. The AMP-activated protein kinase pathway--new players upstream and downstream. J

Cell Sci 2004; 117(Pt 23): 5479-87.
[http://dx.doi.org/10.1242/jcs.01540] [PMID: 15509864]

[367] Jalving M, Gietema JA, Lefrandt JD, *et al.* Metformin: taking away the candy for cancer? Eur J Cancer 2010; 46(13): 2369-80.
[http://dx.doi.org/10.1016/j.ejca.2010.06.012] [PMID: 20656475]

[368] Koh M, Lee JC, Min C, Moon A. A novel metformin derivative, HL010183, inhibits proliferation and invasion of triple-negative breast cancer cells. Bioorg Med Chem 2013; 21(8): 2305-13.
[http://dx.doi.org/10.1016/j.bmc.2013.02.015] [PMID: 23490148]

[369] Vazquez-Martin A, Oliveras-Ferraros C, Cufí S, *et al.* The anti-diabetic drug metformin suppresses the metastasis-associated protein CD24 in MDA-MB-468 triple-negative breast cancer cells. Oncol Rep 2011; 25(1): 135-40.
[PMID: 21109968]

[370] Deng XS, Wang S, Deng A, *et al.* Metformin targets Stat3 to inhibit cell growth and induce apoptosis in triple-negative breast cancers. Cell Cycle 2012; 11(2): 367-76.
[http://dx.doi.org/10.4161/cc.11.2.18813] [PMID: 22189713]

[371] Vazquez-Martin A, Oliveras-Ferraros C, Menendez JA. The antidiabetic drug metformin suppresses HER2 (erbB-2) oncoprotein overexpression *via* inhibition of the mTOR effector p70S6K1 in human breast carcinoma cells. Cell Cycle 2009; 8(1): 88-96.
[http://dx.doi.org/10.4161/cc.8.1.7499] [PMID: 19106626]

[372] Pollak M. Metformin and other biguanides in oncology: advancing the research agenda. Cancer Prev Res (Phila) 2010; 3(9): 1060-5.
[http://dx.doi.org/10.1158/1940-6207.CAPR-10-0175] [PMID: 20810670]

[373] Kourelis TV, Siegel RD. Metformin and cancer: new applications for an old drug. Med Oncol 2012; 29(2): 1314-27.
[http://dx.doi.org/10.1007/s12032-011-9846-7] [PMID: 21301998]

[374] Gonzalez-Angulo AM, Meric-Bernstam F. Metformin: a therapeutic opportunity in breast cancer. Clin Cancer Res 2010; 16(6): 1695-700.
[http://dx.doi.org/10.1158/1078-0432.CCR-09-1805] [PMID: 20215559]

[375] Ben Sahra I, Le Marchand-Brustel Y, Tanti JF, Bost F. Metformin in cancer therapy: a new perspective for an old antidiabetic drug? Mol Cancer Ther 2010; 9(5): 1092-9.
[http://dx.doi.org/10.1158/1535-7163.MCT-09-1186] [PMID: 20442309]

[376] Anisimov VN. Metformin for aging and cancer prevention. Aging (Albany NY) 2010; 2(11): 760-74.
[http://dx.doi.org/10.18632/aging.100230] [PMID: 21084729]

[377] Zhu Z, Jiang W, Thompson MD, McGinley JN, Thompson HJ. Metformin as an energy restriction mimetic agent for breast cancer prevention. J Carcinog 2011; 10: 17.
[http://dx.doi.org/10.4103/1477-3163.83043] [PMID: 21799661]

[378] Hadad SM, Fleming S, Thompson AM. Targeting AMPK: a new therapeutic opportunity in breast cancer. Crit Rev Oncol Hematol 2008; 67(1): 1-7.
[http://dx.doi.org/10.1016/j.critrevonc.2008.01.007] [PMID: 18343152]

[379] Bayraktar S, Hernadez-Aya LF, Lei X, *et al.* Effect of metformin on survival outcomes in diabetic patients with triple receptor-negative breast cancer. Cancer 2012; 118(5): 1202-11.
[http://dx.doi.org/10.1002/cncr.26439] [PMID: 21800293]

[380] Phillips SA, Kung JT. Mechanisms of adiponectin regulation and use as a pharmacological target. Curr Opin Pharmacol 2010; 10(6): 676-83.
[http://dx.doi.org/10.1016/j.coph.2010.08.002] [PMID: 20810317]

[381] Tonelli J, Li W, Kishore P, *et al.* Mechanisms of early insulin-sensitizing effects of thiazolidinediones in type 2 diabetes. Diabetes 2004; 53(6): 1621-9.
[http://dx.doi.org/10.2337/diabetes.53.6.1621] [PMID: 15161771]

[382] Kriketos AD, Gan SK, Poynten AM, Furler SM, Chisholm DJ, Campbell LV. Exercise increases adiponectin levels and insulin sensitivity in humans. Diabetes Care 2004; 27(2): 629-30. [http://dx.doi.org/10.2337/diacare.27.2.629] [PMID: 14747265]

[383] Silva FM, de Almeida JC, Feoli AM. Effect of diet on adiponectin levels in blood. Nutr Rev 2011; 69(10): 599-612. [http://dx.doi.org/10.1111/j.1753-4887.2011.00414.x] [PMID: 21967160]

[384] McMillan DC, Sattar N, McArdle CS, McArdle CS. ABC of obesity. Obesity and cancer. BMJ 2006; 333(7578): 1109-11. [http://dx.doi.org/10.1136/bmj.39042.565035.BE1] [PMID: 17124223]

[385] Brown JC, Winters-Stone K, Lee A, Schmitz KH. Cancer, physical activity, and exercise. Compr Physiol 2012; 2(4): 2775-809. [PMID: 23720265]

[386] Swarbrick MM, Havel PJ. Physiological, pharmacological, and nutritional regulation of circulating adiponectin concentrations in humans. Metab Syndr Relat Disord 2008; 6(2): 87-102. [http://dx.doi.org/10.1089/met.2007.0029] [PMID: 18510434]

[387] Yang W-S, Lee W-J, Funahashi T, *et al.* Weight reduction increases plasma levels of an adipose-derived anti-inflammatory protein, adiponectin. J Clin Endocrinol Metab 2001; 86(8): 3815-9. [http://dx.doi.org/10.1210/jcem.86.8.7741] [PMID: 11502817]

[388] Abbenhardt C, McTiernan A, Alfano CM, *et al.* Effects of individual and combined dietary weight loss and exercise interventions in postmenopausal women on adiponectin and leptin levels. J Intern Med 2013; 274(2): 163-75. [http://dx.doi.org/10.1111/joim.12062] [PMID: 23432360]

[389] Lashinger LM, O'Flanagan CH, Dunlap SM, *et al.* Starving cancer from the outside and inside: separate and combined effects of calorie restriction and autophagy inhibition on Ras-driven tumors. Cancer Metab 2016; 4: 18. [http://dx.doi.org/10.1186/s40170-016-0158-4] [PMID: 27651895]

[390] Flachs P, Mohamed-Ali V, Horakova O, *et al.* Polyunsaturated fatty acids of marine origin induce adiponectin in mice fed a high-fat diet. Diabetologia 2006; 49(2): 394-7. [http://dx.doi.org/10.1007/s00125-005-0053-y] [PMID: 16397791]

[391] Banga A, Unal R, Tripathi P, *et al.* Adiponectin translation is increased by the PPARgamma agonists pioglitazone and omega-3 fatty acids. Am J Physiol Endocrinol Metab 2009; 296(3): E480-9. [http://dx.doi.org/10.1152/ajpendo.90892.2008] [PMID: 19088251]

[392] Dinca M, Serban MC, Sahebkar A, *et al.* Does vitamin D supplementation alter plasma adipokines concentrations? A systematic review and meta-analysis of randomized controlled trials. Pharmacol Res 2016; 107: 360-71. [http://dx.doi.org/10.1016/j.phrs.2016.03.035] [PMID: 27038530]

[393] Xu A, Wang H, Hoo RL, *et al.* Selective elevation of adiponectin production by the natural compounds derived from a medicinal herb alleviates insulin resistance and glucose intolerance in obese mice. Endocrinology 2009; 150(2): 625-33. [http://dx.doi.org/10.1210/en.2008-0999] [PMID: 18927219]

[394] Guo Q, Xu L, Liu J, *et al.* Fibroblast growth factor 21 reverses suppression of adiponectin expression *via* inhibiting endoplasmic reticulum stress in adipose tissue of obese mice. Exp Biol Med 2016; •••: 1535370216677354. [PMID: 27811171]

[395] Boden G, Duan X, Homko C, *et al.* Increase in endoplasmic reticulum stress-related proteins and genes in adipose tissue of obese, insulin-resistant individuals. Diabetes 2008; 57(9): 2438-44. [http://dx.doi.org/10.2337/db08-0604] [PMID: 18567819]

[396] Hampe L XC, Harris PW, *et al.* Modulation of adiponectin assembly and secretion by designated

peptides to counter obesity-related metabolic disorders. Manuscript in submission and revision.

[397] Menzaghi C, Salvemini L, Paroni G, *et al.* Circulating high molecular weight adiponectin isoform is heritable and shares a common genetic background with insulin resistance in nondiabetic White Caucasians from Italy: evidence from a family-based study. J Intern Med 2010; 267(3): 287-94. [http://dx.doi.org/10.1111/j.1365-2796.2009.02141.x] [PMID: 19761474]

[398] Hara K, Horikoshi M, Yamauchi T, *et al.* Measurement of the high-molecular weight form of adiponectin in plasma is useful for the prediction of insulin resistance and metabolic syndrome. Diabetes Care 2006; 29(6): 1357-62. [http://dx.doi.org/10.2337/dc05-1801] [PMID: 16732021]

[399] Peterson SJ, Drummond G, Kim DH, *et al.* L-4F treatment reduces adiposity, increases adiponectin levels, and improves insulin sensitivity in obese mice. J Lipid Res 2008; 49(8): 1658-69. [http://dx.doi.org/10.1194/jlr.M800046-JLR200] [PMID: 18426778]

[400] Fowler JA, Lwin ST, Drake MT, *et al.* Host-derived adiponectin is tumor-suppressive and a novel therapeutic target for multiple myeloma and the associated bone disease. Blood 2011; 118(22): 5872-82. [http://dx.doi.org/10.1182/blood-2011-01-330407] [PMID: 21908434]

[401] Jonas D, Van Scoyoc E, Gerrald K, *et al.* Drug Class Review: Newer Diabetes Medications, TZDs, and Combinations: Final Original Report. Portland (OR) 2011.

[402] Chin J, Hong JY, Lee J, *et al.* Selective peroxisome proliferator-activated receptor δ isosteric selenium agonists as potent anti-atherogenic agents in vivo. Bioorg Med Chem Lett 2010; 20(24): 7239-42. [http://dx.doi.org/10.1016/j.bmcl.2010.10.103] [PMID: 21074432]

[403] Murakami H, Ono A, Takahashi T, *et al.* Phase I study of Efatutazone, an oral PPARγ agonist, in patients with metastatic solid tumors. Anticancer Res 2014; 34(9): 5133-41. [PMID: 25202104]

[404] Catalano S, Mauro L, Bonofiglio D, *et al.* In vivo and in vitro evidence that PPARγ ligands are antagonists of leptin signaling in breast cancer. Am J Pathol 2011; 179(2): 1030-40. [http://dx.doi.org/10.1016/j.ajpath.2011.04.026] [PMID: 21704006]

[405] Taliaferro-Smith L, Nagalingam A, Knight BB, Oberlick E, Saxena NK, Sharma D. Integral role of PTP1B in adiponectin-mediated inhibition of oncogenic actions of leptin in breast carcinogenesis. Neoplasia 2013; 15(1): 23-38. [http://dx.doi.org/10.1593/neo.121502] [PMID: 23358729]

[406] Pishvaian MJ, Marshall JL, Wagner AJ, *et al.* A phase 1 study of efatutazone, an oral peroxisome proliferator-activated receptor gamma agonist, administered to patients with advanced malignancies. Cancer 2012; 118(21): 5403-13. [http://dx.doi.org/10.1002/cncr.27526] [PMID: 22570147]

[407] Komatsu Y, Yoshino T, Yamazaki K, *et al.* Phase 1 study of efatutazone, a novel oral peroxisome proliferator-activated receptor gamma agonist, in combination with FOLFIRI as second-line therapy in patients with metastatic colorectal cancer. Invest New Drugs 2014; 32(3): 473-80. [http://dx.doi.org/10.1007/s10637-013-0056-3] [PMID: 24337768]

[408] Shimazaki N, Togashi N, Hanai M, *et al.* Anti-tumour activity of CS-7017, a selective peroxisome proliferator-activated receptor gamma agonist of thiazolidinedione class, in human tumour xenografts and a syngeneic tumour implant model. Eur J Cancer 2008; 44(12): 1734-43. [http://dx.doi.org/10.1016/j.ejca.2008.04.016] [PMID: 18511262]

[409] Smith MR, Lee H, Fallon MA, Nathan DM. Adipocytokines, obesity, and insulin resistance during combined androgen blockade for prostate cancer. Urology 2008; 71(2): 318-22. [http://dx.doi.org/10.1016/j.urology.2007.08.035] [PMID: 18308111]

[410] Harris PW, Hampe L, Radjainia M, Brimble MA, Mitra AK. An investigation of the role of the adiponectin variable domain on the stability of the collagen-like domain. Biopolymers 2014; 102(4):

313-21.
[http://dx.doi.org/10.1002/bip.22501] [PMID: 24752567]

[411] Takuwa A, Yoshida T, Maruno T, *et al.* Ordered self-assembly of the collagenous domain of adiponectin with noncovalent interactions *via* glycosylated lysine residues. FEBS Lett 2016; 590(2): 195-201.
[http://dx.doi.org/10.1002/1873-3468.12034] [PMID: 26823167]

CHAPTER 6

Adjuvants in Cancer Vaccines - Progress and Challenges

Abid H. Banday[1,2,*], **Syed Irtiqa Urfi**[2] and **Victor J. Hruby**[1]

[1] *Department of Chemistry and Biochemistry, University of Arizona, Tucson 85721, USA*

[2] *Department of Chemistry, Islamia College of science and Commerce, Hawal, Srinagar 190002, India*

Abstract: Ever since the use of aluminium hydroxide (alum) as a potent adjuvant, there has been a colossal rise in the available adjuvants of varied nature for vaccine development. We now have come to a point where we know that some of the immune components such as cytokines are themselves associated with adjuvant properties. However, despite hundreds of well-designed pre-clinical candidates obtained after decades of research, only a handful of vaccine adjuvants have been approved and licensed for human use. This may be attributed to the complexity and the difficulty in designing adjuvants that are safe, potent and economically viable. However recent advances in our understanding of innate immunity, the immune response pathways and the mechanism of adjuvant interactions has led to better adjuvant formulations. For cancer, the success stories of treatment through immunotherapy were not as significant as that of chemo- or radiotherapy. However, vaccine development research has got an enormous attention after the US FDA approval of various preventive cancer vaccines such as Gardasil (Merck), Cervarix (Glaxosmithkline) and the therapeutic vaccine Sipulencel-T (Provenge). Approval of these vaccines has invigorated the use of cellular immunotherapy as an effective alternate method for the treatment of cancer and has provided support for renewed interest and attention which the development of new therapeutic cancer vaccines deserves. The present chapter discusses the role, mechanism of action, current clinical scenario of new cancer vaccine adjuvants, application of their molecularly defined formulations to new generation cancer vaccines and the future challenges, for the benefit of a varied audience.

Keywords: Adjuvant formulation, Alum, AS03, AS04, BCG, BSA, Cancer vaccine, Cellular response, CLRs, Cytokines, Dendritic cells, Humoral response, Immunotherapy, Immunization, Inflammasome, MF59, New generation vaccines, NLRs, Pattern recognition receptors, Potential adjuvants, TLRs, Virus-like particles.

* **Corresponding author Abid H. Banday:** Department of Chemistry, Islamia College of Science and Commerce, Hawal, Srinagar, India-190002; Tel: +91 969 7301630; Fax: +91 194 2429014; E-mail: abidrrl@gmail.com

INTRODUCTION

In modern medicine vaccines as pharmacological agents represent an outstanding success story and are responsible for a significant reduction in morbidity and mortality worldwide. Vaccination has been the most successful medical intervention from a public health perspective for the treatment and prevention of infectious as well as noninfectious diseases. However, vaccine development demands immense improvements against some robust pathogens such as human immunodeficiency virus (HIV), hepatitis C virus (HCV) and respiratory syncytial virus (RSV), for which the current methods of treatment are not effective. Improvements in vaccine development may be achieved by the addition of new adjuvants to vaccine formulations. Adjuvants are compounds that augment the quantum, breadth, quality and performance of specific immune responses to antigens, but have minimal toxicity or lasting immune effects on their own [1]. Substantial reduction in the amount of antigen and/or the necessary number of immunizations for a desired immune response may be accomplished by addition of adjuvants to some vaccines. These substances may be foreign to the body or sometimes may be host-derived components such as heat shock proteins, cytokines, hormones, glycans *etc.* Adjuvants have diverse chemical natures ranging from delivery systems (mineral salts, surface active agents, synthetic micro particles, oil-in-water emulsions, liposomes, *etc.*) to compounds that are recognized by the innate immune system (microbial products, toxins and lipids, nucleic acids, peptidoglycans, carbohydrates, peptides, cytokines/hormones, and small-molecule receptor agonists). Like the diversity in the structure of adjuvants, there is also diversity in the modes of action of adjuvants. Augmentation of immune boosting through addition of adjuvants is theoretically believed to be achieved by mimicking specific sets of evolutionarily conserved molecules such as lipopolysaccharide (LPS), liposomes, components of bacterial cell walls, endocytosed nucleic acids such as double-stranded RNA (dsRNA), single-stranded DNA (ssDNA), and unmethylated CpG dinucleotide-containing DNA which are collectively known as pathogen associated molecular patterns (PAMPs) [2]. This hypothesis seems reasonably rational because an immune response is elicited by the immune system that has evolved to recognize these specific antigenic moieties wherever encountered. The presence of an adjuvant in conjunction with the vaccine can greatly increase the innate immune response to the antigen by augmenting the activities of various immune components such as dendritic cells (DCs), lymphocytes, and macrophages by mimicking a natural infection [3].

In this chapter, we provide an overview of recent advances in the understanding of receptor based mechanism of adjuvant action. We first survey advances in the discovery of main receptor classes that have been linked to adjuvant action, and then discuss the special importance of toll like receptors (TLRs). The current clinical status of various vaccine adjuvants, receptor based design and issues related to safety are also discussed. The later aspect is of particular timeliness considering that over the past several years, concerns regarding the safe use of vaccine adjuvants in humans are coming up across volumes of published data. We then discuss the latest advancements in adjuvant and vaccine technologies and conclude with remarks about the further potential and future prospects of adjuvant technology mainly towards the development of new generation vaccine adjuvants.

DIVERSITY IN ACTION

As already mentioned, there is diversity in the mechanism through which adjuvants exert their effects [4]. A continued stimulation of the immune system is attained by some adjuvants like alum and oil-in-water or water-in-oil type emulsions through generation of antigen trapping depots at the site of injection. As such these adjuvants act as delivery systems providing slow and sustained release of the antigen [5]. Further, stimulation of inflammasomes is potentially enhanced by some of these adjuvants [6]. Such adjuvants increase recruitment and activation of antigen presenting cells (APCs) by boosting the antigen persistence at the injection site [7]. Particulate adjuvants (*e.g.* alum) have the capability to bind antigens to form multi-molecular aggregates which encourage uptake by APCs [8]. Some adjuvants direct the major histocompatibility complexes (MHCs) for antigen presentation [9]. However, the most interesting class of adjuvants, the ligands for pattern recognition receptors (PRR) [10], acts by inducing the innate immunity, predominantly targeting the APCs and consequently influencing the adaptive immune response (Table **1**).

This class of adjuvants produces an indirect component linked immune augmentation. Members of nearly all of the PRR families are potential targets for adjuvants [11 - 13] and provide highly specific recognition of a vast range of microbes [14]. Recognition of a PAMP by a PRR results in the activation of the innate immune system *via* several intracellular signaling pathways that are outside the scope of this chapter but can be consulted for a broader outlook elsewhere [15]. The main PRR subtypes include retinoic acid inducible gene-I (RIG-I)-like receptors (RLRs), Toll-like receptors (TLRs), nucleotide binding oligomerization domain (NOD)-like receptors (NLRs) and C-type lectin receptors (CLRs). TLRs and CLRs are present on the cell membrane and in endosomes, while NLRs and RIG-I helicases are intracellular microbial sensors.

Table 1. Properties of Pattern Recognition Receptors.

Receptor class	Localization	Ligand	Ligand Source
TLR1	Plasma Membrane	JBT3002, Myr3CSK4, N-Pam-S-Lau2CSK4, Pam2CSK4 and Pam3CSK4	Bacteria, Viruses, Parasites, Self
TLR4	-do-	Lipid A, RC529, RC544, RC528, ER-112022 and Taxol	-do-
TLR6	-do-	FSL-1 and MALP-2	-do-
TLR7/8	-do-	7-deaza-dG, Bropirimine, CL075, CL087, CL097, Gardiquimod, Imiquimod, Loxoribine, R-848 and TOG	-do-
Nod1	Cytoplasm	Cl2-DAP, FK156, iE-DAP, m-tri-DAP, Murabutide, Muramyl tetrapeptide and Muramyl tripeptide.	Bacteria, Self
Nod2	-do-	FK565, Muramyl dipeptide and Ml8-Muramyl dipeptide	-do-
CLRs	Plasma Membrane	Beta-glucans	Fungi
RLRs	Cytoplasm	Double-stranded RNAs	RNA Viruses

The PRR ligands signal through pathways that involve distinct adaptor molecules leading to the activation of different transcription factors such as NF-κB and IRF3 which induce the production of cytokines and chemokines that play a key role in the priming, expansion and polarization of the immune responses [16, 17]. MyD88, TRIF, TRAM, and MAL/TIRAP are the important adaptor molecules of the TLR intracellular pathways [18, 19]. However, there are some receptor specific adaptor molecules such as Syk and Raf-1 for CLRs or Rip2/RICK for NLRs such as NOD1/NOD2 which are also involved in intracellular signaling [20]. Activation of some members of the NOD-like receptor (NLR) family, such as NLRP3 and NLRC4, triggers the formation of a protein complex, called inflammasome, which is implicated in the induction of the pro-inflammatory cytokines IL-1β and IL-18 [21]. The NLRP3 and NLRC4 inflammasomes have been found to be involved in the innate immunity induced by certain adjuvants but their mechanism of action remains unclear (Fig. **1**). PRRs recruit one or more of the above adaptor molecules in order to provide specific signaling [22]. Previous studies have indicated that some PRRs are able to interact with each other, thereby modulating the magnitude and/or type of cytokine production [23], with synergistic or inhibitory effects or both. The incorporation of pathogen associated molecular patterns (PAMPs) in vaccine formulations can improve and accelerate the induction of vaccine-specific responses. A number of these agonists are now in clinical trials or in late preclinical stages of development for hepatitis and human papillomavirus vaccines [24, 25]. Though traditional PRR agonists elicit Th2 (weak) immune response, however in combination with alum or classical emulsion adjuvants, the immune response can be biased towards a Th1 response [26 - 28]. The understanding of receptor based interaction remains a challenging area of vaccine development and more insights into the area will definitely result in the production of more effective vaccine adjuvants.

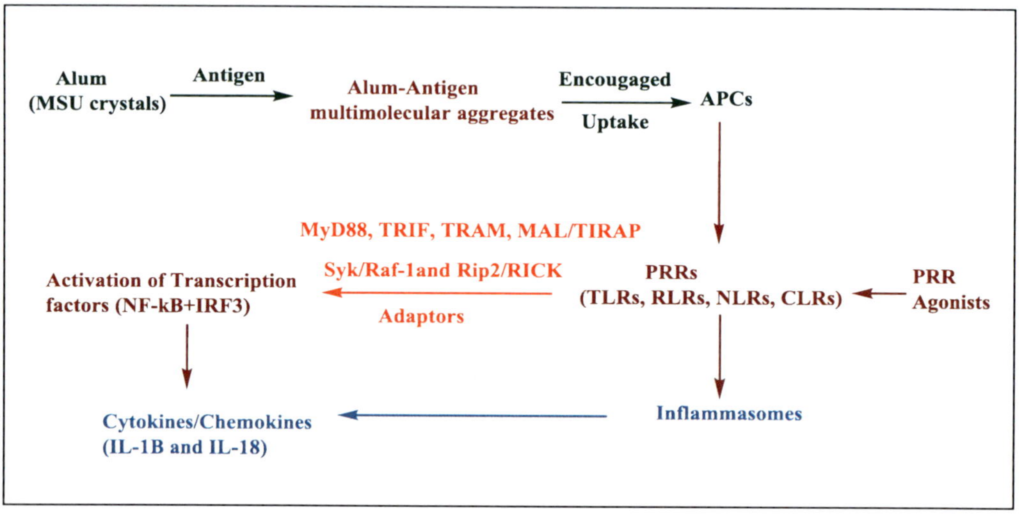

Fig. (1). Schematic representation of mode of action of Alum and PRR agonists.

The choice of any of the adjuvants reflects a compromise between a requirement for adjuvanticity and an acceptable low level of adverse reactions. Normally adjuvants should primarily increase the humoral or cellular immune response to an antigen. In the conventional vaccines, adjuvants are used to elicit an early, high and long-lasting immune response. The new generation subunit or synthetic vaccines, developed through biosynthetic or recombinant technologies, are poor immunogens and require adjuvants to evoke the immune response. The use of adjuvants enables the use of fewer antigens to achieve the desired immune response, and this reduces vaccine production costs. With a few exceptions, adjuvants are foreign to the body and cause adverse reactions. The evaluation and development of non-toxic immunomodulators, as the alternate adjuvants for providing maximum and lasting protective immune responses with existing vaccines, is beneficial because of their safety profile compared to the synthetic variants. One such adjuvant, plant derived QS21 (a saponin) possessing significant immunomodulatory activity, is already being used as adjuvant with various subunit vaccine candidates. It is in the phase-II clinical trials in combination with alum adjuvants [29]. Adaptive immunity adjuvants are either endogenous cytokines or natural and synthetic compounds such as Quillaja saponins and tucaresol, which bind to highly specific receptors on T cells and activate them through co-stimulatory signals and induction of cytokine regulated genes [30]. Even though many adjuvants stimulate Th1 immunity, only QS21 and some of its analogs facilitate cytotoxic T lymphocyte (CTL) production against exogenous antigens which is essential for high potency. Although these saponins are effective adjuvants, they have two main drawbacks *i.e.*, toxicity and structural instability resulting in a change in the type of immunity stimulated by the vaccines [31]. Maintaining the useful attributes and reducing the drawbacks associated with various adjuvants being used will be an interesting field of research.

The immunological aspects through which adjuvants may exert their immune-enhancing effects can be summarized as follows [32].

• First, adjuvants cause greater T cell activity through the translocation of antigens to the lymph nodes where they are recognized by these cells. This ultimately results in increased clearance of pathogen throughout the organism.

• Second, adjuvants may augment the longevity of antigens and make their delivery prolonged by providing physical protection to them. This will result in a more robust immune response as the organism will be exposed to the antigen for a longer duration. Further the immune system will get additional time for up-regulating the production of B and T cells needed for greater immunological memory in the adaptive immune response.

• Third, adjuvants possibly prove helpful in increasing the capability to localize the reactions at the injection site after vaccination. This in turn stimulates enhanced release of danger signals by chemokine releasing cells *e.g.* helper T cells and mast cells.

• Fourth, adjuvants interact with the pattern recognition receptors (PRRs) on the surface or inside the accessory cells, which is believed to enhance innate immune response to antigen.

• Finally, adjuvants could possibly trigger the release of inflammatory cytokines resulting in the recruitment of B and T cells at the infection sites. It may further increase the number of transcriptional events which cumulatively leads to the elicitation of stronger immune response.

ADJUVANTS IN APPROVED PREVENTIVE AND THERAPEUTIC CANCER VACCINES

Recent success stories of cancer vaccines and their optimistic clinical results have reinvigorated the enthusiasm towards their potential to slow down or to eradicate various tumors. The development of cancer vaccines initially started with whole tumor cells but eventually it advanced to sub-unit vaccines involving only a part of the tumor component system. Most of the sub-unit vaccines involve an antigen, dendritic cell, DNA and/or a vector. Most of the vaccines that have been approved or are under clinical trials are sub-unit vaccines. Following are the main classes of cancer vaccines being currently studied.

a. Whole-cell tumor vaccines

b. Antigen vaccines

c. Dendritic cell vaccines

d. DNA vaccines

e. Vector-based vaccines

The use of weakened tumor cells for the development of cancer vaccines was first proposed by Paul Ehlrich more than 100 years ago. However, in the recent years the clinical efficacy has significantly augmented for some types of immunotherapy apart from vaccines. After the initial clinical disappointment, cancer immunotherapy was significantly boosted by the U.S. FDA approval of Gardasil and Cervarix, the two prophylactic cancer vaccines effective against various functional types of human papilloma-virus (HPV). The field of cancer vaccine development received another significant impetus with the approval of Spileucel-T [33] as the first therapeutic cancer vaccine against prostate cancer in 2010. These success stories reinvigorated the momentum that the field of cancer vaccine development deserved and today we see scores of research group working exclusively in the area to come up with the best vaccine conjugates against the deadly disease. Following is a brief discussion about the clinically approved cancer vaccines.

Gardasil and Cervarix

Gardasil and Cervarix are effective vaccines based on purified virus-like particles of the major capsid (L1) protein of HPV as adjuvants [34 - 36]. Both these vaccines have been approved against various functional types of human papilloma-virus (HPV) types 6, 11, 16 and 18. These viruses are responsible for 70% of cervical cancers globally. The VLPs are prepared through self-assembly of L1 proteins produced in recombinant saccharomyces cerevissiae through separate fermentation processes. The yeast is grown from chemically pure fermentation media of carbohydrates, amino acids, vitamins and mineral salts.

The procedure involves the cell wall disruption of yeast cells where upon the VLPs are released which are then purified through suitable physical and chemical methods. The VLPs so purified are adsorbed on aluminium hydroxyl sulphate which then act as adjuvants. Cervarix consists of monophosphoryl Lipid-A in combination with HPV types 16 and 18 as adjuvants thereby broadening the immune response. Cervarix is thus a bivalent vaccine approved against cervical cancer caused by HPV 16 and 18 in females aged between 9 and 25. It is safe and highly tolerable.

Gardasil

Gardasil (Merck) consists of VLPs based on human papillomavirus 6,11,16 and 18 adsorbed on the surface of aluminium hydroxyl sulphate with added amounts of surface free adjuvants. It is a non-infectious quadrivalent recombinant vaccine effective against all four types of HPV [37, 38]. Gardasil is effective against vaginal, cervical and vulvar cancers caused by human papilloma-virus 16 and 18 in females aged between 9 and 26. It is also effective against genital warts in both males and females aged between 9 and 26.

Spileucel-T

Spileucel-T (Provenge, Dendrion) [39] is the first therapeutic cancer vaccine approved by US FDA for use against advanced metaststic prostate cancer in men. This vaccine consists of the cytokine based granulocyte-macrophage colony stimulating factor (GM-CSF) as adjuvant and prostatic acid phosphatase (PAP) as antigen. PAP is abundantly displayed by most prostatic cells on their surface. Spileucel-T activates dendritic cells (DCs) and along with PAP it induces a strong and robust immune response against the prostatic antigen.

POTENTIAL ADJUVANTS FOR USE IN CANCER VACCINES

There are only handful of efficacious adjuvants that have been approved for human use despite much advances in studies related to their design, synthesis and clinical evaluation. This can be attributed to the partial recognition of cancer cells by the immune cells as they are primarily 'self' and indistinguishable from the normal cells. The second difficulty lies in the suppression of effects of B- and Killer T cells by signals originating from cancer cells. It thus necessitates the use of non-specific immune boosters which can rip through the defense lines that cancer cells build to evade attacks from B and T cells. These boosts, called adjuvants, also augment the immune system to fight against specific targets in an effective manner [40]. Adjuvants may include derivatives of natural molecules, synthetically designed small molecules, cytokines, biomolecules such as purified proteins from microbes or non-biological substances. Most of the adjuvants are effective in stimulating components of both the adaptive and the innate immune system. They activate dendritic cells and macrophages to effectively process and present the antigen to T-cells. The most widely used adjuvant till date are the alum adjuvants and the oil in water type emulsions. Such adjuvants are licensed for use in various approved vaccines (Tables **2** and **3**). There is a series of adjuvants that have been approved for use in prophylactic vaccines against some infectious diseases [41, 42] and these may include:

a. Alum (Aluminium hydroxide or phosphate)
b. AS03, AS04 (Oil-in-water type emulsion)
c. MF59 (Oil-in-water type emulsion) or
d. Monophophoryl lipid A (MPL)
e. Virus-like Particles
f. BacilleCalmette-Guérin (BCG)
g. Bovine serum albumin (BSA), heat-shock proteins (HSPs) and keyhole limpet hemocyanin (KLH)
h. TLR agonists
i. Cytokines

Table 2. Adjuvants approved for vaccines in human use.

Adjuvant name	Class	Indications
Alum	Mineral salts	General use for vaccines against various diseases
MF59	Oil in water emulsion	Influenza (Fluad) and Pandemic flu [93]
AS03	Oil in water emulsion and tocopherol	Pandemic Flu [94]
AS04	Monophosphoryl Lipid A and Alum	Hepatitis B virus (HBV) and Human Papilloma Virus (HPV) [95]
Virosomes	Liposomes	Hepatitis A virus (HAV) and Flu
Pam_3Cys	Lipopeptide, TLR-2 agonist	Lyme, Cancer and HPV [96]

Table 3. Adjuvants tested in humans but not licensed yet.

Adjuvant Name	Class	Indications	Stage of clinical development
Montanide	Oil in water emulsion	Malaria and Cancer	Phase III [97]
Advax	Delta inulin based synthetic polysaccharide adjuvant	Broad use mainly against HBV and Influenza	Phase II [98]
Imidzaquinolines	Small moleculeTLR7 and TLR8 agonists	Broad use	Phase II

(Table 3) contd.....

Adjuvant Name	Class	Indications	Stage of clinical development
RC529	Synthetic MPL + Alum	HBV	Phase II [99]
ISS	Oligonucleotide + Alum	HBV	Phase II [100]
AS01	MPL + Liposomes + QS21	Malaria and Tuberculosis	Phase II [101]
AS02	MPL+ Oil in water emulsion + QS21	Malaria	Phase II [101]
QS21	Saponin	Broad use	Phase II [102]
MF59+MTP-PE	Lipidated MDP + Oil in Water emulsion	HIV, Flu	Phase I [103, 104]
CpG	Oligonucleotide + Alum + MF59	HBV, Malaria, Cancer	Phase I [105, 106]
IC31	Peptide + Oligonucleotide	Tuberculosis	Phase I [107]
Iscoms	Saponins + Cholesterol+ Phospholipids	Broad use	Phase I [108, 109]
PLG	Polymeric microparticles	DNA vaccine (HIV)	Phase I [110]
Flagellin	Flagellin linked to Antigen	Flu	Phase I [111, 112]

Alum (Aluminium Hydroxide or Phosphate)

Alum adjuvants remain the mostly used class of adjuvants in various approved vaccines against Hepatitis-B virus (HBV), human papilloma virus (HPV), Influenza-B, Diptheria-tetanus (DT) and Streptococcus pneumonia [43] *etc.* Alum adjuvants have been shown to act through different pathways. The main mechanisms include; 1) the formation of particulate structures promoting phagocytosis of the antigen by APCs such as dendritic cells, macrophages and B-ells; 2) formation of depots facilitating sustained and slow release of antigen; and 3) increasing antigen presentation and MHC class-II expression. There are reports which establish the activation of inflammasomes by alum adjuvants which induces the secretion of chemokines by macrophages and monocytes. However, the safety, tolerability, toxicity and other issues limit the utility of alum adjuvants [44]. These adjuvants cause severe aluminium accumulation in patients with reduced renal efficiency and slower aluminium excretion [45]. It has been shown that such patients suffer from severe bone and brain disorders leading to neurological disorders such as Alzheimer's disease and amyotrophic lateral syndrome. Another major issue with the use of alum adjuvants is their limited applicability towards therapeutic cancer vaccination because of reduced T-cell response (strong response) as compared to Th-2 response (weaker response). Further these adjuvants induce granulomas when introduced through subcutaneous or the intradermal route [46]. These drawbacks, however, don't hamper the widespread use of alum adjuvants owing to their economical production, antigen stabilization through multimolecular aggregation (see Fig. **1**) and a sustained antibody response.

AS03 and AS04

Squalene based oil in water type emulsion adjuvant having a trade name AS03 (adjuvant system 3) is mainly used in the pandemic flu A/(H1N1) vaccines manufactured by GlaxoSmithKline. Recently launched Q-pan vaccine potent against H5N1 influenza also utilizes AS03. Despite these uses it was recently reported to be a cause for narcolepsy [47]. The successor of AS03 is another adjuvant system AS04. It consists of a combination of aluminium phosphate and monophosphoryl lipid A (MPL). Various GlaxoSmithKline vaccine products such as cervarix for HPV and Fendrix for HBV utilize AS04. MPL is considered to be a well-tolerated safe and potent synthetic analog of lipid A, the immunologically active component of cell walls lipopolysaccharide (LPS) of gram negative bacteria. MPL has found its use as a potential agonist for toll like receptor-4 (TLR-4) which has a potential for stimulating a robust B and T cell response. FDA, USA has approved MPL as first TLR agonist for use in the cervical cancer vaccine- cervarix. In Europe it is licensed for the treatment of allergens because it has the ability to down regulate Th2 responses to allergens. TLR-4 agonists are the latest target to be developed as a new generation vaccine adjuvant, owing to their structure activity relationship data. These vaccines include glucopyranosyl lipid A (GPA) and aminoalkylglucosamimide phosphate (AGPs). One of the aminoalkylglucosamimide phosphate derivatives RC-529 exhibited enormous potential and thus is a licensed adjuvant in HBV vaccine (Dynavax) (Table **3**). The mechanism of action & role for human use of various adjuvants may be described in a review that might be consulted for further reading [48].

MF59

Another AS03 like Squalene based oil in water type nano emulsion with droplet size less than 250nm is propriety of Novarites called MF59. It is the first emulsion of this kind to be commercialized for use against influenza seasonal influenza & not by U.S.A. Tests have been conducted to establish its use against hepatitis B virus (HBV), herpes simplex virus (HSV) and human immunodeficiency virus (HIV) in their corresponding formulations. It has been reported to have balanced Th1/Th2 responses hence giving it overall better safety parameters. MF59 directly stimulates monocytes, granulocytes producing cytokines, macrophages and chemokines, after initial depot formation thus making it different from alum [49].

3-O-Desacyl-4'-Monophophoryl Lipid A (MPL)

Lipid A has been studied for long as adjuvant with strong immunostimulatory response against mildly immunogenic antigens such as tumour associated antigens [50 - 55]. Lipid A and its derivatives have been studied to act as toll like receptors, mainly TLR-4 and stimulate both innate and adaptive response immune system.

MPL is an adjuvant of choice because of its toxicity and has thus been used in combination with various other immune boosters mainly saponins. Various new adjuvants such as Detox, AS01B, AS02B & AS15 all saponin based & MPL based adjuvants have been studied for producing specific immune response against TAA's. These adjuvants have also been found to impair the ability of tumor cells to escape the attack of immune system. Further these adjuvants induce strong immune stimulation through elicitation of full complement of immune activities [56 - 73].

3-O-desacyl-4'-monophophoryl lipid A (MPL) is a synthetic derivative of Lipid A which is the biologically active portion of lipopolysaccharide (LPS) extracted from the cell wall of gram negative bacteria. MPL is the only synthetic derivative of lipid A that has been evaluated and approved by U.S FDA as an adjuvant. The preventive cancer vaccine cervarix is based on MPL [74 - 76].

Virus Like Particles (VLPs)

When viral antigens are to be targeted, virus like particles are the adjuvants of choice as they provide the required immunogenic responses act as delivery systems for the cargo by encapsulating the antigen. The U.S FDA approved HPV vaccine Gardasil [77, 78] contains VLPs as adjuvant which are prepared through purification of major capsid protein (L1) of HPV types 6,11,16 &18.

Bacille Calmette-Guerin (BCG)

The main challenge in the development of cancer vaccine is the difficulty in stimulating a tumor specific cytotoxic T lymphocyte (CTL) response and this is the only reason that has led to development of diverse cancer vaccine conjugates. Due to lack of antigen diversity, various trials have been conducted with adjuvants for increased stimulation of tumor specific CTLs. These adjuvants cover a wide range of immune stimulants from live- attenuated tuberculosis vaccine BacilleCalmette-Guerin (BCG) derived from mycobacterium bovis to specific receptor stimulating defined molecular compositions. An insight into the mechanism of the interaction between pathogen associated molecular patterns (PAMPs) and pattern recognition receptors (PRPs) generating signals that lead to activation of IRF and NF-kB has been of great help in designing improved immunological adjuvants. The basis of innate immunity is the downstream signaling towards expression of pro inflammatory cytokines & IRF & NF-kB activation [79]. The interaction of BCG with various immune receptors such as TLR2, TLR4 and the nucleotide binding oligomerization domain like receptors (NLRs) helped it to become the only therapeutic cancer vaccine to be approved by FDA in 1990 until the approval of sipuleucel-T (Provenge). VCG and other TLR agonists, like TLR9 and CPG generate tumor associate antigen specific adaptive immunity by stimulating innate immunity and natural killer (NK) cells. BCG also acts an adjuvant in nucleotide binding oligomerization domain (NOD) agonist, muramyl dipeptide (MDP) containing cell based adjuvant, and Melacine an approved treatment for melanoma in Canada.

Bovine Serum Albumin (BSA), Heat-Shock Proteins (HSPs) and Keyhole Limpet Hemocyanin (KLH)

Adjuvants like bovine serum albumin (BSA), heat- shock proteins (HSPs) and keyhole limpet hemocyanin (KLH) have been in use as non-specific adjuvant conjugates along with peptide based vaccines or recombination proteins to improve efficiency by taking peptide to MHC class–I molecules at cell surface to be presented to lymphocytes. HSPs are also reported to be capable of activating innate immune response without antigen presence which may be because HSP-12 preparations are contaminated with agonists like flagellin (TLR-5) or bacterial LPS (TLR-4) molecules [80].

Owing to its great immune stimulation properties BSA finds immense use in synthetic carbohydrate based cancer vaccines as carrier as well as an adjuvant. Strong Mitogen- activated protein kinase (MAPK)–dependent hypersensitivity is generated by KLH which is a high molecular weight metalloprotein [81].

Increased ability of a protein or peptide based vaccine to bind to a T cell receptor antibody specifically for anti- idiotype and EGFR variant III can be achieved by using BSA as carrier. This is particularly useful in non-Hodgkins lymphoma and glioblastoma respectively [82, 83]. Furthermore gene based cancer vaccines are prepared with cytokines as recombinant proteins, their innate immunity can be enhanced by fusing with selected TAAs & co- expressing along with TAAs *e.g.* interferon alpha 2b (IFN-alpha 2b) is proven to be beneficial therapeutically for high risk melanoma [84].

TLR Agonists

Cancer vaccine formulations have taken a great leap forward in cancer immunotherapy with the addition of TLR agonists as adjuvants. The TLR agonists like TLR-3 (poly I:C), TLR-4 (monophosphoryl lipid MPL), TLR-5(flagellin), TLR-7(immiquimod), TLR-7/8 and TLR9 (CpG)have been so far used in cancer vaccines [85 - 88].

TLR agonists on binding to the particular receptor class produce a response that is helpful in improving the potency of a vaccine to a large extent, through individual or collaborative mechanisms. General formulations of TLR adjuvants are as micro or nanoparticle emulsions, the examples of which include oil in water emulsions or QS-21, ISCOMATRIX and other saponic based formulations and liposome along with TAAs [89]. TLR-3 and rig like receptor (RLR) agonists have the ability to mimic polyI:C signals of double stranded RNA viruses, so they are being formulated as cancer vaccine adjuvants. Thus they are quite useful against those RNA viruses that were able to avert signaling through the RIG-I-like receptor pathway [90].

Adjuvant formulations based on monophosphoryl Lipid-A are used in many cancer vaccines such as cervarix and Melacine (approved in Canada against melanoma). Such formulations are also under study for some vaccines which are currently in late clinical stages. Various synthetic TLR-4 agonists such as glucopyranosyl lipid-A (GLA) and RC-529 (Glaxo-Smithkline, Dynavax) are being studied for use in cancer vaccine formulations [91]. The latter is already licensed for Hepatitis-B virus in Europe. Of late, CpG oligodeoxynucleotides (ODN) based adjuvants have been shown to trigger strong and robust immune responses against tumour associated antigens along with chemotherapeutic agents. Such formulations were evaluated as effective against non-small cell lung cancer (NSCLC) [92]. However, because of their serious side effects, these formulations couldn't cross phase-III clinical trials. Adjuvants based on DNA have extensively been used in vaccines like Juvaris, Dynavax and Intercell (DNA/Lipid cationic complexes). Recently much attention has been paid towards formulation and clinical evaluation of TLR, NLR and RLR specific small molecule adjuvants because of their enormous potential to safely activate the innate immune system.

Cytokines

In some cancer vaccine formulations which are under early stages of clinical developments, other non-specific immune adjuvants such as interleukins (IL-2, IL-7, IL-12, and IL-21), interferons (IFN-α, IFN-β, and IFN-γ) and granulocyte-macrophage colony-stimulating factor (GM-CSF), a cytokine responsible for production of more immune and blood cells in the bone marrow, are being studied. Sipuleucel-T, a therapeutic cancer vaccine based on GM-CSF has already been approved as an adjuvant.

FUTURE PERSPECTIVE

Advances in the scientific and technological aspects related to vaccinology especially related to the identification, expression and delivery of vaccine adjuvants has cleared the perspective that twenty-first century shall see substantial improvements in vaccine development and in their applications to prevent and treat infectious diseases, chronic diseases, autoimmune disorders, addictions, cancer and other ailments. However, the focus in the research and development field should be treating cancer which still remains the most challenging disease with high mortality world over. The modest stimulation of known tumour associated antigens has been an area of concern as it prevents their use in the development of cancer vaccines. However, the identification of stronger and robust antigens and their augmentation through the use of appropriate adjuvants shall be the prime concern of research towards the elicitation of desired responses mainly the IgG and CTL responses.

The use of multiantigenic vaccines improves the stimulation whereas the augmentation can be achieved through the exploitation of new-generation improved vaccine adjuvants (Fig. **2**). Safety of vaccine adjuvants has been another area of concern. The use of purified recombinant proteins in place of sub-unit or whole cell pathogens has improved vaccine safety. However, it results in reduced immunogenicity pertaining to the absence of recognizable pathogens. The development of new generation vaccines using recombinant protein antigens or purified synthetic adjuvants and targeted delivery systems may ensure efficient presentation of the conjugates to antigen presenting cells. The use of targeted delivery systems ensures the safe delivery of the adjuvants to key immune components reducing their systemic distribution. Such safe and targeted vaccine formulations shall be especially useful against infectious diseases and cancer. Further improvements may include comparison of different adjuvants within similar vaccine formulation, identification of different vaccines acting synergistically in combination and the evaluation of better animal models for studies because the PRR expression and function differs between different models.

The recent success of HIV vaccines based on recombinant proteins has reinvigorated and encouraged the use of similar strategies towards the treatment of cancer especially through the use of combination therapy involving adjuvant vaccines and gene-based recombinant vectors. The use of recombinant vectors may provide the most efficient strategy for cancer therapy elicitation of tumour specific CD4+ and CD8+ T-cell responses. Further, identification and evaluation of new targets and pathways towards exploring new routes for design of more effective and efficient formulations holds a great promise against the dreadful disease of 'cancer'.

The present chapter has been written to provide an overview of recent advances in the understanding of receptor based mechanism of adjuvant action and developments in design of more effective formulations. Substantial help has been taken from different sources for its composition which may be of great help to the readers [113 - 116].

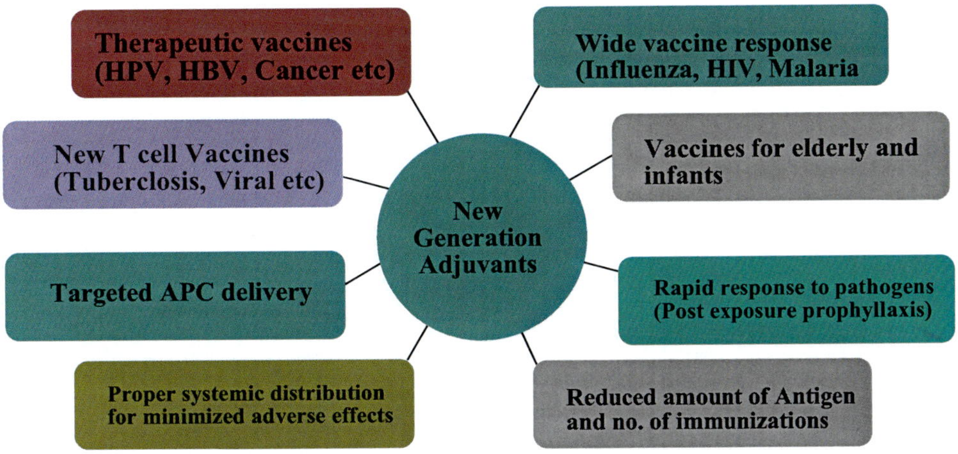

Fig. (2). Preferred attributes of a typical adjuvant for new generation vaccines.

CONCLUDING REMARK

The dreadful disease of cancer remains a global challenge in spite of the notable advances achieved in the related research and development. The conventional strategies of cancer treatment have shown good successes in the recent past but cancer immunotherapy still remains a distant dream. The challenge that is being faced primarily is the target specific action and delivery of efficient vaccine formulations. Further, elicitation of stronger responses and stimulation of desired immune components has been a great challenge. Though alum and oil in water type adjuvants have been commonly used, the lack of targeted action has limited their applicability. It has thus necessitated the use of multi-adjuvant cum multi-antigenic formulations which can be directed against particular PRRs such as toll-like receptors. The successful clinical outcomes of adjuvants in vaccine formulations against human papilloma-virus, Hepatitis-B virus and flu has encouraged the development of safer and well-tolerated new generation vaccine formulations for broader human use especially against cancer. Such vaccines may comprise of improved new vaccine adjuvants, differential combinations of existing adjuvants or totally novel adjuvants targeting new and emerging molecular targets. The new generation vaccine adjuvants should be able to produce strong and targeted immune responses. Such adjuvants should preferably be derivatives of some natural molecules such as sugars or glycoproteins thereby reducing the issues of autoimmune stimulation associated with various adjuvants. The primary challenge pertaining to design and development of such adjuvants and vaccine formulations is their ability to produce effective responses against cancer and other infectious diseases.

CONFLICT OF INTEREST

The author (editor) declares no conflict of interest, financial or otherwise.

ACKNOWLEDGMENT

Declared none.

REFERENCES

[1] Leroux RG. Unmet needs in modern vaccinology adjuvants to imporve the immune response. Vaccine 2010; 28(3): 25-3.
[http://dx.doi.org/10.1016/j.vaccine.2010.07.021]

[2] Gavin AL, Hoebe K, Duong B, *et al.* Adjuvant-enhanced antibody responses in the absence of toll-like receptor signaling. Science 2006; 314(5807): 1936-8.
[http://dx.doi.org/10.1126/science.1135299] [PMID: 17185603]

[3] Majde JA. Progress in leukocyte biology. 1987.

[4] Awate S, Babiuk LA, Mutwiri G. Mechanisms of action of adjuvants. Front Immunol 2013; 4(4): 114-38.
[PMID: 23720661]

[5] Chedid L. Adjuvants of immunity. Ann Inst Pasteur Immunol 1985; 136D(3): 283-91.
[http://dx.doi.org/10.1016/S0769-2625(85)80113-6] [PMID: 3913381]

[6] Li H, Willingham SB, Ting JP, Re F. Inflasonme activation by Alum and Alum's adjuvant effects are mediated by NLRP3. J Immunol 2008; 181: 17-21.
[http://dx.doi.org/10.4049/jimmunol.181.1.17] [PMID: 18566365]

[7] De Gregorio E, Tritto E, Rappuoli R. Alum adjuvanticity: unraveling a century old mystery. Eur J Immunol 2008; 38(8): 2068-71.
[http://dx.doi.org/10.1002/eji.200838648] [PMID: 18651701]

[8] Clements CJ, Griffiths E. The global impact of vaccines containing aluminium adjuvants. Vaccine 2002; 20(3) (Suppl. 3): S24-33.
[http://dx.doi.org/10.1016/S0264-410X(02)00168-8] [PMID: 12184361]

[9] Janeway CA, Travers P, Walport M. Immunobiology: The Immune System in Health and Disease. 5th ed., New York: Garland Science 2001.

[10] Beutler B. Inferences, questions and possibilities in Toll-like receptor signalling. Nature 2004; 430(6996): 257-63.
[http://dx.doi.org/10.1038/nature02761] [PMID: 15241424]

[11] Olive C. Pattern recognition receptors: sentinels in innate immunity and targets of new vaccine adjuvants. Expert Rev Vaccines 2012; 11(2): 237-56.
[http://dx.doi.org/10.1586/erv.11.189] [PMID: 22309671]

[12] Schmidtke JR, Johnson AG. Regulation of the immune system by synthetic polynucleotides. I. Characteristics of adjuvant action on antibody synthesis. J Immunol 1971; 106(5): 1191-200.
[PMID: 4102398]

[13] Wang Y, Shan C, Ming S, Liu Y, Du Y, Jiang G. Immunoadjuvant effects of bacterial genomic DNA and CpG oligodeoxynucleotides on avian influenza virus subtype H5N1 inactivated oil emulsion vaccine in chicken. Res Vet Sci 2009; 86(3): 399-405.
[http://dx.doi.org/10.1016/j.rvsc.2008.09.006] [PMID: 18977008]

[14] Youmans AS, Youmans GP. Factors affecting immunogenic activity of mycobacterial ribosomal and ribonucleic acid preparations. J Bacteriol 1969; 99(1): 42-50.
[PMID: 4979447]

[15] Yamamoto M, Sato S, Hemmi H, *et al.* Essential role for TIRAP in activation of the signalling cascade shared by TLR2 and TLR4. Nature 2002; 420(6913): 324-9.
[http://dx.doi.org/10.1038/nature01182] [PMID: 12447441]

[16] Akira S, Uematsu S, Takeuchi O. Pathogen recognition and innate immunity. Cell 2006; 124(4): 783-801.
[http://dx.doi.org/10.1016/j.cell.2006.02.015] [PMID: 16497588]

[17] Geijtenbeek TB, Gringhuis SI. Signalling through C-type lectin receptors: shaping immune responses. Nat Rev Immunol 2009; 9(7): 465-79.
[http://dx.doi.org/10.1038/nri2569] [PMID: 19521399]

[18] a) O'Neill LA, Fitzgerald KA, Bowie AG. The Toll-IL-1 receptor adaptor family grows to five members. Trends Immunol 2003; 24(6): 286-90.
[http://dx.doi.org/10.1016/S1471-4906(03)00115-7] [PMID: 12810098] b) Fitzgerald KA, Rowe DC, Barnes BJ, *et al.* LPS-TLR4 signaling to IRF-3/7 and NF-kappaB involves the toll adapters TRAM and TRIF. J Exp Med 2003; 198(7): 1043-55.
[http://dx.doi.org/10.1084/jem.20031023] [PMID: 14517278]

[19] Datta SK, Redecke V, Prilliman KR, *et al.* A subset of Toll-like receptor ligands induces cross-presentation by bone marrow-derived dendritic cells. J Immunol 2003; 170(8): 4102-10.
[http://dx.doi.org/10.4049/jimmunol.170.8.4102] [PMID: 12682240]

[20] Strober W, Murray PJ, Kitani A, Watanabe T. Signalling pathways and molecular interactions of NOD1 and NOD2. Nat Rev Immunol 2006; 6(1): 9-20.
[http://dx.doi.org/10.1038/nri1747] [PMID: 16493424]

[21] Martinon F, Burns K, Tschopp J. The inflammasome: a molecular platform triggering activation of inflammatory caspases and processing of proIL-beta. Mol Cell 2002; 10(2): 417-26.
[http://dx.doi.org/10.1016/S1097-2765(02)00599-3] [PMID: 12191486]

[22] Doyle SL, O'Neill LA. Toll-like receptors: from the discovery of NFkappaB to new insights into transcriptional regulations in innate immunity. Biochem Pharmacol 2006; 72(9): 1102-13.
[http://dx.doi.org/10.1016/j.bcp.2006.07.010] [PMID: 16930560]

[23] Lee MS, Kim YJ. Signaling pathways downstream of pattern-recognition receptors and their cross talk. Annu Rev Biochem 2007; 76: 447-80.
[http://dx.doi.org/10.1146/annurev.biochem.76.060605.122847] [PMID: 17328678]

[24] Thoelen S, Van Damme P, Mathei C, *et al.* Safety and immunogenicity of a hepatitis B vaccine formulated with a novel adjuvant system. Vaccine 1998; 16(7): 708-14.
[http://dx.doi.org/10.1016/S0264-410X(97)00254-5] [PMID: 9562690]

[25] Giannini SL, Hanon E, Moris P, *et al.* Enhanced humoral and memory B cellular immunity using HPV16/18 L1 VLP vaccine formulated with the MPL/aluminium salt combination (AS04) compared to aluminium salt only. Vaccine 2006; 24(33-34): 5937-49.
[http://dx.doi.org/10.1016/j.vaccine.2006.06.005] [PMID: 16828940]

[26] Heath WR, Carbone FR. Cross-presentation, dendritic cells, tolerance and immunity. Annu Rev Immunol 2001; 19: 47-64.
[http://dx.doi.org/10.1146/annurev.immunol.19.1.47] [PMID: 11244030]

[27] Napolitani G, Rinaldi A, Bertoni F, Sallusto F, Lanzavecchia A. Selected Toll-like receptor agonist combinations synergistically trigger a T helper type 1-polarizing program in dendritic cells. Nat Immunol 2005; 6(8): 769-76.
[http://dx.doi.org/10.1038/ni1223] [PMID: 15995707]

[28] Krumbiegel D, Zepp F, Meyer CU. Combined Toll-like receptor agonists synergistically increase

production of inflammatory cytokines in human neonatal dendritic cells. Hum Immunol 2007; 68(10): 813-22.
[http://dx.doi.org/10.1016/j.humimm.2007.08.001] [PMID: 17961769]

[29] Hyslop NS, Morrow AW. The influence of aluminium hydroxide content, dose volume and the inclusion of saponin on the efficacy of inactivated foot-and-mouth disease vaccines. Res Vet Sci 1969; 10(2): 109-20.
[PMID: 5818030]

[30] FernA ndez-Tejada A, Chea EK, George C, *et al.* Development of a minimal saponin vaccine adjuvant based on QS-21. Nat Chem 2014; 6(7): 635-43.
[http://dx.doi.org/10.1038/nchem.1963] [PMID: 24950335]

[31] Marciani DJ, Pathak AK, Reynolds RC, Seitz L, May RD. Altered immunomodulating and toxicological properties of degraded Quillaja saponaria Molina saponins. Int Immunopharmacol 2001; 1(4): 813-8.
[http://dx.doi.org/10.1016/S1567-5769(01)00016-9] [PMID: 11357894]

[32] Schijns VE. Immunological concepts of vaccine adjuvant activity. Curr Opin Immunol 2000; 12(4): 456-63.
[http://dx.doi.org/10.1016/S0952-7915(00)00120-5] [PMID: 10899018]

[33] Longo DL. New therapies for castration-resistant prostate cancer. N Engl J Med 2010; 363(5): 479-81.
[http://dx.doi.org/10.1056/NEJMe1006300] [PMID: 20818868]

[34] FDA licensure of bivalent human papillomavirus vaccine (HPV2, Cervarix) for use in females and updated HPV vaccination recommendations from the Advisory Committee on Immunization Practices (ACIP). MMWR Morb Mortal Wkly Rep 2010; 59(20): 626-9.
[PMID: 20508593]

[35] McNeil C. Who invented the VLP cervical cancer vaccines? J Natl Cancer Inst 2006; 98(7): 433.
[http://dx.doi.org/10.1093/jnci/djj144] [PMID: 16595773]

[36] Paavonen J, Naud P, Salmeron J, *et al.* HPV PATRICIA Study Group. Efficacy of human papillomavirus (HPV)-16/18 AS04-adjuvanted vaccine against cervical infection and precancer caused by oncogenic HPV types (PATRICIA): final analysis of a double-blind, randomised study in young women. Lancet 2009; 374(9686): 301-14.
[http://dx.doi.org/10.1016/S0140-6736(09)61248-4] [PMID: 19586656]

[37] Markowitz LE, Dunne EF, Saraiya M, Lawson HW, Chesson H, Unger ER. Centers for Disease Control Prevention (CDC); Advisory Committee on Immunization Practices (ACIP). Quadrivalent Human Papillomavirus Vaccine: Recommendations of the Advisory Committee on Immunization Practices (ACIP). MMWR. Recommendations and reports: Morb Mortal Wkly. Recommendations and reports/ Centers for Disease Control 2007; 56: 1-24.

[38] Lowy DR, Schiller JT. Prophylactic human papillomavirus vaccines. J Clin Invest 2006; 116(5): 1167-73.
[http://dx.doi.org/10.1172/JCI28607] [PMID: 16670757]

[39] Kantoff PW, Higano CS, Shore ND, *et al.* Sipuleucel-T immunotherapy for castration-resistant prostate cancer. N Engl J Med 2010; 363(5): 411-22.
[http://dx.doi.org/10.1056/NEJMoa1001294] [PMID: 20818862]

[40] Dubensky TW Jr, Reed SG. Adjuvants for cancer vaccines. Semin Immunol 2010; 22(3): 155-61.
[http://dx.doi.org/10.1016/j.smim.2010.04.007] [PMID: 20488726]

[41] O'Hagan DT, De Gregorio E. The path to a successful vaccine adjuvant—'the long and winding road'. Drug Discov Today 2009; 14(11-12): 541-51.
[http://dx.doi.org/10.1016/j.drudis.2009.02.009] [PMID: 19508916]

[42] Tritto E, Mosca F, De Gregorio E. Mechanism of action of licensed vaccine adjuvants. Vaccine 2009; 27(25-26): 3331-4.

[http://dx.doi.org/10.1016/j.vaccine.2009.01.084] [PMID: 19200813]

[43] Kang SJ, Locksley RM. The inflammasome and alum-mediated adjuvanticity. F1000 Biol Rep 2009; 1: 15.
[PMID: 20948671]

[44] Gherardi RK, Eidi H, Crépeaux G, Authier FJ, Cadusseau J. Biopersistence and brain translocation of aluminum adjuvants of vaccines. Front Neurol 2015; 6: 4.
[http://dx.doi.org/10.3389/fneur.2015.00004] [PMID: 25699008]

[45] Kramer MF, Heath MD. Aluminium in allergen-specific subcutaneous immunotherapy--a German perspective. Vaccine 2014; 32(33): 4140-8.
[http://dx.doi.org/10.1016/j.vaccine.2014.05.063] [PMID: 24892252]

[46] Petrovsky N, Aguilar JC. Vaccine adjuvants: current state and future trends. Immunol Cell Biol 2004; 82(5): 488-96.
[http://dx.doi.org/10.1111/j.0818-9641.2004.01272.x] [PMID: 15479434]

[47] Stowe J, Andrews N, Kosky C, *et al.* Risk of Narcolepsy after AS03 Adjuvanted Pandemic A/H1N1 2009 Influenza Vaccine in Adults: A Case-Coverage Study in England. Sleep 2016; 39(5): 1051-7.
[http://dx.doi.org/10.5665/sleep.5752] [PMID: 26856903]

[48] a) Lamine MM. Ennio DGio, Valiante NM, Rappuoli R. New adjuvants for human vaccines. Curr Opin Immunol 2010; 22: 41-416.
[PMID: 20061127] b) Chiang CL, Kandalaft LE, Coukos G. Adjuvants for enhancing the immunogenicity of whole tumor cell vaccines. Int Rev Immunol 2011; 30(2-3): 150-82.
[http://dx.doi.org/10.3109/08830185.2011.572210] [PMID: 21557641]

[49] Carter D, Reed SG. Role of adjuvants in modeling the immune response. Curr Opin HIV AIDS 2010; 5(5): 409-13.
[http://dx.doi.org/10.1097/COH.0b013e32833d2cdb] [PMID: 20978382]

[50] Tomai MA, Johnson AG. T cell and interferon-gamma involvement in the adjuvant action of a detoxified endotoxin. J Biol Response Mod 1989; 8(6): 625-43.
[PMID: 2513379]

[51] Schwarz TF. Clinical update of the AS04-adjuvanted human papillomavirus-16/18 cervical cancer vaccine, Cervarix. Adv Ther 2009; 26(11): 983-98.
[http://dx.doi.org/10.1007/s12325-009-0079-5] [PMID: 20024678]

[52] Brichard VG, Lejeune D. GSK's antigen-specific cancer immunotherapy programme: pilot results leading to Phase III clinical development. Vaccine 2007; 25 (Suppl. 2): B61-71.
[http://dx.doi.org/10.1016/j.vaccine.2007.06.038] [PMID: 17916463]

[53] Reed SG, Bertholet S, Coler RN, Friede M. New horizons in adjuvants for vaccine development. Trends Immunol 2009; 30(1): 23-32.
[http://dx.doi.org/10.1016/j.it.2008.09.006] [PMID: 19059004]

[54] Baldridge JR, McGowan P, Evans JT, *et al.* Taking a Toll on human disease: Toll-like receptor 4 agonists as vaccine adjuvants and monotherapeutic agents. Expert Opin Biol Ther 2004; 4(7): 1129-38.
[http://dx.doi.org/10.1517/14712598.4.7.1129] [PMID: 15268679]

[55] Didierlaurent AM, Morel S, Lockman L, *et al.* AS04, an aluminum salt- and TLR4 agonist-based adjuvant system, induces a transient localized innate immune response leading to enhanced adaptive immunity. J Immunol 2009; 183(10): 6186-97.
[http://dx.doi.org/10.4049/jimmunol.0901474] [PMID: 19864596]

[56] Baudner BC, Ronconi V, Casini D, *et al.* MF59 emulsion is an effective delivery system for a synthetic TLR4 agonist (E6020). Pharm Res 2009; 26(6): 1477-85.
[http://dx.doi.org/10.1007/s11095-009-9859-5] [PMID: 19255727]

[57] Kato H, Takeuchi O, Sato S, *et al.* Differential roles of MDA5 and RIG-I helicases in the recognition

of RNA viruses. Nature 2006; 441(7089): 101-5.
[http://dx.doi.org/10.1038/nature04734] [PMID: 16625202]

[58] Schulz O, Diebold SS, Chen M, *et al.* Toll-like receptor 3 promotes cross-priming to virus-infected cells. Nature 2005; 433(7028): 887-92.
[http://dx.doi.org/10.1038/nature03326] [PMID: 15711573]

[59] Adams M, Navabi H, Jasani B, *et al.* Dendritic cell (DC) based therapy for cervical cancer: use of DC pulsed with tumour lysate and matured with a novel synthetic clinically non-toxic double stranded RNA analogue poly [I]:poly [C(12)U] (Ampligen R). Vaccine 2003; 21(7-8): 787-90.
[http://dx.doi.org/10.1016/S0264-410X(02)00599-6] [PMID: 12531360]

[60] Choe J, Kelker MS, Wilson IA. Crystal structure of human toll-like receptor 3 (TLR3) ectodomain. Science 2005; 309(5734): 581-5.
[http://dx.doi.org/10.1126/science.1115253] [PMID: 15961631]

[61] Ewel CH, Urba WJ, Kopp WC, *et al.* Polyinosinic-polycytidylic acid complexed with poly-L-lysine and carboxymethylcellulose in combination with interleukin 2 in patients with cancer: clinical and immunological effects. Cancer Res 1992; 52(11): 3005-10.
[PMID: 1591717]

[62] Giantonio BJ, Hochster H, Blum R, *et al.* Toxicity and response evaluation of the interferon inducer poly ICLC administered at low dose in advanced renal carcinoma and relapsed or refractory lymphoma: a report of two clinical trials of the Eastern Cooperative Oncology Group. Invest New Drugs 2001; 19(1): 89-92.
[http://dx.doi.org/10.1023/A:1006458232384] [PMID: 11291838]

[63] Gitlin L, Barchet W, Gilfillan S, *et al.* Essential role of mda-5 in type I IFN responses to polyriboinosinic:polyribocytidylic acid and encephalomyocarditis picornavirus. Proc Natl Acad Sci USA 2006; 103(22): 8459-64.
[http://dx.doi.org/10.1073/pnas.0603082103] [PMID: 16714379]

[64] Kawai T, Akira S. Innate immune recognition of viral infection. Nat Immunol 2006; 7(2): 131-7.
[http://dx.doi.org/10.1038/ni1303] [PMID: 16424890]

[65] Mailliard RB, Wankowicz-Kalinska A, Cai Q, *et al.* Alpha-type-1 polarized dendritic cells: a novel immunization tool with optimized CTL-inducing activity. Cancer Res 2004; 64(17): 5934-7.
[http://dx.doi.org/10.1158/0008-5472.CAN-04-1261] [PMID: 15342370]

[66] Uematsu S, Akira S. Toll-like receptors and Type I interferons. J Biol Chem 2007; 282(21): 15319-23.
[http://dx.doi.org/10.1074/jbc.R700009200] [PMID: 17395581]

[67] Yoneyama M, Kikuchi M, Natsukawa T, *et al.* The RNA helicase RIG-I has an essential function in double-stranded RNA-induced innate antiviral responses. Nat Immunol 2004; 5(7): 730-7.
[http://dx.doi.org/10.1038/ni1087] [PMID: 15208624]

[68] Marshall-Clarke S, Downes JE, Haga IR, *et al.* Polyinosinic acid is a ligand for toll-like receptor 3. J Biol Chem 2007; 282(34): 24759-66.
[http://dx.doi.org/10.1074/jbc.M700188200] [PMID: 17573354]

[69] Lee J, Chuang TH, Redecke V, *et al.* Molecular basis for the immunostimulatory activity of guanine nucleoside analogs: activation of Toll-like receptor 7. Proc Natl Acad Sci USA 2003; 100(11): 6646-51.
[http://dx.doi.org/10.1073/pnas.0631696100] [PMID: 12738885]

[70] Sauder DN, Smith MH, Senta-McMillian T, Soria I, Meng TC. Randomized, single-blind, placebo-controlled study of topical application of the immune response modulator resiquimod in healthy adults. Antimicrob Agents Chemother 2003; 47(12): 3846-52.
[http://dx.doi.org/10.1128/AAC.47.12.3846-3852.2003] [PMID: 14638493]

[71] Akira S, Takeda K, Kaisho T. Toll-like receptors: critical proteins linking innate and acquired immunity. Nat Immunol 2001; 2(8): 675-80.

[http://dx.doi.org/10.1038/90609] [PMID: 11477402]

[72] Weiner GJ, Liu HM, Wooldridge JE, Dahle CE, Krieg AM. Immunostimulatory oligodeoxynucleotides containing the CpG motif are effective as immune adjuvants in tumor antigen immunization. Proc Natl Acad Sci USA 1997; 94(20): 10833-7. [http://dx.doi.org/10.1073/pnas.94.20.10833] [PMID: 9380720]

[73] Krieg AM. Antitumor applications of stimulating toll-like receptor 9 with CpG oligodeoxynucleotides. Curr Oncol Rep 2004; 6(2): 88-95. [http://dx.doi.org/10.1007/s11912-004-0019-0] [PMID: 14751085]

[74] Harper DM. Current prophylactic HPV vaccines and gynecologic premalignancies. Curr Opin Obstet Gynecol 2009; 21(6): 457-64. [PMID: 19923989]

[75] Gilewski TA, Ragupathi G, Dickler M, *et al.* Immunization of high-risk breast cancer patients with clustered sTn-KLH conjugate plus the immunologic adjuvant QS-21. Clin Cancer Res 2007; 13(10): 2977-85. [http://dx.doi.org/10.1158/1078-0432.CCR-06-2189] [PMID: 17504999]

[76] Ragupathi G, Gathuru J, Livingston P. Antibody inducing polyvalent cancer vaccines. Cancer Treat Res 2005; 123: 157-80. [http://dx.doi.org/10.1007/0-387-27545-2_7] [PMID: 16211870]

[77] a) Chroboczek J, Szurgot I, Szolajska E. Virus-like particles as vaccine. Acta Biochim Pol 2014; 61(3): 531-9. [PMID: 25273564] b) Gilbert SC. Virus-like particles as vaccine adjuvants. Mol Biotechnol 2001; 19(2): 169-77. [http://dx.doi.org/10.1385/MB:19:2:169] [PMID: 11725486]

[78] a) Buonaguro FM, Tornesello ML, Buonaguro L. Virus-like particle vaccines and adjuvants: the HPV paradigm. Expert Rev Vaccines 2009; 8(10): 1379-98. [http://dx.doi.org/10.1586/erv.09.81] [PMID: 19803760] b) Guillén G, Aguilar JC, DueA as S, *et al.* Virus-Like Particles as vaccine antigens and adjuvants: application to chronic disease, cancer immunotherapy and infectious disease preventive strategies. Procedia Vaccinol 2010; 2(2): 128-33. [http://dx.doi.org/10.1016/j.provac.2010.07.004]

[79] Iwasaki A, Medzhitov R. Regulation of adaptive immunity by the innate immune system. Science 2010; 327(5963): 291-5. [http://dx.doi.org/10.1126/science.1183021] [PMID: 20075244]

[80] Tsan MF, Gao B. Heat shock proteins and immune system. J Leukoc Biol 2009; 85(6): 905-10. [http://dx.doi.org/10.1189/jlb.0109005] [PMID: 19276179]

[81] Engstrom L, Pinzon-Ortiz MC, Li Y, *et al.* Characterization of a murine keyhole limpet hemocyanin (KLH)-delayed-type hypersensitivity (DTH) model: role for p38 kinase. Int Immunopharmacol 2009; 9(10): 1218-27. [http://dx.doi.org/10.1016/j.intimp.2009.07.003] [PMID: 19616132]

[82] Heimberger AB, Sampson JH. The PEPvIII-KLH (CDX-110) vaccine in glioblastoma multiforme patients. Expert Opin Biol Ther 2009; 9(8): 1087-98. [http://dx.doi.org/10.1517/14712590903124346] [PMID: 19591631]

[83] Timmerman JM, Vose JM, Czerwinski DK, *et al.* Tumor-specific recombinant idiotype immunisation after chemotherapy as initial treatment for follicular non-Hodgkin lymphoma. Leuk Lymphoma 2009; 50(1): 37-46. [http://dx.doi.org/10.1080/10428190802563355] [PMID: 19125383]

[84] Tarhini AA, Kirkwood JM. Clinical and immunologic basis of interferon therapy in melanoma. Ann N Y Acad Sci 2009; 1182: 47-57. [http://dx.doi.org/10.1111/j.1749-6632.2009.05073.x] [PMID: 20074274]

[85] Adams S, O'Neill DW, Nonaka D, *et al.* Immunization of malignant melanoma patients with full-length NY-ESO-1 protein using TLR7 agonist imiquimod as vaccine adjuvant. J Immunol 2008; 181(1): 776-84.
[http://dx.doi.org/10.4049/jimmunol.181.1.776] [PMID: 18566444]

[86] Igartua M, Pedraz JL. Topical resiquimod: a promising adjuvant for vaccine development? Expert Rev Vaccines 2010; 9(1): 23-7.
[http://dx.doi.org/10.1586/erv.09.135] [PMID: 20021302]

[87] Kumar H, Kawai T, Akira S. Pathogen recognition in the innate immune response. Biochem J 2009; 420(1): 1-16.
[http://dx.doi.org/10.1042/BJ20090272] [PMID: 19382893]

[88] Zhu Q, Egelston C, Vivekanandhan A, *et al.* Toll-like receptor ligands synergize through distinct dendritic cell pathways to induce T cell responses: implications for vaccines. Proc Natl Acad Sci USA 2008; 105(42): 16260-5.
[http://dx.doi.org/10.1073/pnas.0805325105] [PMID: 18845682]

[89] Sun HX, Xie Y, Ye YP. ISCOMs and ISCOMATRIX. Vaccine 2009; 27(33): 4388-401.
[http://dx.doi.org/10.1016/j.vaccine.2009.05.032] [PMID: 19450632]

[90] Jasani B, Navabi H, Adams M. Ampligen: a potential toll-like 3 receptor adjuvant for immunotherapy of cancer. Vaccine 2009; 27(25-26): 3401-4.
[http://dx.doi.org/10.1016/j.vaccine.2009.01.071] [PMID: 19200817]

[91] Baldwin SL, Shaverdian N, Goto Y, *et al.* Enhanced humoral and Type 1 cellular immune responses with Fluzone adjuvanted with a synthetic TLR4 agonist formulated in an emulsion. Vaccine 2009; 27(43): 5956-63.
[http://dx.doi.org/10.1016/j.vaccine.2009.07.081] [PMID: 19679214]

[92] Vollmer J, Krieg AM. Immunotherapeutic applications of CpG oligodeoxynucleotide TLR9 agonists. Adv Drug Deliv Rev 2009; 61(3): 195-204.
[http://dx.doi.org/10.1016/j.addr.2008.12.008] [PMID: 19211030]

[93] Atmar RL, Keitel WA, Patel SM, *et al.* Safety and immunogenicity of nonadjuvanted and MF59-adjuvanted influenza A/H9N2 vaccine preparations. Clin Infect Dis 2006; 43(9): 1135-42.
[http://dx.doi.org/10.1086/508174] [PMID: 17029131]

[94] Lambrecht BN, Kool M, Willart MA, Hammad H. Mechanism of action of clinically approved adjuvants. Curr Opin Immunol 2009; 21(1): 23-9.
[http://dx.doi.org/10.1016/j.coi.2009.01.004] [PMID: 19246182]

[95] Garcon N, Chomez P, Van Mechelen M. GlaxoSmithKline Adjuvant Systems in vaccines: concepts, achievements and perspectives. Expert Rev Vaccines 2007; 6(5): 723-39.
[http://dx.doi.org/10.1586/14760584.6.5.723] [PMID: 17931153]

[96] Poland GA, Jacobson RM. The prevention of Lyme disease with vaccine. Vaccine 2001; 19(17-19): 2303-8.
[http://dx.doi.org/10.1016/S0264-410X(00)00520-X] [PMID: 11257352]

[97] Aucouturier J, Dupuis L, Deville S, Ascarateil S, Ganne V. Montanide ISA 720 and 51: a new generation of water in oil emulsions as adjuvants for human vaccines. Expert Rev Vaccines 2002; 1(1): 111-8.
[http://dx.doi.org/10.1586/14760584.1.1.111] [PMID: 12908518]

[98] Petrovsky N. Vaccine adjuvants: in search of new paradigms. Interview by Jenaid Rees. Expert Rev Vaccines 2013; 12(7): 723-6.
[http://dx.doi.org/10.1586/14760584.2013.811197] [PMID: 23885817]

[99] Thompson BS, Chilton PM, Ward JR, Evans JT, Mitchell TC. The low-toxicity versions of LPS, MPL adjuvant and RC529, are efficient adjuvants for CD4+ T cells. J Leukoc Biol 2005; 78(6): 1273-80.
[http://dx.doi.org/10.1189/jlb.0305172] [PMID: 16204643]

[100] Barry M, Cooper C. Review of hepatitis B surface antigen-1018 ISS adjuvant-containing vaccine safety and efficacy. Expert Opin Biol Ther 2007; 7(11): 1731-7. [http://dx.doi.org/10.1517/14712598.7.11.1731] [PMID: 17961095]

[101] Schleiss MR, Choi KY, Anderson J, *et al.* Glycoprotein B (gB) vaccines adjuvanted with AS01 or AS02 protect female guinea pigs against cytomegalovirus (CMV) viremia and offspring mortality in a CMV-challenge model. Vaccine 2014; 32(23): 2756-62. [http://dx.doi.org/10.1016/j.vaccine.2013.07.010] [PMID: 23867012]

[102] Garcon N, Van Mechelen M. Recent clinical experience with vaccines using MPL- and QS-2--containing adjuvant systems. Expert Rev Vaccines 2011; 10(4): 471-86. [http://dx.doi.org/10.1586/erv.11.29] [PMID: 21506645]

[103] Podda A, Giudice GD, O'Hagan DT. 9 - MF59: a safe and potent adjuvant for human use.Immunopotentiators in Modern Vaccines. London: Academic Press 2006; pp. 149-59. [http://dx.doi.org/10.1016/B978-012088403-2/50010-1]

[104] Durando P, Fenoglio D, Boschini A, *et al.* Safety and immunogenicity of two influenza virus subunit vaccines, with or without MF59 adjuvant, administered to human immunodeficiency virus type 1-seropositive and -seronegative adults. Clin Vaccine Immunol 2008; 15(2): 253-9. [http://dx.doi.org/10.1128/CVI.00316-07] [PMID: 18003811]

[105] Klinman DM. Adjuvant activity of CpG oligodeoxynucleotides. Int Rev Immunol 2006; 25(3-4): 135-54. [http://dx.doi.org/10.1080/08830180600743057] [PMID: 16818369]

[106] Bode C, Zhao G, Steinhagen F, Kinjo T, Klinman DM. CpG DNA as a vaccine adjuvant. Expert Rev Vaccines 2011; 10(4): 499-511. [http://dx.doi.org/10.1586/erv.10.174] [PMID: 21506647]

[107] Kaufmann SH. Novel tuberculosis vaccination strategies based on understanding the immune response. J Intern Med 2010; 267(4): 337-53. [http://dx.doi.org/10.1111/j.1365-2796.2010.02216.x] [PMID: 20433580]

[108] Cox JC, Sjölander A, Barr IG. ISCOMs and other saponin based adjuvants. Adv Drug Deliv Rev 1998; 32(3): 247-71. [http://dx.doi.org/10.1016/S0169-409X(98)00013-1] [PMID: 10837647]

[109] Sjolander A, Cox JC, Barr IG. ISCOMs: an adjuvant with multiple functions. J Leukoc Biol 1998; 64(6): 713-23. [PMID: 9850152]

[110] Vajdy M, Srivastava I, Polo J, Donnelly J, O'Hagan D, Singh M. Mucosal adjuvants and delivery systems for protein-, DNA- and RNA-based vaccines. Immunol Cell Biol 2004; 82(6): 617-27. [http://dx.doi.org/10.1111/j.1440-1711.2004.01288.x] [PMID: 15550120]

[111] Turley CB, Rupp RE, Johnson C, *et al.* Safety and immunogenicity of a recombinant M2e-flagellin influenza vaccine (STF2.4xM2e) in healthy adults. Vaccine 2011; 29(32): 5145-52. [http://dx.doi.org/10.1016/j.vaccine.2011.05.041] [PMID: 21624416]

[112] Ben-Yedidia T, Arnon R. Towards an epitope-based human vaccine for influenza. Hum Vaccin 2005; 1(3): 95-101. [http://dx.doi.org/10.4161/hv.1.3.1851] [PMID: 17012851]

[113] Banday AH, Jeelani S, Hruby VJ. Cancer vaccine adjuvants--recent clinical progress and future perspectives. Immunopharmacol Immunotoxicol 2015; 37(1): 1-11. [http://dx.doi.org/10.3109/08923973.2014.971963] [PMID: 25318595]

[114] http://www.invivogen.com/review-vaccine-adjuvants

[115] Timmermans K, Plantinga TS, Kox M, *et al.* Blueprints of signaling interactions between pattern recognition receptors: implications for the design of vaccine adjuvants. Clin Vaccine Immunol 2013;

20(3): 427-32.
[http://dx.doi.org/10.1128/CVI.00703-12] [PMID: 23345580]

[116] Marciani DJ. Vaccine adjuvants: role and mechanisms of action in vaccine immunogenicity. Drug Discov Today 2003; 8(20): 934-43.
[http://dx.doi.org/10.1016/S1359-6446(03)02864-2] [PMID: 14554157]

CHAPTER 7

Dicycloplatin, a Platinum - Based Supramolecule with 4 Hydrogen Bonds, is a Promising Chemotherapy Drug

Xuqing Yang[1], **Jing Jie Yu**[2,*], **Yi Guo**[2] and **Michael D. Mueller**[2]

[1] *Department of Chemistry, Peking University and Beijing Xingda Scientific System Company, Beijing, China*

[2] *WVU Cancer Institute, School of Medicine, School of Pharmacy, West Virginia University, Morgantown, WV, USA*

Abstract: Improving therapeutic efficacy and reducing side effects of chemotherapeutic agents are primary goals of cancer research. Dicycloplatin (DCP), a promising platinum drug for cancer chemotherapy approved by the Chinese FDA in 2012, shows better solubility and stability than cisplatin and carboplatin; preclinical *in vitro* and *in vivo* studies and a phase I clinical trial found greater antitumor activity and fewer adverse events than carboplatin. Phase II clinical trials in advanced non-small cell lung cancer found efficacy and safety of DCP-plus-paclitaxel comparable to carboplatin-plus-paclitaxel, but with better drug tolerability. DCP bio-pharmaceutical analysis showed different spectrum profiles of DCP and carboplatin in plasma. The bearable side effects of dicycloplatin might be due to the physicochemical properties of dicycloplatin, a supramolecule containing 4 hydrogen bonds. Of note, Histoculture Drug Response Assay and the successful treatment of a lung cancer patient with brain metastatic tumor indicate possible BBB penetration. Molecular studies of the dicycloplatin-induced activation pathway of DNA damage response in cancer cells - including investigation of Chk2, BRCA1 and p53 - demonstrated a signaling pathway similar to cisplatin-induced activation. This chapter concludes with a discussion of the first American cancer patient to receive dicycloplatin chemotherapy. The patient was treated in Beijing in the summer of 2016.

Keywords: Cancer Chemotherapy, Dicycloplatin, Hydrogen Bond, New Platinum Drug, Supramolecule.

INTRODUCTION

Cancer is one of the leading causes of morbidity and mortality worldwide, with approximately 14.1 million new cases in 2012; of these, 7.4 million cases were

* **Corresponding author Jing Jie Yu:** WVU Cancer Institute, School of Medicine and School of Pharmacy, West Virginia University, Morgantown, WV, USA; Tel: 304-293-8661; Fax: 304-293-4667; E-mail: jyu@hsc.wvu.edu

Atta-ur-Rahman (Ed.)

men and 6.7 million were women [1]. The annual number of new cases is expected to increase to 24 million by 2035 [2].

The rate of cancer death in the United States for men and women combined fell 25% from its peak in 1991 to 2014. This decline translates to the prevention of millions of cancer deaths. That is significant progress; however, 1,685,210 new cancer cases and 595,690 cancer deaths are projected in the USA in 2016 [1, 3].

Cancer remains a global killer despite breakthroughs in research, better treatments, and stronger prevention programs. Furthermore, the side effects of platinum anticancer drugs and other chemotherapeutic agents are often brutal, with many patients dropping out of treatment. Undesirable side effects, such as myelosuppression, nephrotoxicity, hepatotoxicity, neurotoxicity, ototoxicity, and nausea and vomiting, are constraints to dosage and long-term use.

Cisplatin, carboplatin and oxaliplatin were approved by the USFDA in 1978, 1989, and 2004, respectively; each went on to gain worldwide approval. Three other platinum chemotherapy drugs have been approved in single markets: nedaplatin in Japan, lobaplatin in China, and heptaplatin in Korea [4].

Still, platinum remains a cornerstone of chemotherapy for various cancers. The need for highly-potent anticancer drugs derived from platinum analogues with low toxicity and broad-spectrum effects led to the development of dicycloplatin. The pharmacologically active structure and physicochemical properties of dicycloplatin, a supramolecular compound designed by Professor Xuqing Yang, are based on the molecular biology findings of physicists and chemists.

In 1951, Linus Pauling, Robert Corey, and Herman Branson proposed that the α-helix and β-sheet form the backbones of proteins. The Pauling–Corey–Branson models of proteins showed two hydrogen-bonded helical structures forming a polypeptide chain [5].

These models preceded James Watson and Francis Crick's discovery that DNA is composed of a phosphate group linked to a deoxyribose sugar, which in turn is linked to one of four nitrogenous bases - adenine (A), guanine (G), cytosine (C), and thymine (T) - the nucleotides pairing-up via hydrogen bonds. As we know, guanine establishes 3 hydrogen bonds with cytosine; adenine forms 2 hydrogen bonds with thymine. These specific pairings suggested a copying mechanism for the genetic material [6].

Hydrogen bonds are a key element of life. They are weaker than covalent bonds and require/release less energy when formed/broken, thus react easily at normal body temperature. Hydrogen bonds play an important role in determination of a molecule's physiological or biochemical nature and maintenance of stability [5, 6].

A supramolecule is a non-covalent association of complexes or molecules. DCP is a supramolecule assembled through four strong hydrogen bonds (1.7 Å~1.9 Å), with a configuration that seems to possess unique characteristics relating to target recognition and selectivity, increased hydrophilia and greater stability in an acidic medium, and more bondable to cellular DNA [7].

Nobel laureates Donald J. Cram, Jean-Marie Lehn and Charles J. Pedersen, who discovered molecules that mimic important biological processes, including the mode of action of enzymes, laid the foundation for an important interdisciplinary area of research within chemistry termed host-guest chemistry (Cram) or supramolecular chemistry (Lehn) [8]. This influenced the design of DCP. Dicycloplatin mainly consists of two parts - carboplatin and 1,1-cyclobutane dicarboxylate - in a host-guest complex [7].

DCP's structure is consistent with the Exclusion Principle of Wolfgang Pauli, who won the Nobel Prize in Physics in 1945 [9], and with Hund's Rule [10].

The design of DCP also follows Lipinski's 1997 Rule of Five. The Rule of Five for an orally-active drug is to be no more than one violation of the following criteria: no more than 5 hydrogen bond donors (the total number of nitrogen-hydrogen and oxygen-hydrogen bonds); no more than 10 hydrogen bond acceptors (all nitrogen or oxygen atoms); a molecular mass less than 500 Daltons; and an octanol-water partition coefficient log P not greater than 5 [11].

For DCP:

1. Hydrogen bond donors = 4;

2. Hydrogen bond acceptors = 4;

3. Molecular mass = 515.4;

4. Log P = 1; and

5. Rotatable bonds = 4.

An-Tuo-Ke-Jin (安妥可金, the capsule form of dicycloplatin), was developed for oral administration in 1992. Subsequently, dicycloplatin intravenous injection was developed for a phase I human clinical trial, completed in 2006. Phase II clinical trials using DCP+paclitaxel in non-small-cell lung cancer patients were completed in 2009 [12, 13]. In March 2012, dicycloplatin injection was approved for chemotherapy by the Chinese FDA (Certificate Nos. H20120020 and H20120021).

The following sections describe the structure of dicycloplatin and report the pre-clinical and clinical studies, cellular and molecular investigations, mechanism research, along with accounts of special cases.

PHYSICOCHEMICAL PROPERTIES

DCP's chemical name is cis-diammine (1,1-cyclobutane-dicarboxylate) platinum (II), 1,1-cyclobutancdicarboxylic acid complex. Its molecular formula is $C_{12}H_{20}N_2O_8Pt$; molecular weight, 515.39. DCP's solubility in water is 4 g/100 ml, compared to 1.5 g/100 ml for carboplatin and 0.2 g/100 ml for cisplatin). DCP is stable in water long-term. Worth noting, experiments indicate that physiological saline cannot serve as a solvent for DCP, as DCP dissolved in physiological saline easily degrades into carboplatin acid and 1,1-cyclobutane dicarboxylic acid [7]. The chemical structure of DCP is shown in Fig. (**1**).

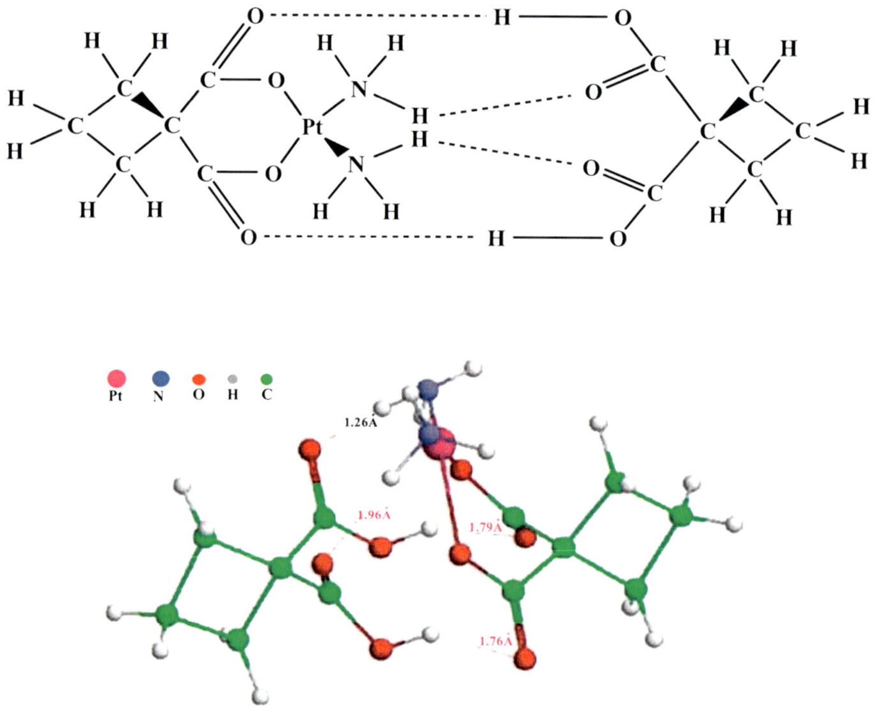

Fig. (1). Chemical structure of dicycloplatin generated by modeling. *Illustrative representation of the data originating from Yang et al.*

Dicycloplatin is synthesized from platinum powder through a 10-step process. X-ray single crystal diffraction shows that the end product is a supramolecule composed of carboplatin (host) and 1,1-cyclobutane dicarboxylate ligand (guest) interconnected via four hydrogen bonds Fig. (**1**).

Unlike carboplatin - a covalent-bond molecule - dicycloplatin is a hydrogen bond supramolecule. This may explain why DCP appears to have a more stable chemical structure, good water solubility, and a better safety profile [7]. The solubility and stability of platinum complexes have a direct bearing on their activity, toxicity, and pharmacokinetics. Clinical observations in China show DCP possesses superior anticancer efficacy through its broad-spectrum effects, high performance, low toxicity, and good penetration.

Electrospray ionization mass spectrometry (ESI-MS) found that a model of the structure of DCP to be soluble and stable [7]. Experimental tests for the solubility and stability of dicycloplatin and carboplatin in water were performed in 1998. Aqueous solutions of each were sealed in ampules and stored at room temperature. One week later, the carboplatin ampule began to show platinum-precipitation; the DCP ampule remained clear and transparent. No chemical decomposition of DCP in aqueous solution was observed after 10 years of storage at room temperature (Fig. **2A**). Using MS and NMR, no differences in quality were found between the ten-year-old DCP solution and newly prepared DCP products [7, 14]. Thus, DCP overcomes carboplatin instability in aqueous solution while maintaining its therapeutic effects.

(A)

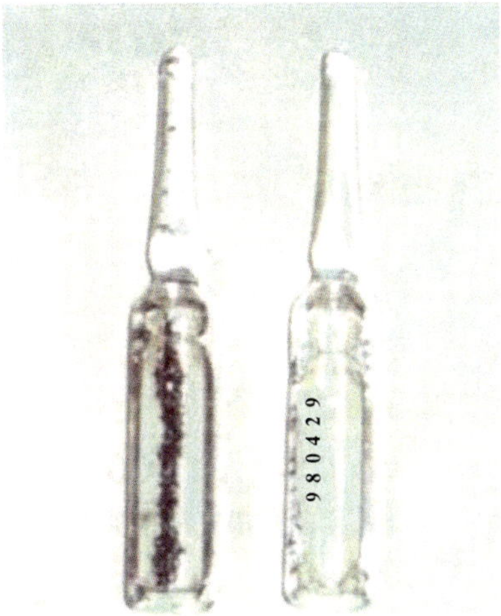

Carboplatin DCP

Fig. 2 contd.....

(B)

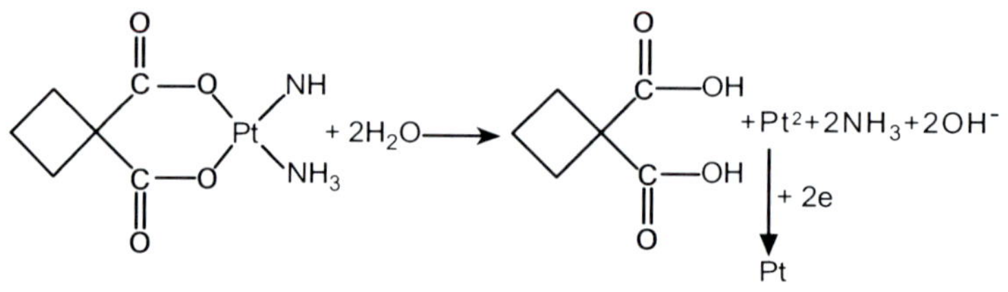

(C)

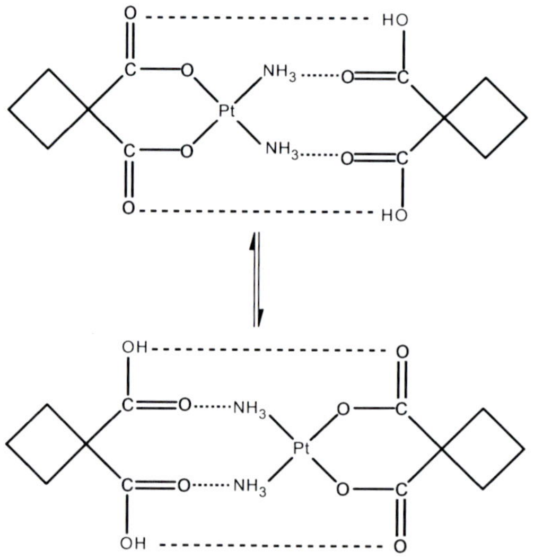

Fig. (2). (A) Comparison of DCP and carboplatin in aqueous solution after 10 years storage, **(B)** Carboplatin Hydrolysis in water, **(C)** H+ internally exchanges within DCP supramolecule.

Dicycloplatin's stability may be due to its hydrolytic behavior. For carboplatin in aqueous solution, the leaving group (1,1-cyclobutane dicarboxylic acid) accepts H^+ of water; the remaining $Pt(NH_3)_2$ is unstable, leading to molecular breakdown (Fig. **2B**).

As noted, DCP contains four O-H···O strong hydrogen bonds. In hydrolysis in aqueous solution, the leaving group (1,1-cyclobutane dicarboxylic acid) of the host molecule (carboplatin acid) - does not accept H^+ of water, which is weak. Within the supramolecular structure, the 1,1-cyclobutane dicarboxylic acid accepts H^+ internally only from the O-H···O strong hydrogen bond. It is hypothesized that, the leaving group switches over and reforms with the remaining $Pt(NH_3)_2$ - maintaining the structure of DCP (Fig. **2C**), thus preserving the therapeutic integrity of the drug molecule. This is the hypothetical explanation [14 - 16].

To differentiate dicycloplatin from physical mixtures of carboplatin and cyclobutane dicarboxylate, a method of quality control was developed using X-ray powder diffraction and high-performance liquid chromatography. The test results showed that the dicycloplatin supramolecule is substantially different from a physical mixture of carboplatin and cyclobutane dicarboxylate [14].

Furthermore, as the structure of DCP consists of carboplatin and cyclobutane dicarboxylate, it is important to determine if DCP is hydrolyzed to carboplatin inside the body and if the effect-driving substance of this drug is carboplatin instead of dicycloplatin. To answer this question, prototype concentration of dicycloplatin in patient blood samples after DCP administration was measured using AB-4000™ LC-MS/MS instrumentation, with carboplatin plasma concentration as comparison. The results demonstrated that the spectrum-profiles of DCP and carboplatin are different; the majority of DCP is still present as prototype DCP in plasma 2-h after administration [17].

PRE-CLINICAL RESULTS

Pharmacology, Pharmacokinetics and Toxicology

Dicycloplatin is a cytotoxic platinum complex. Its characteristics include low toxicity, high efficiency, wide-spectrum effects, and good water solubility. Early *in vitro* cytotoxicity studies demonstrated significant antitumor activity against a variety of human cancer cell lines, such as BCG-823, BEL-7402 and EJ.

DCP also demonstrated *in vivo* anti-tumor effects in the experimental tumor xenograft model of A549 human non-small cell lung cancer, with 67-90% inhibition of tumor growth, compared to 51-63% for carboplatin. Other trials in animal models found inhibitory effects in cancers of stomach, colon, lung and breast, with lower toxicity compared to similar platinum compounds.

Pharmacokinetic studies of DCP in comparison to cisplatin and carboplatin in rats suggest that DCP and carboplatin have faster renal excretion than cisplatin. P-glycoprotein [P-gp] organic anion transporter (OAT) or organic anion-transporting polypeptide (OATP) may be involved in excretion of DCP [18 - 25].

In a pharmacokinetic study of DCP in dogs, maximum plasma concentration (C_{max}) area under concentration (AUC) and dosage appear to be linearly correlated; the summarized plasma pharmacokinetic parameters in rats and dogs following DCP treatment are shown in Table **1**. Zhao *et al.* reported that plasma concentration of platinum originating from DCP decreased rapidly within the first 4 hours and declined slowly long-term in rats and dogs. This study also showed distribution of DCP, with the highest concentration in plasma and some in intestine and prostate [26].

Table 1. Plasma pharmacokinetic parameters in rats and dogs following dicycloplatin administration.

Preclinical pharmacokinetic parameters of dicycloplatin						
Dose (mkg)	**Cmax (μg/mL)**	**AUC_{inf} (μg×h/mL)**	**T1/2α (h)**	**T1/2β (h)**	**Vc (L/kg)**	**Cl (L/kg/h)**
10	54	38.8	0.27	11.9	0.17	0.26
20	130	80.4	0.24	10.8	0.16	0.25
40	255.3	176.6	0.25	11.6	0.16	0.22
5	20	41.7	0.56	18.1	0.28	0.123
10	37.4	81.1	0.72	22.8	0.326	0.14
20	81.6	146.2	0.58	21.6	0.318	0.154

Note: Rat data (first 3 lines); dog data (last 3 lines). C_{max}: maximum plasma concentration. AUC_{inf}: area under the curve from time 0 to infinite; T ½ α: early distribution half-lives; T ½ β: late elimination half-lives; Vc: volume of distribution; Cl: clearance. *Data provided by Sopo-Xingda Pharmaceutical, Inc.*

Acute toxicity was investigated by intravenous and intraperitoneal single administration of different doses. The results indicated that the toxicity profile of DCP [Lethal dosage (LD_{50})=210 mg/kg] is higher than that of carboplatin (164 mg/kg) and cisplatin (14.27 mg/kg), indicating greater toxicity of cisplatin and carboplatin. The inhibition rate of carboplatin was lower than that of dicycloplatin, so the safety range of dicycloplatin (*i.e.* the gap between therapeutic dose and toxic dose) was wider than that of carboplatin, an important indicator of relative drug safety [21 - 25].

Studies of cytotoxicity and hematological toxicity of DCP, cisplatin, and carboplatin found that DCP induced renal toxicity in rats significantly less often than cisplatin, with similar episodes of myelosuppression as those found with carboplatin. *In vivo* evaluation of DCP toxicity suggests that DCP causes the least amount of damage to major target organs in rats. These reports provided a generally favorable safety profile for DCP.

In summary, intravenous administration of DCP at doses of 10, 20 and 40 mg/kg had no significant effect on the central nervous system of mice, and the intravenous administration of DCP at doses of 1.5, 3.0 and 6.0 mg/kg had no significant effect on blood pressure, ECG, and respiration in anesthetized rats [27 - 29].

A long-term DCP toxicity trial in rats that examined peripheral blood and bone marrow in high- and medium-dose groups (21 and 15 mg/kg) found a significant decline in red blood cell count, hemoglobin, white blood cell count, and platelets, which was more pronounced in the high-dose group. These effects were not significant in the low-dose group (8.7 mg/kg). All the indicators returned to normal levels one month after drug withdrawal.

Routine blood work showed that ASP, ALP, AKP, BUN, and CR in the DCP dose groups were higher than those of the control group, especially in the high-dose group. The main target organs affected by DCP are kidneys, liver, lungs and heart, with more obvious dose-effects in the high-dose group. In the medium- and low-dose groups, no myocardial injuries were observed [24 - 30].

Mechanism of Action

In general, platinum-based chemotherapeutic agents such as cisplatin and carboplatin are the mainstream drugs for treating advanced cancers. These drugs cause cell death by forming covalent crosslinks in DNA that interfere with DNA replication and transcription, which leads to cell death [31 - 33].

There are four drug-action steps in platinum chemotherapy: 1) **Uptake** (Cell Entry), which is not a matter of simple diffusion but involves plasma membrane proteins or transporter proteins, such as CTR1 [34, 35]; 2) **Activation**, by which the parental drug is converted to active species that bind to the target or meant to attack the cancer cell DNA, usually achieved by hydrolysis inside the cell due to a much lower concentration of chloride ion; 3) **DNA binding,** when the active species bind with the DNA and cause DNA crosslinks, the potency of the drug directly related to ability to form bi-functional interstrand and/or intrastrand crosslinks on DNA, thus disrupting cellular processes such as transcription and replication, leading to apoptosis; and 4) **Cellular Response**. With regard to the last step, platinum drugs provoke a complex series of cellular reactions. Lethal drug doses kill cells primarily by forming DNA adducts, resulting in cell-cycle arrest at the G2 phase. Platinum-DNA adducts are recognized by cellular proteins, including some that enhance the survival of the cancer cell through the DNA repair system, and others that hasten death by conferring sensitivity to the drug and triggering apoptosis. Each drug-action step is important in successful platinum chemotherapy [36, 37].

Molecular Mechanism Studies

Researchers at West Virginia University investigated cisplatin- and dicycloplatin-induced DNA damage response in cisplatin-sensitive ovarian cancer A2780 cells (a wild-type p53 cell model). After 1-hr drug exposure at IC_{50} doses (cisplatin 3 μM; dicycloplatin 3.79 μM), several kinases of the DNA-damage repair pathway were activated by both platinum drugs. The activation included phosphorylation of p53, phosphorylation of Chk2, and increased proteins of ATM, p53, p48 and p21. Thus, dicycloplatin activates pathways similar to other platinum compounds (Fig. **3**) [17].

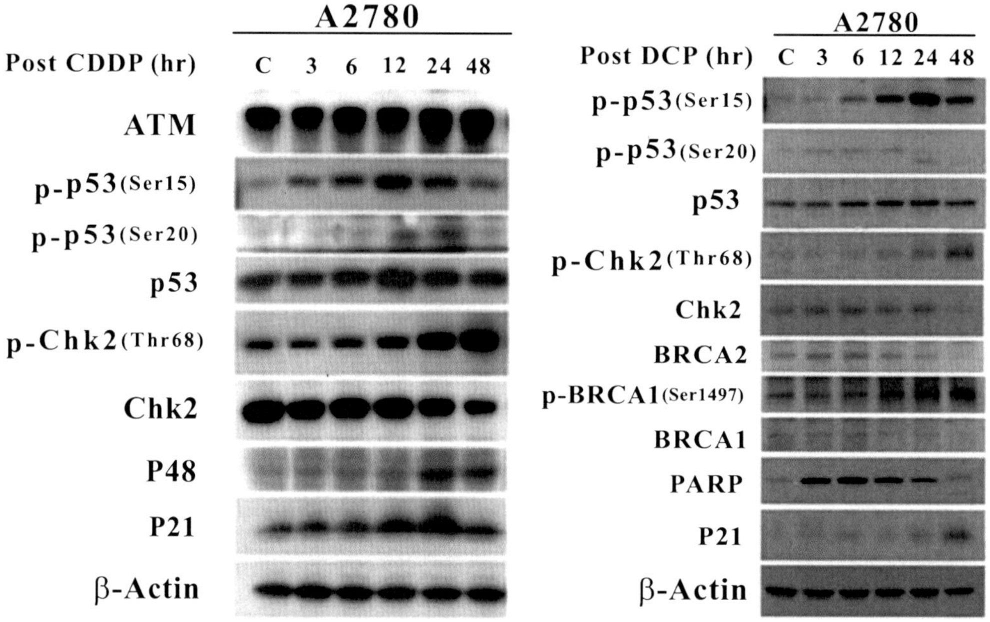

Fig. (3). DCP & CDDP in A2780 cells: DNA-damage response induced by cisplatin (CDDP) and dicycloplatin (DCP) in ovarian cancer A2780 cells. Cells were treated with cisplatin or dicycloplatin for 1 h, then washed and re-incubated in drug-free media for the indicated times. Proteins were extracted, separated by polyacrylamide gel electrophoresis (PAGE), transferred onto PVDF membrane, and probed with the indicated antibodies. *Reproduced with permission from Anticancer Research 34: 455-464 (2014).*

An *in vitro* study of dicycloplatin inhibition of growth of human hepatoma cells BEL-7402 found alterations of morphology and ultrastructure of such cells. Obvious morphological changes and shedding of some cells were observed after five days of treatment with dicycloplatin at a dose of 20 μg/ml. Under electron microscope, cells were observed to die after five days of treatment with dicycloplatin at a dose of 50 μg/ml, mainly due to steatosis and vacuolar degeneration [38].

This study also reported DCP effects on gene expression of C-myc, N-ras and p53. Dicycloplatin seemed to inhibit BEL-7402 cell proliferation. RT-PCR analysis showed that it down-regulates gene expression of C-myc in hepatoma BEL-7402 cells. Northern Blot analysis showed that p53 gene expression declined after treatment with dicycloplatin, and N-ras gene expression slightly declined in the group receiving a dose of 10 μg/ml, but increased with a dose of 20 μg/ml. Experimental results showed that dicycloplatin had killing- and growth-inhibition effects on human hepatoma BEL-7402 cells, with cytotoxicity as the primary mode of action. This suggests that the decline in C-myc and p53 gene expression after treatment with dicycloplatin may be a result of gene transcription being affected after cancer cell DNA was damaged by dicycloplatin [38].

In a comparative study of platinum DNA damage caused by cisplatin, carboplatin and dicycloplatin, researchers found that all three can induce breaking of pBR322 plasmid DNA. The half-breaking dose is 33.71 μmol/L, 3.31 mmol/L and 1.61 mmol/L, respectively; the breaking capacity is cisplatin > dicycloplatin > carboplatin [39].

Investigation of dicycloplatin-induced apoptosis in prostate cancer PC3 cells and lung cancer NCI/H446 cells showed that DCP induces cell-cycle arrest, inhibits proliferation, and leads to apoptosis [20, 40, 41].

CLINICAL STUDIES

Phase I Clinical Trial

From February 2004 to October 2005, a Phase I clinical and pharmacokinetic study to determine efficacy and safety was conducted at State Key Laboratory of Oncology in Southern China with 29 terminal cancer patients on 8 dose-levels of dicycloplatin injection. All recruited patients signed informed consent forms according to Chinese regulatory guidelines, and all patients completed the clinical observations [42].

This trial investigated patient tolerance of dicycloplatin and toxicity to target organs to determine acute toxicity, dose-limiting toxicity, and maximum-tolerated dose. At the same time, DCP pharmacokinetic behavior - such as concentration-time curve and pharmacokinetic parameters - were observed. The data obtained provided a theoretical basis and administration protocol for rational treatment in subsequent Phase II clinical trials.

***Eligibility and Recruitment**:* All patients had advanced malignant tumors diagnosed by pathology or cytology; conventional treatment (or standard regimen) had failed or tumor had recurred and no effective treatments were available to these patients. General condition: Eastern Cooperative Oncology Group (ECOG) point was ≤ 2, and expected lifetime was ≥ 3 months. A total of 29 patients (6 females and 23 males) between 30 and 70 years of age were recruited. Patients included 10 cases of nasopharynx cancer (34.48%), 8 of colorectal cancer (27.59%), 7 of non-small cell lung cancer (24.14%), 1 each of synovial sarcoma, malignant schwannoma, breast cancer, and Hodgkin's lymphoma (3.45% of each). Fourteen cases (48.28%) had received surgery and 16 cases (55.17%) had a history of radiotherapy; 29 cases (100%) had received chemotherapy, of which 27 (93.10%) had received platinum chemotherapy. All the subjects were hospitalized during the trial [42].

This clinical Phase I trial was conducted with a series dosing (mg/m^2) of 50, 100, 175, 250, 350, 450, 650 (n=3/group), and the 550 mg/m^2 dose (n=8) assumed as phase II dose) for the DCP treatment. The initial dose was calculated based on 1/10 of mice LD_{10} and determined as 50 mg/m^2. Dose escalation was determined according to modified Fibonacci method as per 100%, 66.7%, 50%, 40%, 33.3%, 16.7%.

DCP Administration: Dicycloplatin injection was dissolved in 250 ml of 5% glucose solution and Intravenously (IV) administered to the patient within 1 hour. Each patient received one grade of dose-level for DCP treatment. Observation duration was at least 2 weeks; if any hematology or non-hematology organ toxicity occurred then it was observed until the toxicity disappeared.

Results: Primary **efficacy** analysis was obtained for 20 cases (of 29 patients): partial response (PR) for 2 cases (non-small cell lung cancer patients in 550 mg/m^2 dose group); stable disease (SD) for 13 cases; progressive disease (PD) for 5 cases; no data (ND) for 9 cases. Trial observations indicated non-small cell lung cancer, head and neck cancer or colorectal carcinoma as primary targets for Phase II clinic trials.

The main **adverse effects** of dicycloplatin observed were nausea/vomiting and myelosuppression. Other adverse reactions seemed to relate to drug dose and incidences were very low, suggesting that the drug is well tolerated. In the 650 mg/m^2 dose group, adverse reactions occurred for 3 subjects: III°-IV° vomiting, III° hair loss, and III° PLT decrease, indicating that the main toxicities occur in the gastrointestinal system and bone marrow. The human maximum tolerated dose (MTD) of DCP was determined to be 650 mg/m^2. Thus, 550 mg/m^2 once every three weeks would be recommended as the starting dose for subsequent Phase II clinical trials [42].

Pharmacokinetic analysis to determine the total and ultra-filtered platinum concentrations in plasma demonstrated linear pharmacokinetics characterized by distribution half-life of 1.49 ± 0.39 h and low steady-state apparent volume of distribution (177.88 ± 44.92 L/m^2).

Specifically, 1-h after intravenous administration, dicycloplatin appeared at a constant rate; the drug concentration-time curve showed a two-compartment model disposition characteristic. For a 450 mg/m^2 dose of DCP, the clearance rate of free drug CL was 1.54 ± 0.65 L/h, $t_{1/2\beta}$: 90.40 ± 26.30 h, and protein binding rate: 0.31 ± 0.11. At 550 mg/m^2 dose level, the clearance rate of free drug CL was 2.98 ± 1.30, $t_{1/2\beta}$: 62.05 ± 12.80, and protein binding rate: 0.508 ± 0.195. Seventy-two hours after drug administration, the excretion rate of dicycloplatin via urine was 15.3-25.4%. Within the 250-to-650 mg/m^2 dose range, $AUC_{0\rightarrow\infty}/D$ was 1.14-1.34, AUC was linear to dose, $t_{1/2}$ was almost the same, which indicated that platinum in dicycloplatin essentially had a linear pharmacokinetic characteristic in the body within that dose range. When the dose increased, the area under the concentration-time curve increased correspondingly. The AUC was linear to the dose (r=0.9898), indicating that dicycloplatin has a linear pharmacokinetic process in the human body in the 100 to 550 mg/m^2 dose range. The logarithm of elimination phase's concentration against time was a line, demonstrating that the drug is eliminated via first-order kinetics [42].

Finally, renal excretion (one path for dicycloplatin excretion) accounted for 28.57-45.71% of total excretion; the renal clearance rate was (0.69 ± 0.20) L/h. Dicycloplatin is eliminated slowly in the body ($t_{1/2\beta}$: 80.90±17.48 h), suggesting a 3-4 week course of treatment for Phase II clinic trials.

In conclusion, DCP displayed a favorable safety profile at dose range from 50 mg/m^2 to 550 mg/m^2 and the first efficacy signal was observed at the 550 mg/m^2 dose level [42].

Phase II Clinical Trial

From January 2007 to February 2008, a double-blinded, randomized multicenter phase II clinical trial comparing dicycloplatin and paclitaxel against carboplatin and paclitaxel in non-small cell lung cancer was conducted in China. Two hundred and forty chemotherapy-naïve patients with advanced NSCLC were recruited; 15 multi-center/hospitals nationwide participated. All patients signed informed consent form according to Chinese regulatory guidelines [43, 44].

Eligibility: Eligible patients were >18 years old with cytological or histological confirmation of stage IIIB (with pleural effusion) and stage IV NSCLC. Patients had measurable disease and an ECOG performance status of 0 to 2, with an expected life-time ≥ 3 months. Additional eligibility criteria included blood counts, enzyme, and kidney function reaching required levels. Excluded patients included those who received prior chemotherapy, radiotherapy or biologic therapy for NSCLC; had documented brain and meninx metastases; had a major operation within 4 weeks before the trial; or a significant history of cardiac disease and/or other active uncontrolled diseases.

However, relapsed tumor after an operation and adjuvant chemotherapy with non-platinum drugs was permitted if completed at least 6 months before registration. Patients who previously had a malignancy within 5 years were also excluded except for basal cell carcinoma, cervical carcinoma *in situ* or prostate carcinoma *in situ*. Patients who were pregnant or in the lactation period, allergic to paclitaxel or platinum, abused alcohol and other substances, or had primary organ failure were not eligible [43, 44].

Treatment and Analyses: The aim of this trial was to compare the efficacy and toxicity of dicycloplatin-plus-paclitaxel to those of carboplatin-plus-paclitaxel as a first-line treatment for patients with advanced NSCLC. Accordingly, 240 NSCLC patients were randomly assigned (1:1) into two platinum-plus-paclitaxel arms to receive either dicycloplatin at 450 mg/m^2 plus paclitaxel at 175 mg/m^2 or carboplatin AUC = 5 plus paclitaxel at 175 mg/m^2 every 3 weeks (q3w) for up to 4 to 6 cycles. The two arms were well-balanced with regard to gender, age, ECOG status, disease stage, and histological classification. The primary endpoint was response rate. Secondary endpoints included progression-free survival (PFS), overall survival (OS), and adverse events. The 240 patients disposition is described according to CONSORT criteria. Safety was assessed in 240 patients and efficacy was evaluated in 236 patients (118 patients in each arm), due to 4 patients later found to be ineligible [43, 44].

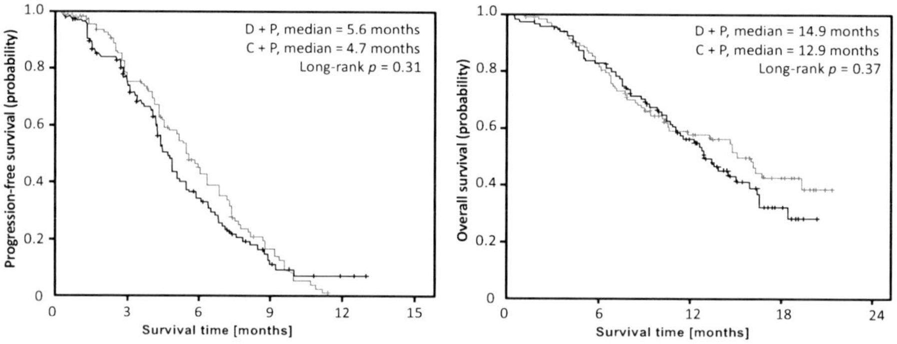

Fig. (4). Kaplan-Meier Curves for Progression-free & Overall Survivals. *Liu KJ et al, Arch Med Sci 2014; 10, 4: 717–724.*

Results: The response rates were 36.44% for dicycloplatin-paclitaxel and 30.51% for carboplatin-paclitaxel (p = 0.33). The median PFS was 5.6 months in the dicycloplatin-paclitaxel arm and 4.7 months in the carboplatin-paclitaxel arm (p = 0.31). The median OS was 14.9 months for dicycloplatin-paclitaxel and 12.9 months for carboplatin-paclitaxel (p = 0.37) (Fig. **4**). Adverse events in the two arms were well-balanced. The most common grade III/IV adverse event was hematologic toxicity [43, 44].

Conclusions: Trial results indicated no significant difference in patient outcomes between the two arms. Patients treated with dicycloplatin-paclitaxel had similar response (efficacy) and survival rates as those treated with carboplatin-paclitaxel, and toxicities of both treatments were generally tolerable. There were no statistically significant differences between treatment arms in baseline characteristics (Table **2**). The median cycles of therapy received in each treatment arm were four per patient. No major differences existed between the two arms concerning dose reduction [43, 44].

Table 2. Efficacy Results of DCP+paclitaxel and carboplatin+paclitaxel.

Variable	D + P (*N* = 118)		C + P (*N* = 118)		Value of *p*
	n	%	*n*	%	
Response					
PR [%]	43	36.44	36	30.51	0.33
SD [%]	58	49.15	59	50.00	0.43
PD [%]	7	5.93	15	12.71	0.19
Non-evaluable	10	8.47	8	6.78	0.76
Response rates [%]	36.44		30.51		0.04
95% CI	27.76–45.12		22.20–38.82		
Disease control rates [%]	85.59		80.51		0.02
95% CI	79.26–91.93		73.36–87.66		
PFS, median [months]	5.6		4.7		0.31
95% CI	4.9–6.3		4.2–5.2		
6 months [%]	43		33		0.19
Survival, median [months]	14.9		12.9		0.37
95% CI	12.4–17.7		10.9–15.1		
1 year [%]	58		56		0.90

Response rates were based on intention to treat (ITT). D + P: dicycloplatin plus paclitaxel, C + P: carboplation plus paclitaxel, PR: partial remission, SD: stable disease, PD: progression disease, PFS: progression-free survival, CI: confidence interval. *Liu KJ et al, Arch Med Sci 2014; 10, 4: 717–724.*

CASE REPORTS

DCP Treatment of a Prostate Cancer Patient Indicates Therapeutic Efficacy

A 78-year-old male patient presented at Shanghai Sixth People's Hospital with high fever, frequent urination, urinary urgency, dysuria, and difficulty walking on March 17, 2006. Three months earlier, he was diagnosed with acute urinary tract infection based on blood test results (CEA positive, WBC count 14.3x109/L, neutrophil 89.9%) and treated with an anti-inflammation medicine.

On March 22, laboratory tests showed his prostate specific antigen (PSA) to be 55.813 ng/ml (normal range 0.000-4.00 ng/ml); his free prostate specific antigen (fPSA) was 11.214 ng/ml (normal 0.00-1.90 ng/ml). Ultrasonography showed a full-shaped hyperplastic prostate with hyperplastic nodes in inner gland as large as 18 x19 mm. There were several cysts in the prostate. Using prostate needle biopsy under ultrasonography, 6 sites were sampled for pathology examination. The results, together with HE slicing, confirmed prostate cancer of inner peripheral zone in the right-side prostate with a 5 Gleason Score (3+2), atypical adenomatous hyperplasia (AAH) of right side transition zone, and hyperplasia with chronic interstitial inflammation of other prostate tissue.

Physicians recommended orchiectomy, followed by treatment with zoladex. Instead, the patient elected to receive DCP chemotherapy. He began DCP treatment in mid-April of 2006. The initial dose was 450 mg by IV; he had no side effects. Two weeks later, the patient began a schedule of four 500 mg DCP treatments over a two-month period. During treatment, the patient had blood tests regularly. On May 12, his PSA was 8.797 ng/ml and fPSA was1.819 ng/ml; αFP was 2.52 ng/ml and CEA was 2.61 ng/ml (both negative). On May 16, his PSA was 8.35 ng/ml and fPSA was 1.62 ng/ml. On August 18, his PSA dropped to 4.298 ng/ml and his fPSA was 0.900 ng/ml. Finally, on October 7, the patient's PSA – at 3.282 ng/ml – had returned to normal.

After 5 cycles of DCP treatment (once every half-month), the patient's symptoms (frequent urination, urinary urgency) gradually alleviated. He started walking again. He gained weight and was able to take care for himself. To reinforce his treatment, two cycles of 600 mg of DCP were given by IV, one cycle every six months.

Routine laboratory tests on January 18, 2007 showed his PSA to be 3.051 ng/ml; two months later, it was 2.786 μg/L and his fPSA was 0.676 μg/L. On September 6, blood test PSA was 3.129 μg/L and fPSA was 0.709 μg/L; two months later, his PSA was 3.476 μg /L and fPSA as 0.878 μg/L (CEA and αFP were both negative). On April 22, 2008, the patient's PSA was normal (3.04 μg /L).

During DCP treatment, the patient's white and red blood cell counts decreased slightly. He experienced mild nausea but no other discomfort. After DCP treatment, the patient's PSA level stayed within normal range. Two years after DCP treatment, he stated that he felt well, could take care of himself and even performed some housework. He is still under follow-up.

DCP May Cross the Blood–Brain Barrier (BBB)

A 42-year-old male with lung cancer (upper apical region of right lung) was diagnosed in June 2006 at Beijing Hospital based on Positron Emission Tomography/Computerized Tomography (PET/CT). Clinicians found significantly increased metabolic activity of the mass and high SUV (standardized uptake value) at delayed imaging. The patient's cancer soon spread to lymph nodes, mediastina, and meninges. An encephalic tumor was also found [17].

A regimen of large-dose DCP, 1200 mg by IV every 3 weeks (q3w) was prescribed, and oral Iressa was begun as adjuvant treatment to support cranial radiotherapy at PLA General Hospital in Beijing. This regimen continued for four months, with five high-dose treatments of DCP and whole-brain irradiation to 38 Gy in 19 daily fractions, followed by a boost to the lesion area with additional 22 Gy delivered in 11 daily fractions. During treatment, the patient developed slight leucopenia and thrombocytopenia which were resolved by symptomatic treatment. The patient was discharged on February 1, 2008 with normal white blood cell count, platelet count, biochemistry, and blood coagulation tests. Magnetic resonance imaging (MRI) on January 25, 2008 showed the brain lesion disappeared (Fig. **5**).

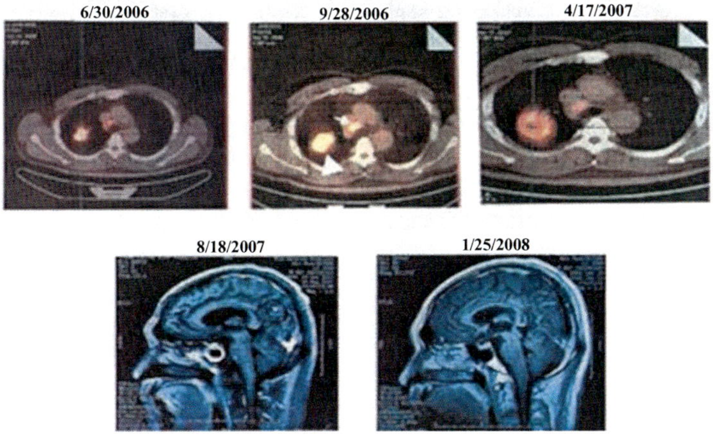

Fig. (5). Dicycloplatin in the treatment of lung cancer with brain metastasis. *Reproduced with permission from Anticancer Research 34: 455-464 (2014).*

That successful treatment suggested the possibility that DCP crosses the Blood-Brain Barrier. *In vitro* data of histoculture drug response assay in various tumor specimens demonstrated DCP chemosensitivity in brain tissue. As shown in Fig. (**6**) (right panel), DCP sensitivity in glioma tissues is much higher compared to other chemotherapeutic agents (data provided by Sopo-Xingda Pharmaceutical, Inc.). Worth noting, Zhao *et al.* reported that a visible distribution of platinum originating from DCP was seen in brains of rats [26].

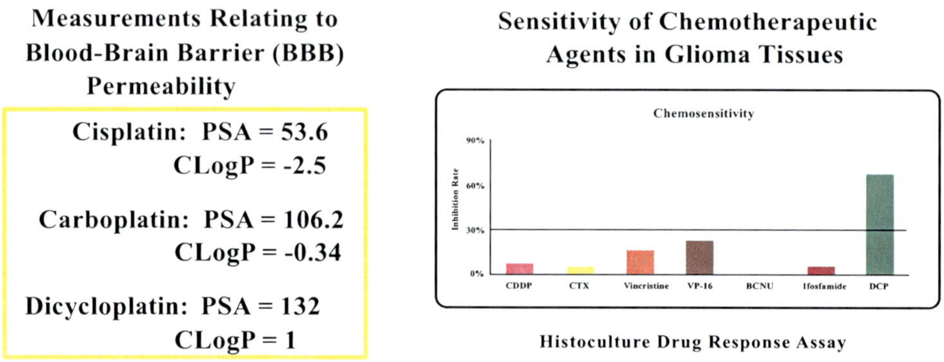

Fig. (6). Dicycloplatin may pass the blood-brain barrier, as suggested by chemosensitivity assay in brain tissue - data provided by Sopo-Xingda Pharmaceutical, Inc. and by theoretical calculations. *Reproduced with permission from Anticancer Research 34: 455-464 (2014).*

Theoretically, a larger polar surface area (PSA) hints at greater solubility, and a higher calculated logP (ClogP) predicts better permeability, as opposed to a smaller PSA and logP [45 - 47]. Therefore, we calculated the PSA and logP for cisplatin, carboplatin and DCP. As shown in Fig. (**6**) (left panel), DCP has the largest PSA and the largest ClogP. Thus, DCP seems to have better solubility and permeability based on these characteristics, with greater likelihood of crossing the BBB [17].

Treatment of American Cancer Patient Suggests Chemotherapeutic Advantages of DCP over Other Platinum Drugs

A 65-year-old Caucasian male with increasing hematuria over a 4-month period, without lower urinary tract symptoms, was seen in June of 2016 at West Virginia University Hospital, Morgantown, USA. The patient was a former smoker. The work-up, with CT-Urography and cystoscopy, showed a 1.5 cm bladder mass. Transurethral resection of the mass was performed on June 30, 2016. The pathologic diagnosis was non-invasive high-grade papillary urothelial carcinoma (Ta) involving the right lateral wall of the bladder. Immunotherapy with BCG and surveillance cystoscopy were recommended to the patient. Patient declined BCG treatment but elected cystoscopic surveillance every 3 months at WVU Hospital after dicycloplatin chemotherapy in Beijing, China.

With regard to family history, the patient's mother died of ovarian cancer at age 50, and his father had advanced prostate cancer when he died at age 81 of other causes. The patient's family history and his familiarity with DCP through his work as a medical writer, led him to pursue DCP chemotherapy. The patient traveled to Beijing in mid-July to receive DCP treatment over eight weeks between July 19 and September 7, 2016.

The patient was evaluated by a Chinese oncologist and signed a consent form at Chinese PLA General Hospital (generally known as Beijing 301 Hospital). Baseline blood counts and chemistry were evaluated prior to chemotherapy and then weekly, the day before each treatment.

To protect the liver and kidneys from chemotherapy adverse effects, the DCP protocol began with omeprazole and reduced glutathione sodium, dissolved separately in 100 ml of 0.9% saline and given by IV prior to each DCP administration. An intramuscular injection of diphenhydramine was given before each treatment. Dicycloplatin 300 mg was dissolved in 250 ml 5% glucose solution and infused over one hour. The patient was treated in the Day Ward of the Tumor Building at Beijing 301 Hospital and was observed for 30 minutes after each treatment.

The American cancer patient received one cycle of 300 mg of DCP via IV weekly for 8 weeks. In addition, An-Tuo-Ke-Jin, the capsule form of dicycloplatin, was administered orally to achieve the standard weekly DCP dosage. One capsule contains 3 mg of dicycloplatin; 4 capsules were taken 3 times a day for 1 ½ months.

Adverse effects included mild nausea, moderate fatigue, and, towards the last weeks, back and leg aches. There was no vomiting or hair loss, and the patient continued sightseeing despite moderate fatigue and aches. Weekly blood counts are shown in Table **3**. Blood values remained within normal limits, though there were notable decreases in red blood cells, white blood cells, and platelets.

After completion of DCP treatment, the patient returned to the United States for follow-up at West Virginia University Hospital. Cystoscopies performed on October 13, 2016, on January 16, 2017, and on April 20, 2017 revealed no recurrence of tumor and the resection area was observed to be clear of new growth. The resection site appeared to be healing (Fig. **7**). Wash solutions collected during each cystoscopy were examined by a cytologist and no malignant cells were observed. In summary, dicycloplatin chemotherapy seems successful to date and adverse effects were bearable.

Table 3. CBC with Differential of the American Cancer Patient Treated at Beijing 301 Hospital, China.

			Pre-DCP	DCP Treatment						
	Reference Range	Unit	7/20/2016	7/26/2016	8/2/2016	8/9/2016	8/16/2016	8/23/2016	8/30/2016	9/6/2016
Hemoglobin	137~179(M); 116~155(F)	g/L	137	142	137	134	132	139	135	134
RBC	4.3-5.9 (M); 3.9~5.2 (F)	10^12/L	4.63	4.81	4.64	4.51	4.49	4.31	4.42	4.34
WBC	3.5 ~ 10	10^9/L	7.37	7.04	6.32	5.32	4.7	4.62	3.9	4.04
Platelets	100 ~ 300	10^9/L	248	267	244	225	244	187	147	177

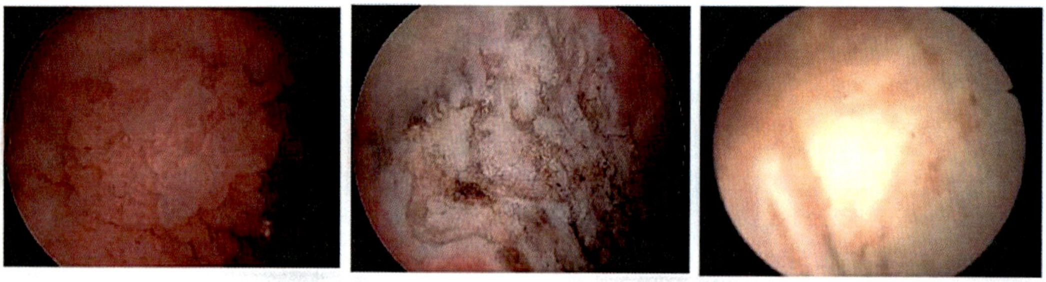

Fig. (7). Images obtained during cystouroscopic procedure before resection (left panel) and after tumor resection (middle panel) on June 30, 2016. The residue of bladder tumor lesion observed during the second follow-up is shown in the right panel (6-month post-DCP treatment).

CONCLUSION

The design of dicycloplatin, a supramolecule built with 4 hydrogen bonds, provides a chemotherapy compound with a novel molecular geometry that offers an improved therapeutic dynamic and a better safety profile.

The DCP chemotherapy received by the first American patient – and follow-up observations of his resection site – may not prove DCP efficacy is superior. However, the patient has received DCP only and there is no evidence to date of tumor recurrence. Also of note, the patient experienced tolerable side effects.

Worth injecting here, *in vitro* investigations indicate that DCP mainly affects tumor cells, largely bypassing normal cells. Researchers using Atomic Force Microscopy (AFM) observed that dicycloplatin does not affect single DNA molecules of normal cells, unlike other platinum drugs, such as oxaliplatin, which alters DNA structure of normal cells (data not shown). In another study, DCP, at certain concentrations, mainly affected cancer cells (lung carcinoma and melanoma cells) but did not affect or only slightly affected normal cells (dermal fibroblast and mammary epithelia cells) (data not shown). These results are consistent with the medical data for – and the observations of – the first American cancer patient to receive DCP chemotherapy. The patient did not suffer from serious myelosuppression or other toxic effects.

In addition, symptoms and clinical data for the American patient suggest important considerations for DCP chemotherapy - and possibly for other platinum anticancer drugs - relating to dose management and administration schedule. That is, it may be possible to treat cancer patients with mid-level doses (300 mg to 550 mg of DCP) instead of higher doses (650 mg or 800 mg of DCP), and more frequent IV infusions (weekly instead of bi-weekly or every three weeks), to reach target dosage and efficacy, with fewer adverse effects.

Finally, the dicycloplatin experience of the American cancer patient - carefully documented in the United States and in China - is very much the same as reported by Chinese physicians, researchers and patients. Studies at West Virginia University illuminate the underlying molecular mechanisms and gene activation process. This collaborative effort is an excellent example of an East-West coming together to better understand and improve platinum drug chemotherapy.

Clearly, a platinum-based anticancer drug as effective as cisplatin and carboplatin – with bearable side effects – is an important therapeutic advance. Clinical investigation of DCP in the United States and Europe is warranted.

CONFLICT OF INTEREST

The authors (editor) declares no conflict of interest, financial or otherwise.

ACKNOWLEDGEMENT

We acknowledge the Molecular Medicine Core Facility, WVU Cancer Institute, West Virginia University, USA for supporting this study and Beijing Xingda Scientific System Company for guidance of dicycloplatin research in coordination with drug development.

REFERENCES

[1] The World Cancer Research Fund International [homepage on the internet]. England: Cancer facts and figures [updated: January 2015] Available at: http://www.wcrf.org/int/cancer-facts-and-figures

[2] Cancer Research UK [homepage on the internet]. England: Worldwide cancer statistics [updated: February 2015] Available at: http://www.cancerresearchuk.org/health-professional/ cancer-statistics/worldwide- cancer

[3] The American Cancer Society [homepage on the internet]. Atlanta: Cancer Facts and Figures 2016 [updated: 5th January 2017; cited: 28th feburary 2017] Available at: https://www.cancer.org/research/cancer-facts-statistics/all-cancer-facts-figures/cancer-facts-figures-2016.html

[4] Apps MG, Choi EH, Wheate NJ. The state-of-play and future of platinum drugs. Endocr Relat Cancer 2015; 22(4): R219-33.
[PMID: 26113607]

[5] Pauling L, Corey RB, Branson HR. The structure of proteins; two hydrogen-bonded helical configurations of the polypeptide chain. Proc Natl Acad Sci USA 1951; 37(4): 205-11.http://www.pnas.org/content/37/4/205
[http://dx.doi.org/10.1073/pnas.37.4.205] [PMID: 14816373]

[6] Pray L. Discovery of DNA structure and function: Watson and Crick. Nature Education 2008; 1(1): 100.

[7] Yang XQ, Jin XL, Song QH, *et al.* Structural studies of dicycloplatin: an antitumor supramolecule. Scientia Sinica Chimica 2010; 40: 485-91.

[8] Royal Swedish Academy of Sciences. Awarded for syntheses of molecules that mimic important biological processes. 14th October 1987. (Press release)

[9] Nobelprize.org. 2014.http://www.nobelprize.org/nobel_prizes/physics/laureates/1945/pauli-facts.html

[10] Katz DM. Physics for scientists and engineers: foundations and connections with modern physicsBoston: Cengage learning 2015; 2: 1395-9.

[11] Lipinski CA, Lombardo F, Dominy BW, Feeney PJ. Experimental and computational approaches to estimate solubility and permeability in drug discovery and development settings. Adv Drug Deliv Rev 2001; 46(1-3): 3-26.
[http://dx.doi.org/10.1016/S0169-409X(00)00129-0] [PMID: 11259830]

[12] Guan ZZ, Yang XQ, Song QH, *et al.* Phase I and phase II clinical trial of novel anticancer drug dicycloplatin in patients with advanced solid malignancies. Xth International Symposium on Platinum Coordination Compounds in Cancer Chemotherapy. 2007 Nov 30-Dec 3; Verona, Italy.

[13] Zeng Y, Shi MQ, Zhou GR, Xia GH, Feng JF. A randomized multicenter phase II trial of dicycloplatin plus paclitaxel versus carboplatin plus paclitaxel in the treatment of patients with advanced non-small cell lung cancer. J Nantong Univ 2011; 31: 327-9. [Med Sci].

[14] Yang X, Zheng J, Song Q, *et al.* Determination methods for the anticancer drug dicycloplatin, a supramolecule assembled through hydrogen bonding. Analyst (Lond) 2015; 140(8): 2704-12.
[http://dx.doi.org/10.1039/C4AN02274B] [PMID: 25686082]

[15] Zhang J, Chen P, Yuan B, Ji W, Cheng Z, Qiu X. Real-space identification of intermolecular bonding with atomic force microscopy. Science 2013; 342(6158): 611-4.
[http://dx.doi.org/10.1126/science.1242603] [PMID: 24072819]

[16] De Marco L, Thämer M, Reppert M, Tokmakoff A. Direct observation of intermolecular interactions mediated by hydrogen bonding. J Chem Phys 2014; 141(3): 034502.
[http://dx.doi.org/10.1063/1.4885145] [PMID: 25053321]

[17] Yu JJ, Yang X, Song Q, Mueller MD, Remick SC. Dicycloplatin, a novel platinum analog in chemotherapy: synthesis of chinese pre-clinical and clinical profile and emerging mechanistic studies. Anticancer Res 2014; 34(1): 455-63.
[PMID: 24403501]

[18] Yan DM, Tu LL, Peng XY, Li WJ, Shen ZR. *In vitro* and *in vivo* antitumour activity of dicycloplatin, a new active platinum complex. Chin J Mod Appl Pharm 2007; 24: 352-6.

[19] Liu T, Su H, Fang JC, Wang NQ, Yang XQ, Cui WC. Studies on the cooperative anticancer effects of dicycloplatin and 8 chemotherapeutic drugs in mice transplanted hepatoma 22 or sarcoma 180. Chin J Mod Clin Med 2003; 2: 1-3.

[20] Yan DM, Tu LL, Zheng XL, Qian BC. Apoptosis Induced by Dicycloplatin in Human Ovarian Cell Line A2780. J Oncol 2009; 15: 667-70.

[21] Yang XQ, Guan ZZ, Song QH, *et al.* Dicycloplatin, a Novel Drug for Cancer Chemotherapy. China: Presentation in Beijing 2007.

[22] Liu T, Su H, Fang JC, Wang NQ, Yang XQ, Cui WC. A comparison between dicycloplatin and carboplatin anticancer effect in animal models. Chin J Basic Clin Med 2003; 9: 36-8.

[23] Yan CH, Shi C, Li YF, Liao MY. Comparison of toxicities of dicycloplatin with cisplatin and carboplatin. Zhongguo Xin Yao Zazhi 2005; 14: 1156-9.

[24] Li XQ, Chen XJ, Ren WC, Zhang H, Gu X, Wang GJ. Effects of probenecid, methotrexate, digoxin and verapamil on the renal excretion of dicycloplatin in rats. Zhongguo Yaoke Daxue Xuebao 2006; 37: 447-9.

[25] Li XQ, Xu FL, Chen XJ, Zhang PY, Wang SL, Gu X. The effects of probeneside and other drugs on the renal excretion of dicycloplatin in rats. J Qinghai Med Coll 2005; 26: 243-6.

[26] Zhao D, Zhang Y, Xu C, *et al.* Pharmacokinetics, tissue distribution, and plasma protein binding study of platinum originating from dicycloplatin, a novel antitumor supramolecule, in rats and dogs by ICP-MS. Biol Trace Elem Res 2012; 148(2): 203-8.
[http://dx.doi.org/10.1007/s12011-012-9364-2] [PMID: 22367705]

[27] Wang C, Yu T. Comparative studies of cytotoxicity and hematologic toxicity induced by cisplatin, carboplatin and dicycloplatin in mice. Med J Liaoning 2004; 18: 273-4.

[28] Wang NQ, Fang JC, Liu T. Toxicity studies of new anticancer drug, dicycloplatin. Proc Chin Pharmacol 2001; 18: 30-1. [in Chinese].

[29] Lin F, Lu L, Yang XQ, Cui WC, Wang FM. The comparison study of major target organ hurt induced by dicycloplatin and its derivates in rats. Carcinogen Teratogen Mutagen 2005; 17: 276-9.

[30] Lin F, Feng FL, Yang XQ, Cui WC, Wang FM. *In vivo* evaluation of dicycloplatin toxicity on pregnancy in rats. Chin J Toxicol 2005; 19: 72-3.

[31] Kelland L. The resurgence of platinum-based cancer chemotherapy. Nat Rev Cancer 2007; 7(8): 573-84.
[http://dx.doi.org/10.1038/nrc2167] [PMID: 17625587]

[32] Reed E. Platinum-DNA adduct, nucleotide excision repair and platinum based anti-cancer chemotherapy. Cancer Treat Rev 1998; 24(5): 331-44.
[http://dx.doi.org/10.1016/S0305-7372(98)90056-1] [PMID: 9861196]

[33] Behrens BC, Hamilton TC, Masuda H, *et al.* Characterization of a cis-diamminedichloroplatinum(II--resistant human ovarian cancer cell line and its use in evaluation of platinum analogues. Cancer Res 1987; 47(2): 414-8.
[PMID: 3539322]

[34] Lee J, Peña MM, Nose Y, Thiele DJ. Biochemical characterization of the human copper transporter Ctr1. J Biol Chem 2002; 277(6): 4380-7.
[http://dx.doi.org/10.1074/jbc.M104728200] [PMID: 11734551]

[35] Gottesman MM, Fojo T, Bates SE. Multidrug resistance in cancer: role of ATP-dependent transporters. Nat Rev Cancer 2002; 2(1): 48-58.
[http://dx.doi.org/10.1038/nrc706] [PMID: 11902585]

[36] Mueller MD. Report on platinum drug symposium. W V Med J 2008; 104(4): 8-9.
[PMID: 18646676]

[37] Yu JJ. Unlocking the molecular mechanisms of DNA repair and platinum drug resistance in cancer chemotherapy. Curr Drug Ther 2009; 4(1): 19-28.
[http://dx.doi.org/10.2174/157488509787081895]

[38] Li GQ, Chen XG, Wu XP, *et al.* Effect of dicycloplatin, a novel platinum chemotherapeutical drug, on inhibiting cell growth and inducing cell apoptosis. PLoS One 2012; 7(11): e48994.

[http://dx.doi.org/10.1371/journal.pone.0048994] [PMID: 23152837]

[39] Ma HZ, Liu ZX, Wang QJ, Yan CH. liao MY. The breaking effects of cisplatin, carboplatin and dicycloplatin on pBR322 plasmid DNA. Carcinog Teratogen Mutagen 2003; 15: 220-2.

[40] Xu MW, Zhang J, Xu CG, Wang W. Apoptosis induced by dicycloplatin in human androgen independent prostate cancer line PC3. J Mod Urol 2012; 17: 541-4.

[41] Wang C, Yan RF. Assay of the lung cancer cell apoptosis induced by dicycloplatin. Pract Oncol J 2005; 19: 21-2.

[42] Li S, Huang H, Liao H, *et al.* Phase I clinical trial of the novel platin complex dicycloplatin: clinical and pharmacokinetic results. Int J Clin Pharmacol Ther 2013; 51(2): 96-105. [http://dx.doi.org/10.5414/CP201761] [PMID: 23127487]

[43] Liu KJ, Guan ZZ, Liang Y, *et al.* A double-blind, randomized phase II study of dicycloplatin plus paclitaxel versus carboplatin plus paclitaxel as first-line therapy for patients with advanced non-smal--cell lung cancers. Arch Med Sci 2014; 10(4): 717-24. [http://dx.doi.org/10.5114/aoms.2014.44862] [PMID: 25276156]

[44] Peng J, Wu HY, Guan ZZ. Chemotherapy-naïve patients with advanced non-small cell lung cancer–single institution experience. Chin J Clin Oncol 2009; 12: 711-4.

[45] Jouyban A, Soltani S. Blood–Brain Barrier Permeation.Toxicity and Drug Testing. InTech 2012; pp. 3-24.http://www.intechopen.com/books/toxicity-and-drug-testing/blood-brain-barrier-penetration [http://dx.doi.org/10.5772/20360]

[46] Ganellin CR. Uses of partition coefficients by brain penetration applied to the design of H2-receptor histamine antagonists.Handbook of Preformulation: Chemical, Biological, and Botanical Drugs. Boca Raton: Taylor and Francis Group CRC Press 1991; pp. 103-10.

[47] Lipinski CA, Lombardo F, Dominy BW, Feeney PJ. Experimental and computational approaches to estimate solubility and permeability in drug discovery and development settings. Adv Drug Deliv Rev 2001; 46(1-3): 3-26. [http://dx.doi.org/10.1016/S0169-409X(00)00129-0] [PMID: 11259830]

SUBJECT INDEX

A

B

C

D

I

K

L

M

R

S

T

V

W

Made in United States
North Haven, CT
04 November 2026

21727461R00212